PHOSPHOLIPIDS

New Comprehensive Biochemistry

Volume 4

General Editors

A. NEUBERGER
London

L.L.M. van DEENEN
Utrecht

ELSEVIER BIOMEDICAL PRESS
AMSTERDAM · NEW YORK · OXFORD

Phospholipids

Editors

J.N. HAWTHORNE and G.B. ANSELL

Nottingham and Birmingham

1982

ELSEVIER BIOMEDICAL PRESS
AMSTERDAM · NEW YORK · OXFORD

ISBN for the series: 0444 80303 3
ISBN for the volume: 0444 80427-7

Published by:
Elsevier Biomedical Press
Molenwerf 1, P.O. Box 1527
1000 BM Amsterdam, The Netherlands

Sole distributors for the U.S.A.and Canada:
Elsevier Science Publishing Company Inc.
52 Vanderbilt Avenue
New York, NY 10017, U.S.A.

Library of Congress Cataloging in Publication Data
Main entry under title:

Phospholipids.

(New comprehensive biochemistry; v. 4)
Includes bibliographical references and index.
1. Phospholipids. 2. phospholipids–Metabolism.
I. Hawthorne, J.N. (John Nigel) II. Ansell, G.B.
(Gordon Brian) III. Series.
QD415.N48 vol. 4 574.19′2s [574.19′214] 82-18382
[QP752.P53]
ISBN 0-444-80427-7 (U.S.)

Printed in The Netherlands

To the memory of Maurice Gray (1930–1980),
a good friend and dedicated lipid
biochemist.

Preface

In the general preface to the original series of volumes entitled *Comprehensive Biochemistry*, Florkin and Stotz stated: "The Editors are keenly aware that the literature of biochemistry is already very large". Even so, the chemistry of the phospholipids formed only part of Vol. 6 (1965) and the whole of lipid metabolism was covered in Vol. 18 published in 1970, of which only a small part was concerned with phospholipid metabolism. For the present series, therefore, we were charged by the General Editors to produce a volume on phospholipids which was to emphasise metabolic aspects since their structural role in membranes was covered in Vol. 3. We had to ensure coverage of developments in the last decade while, at the same time, summarising essential findings of earlier periods.

There are various ways in which the book could have been organised. As will be seen, we finally decided to devote separate chapters to individual or closely related phospholipids in which the essential chemistry is first described followed by an account of the metabolism, due regard being paid to the pioneering work of the past. We have included a chapter on phospholipases in general and one on phospholipase A2 since its structure and the mechanism of its action have been investigated in greater detail than any other phospholipid metabolising enzyme. The increasingly important topic of phospholipid exchange proteins is also treated separately. Furthermore, since the use of biochemically defined mutants shows great promise for the better understanding of phospholipid biosynthesis and function, a chapter has been devoted to genetic control of the enzymes involved.

This book is intended for advanced students and research workers and we believe that it gives a comprehensive, though not exhaustive, account of phospholipid biochemistry. Throughout, the reader will discover how advances in techniques have added to our knowledge of the ever-expanding field. Though it is difficult sometimes to avoid the impression that all research work is confined to the liver we hope that key references to other organs and other organisms will enable those whose interest lies outside the peritoneal cavity to be satisfied.

If the contents of the book belie the general title of the series, the responsibility lies with the editors not the authors and we would appreciate comments on errors and omissions.

We are grateful to Mrs. J. Paxton for her help in the preparation of the subject index.

J.N. Hawthorne
G.B. Ansell

Nottingham and Birmingham, August 1982

Contents

CHAPTER 1

Phosphatidylserine, phosphatidylethanolamine and phosphatidylcholine

G.B. ANSELL and S. SPANNER

Department of Pharmacology, The Medical School, Birmingham B15 2TJ, U.K.

1. Introduction

This chapter deals with the metabolism of phosphatidylcholine, lysophosphatidylcholine, phosphatidylethanolamine and phosphatidylserine in mammalian cells. Although basic mechanisms for the synthesis and catabolism of these major cell components have been known for some years there have been many recent investigations on their metabolism and possible function. Therefore this account, while summarising well-established facts, covers some of the more recent advances in some detail. Although original sources are usually cited, references to reviews rather than a series of papers are sometimes given.

2. Discovery and chemistry

(a) Phosphatidylcholine and lysophosphatidylcholine

Between 1846 and 1847 Gobley isolated from egg-yolk and brain a lipid which he called "lecithin" (Gk. *lekithos*, egg-yolk) [1] and from which he could obtain glycerophosphoric acid and fatty acids. Diakanow [2,3] and Strecker [4] showed that this lipid contained the base choline, originally isolated from hog bile by Strecker [5] (Gk. *cholě*, bile) and the two workers were able to deduce a provisional structure for lecithin. The subsequent history of lecithin was documented by MacLean and MacLean [6], Wittcoff [7] and Ansell and Hawthorne [8]. It was not until 1950 that Baer and Kates [9] by chemical synthesis showed that lecithin was based on L-α-glycerophosphate (L-3-glycerophosphate or D-1-glycerophosphate, deriving from D-glyceraldehyde) like all other naturally occurring glycerophospholipids. Other methods of synthesis are given by Strickland [10]. The nomenclature of phospholipids has undergone numerous modifications in the last two decades [8,10,11] and account has been taken of the fact that glycerol does not possess rotational symmetry. The latest recommendations are those of the IUPAC-IUB Commission on Biochemical Nomenclature [11] and the stereospecific numbering system is now used for all phospholipids. Thus lecithin is 1,2-diacyl-*sn*-glycero-3-phosphocholine or

Hawthorne/Ansell (eds.) Phospholipids

TABLE 1

Phosphatidylcholine, phosphatidylethanolamine, phosphatidylserine and lysophosphatidylcholine concentrations in various tissues

Tissue		Total phospholipid (μmol/g)	Phosphatidyl-choline (% TPL)	Phosphatidyl-ethanolamine (% TPL)	Phosphatidyl-serine (% TPL)	Lysophos-phatidyl-choline (% TPL)	Ref.
Brain,	rat	60.2	25	12	8	–	Unpublished results
grey matter,	man	50.9	39	40 (incl. plasmal)	13	–	
white matter,	man	82.8	31	34	16	–	29
myelin	man	–	24	7	21	–	
Kidney	rat	36.6	34	27	7	1	34
	man	22.2	33	24	1	3	
Lung	rat	–	54	20	6 (with PI)	–	46
	man	–	53	19	8	3	
Spleen	rat	17.5	42	24	8	1	34
	man	24.7	41	25	8	2	
Skeletal muscle	rat	11.3	51	22	4	3	33
	man	16.9	48	26	3	trace	
Pancreas	ox	28.1	53	21	4	–	313
	guinea pig	30.6	50	18	4	–	
Heart	rat	15.2	36	30	3	1	33
	man	21.5	40	26	3	4	
Plasma	rat	1.5	64	1	–	23	38
	man	2.9	70	3	trace	7	36
Erythrocytes	rat	4.2	42	23	11	4	35, 315
	man	3.9	29	28	14	2	6, 7
Platelets (nmol/10^9)	man	436	40	28	9	1	314
Liver	rat	37.9	48	24	3	1	34
	man	41.3	44	28	3	1	
Bile	rat	4.3	90	4	1	<1	47
Amniotic fluid (μmol/100 ml)	man	14.4	68	12	8	<1	12

TABLE 2

Major molecular species of phosphatidylcholine in the rat as % of the total

Fatty acids	Brain	Liver	Lung	Kidney	R.b.c.	Plasma	Gastric mucosa	Intestine	Bile
16:0/16:0	16.0	4.0	25.0, 41.4	8.0	20.3	–	31.0	–	–
18:0/16:0	4.0	–	4	–	–	–	7.0	–	–
16:0/16:1	–	–	12.0, 8.0	–	–	–	–	–	–
16:0/18:1	30.0	4.0	10.0, 16.6	–	14.5	5.5	–	6.5	–
18:0/16:1	–	–	–	–	–	–	–	–	–
18:0/18:1	12.0	–	–	–	–	–	14.0	–	–
16:0/18:2	–	15.7, 20.0	10.0, 6.4, 8.0	8.0	12.4	19.4	–	24.6	53.7
18:0/18:2	–	10.0, 9.0, 9.0	5.6	10.0	5.6	16.0	–	15.8	9.9
18:1/16:0	10.0	–	12.0	7.0	–	–	–	–	–
16:0/18:3	–	–	–	–	–	–	6.0	–	–
16:0/20:4	8.0	18.9, 22.0	15.0	22.0	5.4	5.8	–	14.0	14.7
18:0/20:4	10.0	25.0, 20.9	–	25.0	7.5	13.3	–	8.6	–
16:0/22:6	–	4.8	–	–	–	6.9	–	–	–
18:0/22:6	–	5.6	–	–	–	4.8	–	–	–
18:2/20:4	–	–	–	–	–	5.5	–	–	–
18:2/16:0	–	–	–	6.0	–	–	–	–	–

Values taken from [49–54].

3-*sn*-phosphatidylcholine. In most naturally occurring lecithins the 1-position is esterified with a saturated fatty acid and the 2-position with an unsaturated one but there are notable exceptions (see Table 2).

Lysolecithin (1-, or 2-lysophosphatidylcholine, or 1- or 2-acyl-*sn*-glycero-3-phosphocholine) also occurs naturally in tissues though at very much lower levels than lecithin (Table 1) [12–14]. It is, of course, a significant component of blood plasma (p. 32) as was found conclusively by Gjone et al. [15]. It is likely that all naturally occurring lysolecithins have the 1-acyl structure (note that 1-acyl-*sn*-glycero-3-phosphocholine is 2-lysophosphatidylcholine) and that the fatty acid is saturated as first suggested by Lüdecke in 1905 [16]. The high levels of lysolecithin reported in the older literature (see [7]) arise as a result of phospholipase activity post mortem.

*(b) Phosphatidylethanolamine (3-*sn*-phosphatidylethanolamine)*

Thudichum [17] separated a nitrogen- and phosphorus-containing lipid fraction from brain tissue which he distinguished from lecithin by its relative insolubility in warm ethanol. He called it "kephalin" and obtained ethanolamine from it as a hydrolysis product (though he thought this base was a breakdown product of choline and not naturally occurring). "Kephalin" or "cephalin" is now known to be a mixture of ethanolamine-, serine-, and inositol-containing phospholipids but it was not until 1913 that one of the contributing bases was shown to be ethanolamine [18,19]. Rudy and Page [20] isolated an ethanolamine glycerophospholipid from the cephalin fraction of brain tissue which they reasonably assumed to be phosphatidylethanolamine. However, it is not easy to separate phosphatidylethanolamine from tissues on a preparative scale, especially if they contain ethanolamine plasmalogen as do brain, cardiac muscle and skeletal muscle (see [10,21]) and it is probable that Rudy and Page's preparation contained plasmalogen. It is likely in retrospect, therefore, that the first pure preparations of phosphatidylethanolamine from natural sources were those of Lea et al. [22] from egg-yolk and Klenk and Dohmen [23] from liver. The phospholipid was first synthesised by Baer et al. [24]; other methods of synthesis are given by Strickland [10]. The fatty acid in the 1-position is usually saturated and that in the 2-position unsaturated.

*(c) Phosphatidylserine (3-*sn*-phosphatidylserine)*

Although MacArthur [25] had demonstrated the presence of α-amino nitrogen in Thudichum's "kephalin" it was not until 1941 that Folch and Schneider [26] showed that an α-amino-β-hydroxylic acid was present. In the same year Folch [27] identified the amino acid as L-serine and showed that phosphatidylserine was a component of the kephalin fraction. In 1948 Folch proposed a structure [28] and Baer and Maurukas [28a] showed that this phospholipid, when reduced, was identical to 1,2-distearoyl-*sn*-glycero-3-phospho-L-serine which they synthesised.

The structures of phosphatidylcholine, phosphatidylethanolamine and phosphatidylserine are given in Fig. 1.

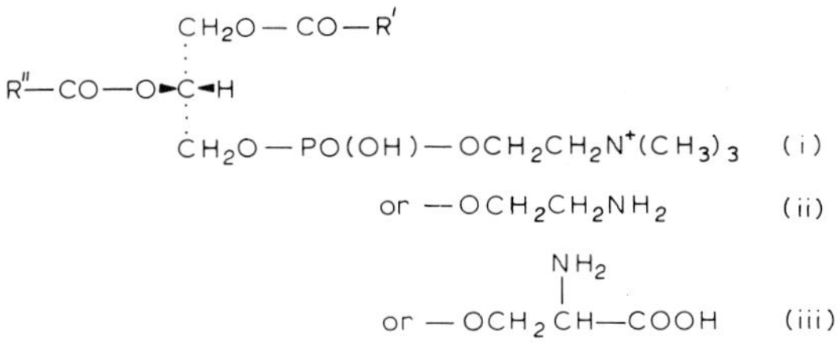

Fig. 1. 1,2-Diacyl-*sn*-glycero-3-phosphocholine (phosphatidylcholine) (i) where R′– and R″– are the fatty acyl substituents. In phosphatidylethanolamine (ii) and phosphatidylserine (iii) the choline is replaced by ethanolamine and serine respectively.

3. Determination and distribution in animal tissues

The phospholipid content of mammalian organs varies from organ to organ and from species to species. Table 1 gives, where possible, the total phospholipid content in μmol/g wet weight of tissue and the phosphatidylcholine [11], phosphatidylethanolamine [12], phosphatidylserine [13] and lysophosphatidylcholine content as a percentage of the total phospholipid in two species, rat and man. For a more comprehensive, but unfortunately now outdated survey the reader is referred to the chapter by White [29] and to journals relevant to the organs concerned.

Although much work has been carried out on the fatty acid content of phospholipids, the values for the molecular species are more difficult to find. This in part is due to the rather complex methodology involved. In Table 2 values are given for the molecular species of phosphatidylcholine in various tissues of the rat and in Tables 3 and 4 values for phosphatidylethanolamine and phosphatidylserine in muscle and kidney of various animals.

TABLE 3

Molecular species of phosphatidylethanolamine in various tissues of different animals

	Skeletal muscle		Kidney		Sarcoplasmic reticulum of skeletal muscle
	Rat	Human	Dog	Pig	Rabbit
16:0/18:1	–	–	12.1	6.7	14.9
18:0/18:1	–	–	–	–	17.9
16:0/18:2	13.1	14.0	10.8	–	11.9
16:0/20:4	11.6	21.0	27.3	10.3	14.9
18:0/20:4	13.9	28.0	25.9	35.6	23.9
18:0/22:6	37.3	9.3	–	–	–
18:1/18:1				6.9	
18:1/20:4			11.2	18.1	

Values taken from [55] and [56].

TABLE 4

Molecular species of phosphatidylserine as % of total in skeletal muscle of the rabbit

	Sarcoplasmic reticulum [a]	Sarcotubular vesicle
16:0/18:1	11.0	–
18:0/18:1	8.0	14.0
18:0/20:3	17.0	8.0
18:0/22:3	5.0	11.0
18:0/20:4	50.0	20.0
18:0/22:5	4.0	8.0
18:0/22:6	–	25.0

[a] Includes phosphatidylinositol. Values taken from [55] and [57].

The methods for the isolation of phospholipids and their subsequent separation into classes dependent upon the nature of the base are now well established. Phospholipids containing choline, ethanolamine and serine are readily extracted into chloroform–methanol (2:1, v/v) from mammalian tissues [30] and this is the method adopted by most workers. Early methods of isolation of individual lipids by column chromatography or paper chromatography following the removal of the fatty acids have, on the whole, given way to two-dimensional thin-layer chromatography. The lipids can then be quantified by the determination of the phosphorus content or in experiments involving radiolabelled compounds, be assayed for radioactivity. The individual lipids may also be separated on silica gel and assayed for their fatty acid content by gas-liquid chromatography. For details of these methods the reader is referred to the reviews by Spanner [21] and Nelson [31].

The determination of the molecular species of the lipid is more complex. For one method and for relevant references the reader is referred to the paper by Kawamoto et al. [32].

4. Biosynthesis

In section 2 the phospholipids were described in the order in which they were discovered but it seems more logical to reverse this order when discussing the biosynthetic pathways because it is known that phosphatidylserine can give rise to phosphatidylethanolamine which in turn can be converted to phosphatidylcholine though these are not the only pathways.

(a) Phosphatidylserine

There are two established pathways for the biosynthesis of phosphatidylserine, a phospholipid which accounts for 5–10% of the total phospholipid in eukaryotic cells. One is a Ca^{2+}-mediated exchange reaction of L-serine with another phospholipid, probably phosphatidylcholine, and until very recently, was thought to be the sole method by which phosphatidylserine is synthesised in animal tissues. The other pathway is the reaction between CDP-diacyl glycerol and L-serine, confined apparently to bacteria and plants.

(i) Base-exchange reaction

In 1959 Hübscher et al. [58] noted that L-serine could be incorporated into phosphatidylserine in a liver mitochondrial preparation by a reaction likely to be independent of energy supply but dependent on Ca^{2+} ions. Subsequently, the energy-independent reaction was shown largely to be confined to the microsomal fraction (endoplasmic reticulum) [59], with a K_m for serine of about 0.5 mM and optimum pH of 8.3 with 25 mM Ca^{2+}. In Ehrlich ascites cells, however, the exchange reaction is found in mitochondria, possibly due to malignant transformation [60]. The reaction has been extensively investigated, particularly in brain tissue, by Porcellati and his co-workers [61–63]. The K_m for serine seems to depend on the Ca^{2+} concentration [63]. Base exchange also occurs with choline (p. 16) and ethanolamine (p. 12) and the question arises whether a single enzyme is responsible and what the preferred lipid acceptor is. The work of Kanfer [64] and Gaiti et al. [63] indicated that more than one enzyme is involved and the partial purification of the L-serine base-exchange enzyme has been reported [65]. A microsomal fraction from brain tissue was solubilised with a mixture of detergents and the protein precipitate obtained by 35–60% saturation with ammonium sulphate fractionated on Sephadex B and DEAE-cellulose columns. Ethanolamine plasmalogen and phosphatidylethanolamine were good acceptor lipids (K_m serine 0.4 mM); other phospholipids could not serve as acceptors and the incorporation of serine was not inhibited by ethanolamine or choline. In cultured brain cells there is a preferential incorporation of serine into 1-alkyl-2-acyl-glycerophosphoserine as well as diacylglycerophosphoserine [66]. The purified enzyme of Taki and Kanfer [65] had no phospholipase D activity though earlier studies [67] had suggested that this phospholipase was identical with the base-exchange enzyme. The molecular mechanism for the exchange is unknown. For an extensive account of the base-exchange reaction for serine and the other phospholipid bases the reader is referred to a recent review by Kanfer [68] in which the relationship of the responsible enzymes to phospholipase D is also discussed.

(ii) Other reactions

In bacteria it is well established that phosphatidylserine is formed by the reaction of L-serine with CDP-diacylglycerol catalysed by phosphatidylserine synthase (CDP-diacylglycerol:L-serine *O*-phosphatidyltransferase, EC 2.7.8.8)

$$\text{L-serine} + \text{CDP-diacylglycerol} \rightarrow \text{phosphatidylserine} + \text{CMP}$$

and first observed in a cell-free system from *E. coli* [69,70]. It also occurs in plants and protozoa [71] but as far as is known does not occur in animal tissues.

Over 20 years ago Hübscher et al. [58] noted an incorporation of L-serine into the phosphatidylserine of mitochondria isolated from liver which was dependent on Mg^{2+} and ATP and stimulated by CMP. This was confirmed by Bygrave and Bücher [72] but the finding was puzzling since other work had more or less eliminated mitochondria as organelles capable of synthesising phospholipids containing nitrogen bases de novo [73,73a] except for the formation of phosphatidylethanolamine by the decarboxylation of phosphatidylserine (p. 9). More recent work has, however, demonstrated an ATP- and CMP-stimulated incorporation of L-serine into the phosphatidylserine of mitochondria of Ehrlich ascites cells [60]. Although the incorporation was also stimulated by the presence of phosphatidic acid, which suggested the involvement of CDP-diacylglycerol, the addition of the latter had no effect. Furthermore, serine did not stimulate the incorporation of ^{32}P from [^{32}P]phosphatidic acid into phosphatidylserine in the presence of CTP or the release of [^{14}C]CMP from [^{14}C]CDP-diacylglycerol and so the involvement of CDP-diacylglycerol had to be rejected. Kiss [74] had noted earlier that stimulation of phosphatidylserine formation in heart slices by phosphatidic acid was unaffected by CTP and thought that it might be caused by a reversal of the action of a phospholipase D.

The ATP-stimulated incorporation of serine into mitochondrial phosphatidylserine therefore has been difficult to explain. It is possible that there are two base-exchange mechanisms, one dependent on, and one independent of, ATP, a view originally put forward by Bygrave and Bücher [72] and supported by the experiments of Yavin and Zeigler [66] with cultured brain cells. Recently a new explanation of the ATP-stimulated synthesis has been put forward though it was studied in brain microsomes and not liver mitochondria [75] but is interesting in that it will explain phosphatidic acid-stimulation as well. In the presence of Ni^{2+} ions, ATP promotes the conversion of phosphatidic acid to pyrophosphatidic acid (*P,P'*-bis(1,2-diacyl-*sn*-glycero-3-pyrophosphate). This then appears to react directly with L-serine to give phosphatidylserine. The reaction is inhibited by –SH inhibitors such as *p*-hydroxymercuribenzoate which do not affect the Ca^{2+}-dependent pathway.

Presumably the final fatty acid composition of phosphatidylserine is determined by the acyltransferase reactions which operate for phosphatidylethanolamine and -choline (pp. 12 and 17) but only one study appears to have been carried out [76].

(b) Phosphatidylethanolamine

There are four established pathways for the biosynthesis of this phospholipid.

(i) The decarboxylation of phosphatidylserine

$$\text{phosphatidylserine} \rightarrow \text{phosphatidylethanolamine} + CO_2$$

(ii) The transfer of ethanolamine by its phosphorylation and transfer to a diacylglycerol acceptor from CDP-ethanolamine

$$\text{ethanolamine} + \text{ATP} \rightarrow \text{phosphoethanolamine} + \text{ADP}$$

phosphoethanolamine + CTP → CDP-ethanolamine + P_i

CDP-ethanolamine + D-1,2-diacylglycerol → phosphatidylethanolamine + CMP

(iii) The Ca^{2+}-dependent base exchange reaction analogous to the one exchanging L-serine

(iv) The acylation of lysophosphatidylethanolamine.

(i) Decarboxylation of phosphatidylserine

It has long been known that L-serine gives rise to ethanolamine in vivo [76] and Arnstein [77] showed that C_2 and C_3 of serine provide the carbon atoms of ethanolamine. There is no evidence that decarboxylation of free serine occurs in animal tissues but the formation of phosphatidylethanolamine by the decarboxylation of phosphatidylserine in the liver was deduced from experiments in vivo using L-[3-^{14}C]serine [78]. Supporting experiments were made by Wilson et al. [79] who incubated labelled serine with liver mitochondria and brain homogenates and showed that the appearance of labelling in the ethanolamine moiety of phosphatidylethanolamine was not reduced by the presence of added ethanolamine or phosphoethanolamine. The work of Kennedy and his collaborators [80,81] clearly demonstrated the decarboxylation of phosphatidylserine by a mitochondrial enzyme in liver. The enzyme responsible, phosphatidylserine decarboxylase (phosphatidylserine carboxy-lyase, EC 4.1.1.65) has a K_m for phosphatidylserine of 6.5 mM and is dependent upon pyridoxal phosphate [82]. From the results of experiments on differentiating cells from cerebral hemispheres incubated with L-[3-^{14}C]serine in culture, Yavin and Zeigler [66] concluded that 13% of the ethanolamine-containing diacylglycerophospholipids derived from the corresponding serine phospholipids and no water-soluble ethanolamine-labelled intermediates could be detected. This particular study also demonstrated the formation of ethanolamine plasmalogen (1-alk-1′-enyl-2-acyl-*sn*-glycero-3-phosphoethanolamine) by a series of reactions one of which was the decarboxylation of serine plasmalogen (see Chapter 2).

It is puzzling why there appear to be two pathways for the formation of ethanolamine-containing glycerophospholipids in tissues, one by the decarboxylation of serine glycerophospholipids and the other by the cytidine pathway utilising free ethanolamine. All ethanolamine in animal tissues derives from serine and the latter can be decarboxylated only when in a lipid-bound form. It can be argued, therefore, that the decarboxylation pathway is the true system for the formation of ethanolamine de novo. This conclusion may be incorrect, however, because phosphoethanolamine can also derive from the catabolism of dihydrosphingosine phosphate, a reaction first observed in vitro in 1968 [83]. Although, as will be seen in Fig. 1, this ethanolamine also derives from serine it could be used for the synthesis of phosphatidylethanolamine by the cytidine pathway de novo.

(ii) The cytidine pathway

This pathway has been one of the most extensively studied since its original discovery by Kennedy and Weiss [84]. Although the first step, the phosphorylation of ethanolamine, was originally thought to be catalysed by choline kinase (ATP:choline phosphotransferase, EC 2.7.1.32) (q.v.) it is now clear that some mammalian tissues contain a separate ethanolamine kinase (ATP:ethanolamine *O*-phosphotransferase, EC 2.7.1.82). Thus, liver contains two such kinases in the cytoplasm, ethanolamine kinase I (M_r 36000) which has no activity towards, and is not inhibited by choline and II (M_r 160000) which has activity towards, and is inhibited by choline [85]. There is, however, a tendency for the two kinases to co-purify [86–88]. It may be that the two activities are mediated by two distinct sites on a single protein (see also [89]). The pH optimum is in the region of 8.5 and the K_m for liver ethanolamine kinase I is 0.4 mM and 1.7 mM for II whereas it is 2.2 mM in brain synaptosomes [89]. The K_m for Mg^{2+}-ATP is 14.3 mM for kinase I and 0.5 mM for kinase II [85].

Relatively large pools of phosphoethanolamine are present in tissues, e.g., 1 μmol/g fresh weight in brain [90] and 0.4 μmol/g in liver [91]. There is, however, no certainty that it all derives from the phosphorylation of ethanolamine since the ester can also derive from the cleavage of the phosphate esters of sphingosine bases, particularly in the liver. The extensive work of Stoffel and his collaborators [83,92–94] has shown that sphingosine ((4D)sphingenine) (though most of the experiments have been done with dihydrosphingosine (sphinganine)) is phosphorylated and the phosphate ester cleaved according to the scheme shown in Fig. 2.

In vivo the phosphoethanolamine produced from sphingosine is not in the same metabolic pool as that formed by the phosphorylation of ethanolamine [95] which is not surprising since these processes occur in different subcellular compartments. In isolated hepatocytes ethanolamine seems to be the most important precursor of phosphoethanolamine [96].

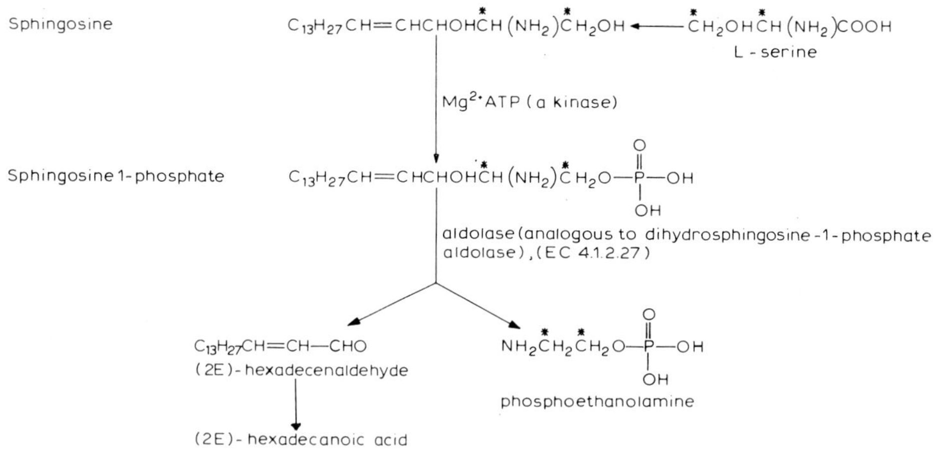

Fig. 2. The formation of phosphoethanolamine from L-serine via sphingosine. C-atoms 2 and 3 of serine (as asterisked) give rise to C-atoms 1 and 2 of ethanolamine.

The formation of CDP-ethanolamine is catalysed by the enzyme ethanolamine phosphate cytidylyltransferase (CTP:ethanolamine phosphate cytidylyltransferase, EC 2.7.7.14) which is distinct from the analogous enzyme forming CDP-choline [97]. The properties of the enzyme whose action is freely reversible have been summarised [98–101]. The M_r-value of the liver enzyme is 100000–120000 (though Chojnacki et al. [99] gave a value of 40000), and its optimal pH is 7.6 with another lower peak of activity at pH 6 according to Sundler [101] who also noted a requirement for a reducing agent e.g. dithiothreitol. Mg^{2+} is an essential co-factor and the K_m for CTP is 50 μM and for phosphoethanolamine 65 μM. The reaction is an ordered sequential one in which CTP is added to the enzyme first and CDP-ethanolamine released from the enzyme last [101]. The specificity for phosphoethanolamine is high though phosphomonomethylaminoethanol can serve as a substrate and deoxy-CTP can substitute for CTP. Levels of CDP-ethanolamine in tissues are extremely low (25 nmol/g liver [102]) and the activity of the cytidylyltransferase may be rate-limiting as has been demonstrated in isolated hepatocytes by Sundler and Åkesson [103]. If it is rate-limiting then the observation by Plantavid et al. [103a] that its activity in vitro is inhibited by *S*-adenosylmethionine, the methyl donor for phosphatidylcholine synthesis, may be important.

The final step in the synthesis of phosphatidylethanolamine is the transfer of phosphoethanolamine from CDP-ethanolamine to a diacylglycerol acceptor, which is catalysed by ethanolamine phosphotransferase (CDP-ethanolamine:1,2-diacylglycerol phosphoethanolamine transferase, EC 2.7.8.1). The diacylglycerol acceptor can be replaced by 1-alkenyl-2-acyl glycerol [104] or 1-alkyl-2-acyl glycerol [105–110]. Deoxy-CDP-ethanolamine can also serve as the donor and it is likely that the same enzyme can donate to all lipid acceptors [107]. Mg^{2+} is the accepted co-factor but Mn^{2+} is also effective.

For a long time after the initial discovery of the enzyme it was uncertain whether it is distinct from the corresponding choline phosphotransferase but Radominska-Pyrek et al. [110] recently succeeded in solubilising the ethanolamine phosphotransferase free from choline phosphotransferase activity. However, since all the choline phosphotransferase disappeared during the preparative procedure it is still just possible that the two activities are located on the same enzyme (see also [111]). The K_m for CDP-ethanolamine was 0.14 mM when the acceptor diacylglycerol was incorporated into a liposome [110] though Coleman and Bell [111a] gave a K_m of 18 μM for the reaction in fat cells.

The reversibility of the action of this enzyme has been demonstrated for liver and some other tissues though this aspect has been studied more extensively for the choline phosphotransferase. Kanoh and Ohno [112] showed that CMP, when incubated with a liver microsomal fraction whose ethanolamine lipids had been labelled with [1,2-^{14}C]ethanolamine led to the formation of CDP-[1,2-^{14}C]ethanolamine. The K_m for CMP was 0.14 mM and the K_i for CDP-ethanolamine which inhibited the back reaction, was 0.05 mM. The reaction proceeded at half the rate observed with CDP-choline (q.v.). For the back reaction 1-stearoyl-2-acylglycero-3-phosphoethanolamine was preferentially used as a substrate [113].

However, when diacylglycerols, liberated by the *back* reaction were isolated and used for the *forward* reaction with microsomal fragments, the transferase preferentially used hexaenoic diacylglycerol as substrate (but see lung, p. 18).

(iii) The base-exchange reaction

The base-exchange reaction for the incorporation of ethanolamine into phosphatidylethanolamine was first demonstrated by Borkenhagen et al. [80]. Bjerve [114] showed that, with a liver-microsomal fraction, free ethanolamine could displace choline or serine from the appropriate diacylglycerophospholipid; ethanolamine could be displaced from phosphatidylethanolamine by serine but not by choline. Bjerve [115] also noted that the incorporation of ethanolamine was inhibited by L-serine *non*-competitively while choline did so *un*competitively, that there was a reciprocal relationship between Ca^{2+} concentration and pH in terms of incorporation and that ethanolamine was predominantly incorporated into hexaenoic acid-containing species of phosphatidylethanolamine. Incorporation of ethanolamine into phosphatidylethanolamine in brain microsomes was ten times faster than into ethanolamine plasmalogen [115a] and incorporation of ethanolamine was faster than that of serine and choline [116]. According to Kanfer [64] the optimum pH for the incorporation of ethanolamine also appears to be different from that of the other two bases for a brain particulate fraction and the K_m found for ethanolamine was 15 μM. There have been numerous further studies and the consensus of opinion is that only a small pool of phosphatidylethanolamine is involved in base exchange in vitro and that this is confined to the endoplasmic reticulum (microsomal fraction) [68]. The same appears to be true in vivo, at least for liver [102]. Evidence for the significance of the base exchange of ethanolamine and choline has been difficult to obtain [68].

(iv) The acylation of lysophosphatidylethanolamine

Since the liver contains phosphatidylethanolamine species which are not readily synthesised by the cytidine pathway it is generally believed that deacylation and reacylation are significant mechanisms for their formation. The acylation of 1- and 2-acylglycerophosphoethanolamine was first described by Lands and his colleagues [117,118]. Using liver slices Van Golde et al. [119] demonstrated that, though de novo synthesis was important for the synthesis of monoenoic and dienoic species, some of the more highly unsaturated fatty acids were introduced by acylation. Experiments in vivo showed that 95% of the linoleic acid in 1-stearoyl-2-linoleoyl phosphatidylethanolamine entered the molecule by means of acylation [120] as did all the arachidonic acid (derived from linoleic acid). Thus the acyl transferase preferentially utilises highly unsaturated fatty acids [51].

(v) General comments on phosphatidylethanolamine synthesis

The cytidine pathway is almost certainly the most important pathway for the synthesis of phosphatidylethanolamine de novo in the liver and the base exchange pathway plays only a minor role. This follows from the observations that the distribution of radioactivity amongst different species of phosphatidylethanolamine after the intraportal injection of labelled ethanolamine was very similar to that obtained when labelled phosphate or glycerol were used [102]. A rate of synthesis of 0.35 μmol phosphatidylethanolamine/min/whole (rat) liver was calculated by Sundler and Åkesson [121]. The cytidine pathway appears to be particularly important in the synthesis of molecular species rich in hexaenoic acid. Thus Åkesson et al. [122] found that after the intraportal injection of [^{3}H]glycerol into rats 60% of the radioactivity appeared in hexaenoic acid-containing species within 5 min. Sundler and Åkesson [121] using [^{3}H]ethanolamine in vivo calculated that the rate of synthesis of the 1-palmitoyl-2-docosahexaenoyl species was six times that of the 1-palmitoyl-2-arachidonyl species. The phosphatidylethanolamine in liver is richer in arachidonic acid, however, which points to the significance of acylation in determining the final composition of the tissue.

One interesting observation made after the intraperitoneal injection of labelled ethanolamine [123] was that the specific radioactivity of the CDP-ethanolamine was higher than that of the precursor phosphoethanolamine. This was confirmed by Sundler [102] and Sundler and Åkesson [121] even for very short time intervals and they suggested that this implies two pools of phosphoethanolamine only one of which can be labelled by exogenous ethanolamine. The other pool is presumably that produced from the catabolism of sphingosine (Fig. 2) which would not be labelled and would therefore reduce the measured specific radioactivity.

(c) Phosphatidylcholine

Five separate mechanisms are now known to lead to the formation of this universal constitutent of mammalian cell membranes:

(i) The stepwise methylation of phosphatidylethanolamine. In this reaction methyl groups are successively introduced into the terminal amino group of phosphatidylethanolamine.

phosphatidylethanolamine + 3 *S*-adenosyl-L-methionine →

phosphatidylcholine + 3 *S*-adenosyl-L-homocysteine

(ii) The transfer of choline to a diacylglycerol acceptor via its phosphorylation and conversion to CDP-choline (cytidine pathway analogous to that for ethanolamine, see p. 9).

(iii) The Ca^{2+}-dependent exchange reaction analogous to that utilising serine or ethanolamine.

(iv) The acylation of lysophosphatidylcholine by acyl CoA:

lysophosphatidylcholine + acyl CoA → phosphatidylcholine + CoA

(v) Transacylation between two molecules of lysophosphatidylcholine:

2 lysophosphatidylcholine → phosphatidylcholine + glycero-3-phosphocholine

(i) Stepwise methylation

It has long been known that ethanolamine is derived from serine (p. 9) and that it is *N*-methylated to form choline [124,125], the methyl groups deriving from *S*-adenosyl-L-methionine. Bremer et al. [78,126] clearly demonstrated by experiments in vivo and in vitro that the methylation of ethanolamine takes place in the liver only when it is in a lipid-bound form (see also [127,128]). Phosphatidylmonomethylaminoethanol and phosphatidyldimethylaminoethanol are intermediates which are found in the liver and dilinoleoylphosphatidyldimethylaminoethanol was shown to be readily taken up by the liver after intravenous injection and methylated to dilinoleoylphosphatidylcholine [129]. However, it proved difficult for some time to demonstrate the introduction of the first methyl group in vitro [130]. These difficulties were resolved in 1978 by Hirata et al. [131] who showed that, for bovine adrenal medulla, two enzymes were involved in methylation. The enzyme catalysing the transfer of a methyl group from *S*-adenosylmethionine to phosphatidylethanolamine had an optimum pH of 6.5, a low K_m for *S*-adenosylmethionine (1.4 μM) and an absolute requirement for Mg^{2+}. The two-stage conversion of phosphatidylmonomethylaminoethanol to phosphatidylcholine was carried out by a second methyltransferase with an optimum pH of 10, a high K_m for *S*-adenosylmethionine and no requirement for Mg^{2+}. Subsequent work showed that these two enzymes are widely distributed [132] and their asymmetric distribution in the cell membrane is discussed on p. 26.

Experiments by LeKim et al. [129] showed that the methylation of phosphatidyldimethylaminoethanol in rat liver microsomes was dependent on the degree of unsaturation of its fatty acids, relative rates being dilinoleoyl-species > 1-stearoyl-2-linoleoyl- > 1-stearoyl-2-oleyl. The capacity of other tissues to carry out this methylation is feeble [132]. The results of Arvidson [133] strongly suggested that the hexaenoic species of phosphatidylcholine were more heavily labelled after an injection of [^{14}C]ethanolamine in vivo and it now appears from a number of studies [51] that the functioning of the methylation pathway in liver results primarily in phosphatidylcholine enriched in 1-palmitoyl-2-docosahexanoyl-, 1-palmitoyl-2-arachidonyl- and 1-stearoyl-2-arachidonyl-containing species. One further important point is that, in quantitative terms, the formation of phosphatidylcholine by the methylation pathway is of significance only in the liver [123] where it amounts to not less than 20% and not more than 40% of that synthesised [103] and the corollary of this is that the body's supply of choline is either synthesised in the liver or brought in with the diet (see p. 31). The suggestion of Morgan et al. [134,135] that dipalmitoyl-phosphatidylcholine, an important surfactant in the lung, is formed by the methylation of the corresponding phosphatidylethanolamine is probably incorrect, since methyltransferase activity is weak in lung [136,137] and lung does not contain dipalmitoylphosphatidylethanolamine in significant amounts [138] (see also p. 18).

(ii) Cytidine pathway

Choline kinase (Mg–ATP: choline phosphotransferase, EC 2.7.1.32) was discovered by Wittenberg and Kornberg [139] and is a Mg^{2+}–ATP-dependent enzyme present in the cytosol of mammalian cells. It was first studied in detail in brain tissue where the highest level of activity is found [140,141]. The K_m for choline varies from 2.6 mM in brain [89] to 44 μM for adult mouse lung (120 μM in foetal lung [142]) and 33 μM for monkey lung [87]. Oldenberg and Van Golde [142] found a K_m for ATP of 8 mM for the enzyme from adult mouse lung. The substrate is actually Mg–ATP but Mg^{2+} in excess of that required for the Mg–ATP complex formation is required by the brain synaptosomal kinase [89]. Optimum activity is usually found between pH 9 and 10 though in mouse it is nearer 8 [142]. Activity is inhibited by Ca^{2+}. Most preparations of choline kinase have activity towards ethanolamine and the activities co-purify [86–88] and often cannot be separated [87]. After a detailed and sophisticated study of the kinetics of choline kinase from rat liver, Infante and Kinsella [143] concluded that, at low concentrations of the reactants they were "added" in the order choline, $Mg–ATP^{2-}$ and Mg^{2+}. This implied in particular a dual role for Mg^{2+}.

CTP:choline phosphate cytidylyltransferase (EC 2.7.7.15) was discovered at the same time as the analogous enzyme forming CDP-ethanolamine and early studies were summarised by Ansell and Chojnacki [98]. It is Mg^{2+}-dependent and the K_m for the liver enzyme is 0.17 mM for phosphocholine and 0.2–0.3 for CTP [144] and in lung 5 mM for phosphocholine and 0.57 mM for CTP [142]. The concentration of phosphocholine in liver is 1.4 mM [145] and therefore much higher than its K_m whereas the concentration of CDP-choline (51 μM) [146] is extremely low. This suggests that the phosphocholine moiety of CDP-choline is rapidly transferred to phosphatidylcholine (or to a much lesser extent to sphingomyelin) and that the cytidylyltransferase reaction is rate-limiting (though Infante [146a] is of the opinion that the kinase reaction is also rate-limiting). The cytidylyltransferase unlike the kinase which is in the cell cytoplasm and the phosphotransferase which is in the endoplasmic reticulum is found in both cell fractions in the liver [144,147] and adult but not foetal lung [142]. According to Choy et al. [144] it exists in the liver cytosol in two different molecular forms one of which (the L-form) has an M_r-value of 200000 and which can aggregate to form polymers (H-form) in the presence of diacylglycerol at 4°C [148]. Fiscus and Schneider [149] found that phospholipids stimulated cytidylyltransferase activity and more recently it has been observed that the L-form of the enzyme is activated 10-fold by lysophosphatidylethanolamine and to a lesser extent by phosphatidylserine, phosphatidylinositol and phosphatidylglycerol but inhibited by some species of lysophosphatidylcholine [150,151]. Feldman et al. [151a] are of the opinion that phosphatidylglycerol is the most potent activator of the L-form.

There have been extensive studies on phosphocholine phosphotransferase (CDP-choline: 1,2-diacyl-glycerol choline phosphotransferase, EC 2.7.8.2) which like its ethanolamine counterpart can transfer the phosphomonoester to 1-alkyl-2-acylglycerols and 1-alk-1′-enyl-2-acylglycerols as well as diacylglycerol. The numerous studies to investigate preference of the liver enzyme for certain molecular species of diacylglycerol have been discussed in some detail by Holub and Kuksis [51]. Desaturated substrates, e.g., 1,2-dipalmitoylglycerol appear to be poor substrates in vitro (q.v.) except in lung [152] but 1-palmitoyl-2-acylglycerols are good substrates when the 2-acyl group is an unsaturated fatty acid [153]. When $^{32}P_i$ was used as a precursor to demonstrate the synthetic capacity of the cytidine pathway in the liver in vivo, 1-palmitoyl-2-oleoyl- and 1-palmitoyl-2-linoleoyl- were the predominant species formed. That the cytidine pathway is pre-eminently important for the synthesis of 1-palmitoyl- and 1-stearoyl-2-lineoylglycerophosphocholine in liver can also be deduced from the high rate of formation of these species when [^{14}C]choline is used as a precursor (0.26 μmol/min/liver [121].

The kinetics of the back reaction for choline phosphotransferase in vitro are different from those for the ethanolamine phosphotransferase (K_m for CMP is 0.19 mM, K_i for CDP-choline 1 mM [112]); diacylglycerol so released may be used for phosphatidylethanolamine synthesis [153a]. However, the back reaction is thought not to occur to a significant extent in the liver in vivo [145]. In the microsomal fraction of the brine shrimp *Artemia salina* the back reaction operates at a faster rate than the forward reaction [153b].

(iii) Base exchange

The energy-independent incorporation of choline into phosphatidylcholine was first demonstrated in a mitochondrial preparation from liver by Dils and Hübscher [154] who subsequently [67] showed that the major activity resided in the microsomal fraction. The reaction, optimally active at pH 8.5, depended upon a calcium concentration of 2–3 mM; the K_m for choline was 8 mM and L-serine was a competitive inhibitor with a K_i of 0.77 mM. Numerous subsequent investigations have been summarised by Kanfer [68].

Bjerve [115] noted that choline incorporation into liver microsomes in vitro was competitively inhibited by both serine and ethanolamine. However, once incorporated into brain microsomal phosphatidylcholine, it could not be displaced by the other two bases. There is indeed evidence that the enzymes responsible for the exchange of the three bases are distinct in brain tissue. For example, the heat inactivation curves for choline incorporation with both particulate and solubilised preparations from brain were different from those for ethanolamine or serine [155]. The three activities have now been separated by a combination of gel filtration and ion-exchange chromatography [156].

Numerous experiments have been performed to decide the relative importance of the incorporation of choline by the base-exchange reaction and its incorporation by the cytidine pathway in vivo. Treble et al. [157] were of the opinion that base-exchange is much more important than the cytidine pathway in liver but subsequent investigations, in which very short time intervals were taken after the injection of labelled choline for the measurement of the radioactivity in metabolites, indicated that the reverse is more likely [145,158]. Sundler et al. [145] calculated that the incorporation of choline into phosphatidylcholine of liver by the cytidine pathway was twenty times faster than by base exchange. Further evidence is provided by the fact that different molecular species are labelled by the two pathways. Thus, Bjerve [159] showed that the base exchange reaction preferentially labelled phosphatidylcholine rich in polyunsaturated fatty acids (hexaenoic species) when liver microsomes were used, whereas cholinephosphotransferase is known to "prefer" monoenoic or dienoic species of diacylglycerols (p. 15). Since the experiments of Arvidson [160] and Sundler et al. [145] showed that monoenoic and dienoic species of phosphatidylcholine were preferentially labelled in the liver a short time after the intraportal injection of labelled choline it follows that the cytidine pathway is the major one for choline incorporation into phosphatidylcholine in the liver in vivo. Experiments summarised by Kanfer [68] indicate that in the brain, too, the incorporation of choline into phosphatidylcholine by base exchange is very small in vivo. It may be concluded that base exchange of choline undoubtedly occurs in tissues but its special significance is unknown at the present time.

(iv) Acylation of lysophosphatidylcholine

The acylation of monoacyl glycerophosphocholine was first noted by Lands [161,162] and the early work has been summarised by Thompson [163]. Two enzymes are responsible. Lysolecithin acyltransferase (acylCoA: 1-acyl *sn*-glycero-3-phosphocholine-*O*-acyltransferase, EC 2.3.1.23) transfers a fatty acid to the 2-position of 1-acyl-glycero-3-phosphocholine, whereas 2-acylglycerophosphocholine acyltransferase (acylCoA: 2-acyl *sn*-glycero-phosphocholine-*O*-acyltransferase, EC 2.3.1.62) transfers a fatty acid to the 1-position of 2-acylglycero-3-phosphocholine. These reactions have been shown to be extremely important for the synthesis of phosphatidylcholine containing specific fatty acids but do not, of course, lead to synthesis de novo. These acyltransferases are distinct from those acylating glycerol-3-phosphate to form phosphatidic acid [164].

It is now clear that the type of fatty acid transferred in the acylation of 1-acyl or 2-acyl-*sn*-glycero-3-phosphocholine is largely determined by the position of the free OH-group on the lysophospholipid [165]. Saturated fatty acids are more readily transferred when the free OH-group is in the 1-position and unsaturated fatty acids when the OH-group in the 2-position is available. The K_m for stearoyl CoA, oleoyl CoA, linoleoyl CoA for acylation in the 2-position when liver microsomes are used as an enzyme source is 1–10 μM though the transfer of the saturated stearic acid is very small [165]. There is considerable preference for arachidonic acid in the acylation of 1-acyl-*sn*-glycero-3-phosphocholine [166] and it seems likely that acylation is the major method for the incorporation of this fatty acid into lecithin in mammalian tissues (see [138] for further evidence). There is some evidence that the nature of the saturated fatty acid in the 1-position determines the nature of the fatty acid introduced into the 2-position [166a]. In liver, acylation is the major route for the incorporation of palmitic acid into hexaenoic species of phosphatidylcholine [166b] and also for the incorporation of linoleic acid into the 1-stearoyl-2-linoleoyl-, but not 1-palmitoyl, species of phosphatidylcholine [167]. These and other acylations are discussed in great detail by Holub and Kuksis [51].

(v) Transacylation of lysophosphatidylcholine

A special type of acylation of lysolecithin was discovered by Erbland and Marinetti

[168] and Van den Bosch [169]. The reaction requires two molecules of 1-acyl-glycero-3-phosphocholine and yields 1,2-diacylglycero-3-phosphocholine and glycero-3-phosphocholine (the reaction, a lysolecithin-lysolecithin acyltransferase reaction, is sometimes referred to as lysophosphatidylcholine interesterification and is discussed in some detail by Holub and Kuksis [51]). By using as substrate 1-[^{14}C]acylglycero-3-[^{32}P]phosphocholine, Van den Bosch et al. [169] showed that the ratio of ^{14}C to ^{32}P in the lecithin produced by the reaction was twice that of the parent lysophospholipid. Though these workers consider it of only minor importance in liver, it may be important in the lung.

Since some of the most interesting work on the interrelationships of these pathways in recent years has concerned the lung, the metabolism of phosphatidylcholine in that tissue will now be discussed.

(vi) The metabolism of phosphatidylcholine in the lung

Phosphatidylcholine metabolism in the lung is unique because of the complexity of the pathways, the high degree of its saturation and because of the ability of the lung to produce a surfactant, rich in dipalmitoyl glycerophosphocholine, which is obligatory for the maintenance of the structural integrity of the alveoli [169a]. This surfactant is almost certainly produced by the type II alveolar cells, as first postulated by Macklin [170] and subsequently substantiated by a number of other workers. While containing some unsaturated phosphatidylcholine and other phospholipids (see Ch. 6) the surfactant is primarily composed of dipalmitoylglycerophosphocholine [171]. The phosphatidylethanolamine of both lung tissue and the surfactant has the more normal pattern of fatty acid distribution i.e. a saturated fatty acid in the 1-position and an unsaturated fatty acid in the 2-position. The fatty acids are brought to the lung by the blood in a free form or in the very low-density lipoprotein (VLDL) fraction synthesised predominantly in the liver or as phospholipids in chylomicrons from the intestine. The lung is also capable of the endogenous production of fatty acids, particularly palmitic acid.

The biosynthesis of the high concentration of disaturated and in particular dipalmitoyl-phosphatidylcholine has led to much speculation and many experiments (see [51,172] for excellent reviews) and the various pathways investigated are shown in Fig. 3. It would appear from the work of Epstein and Farnell [173] that, in the lung of the adult monkey, biosynthesis of phosphatidylcholine de novo occurs by the cytidine pathway (reactions 1–3, Fig. 3) with only about 3% by the methylation pathway. This conclusion receives support in a recent paper by Ishidate and Weinhold [174] which gives fairly convincing evidence that in the rat there is a sufficiently high turnover of dipalmitoylglycerol in vivo to account for the formation of dipalmitoyl-phosphatidylcholine by the cytidine pathway. Indirect evidence for the relative unimportance of the methylation pathway comes from the work of Vereyken et al. [138]. They showed that, whereas in liver where the methylation pathway is quantitatively important (p. 14), the fatty acid patterns of phosphatidylcholine and phosphatidylethanolamine are — as might be expected — similar, those of lung are very different. This is particularly obvious in the percentage of

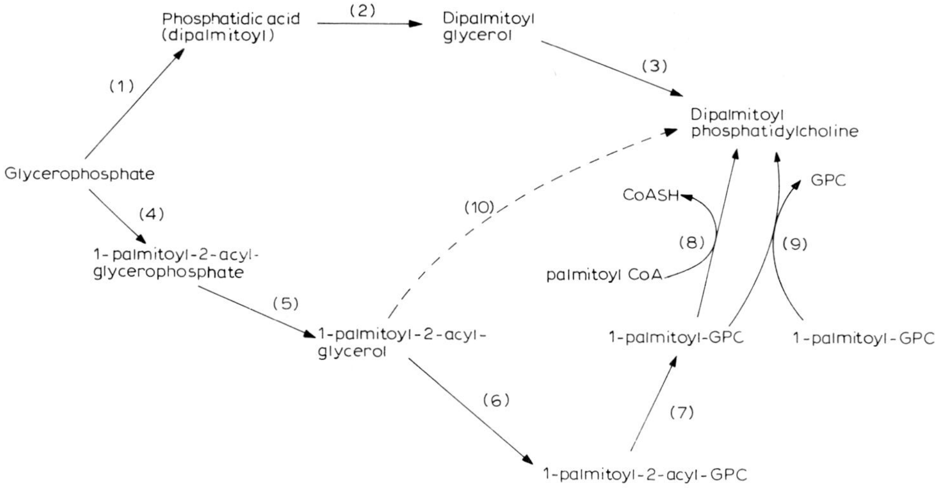

Fig. 3. Possible routes for the formation in lung of dipalmitoylphosphatidylcholine. The numbers in brackets refer to reactions described in the text. GPC, *sn*-glycero-3-phosphocholine (adapted from [172]).

C16:0 and C20:4 (see Table 2). It is, however, interesting that in the rabbit foetus 2–3 days before term a large peak of methyltransferase activity appears in the lung which drops again rapidly at birth.

However, there is considerable evidence for other reactions which are shown in Fig. 3. An important finding by Vereyken et al. [138] supports the occurrence of reaction (8) in lung. They found that, whereas the microsomal acyltransferase of liver showed a preference for unsatured fatty acids particularly C18:2 when 1-palmitoyl-glycerophosphocholine is the acceptor, lung microsomal fraction showed no such preference. On the other hand, Van Heusden et al. [175] noted that lung acyltransferase shows a preference for 1-palmitoyl- over 1-stearoyl-glycerophosphocholine as the acceptor. In the same paper they described the results of experiments in which radiolabelled lysophosphatidylcholines were injected intravenously and deduced that the desaturated phosphatidylcholines were not necessarily produced in the lung by the transacylation of the two lysophosphatidylcholine molecules by the action of lysophosphatidylcholine: lysophosphatidylcholine transacylase (reaction 9) when as in vivo, the lysophosphatidylcholine was taken up from the circulation.

It is quite possible that dipalmitoyl-phosphatidylcholine occurs in two pools in the lung. The microsomal phospholipase A_2, while showing the usual preference for an unsaturated fatty acid in the 2-position, can cleave palmitate from the phospholipid if it is membrane-bound (i.e. presumably formed by the cytidine pathway) [175a] but not from the same dipalmitoyl species if this is formed exogenously by the acylation of 1-palmitoyl-lysophosphatidylcholine received from the circulation, perhaps via reactions (4)–(7). One other possibility is that 1-palmitoyl-2-palmitoleoylglycerophosphocholine (probably synthesised in the lung by the cytidine pathway or taken in from the circulation) is converted to the dipalmitoyl species by biohydrogenation [49] (reaction 10).

5. Catabolic pathways

There are numerous enzymes in mammalian tissues capable of hydrolysing glycerophospholipids. The acylester groups are hydrolysed by phospholipase A_1 and A_2 and lysophospholipase, the phosphoglycerol linkage by phospholipase C and the phosphocholine (ethanolamine) linkage by phospholipase D. Though phospholipases have been known for a long time [7] as constitutents, for example of pancreatic secretions, venoms and bacteria, it is only in the last two decades that the nature and activities of the intracellular phospholipases of mammalian tissues have been examined. Since the phospholipases are dealt with in considerable detail in Chapters 9 and 10, the following account is only a summary of the enzymes which hydrolyse the glycerophospholipids whose metabolism is discussed in this chapter. Most of the investigations have been carried out on liver tissue.

It is clear that the hydrolysis products produced by the five enzymes may be the result either of sequential action or simultaneous action. Van den Bosch and Van Deenen [13] noted that synthetic phosphatidylcholines could be hydrolysed by rat liver homogenates to yield both 1-acyl- and 2-acylglycerophosphocholines both of which are known to be normal constituents of the tissue [13]. A phospholipase A_1 (phosphatid(at)e 1-acylhydrolase, EC 3.1.1.32), preferentially hydrolysing the acyl ester group in the 1-position of glycerophospholipids, was obtained from a particulate fraction of rat brain by Gatt et al. [176]. It had an optimum pH of 4.0 [177] (see also [178,179]). It was clear, however, from the work of Waite and Van Deenen [180] that another intracellular phospholipase A_1 exists in liver which is optimally active at physiological pH and present in the microsomal fraction. A similar enzyme was also found in brain microsomes by Woelk and Porcellati [181] and it is apparent that the enzyme prepared by Gatt and others is lysosomal in origin with optimal activity near pH 4.0. The microsomal enzyme requires Ca^{2+} for activity (though this was not observed in the original study by Waite and Van Deenen [180]) whereas the one in lysosomes does not; it is in fact inhibited by Ca^{2+} [182]. A phospholipase A_1 is also present in the liver cytosol [180] and plasma membranes [183,184] where it can also serve apparently as a transacylase and hydrolyse monoacylglycerol [185]. The different roles of these A_1 enzymes in the activity of the cell is unknown. Since the fatty acid in the 1-position of the glycerophospholipids is usually fully saturated the action of phospholipase A_1 is to release acids of this type and it is generally considered that phosphatidylethanolamine is more readily attacked than phosphatidylcholine [184]. Few studies on the specificity of the intracellular phospholipases A_1 appear to have been carried out. Hydrolysis of the 1-palmitoyl species of phosphatidylcholine by the enzyme from brain neurones was greater than that of the 1-(18:1) or 1-(18-2) analogues [186].

Phospholipase A_2 (phosphatide 2-acylhydrolase, EC 3.1.1.4) is probably the most thoroughly investigated phospholipase and is the phospholipase A of the older literature. As its name implies it specifically hydrolyses the fatty acid from the 2-position of a glycerophospholipid and this is usually occupied by an unsaturated fatty acid [187]. In the liver the typical A_2 enzyme is associated with mitochondria, has an optimum pH near 8.0 and is dependent on Ca^{2+} for activity [182,184]. The enzyme of brain mitochondria has a preference for phosphatidylethanolamine over phosphatidylcholine and has some activity towards phosphatidylserine [181]. The preparation of Waite and Sisson [187] from liver hydrolysed phosphatidylserine as rapidly as phosphatidylethanolamine. A phospholipase A_2 has been detected in the microsomal fraction of liver which can remove the fatty acid from the 2-position of both phosphatidylethanolamine and 1-alkyl-2-acylglycerophosphoethanolamine [188] and the optimum pH for this enzyme is 9.5.

Clearly, in vivo, one of the most important functions of phospholipases A_1 and A_2 must be the removal of fatty acids from the 1- and 2-positions respectively so that they can be replaced via the acyltransferase reactions (pp. 12 and 17) without the necessity for synthesis of the whole molecule de novo i.e. a remodelling process.

Most of the determinations of phospholipase A_1 and A_2 activity have been carried out in the presence of detergents which tend to suppress the activity of another phospholipase, lysophospholipase (lysolecithin acylhydrolase, EC 3.1.1.5). This enzyme has been known for many years and the early work has been well summarised [189,190]. It has been considered that another phospholipase exists in tissues known as phospholipase B which can remove fatty acids from both the 1- and 2-positions of diacylglycerophospholipids. Though the general view is that phospholipase B activity represents the combined actions of A_1 and A_2 followed by lysophospholipase (q.v.) the likelihood of genuine B activity has been resurrected recently (see Chapter 9). There is no doubt, however, that lysophospholipase is a distinct enzyme and is discussed fully by Van den Bosch (Chapter 9). It requires no cations for activity and is unspecific in that it can attack 1-acyl- and 2-acylglycerophosphocholine [191]. Extensive studies on the activity towards 1-palmitoylglycerophosphocholine of a microsomal enzyme from brain by Gatt and his co-workers [190,192] confirmed earlier observations [191,193] that the activity is inhibited by excess substrate. This is apparently caused by the adsorption of lysolecithin micelles onto tissue particles which inhibit the reaction because it is dependent on molecular solutions of substrate (see Van den Bosch, Chapter 9).

More than one lysophospholipase exists in tissues [194] and in liver the activity is largely associated with the microsomal fraction [180]. It is fair to say that most of the work on this enzyme has been carried out with lysolecithin as a substrate though the brain enzyme is known to attack 1-acyl-glycerophosphoethanolamine at about half the rate for the choline analogue [190].

Phospholipase C (phosphatidylcholine cholinephosphohydrolase, EC 3.1.4.3) is not a widely distributed or major phospholipase in mammalian tissue but is common in bacteria (e.g. *B. cereus, C. perfringens*, see Chapter 9). The enzyme liberating diacylglycerol from phosphatidylinositol is the best documented enzyme (see Chapter

7). Williams et al. [195] described a phospholipase C active towards phosphatidylethanolamine but not its lyso-derivative, in brain tissue. No divalent cations were required, though there appeared to be a requirement for some component of the cytosol. Recently a phospholipase C very active towards phosphatidylcholine has been found in lysosomes from liver [196,197]. A phospholipase C active towards phosphatidylserine does not appear to have been described.

The activity of phospholipase D (phosphatidylcholine phosphatidohydrolase, EC 3.1.4.4) may be considered as a special variant of transphosphatidylation:

$$\text{Phosphatidyl-O–R}' + \text{R}''\text{–OH} \overset{Ca^{2+}}{\rightleftharpoons} \text{phosphatidyl-O–R}'' + \text{R}'\text{–OH}$$

When the receptor R″-OH is not a primary alcoholic group but water, the reaction then becomes hydrolysis. There is no evidence at the present time that transphosphatidylation as opposed to hydrolysis is of any physiological significance. Since ethanolamine, serine and choline are primary alcohols it has been considered in the past that phospholipase D activity is identical with base exchange activity as originally proposed by Dils and Hübscher [67]. In fact phospholipase D activity per se was thought to be absent from mammalian tissues but in 1973 Saito and Kanfer [198] showed unequivocally that phosphatidic acid could be released from lysophosphatidylcholine or -ethanolamine by a solubilised preparation from rat brain. The enzyme responsible appeared to depend on Ca^{2+} or Mg^{2+} and, curiously, the pH optimum for its action as hydrolase and transphosphatidylase were different [199]. It has subsequently been purified (approx. M_r 200000), shown to be free from base-exchange activity and with K_m values of 0.75 and 0.91 mM for phosphatidylcholine and -ethanolamine respectively [200]. Brain tissue also contains a lysophospholipase D which can hydrolyse l-acyl-glycerophosphoethanolamine and -choline and is stimulated by Mg^{2+} but inhibited by Ca^{2+} [201,202]. A review of phospholipase D has been written by Heller [203] and its relation to base exchange activity has been discussed by Kanfer [68].

Glycerophosphocholine phosphodiesterase (EC 3.1.4.2) hydrolyses the water-soluble glycero-3-phosphocholine to glycerol-3-phosphate and choline. It was first demonstrated in *Serratia plymethicum* by Hayaishi and Kornberg [203a] and in liver by Dawson [204]. It is likely that the same enzyme hydrolyses glycero-3-phosphoethanolamine [204] but this aspect of the enzyme's activity has been virtually ignored. Baldwin and Cornatzer [205], however, showed that the enzyme from kidney could hydrolyse glycerophosphoethanolamine at about 40% of the rate for the choline compound (K_m for glycerophosphoethanolamine, 11.5 μM, and glycerophosphocholine, 2.2 μM). Since these phosphodiesters seem to be on the major catabolic pathway, at least in liver, the phosphodiesterase is an important enzyme for the release of choline and ethanolamine. Although Dawson [204] found the optimum pH for the liver enzyme to be 7.5, Baldwin and Cornatzer [205] found it to be 9.3 and Webster et al. [206] found the optimal activity of the brain enzyme to be around 9.0. At that pH the rate of hydrolysis by rat spinal cord was as high as 58 μmol/g tissue/h.

Although all authors agree that a metallic bivalent cation is required for activity because chelating agents such as EDTA effectively abolish it, the nature of the cation is unclear. Mg^{2+} ions have been routinely used in enzyme assays but no cation additional to that present in the tissue is necessary for optimal activity in vitro. Baldwin et al. [207] proposed Zn^{2+} as an essential requirement for the kidney enzyme though this is unlikely to be true for the enzyme in liver [204] or brain [206]. Recent observations on the brain enzyme [210] suggest that very low concentrations (approx. 0.1 μM) of Ca^{2+} may be required. Its subcellular distribution in liver [208] and brain [209,210] have been investigated.

From the first observations by Dawson [211] that glycerophosphocholine and glycerophosphoethanolamine are present in liver and in the degradative pathway of choline and ethanolamine glycerophospholipids in that tissue it has become clear that deacylation is a major step in their turnover and accounts in part for the activity of phospholipase A_1 and A_2 and lysophospholipase. Further details, together with the possible role of the lysosomal phospholipase C can be found in Chapter 9. It might be stated that there is no real evidence that phosphatidylserine is catabolised by the mechanism available for phosphatidylcholine and -ethanolamine and no glycerophosphoserine diesterase activity was detected in the kidney [205]. Clearly some phosphatidylserine is converted to phosphatidylethanolamine but whether this is the major catabolic pathway is unknown.

6. Aspects of sub-cellular metabolism

Of the enzymes involved in the metabolism of phosphatidylcholine and -ethanolamine in mammalian tissues only the kinases are definitely located in the cell cytoplasm though some cytidylyltransferase activity may be found there as was first observed by Wilgram and Kennedy [147]. The phosphotransferases and the methyltransferases are associated with the endoplasmic reticulum and it therefore follows that the phosphatidylcholine and -ethanolamine synthesised there are transferred to other organelles such as mitochondria and also to the plasma membrane. Recent work has indicated that the phosphotransferases and the methyltransferases are not randomly distributed in the endoplasmic reticulum but have asymmetric distribution in the membrane bilayer. This is not surprising since the glycerophospholipids themselves are now believed to have an asymmetric distribution in some membranes and this will be discussed first.

In 1977 Rothman and Lenard [212] wrote: “There is now compelling evidence that biological membranes are vectorial structures; that is, their components are asymmetrically distributed between the two surfaces”. Subsequent work has served to confirm this didactic statement and has been admirably reviewed by Op den Kamp [213,214]. Membrane proteins appear to demonstrate absolute asymmetry but lipids, including phospholipids, do not; nevertheless the asymmetry of the latter can be considerable as has been exhaustively shown for the erythrocyte membrane [213,214]. The original studies were by Bretscher [215,216] who used a non-penetrating probe (formylmethionyl methylphosphate) to show that most of the phosphatidylethanolamine and -serine is apparently located on the inside of the erythrocyte membrane and Zwaal and co-workers [217] who used phospholipases to show that phosphatidylcholine and sphingomyelin are preferentially located on the outside of the membrane. The location of phosphatidylcholine on the outside of the membrane has been convincingly confirmed by the use of the phosphatidylcholine exchange protein as a non-penetrating probe [214,218]. The available methods for investigation, including penetrating and non-penetrating probes, digestion by phospholipases and the use of phospholipid exchange proteins are described by Rothman and Lenard [212] and by Op den Kamp [214]. Clearly the differential distribution will affect the charge on either side of the membrane since phosphatidylserine has a net negative charge whereas phosphatidylethanolamine and -choline are essentially zwitterionic at physiological pH.

However, the precise location of phosphatidylethanolamine and phosphatidylserine is more doubtful [213,214] because the experimental evidence is conflicting. It is possible that phosphatidylethanolamine may be more flexibly located and may undergo transbilayer movement (called “flip-flop” by Kornberg and McConnell [219]). Movement of phosphatidylcholine by this process in the erythrocyte membrane is, however, very slow [218] and such a process would require a high activation energy.

The asymmetric distribution of phospholipids in other cell membranes has also been investigated but there is less agreement on the nature of this distribution than there is for that in the erythrocyte membrane [214]. There is a further complication in that the endoplasmic reticulum is the major site for phospholipid synthesis and the plasma membrane and intracellular organelles must receive their phospholipids from the endoplasmic reticulum. This was first shown for rat liver [220] and the extensive research on the process by which transfer takes place is discussed by Kader (Chapter 8).

Since the endoplasmic reticulum is the major site of synthesis (with which this chapter is concerned), studies on this subcellular component are important. It is highly likely also that asymmetry in this membranous cell component, if it exists, will arise during biosynthesis. There have, therefore, been a number of investigations on the nature of this particular asymmetry and how it can be accounted for by the metabolic activity of the endoplasmic reticulum. A pre-requisite of any such investigation in vitro is the knowledge that the isolated microsomal vesicles have the same orientation as the reticulum from which they derive with the cytoplasmic side of the membrane outside and the cisternal side of the membrane inside and this can be achieved with liver microsomes [221]. Higgins and Dawson [221] subjected such vesicles to the action of phospholipase C from *C. perfringens* and showed that their contents were not lost even though 50% of the phospholipids were hydrolysed. The composition of the hydrolysed phospholipids was shown to be 85% phosphatidylcholine, 8% phosphatidylethanolamine, 1% phosphatidylserine and 9% sphingomyelin, which were therefore assigned to the outer leaflet of the vesicle bilayer i.e. the cytoplasmic side. Under circumstances which resulted in the opening of the vesicle, further hydrolysis by phospholipase C occurred, presumably the hydrolysis of the phospholipids of the inner leaflet (cisternal side) whose composition was concluded to be 28% phosphatidylcholine, 34% phosphatidylethanolamine, 9% phosphatidylserine and 5% sphingomyelin; the remainder was assumed to be phosphatidylinositol which was not hydrolysed by the phospholipase C used. The results of these experiments, however, do not agree with those of Sundler et al. [222] and Nillson and Dallner [223,224] who used phospholipase A_2 as a probe and decided that phosphatidylcholine was distributed across the bilayer and that phosphatidylethanolamine was in the outer layer (cytoplasmic side). These experiments in turn, have been criticised more recently by Higgins [225,226] who considers either that phospholipase A_2 opens microsomal vesicles allowing access to the inner layer or that it induces translocation. These discrepancies have been summarised by Op den Kamp [213] who concludes not surprisingly that further work is necessary. In this connection Higgins [226] points out that observations on "inside-out" vesicles would be useful but these have yet to be prepared, though Nillson and Dallner [224] made many attempts.

However, it is becoming clear that, unlike the erythrocyte membrane in which transbilayer mobility is small, the microsomal membrane may exhibit considerable mobility in terms of its phospholipids especially when probes such as phospholipases or exchange proteins are used. Some indications of this mobility which may occur by lateral or transverse diffusion as well as "flip-flop", have been shown by experiments in which the phosphatidylcholine has been labelled in vivo or in vitro before treatment with probes. Zilversmit and Hughes [227] labelled the phospholipids of liver endoplasmic reticulum with ^{32}P in vivo using both normal rats and partially hepatectomised animals in which growth would increase phospholipid synthesis. In the partially hepatectomised animals it is inconceivable that any phospholipids in the inner leaflet would remain unlabelled during the 3-day period of labelling. When microsomal vesicles were prepared from the liver 80–90% of the phosphatidylcholine (and other glycerophospholipids; phosphatidylethanolamine and phosphatidylserine) was exchangeable when incubated with beef heart mitochondria and exchange protein, though the vesicles remained intact. Similar results were obtained by Van den Besselaar et al. [228]. In further experiments Zilversmit [229] used microsomes prepared from the liver of rats 1 h after the administration of [*Me*-^{14}C]choline on the assumption that any asymmetry of labelling would be exaggerated. Incubation of such microsomes with phospholipase A_2 showed that the specific radioactivity of the lysophosphatidylcholine produced after a short period of hydrolysis was the same as that produced after long periods of hydrolysis. If the phosphatidyl [*Me*-^{14}C]choline had been labelled predominantly in the outer layer the specific radioactivity of that initially hydrolysed should have been higher than that hydrolysed after a longer period.

Zilversmit concluded that movement across the bilayer is very rapid and not specifically induced by the presence of exchange protein. Different conclusions were drawn by Higgins [225] who labelled the endoplasmic reticulum of liver in vivo with [*Me*-^{14}C]choline and prepared labelled microsomes. In complementary experiments she labelled the phosphatidylcholine of isolated unlabelled microsomal vesicles from CDP[*Me*-^{14}C]choline. On incubation of the microsomal vesicles with phospholipase C (*C. perfringens*) 89% of the label was removed but only 50% of the total phospholipid was hydrolysed and the vesicles remained intact. These experiments indicated that labelling of the phosphatidylcholine by the cytidine pathway takes place on the outer leaflet and that transfer to the inner leaflet does not take place.

Op den Kamp [213] points out clearly that these conflicting results might be explained in terms of the effects of the various probes used and the ionic environment during incubation with such probes. In other words, under some conditions transbilayer movements could be stimulated and in others suppressed. Therefore, no precise statement of the distribution in the endoplasmic reticulum of glycerophospholipids in general and phosphatidylcholine in particular can be made. In terms of biosynthesis, however, it does seem that the choline phosphotransferase is associated with the outer leaflet since Coleman and Bell [230] showed that 90% of this enzyme could be hydrolysed by chymotrypsin or pronase though the vesicles remained intact. This location has also been demonstrated by Vance et al. [231] and both groups of workers have shown that ethanolamine phosphotransferase has the same location. CDP-choline cannot penetrate microsomal vesicles [232] and since cholinephosphate cytidylyltransferase is either cytoplasmic or on the cytoplasmic surface of the endoplasmic reticulum [231] and further, since the formation of phosphocholine occurs in the cytoplasm, it seems likely that the formation of phosphatidylcholine in the liver by the cytidine pathway takes place on the cytoplasmic surface. The same may be true for phosphatidylethanolamine [231] which would imply that phosphatidylethanolamine is formed on the cytoplasmic surface. An excellent review of this topic has appeared recently [233] and Fig. 4 attempts to summarise current views.

But there is a further complication in that 20% of liver phosphatidylcholine is formed by the stepwise methylation of phosphatidylethanolamine (p. 14) [103], a process which has received considerable attention from Axelrod and his co-workers who, however, have worked with tissues other than liver. The two enzymes responsible for the methylation have been described on p. 14 and were first identified in the microsomal and mitochondrial fractions of the adrenal medulla by Hirata et al. [131]. In extensive studies on erythrocyte ghosts [234] prepared "right-side-out" and "inside-out" it was demonstrated by tryptic digestion that methyltransferase I, which forms phosphatidyl *N*-monomethylaminoethanol, is on the inside of the erythrocyte membrane; the activity of methyltransferase II yielding phosphatidylcholine is on the outside. The formation of phosphatidylmonomethylaminoethanol only took place when *S*-adenosyl-[*Me*-^{3}H]-L-methionine was either introduced into the interior of "right-side-out" ghosts or added to medium in contact with the cytoplasmic side of "inside-out" ghosts, i.e. the first methylation took place on the cytoplasmic or internal layer of the ghost. These and other experiments with phospholipase C confirmed the asymmetry in the ghosts and showed that phosphatidylcholine was formed on the external surface. It was further demonstrated that, if *S*-adenosyl-[*Me*-^{3}H]-L-methionine was incubated with "inside-out" ghosts, and these incubated with

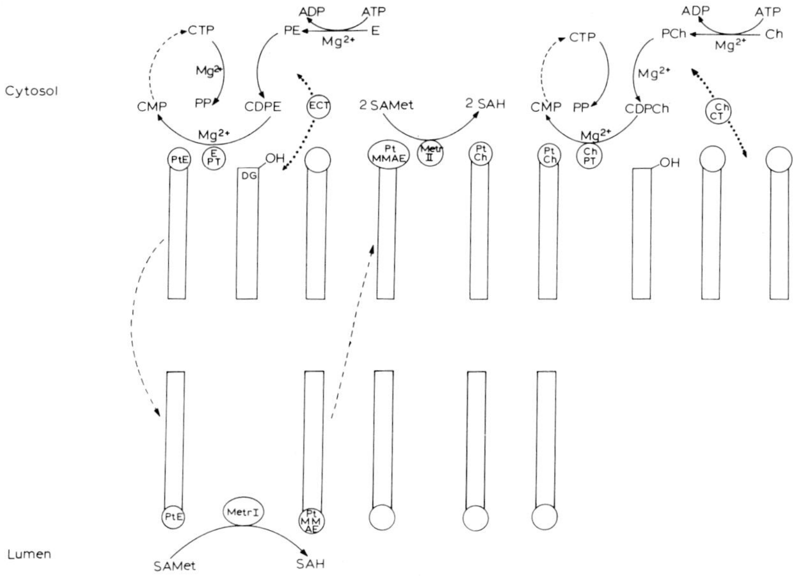

Fig. 4. A diagram showing the likely subcellular locations of the reactions leading to the formation of phosphatidylcholine and phosphatidylethanolamine in the membrane of the endoplasmic reticulum and cytoplasm. See text for details. *Key*: CDPCh, cytidine diphosphate choline; PCh, phosphocholine; Ch, choline; CTP, cytidine triphosphate; CMP, cytidine monophosphate; E, ethanolamine; PE, phosphoethanolamine; CDPE, cytidine diphosphate ethanolamine; PtE, phosphatidylethanolamine; PtMMAE, phosphatidylmonomethylaminoethanolamine; PtCh, phosphatidylcholine; DG, diacylglycerol; SAMet, *S*-adenosylmethionine; SAH, *S*-adenosylhomocysteine; ECT, ethanolamine phosphate cytidylyltransferase; ChCT, choline phosphate cytidylyltransferase; EPT, ethanolamine phosphotransferase; ChPT, choline phosphotransferase; Metr I, methyltransferase I; Metr II, methyltransferase II; ...., distribution between membrane and cytosol.

phospholipase C, then the phospholipid unavailable for hydrolysis increased i.e. it had been transferred to the external surface which now faced inwards. The lag phase was very short (2 min), which indicated that this metabolically induced diffusion was extremely rapid compared with transbilayer movements not so induced.

Erythrocytes are a special case in that they are incapable of net synthesis of phospholipids but are capable of their modification. This modification, which appears to involve only very low levels of methylation, may have a functional significance all its own [235] though this has been challenged [236]. A variation on the methylation processes occurring in the liver has recently been described by Higgins [226] in that phosphatidylcholine formed by stepwise methylation of phosphatidylethanolamine appeared to be formed on both sides of the bilayer of liver microsomal vesicles but the first two products of methylation could be formed only on the inner leaflet. Though the phosphatidylmonomethyl- and phosphatidyldimethylaminoethanol could be translocated to the outer leaflet, the pools of phosphatidylcholine on each side could not equilibrate.

Thus, in the liver endoplasmic reticulum it appears that the cytidine pathway produces phosphatidylcholine in the outer leaflet and the methylation pathway the same phospholipid by sequential methylation in first the inner and then the outer leaflet. The significance of this is obscure at the present time. The methylation pathway considered in isolation is the only one available for the synthesis of choline while the cytidine pathway can presumably re-cycle choline that has escaped oxidation to betaine by the combined action of mitochondrial choline dehydrogenase (EC 1.1.99.1) and betaine aldehyde dehydrogenase (EC 1.2.1.8). To what extent choline liberated from phosphatidylcholine via the action of phospholipases and glycerophosphocholine diesterase is re-used by the cytidine pathway is unknown but after the intraperitoneal injection of [*Me*-^{14}C]choline 70% of the water-soluble radioactivity found in the liver (rat) was in the form of betaine and 30% as phosphocholine within 1 h [237].

7. Transport in the body

The movement of phospholipids in the blood and other body fluids is an extremely complex phenomenon which is incompletely understood. Most investigations have been concerned with phosphatidylcholine and lysophosphatidylcholine which are transported in body fluids in lipoprotein complexes of variable size, primarily dictated by the triglyceride content.

(a) Absorption and the formation of chylomicrons

For an admirable summary of the work up to 1972 the reader is referred to the review by Coleman [238]. Phosphatidylcholine enters the circulation from the lumen of the small intestine from two sources: the diet and the bile. Pancreatic phospholipase A_2 removes the fatty acid from the 2-position of phosphatidylcholine to give lysophosphatidylcholine which is taken up by the endothelial cells of the multitudinous villi which characterise the inner wall of the small intestine and is there re-acylated in the 2-position with an unsaturated fatty acid by the action of lysophosphatidylcholine acyltransferase. There is a certain degree of confusion about this uptake in that, while in man both dietary and biliary phosphatidylcholine are readily hydrolysed by pancreatic phospholipase A_2 [239], in the rat the biliary phosphatidylcholine is resistant to this enzyme. It has therefore been proposed by Boucrot [240] that there is an enterohepatic circulation of biliary phosphatidylcholine in the rat. It may be said in passing that the rat has no gall bladder and this may have some bearing on the different fates of phosphatidylcholine in the two species.

The phosphatidylcholine is transferred from the endothelial cells of the small intestine to the lymphatic system and is carried in the lymph in the form of chylomicrons which are spheres with a triglyceride core and an outer membrane composed of phospholipid, cholesterol and protein. Among the proteins is apoprotein

A I which is synthesised in the intestinal wall and activates the lecithin-cholesterol acyltransferase (LCAT, EC 2.3.1.43)

phosphatidylcholine + cholesterol → 1-acylglycero-3-phosphocholine + cholesterol ester

The chylomicrons are rapidly transferred in the circulation to sites such as the skeletal muscle and adipose tissues. The triglyceride core is largely lost and the resulting much smaller sphere of low-density lipoprotein (LDL) contains only a small residue of triglyceride and cholesterol ester in the core.

On reaching the liver the chylomicrons have been reduced to chylomicron remnants and have lost most of their triglyceride core to adipose tissue and skeletal muscle. The cholesterol esters from all the lipoprotein particles are taken up primarily by the parenchymal cells and the remaining 6% by the Kupffer cells [241]. The apolipoproteins are broken down at a much higher rate in the non-parenchymal cells when compared with the rate in the parenchymal cells. This is particularly true of the VLDL particles [242].

(b) High-density lipoproteins

The high-density lipoproteins (HDL) are the particles richest in phospholipid and apoprotein AI. It has been postulated [243] that these particles possess the original surface remnants of the nascent chylomicrons in their final form while the LDL contain the core remnants. Smith et al. [244] divide the HDL particles into two sub-classes HDL_1 and HDL_2. The HDL particles in the circulation are derived from both the liver and intestine and are primarily the products of lipolysis.

As has been mentioned above, LCAT is activated by apoprotein AI and to a lesser extent by apoprotein CI. Apoprotein AI is found in the HDL fraction and its effect on the LCAT is greatest when the phosphatidylcholine contains unsaturated fatty acids, particularly the C18:2 fatty acid [245]. Apoprotein CI is equally active in the presence of saturated and unsaturated fatty acids [244]. While some of this lysophosphatidylcholine can pass through into the arterial wall and be converted into phosphatidylcholine [246], much of it remains in the HDL particles and is carried to the liver. It has been shown by Smith et al. [244] that the polar head groups on the phospholipid of the HDL particle are not essential for maintaining the supramolecular properties of the lipoprotein. All plasma lipoproteins have, however, a high affinity for lysophosphatidylcholine as do all cellular membranes including plasma membranes. It has been suggested by Portman and Alexander [247] that lysophosphatidylcholine acts as a bond between the lipoprotein particles and the tissue.

(c) The liver and the production of phospholipids for bile and plasma

In liver there is a rapid turnover and metabolism of phospholipids. The liver is capable of acylating lysophosphatidylcholine to phosphatidylcholine and can also synthesise the diacylglycerophospholipids containing choline and ethanolamine by the cytidine pathway (pp. 9, 12, 14, 17). Moreover, it can convert phosphatidylethanolamine to phosphatidylcholine by stepwise methylation, the only organ to do this to a significant extent (p. 14).

Kawamoto et al. [32] have postulated two pools of phosphatidylcholine in liver, a dynamic pool supplying the phospholipid directly to the bile and a second, so-called static pool, which transfers the phospholipid from the liver to the plasma. In the liver endoplasmic reticulum the predominant fatty acid of the phosphatidylcholine is stearic acid while the bile contains predominantly 1-palmitoyl phosphatidylcholine. Bile canalicular membranes have a very low capacity to synthesise phospholipids. The choline phosphotransferase is virtually absent and the activity of lysolecithin acyltransferase, though present, is very low i.e. 17% of the activity of the liver endoplasmic reticulum [47]. It can be seen from the table (Table 5) that the phospholipid content of the liver microsomes from the endoplasmic reticulum and the bile canalicular membranes is very different from that found in the bile [47]. There is also a marked difference in the fatty acid pattern. The bile contains 53.7% of C16:0, C18:2 and 14.7% of C16:0, C20:4 while the bile canalicular membranes have only 19.7% of C16:0, C18:2 and 33.7% of C18:0 combined with either C18:2 or C20:4. Because of the very low enzyme activity of bile canalicular membranes it would appear that the source of bile phospholipids is mainly and possibly exclusively synthesis de novo in the liver and the products of this synthesis may well be the dynamic pool postulated by Kawamoto et al. [32]. This pool must be produced at a high rate as bile contains 3–4 μmol phosphatidylcholine per ml, so a very considerable amount of phosphatidylcholine is poured into the intestine from the bile duct.

The transport of phosphatidylcholine through the liver membrane system is still somewhat obscure, but the production of micelles appears to be obligatory. In experiments in which rat livers were perfused either with sodium taurocholate, a good micelle-forming conjugated bile salt or with a glycine conjugate of dehydrocholate, a poor micelle-forming compound, Young and Hanson [248] found that, although both perfusates increased bile flow, the sodium taurocholate raised the

TABLE 5

The phospholipid content of liver microsomes, bile canalicular membranes and bile

	Liver microsomes	Bile canalicular membranes	Bile
Phospholipids	μmol/mg protein	μmol/mg protein	μmol/ml
Total	0.31	0.65	4.32
	% phospholipid	% phospholipid	% phospholipid
Phosphatidylcholine	54.5	36.0	90.0
Lysophosphatidylcholine	1.5	2.3	0.3
Phosphatidylethanolamine	23.9	22.7	4.4
Phosphatidylserine	3.4	14.8	1.2

[a] These values are very similar to those of liver plasma membranes [47].

phosphatidylcholine content of bile while the dehydrocholate caused a decrease. This would substantiate the theory that phosphatidylcholine must be carried in a micellar form. Swell et al. [249] obtained a similar result from experiments in which dog livers were perfused with or without taurocholate. These workers also measured the uptake of $^{32}P_i$ into plasma and liver and found that taurocholate increased uptake of this radiolabel into phosphatidylcholine and phosphatidylethanolamine in plasma and liver as well as in bile. It would thus appear that plasma and bile have the capacity for selecting the phospholipid species they require from the large and varied lipid pool of liver.

Interesting experiments have been carried out by Jackson et al. [250] on rat liver microsomes, in which the phosphatidylcholine had been prelabelled in vivo with [*Me*-^{14}C]choline, and incubated with lipoproteins from human plasma and phosphatidylcholine exchange protein (see Chapter 8). Examination of the VLDL, LDL and HDL fractions demonstrated that, although their phospholipid content was unchanged after incubation, there was a transfer to the lipoproteins of 40–60% of the labelled phosphatidylcholine from the microsomes in 45 min. This is a rapid exchange.

Though the mode of transfer of phospholipid from the liver to bile is becoming clearer, the mode of transfer of phospholipid from the liver to other tissues via the plasma lipoproteins is still relatively obscure. It has been known for some time that lysophosphatidylcholine and lysophosphatidylethanolamine labelled with ^{32}P and [^{14}C]palmitate when introduced into the plasma as a complex with serum can be taken up by many tissues and converted to their diacyl analogues by acylation [251]. Comparable experiments were carried out by Illingworth and Portman [252] who showed that when doubly labelled lysophosphatidylcholine, i.e. labelled with [^{14}C]palmitate and [*Me*^{3}H]choline, was injected intravenously into squirrel monkeys it was rapidly taken up intact by the brain, acylated and metabolised within that organ as though it had been assimilated into the endogenous pool. Indirect evidence that the brain receives a significant amount of lipid bound choline in this form has been provided by Ansell and Spanner [253,254]. They found that when [^{14}C]ethanolamine was injected intraperitoneally into rats, choline-labelled lipids were found in liver, plasma and brain which could not have been derived from water-soluble precursors. The implication was that the phosphatidyl-[^{14}C]ethanolamine formed in the liver was converted to [^{14}C]choline-labelled glycerophospholipids which entered the plasma and passed to the brain.

The pregnant rat affords an interesting example of the transfer of lysophosphatidylcholine within the body. When [1-^{14}C]palmitoylglycerophosphocholine was injected into pregnant rats there was a rapid disappearance from the circulation and uptake into the liver and placenta of the mother but only a trace of the radioactivity, after a long time interval, appeared in the foetal liver [255]. Eisenberg et al. [255] concluded that the maternal phospholipid did not pass from the placenta to the foetus. The small amount of radioactivity found in the foetus was accounted for as fatty acid from catabolised phospholipid since ^{32}P from ^{32}P-labelled lysophosphatidylcholine was undetectable.

Lysophosphatidylcholine accounts for 27% of the total phospholipid of human plasma (Table 1) [256] and it has been pointed out (p. 29) that plasma membranes have a great affinity for this phospholipid. However, the major phospholipid component of the HDL, LDL and VLDL is phosphatidylcholine [37], and it is not possible to say what exactly is the "preferred" form of transport of choline glycerophospholipids to tissues. No definite information exists on phosphatidylethanolamine or phosphatidylserine, both of which are minor constituents of plasma.

Any investigation is made much more complex by the fact that phosphatidylcholine cannot only donate a fatty acid to cholesterol by the action of LCAT but can take part in other intraconversions. Thus, plasma has been shown capable of converting lysophosphatidylcholine to phosphatidylcholine by the enzyme lysolecithin acyltransferase by incubating human plasma with [1-^{14}C]palmitoyl glycerophosphocholine. Subbaiah and Bagdade [257] demonstrated that labelled phosphatidylcholine was formed at a rate of 6 nmol/ml/h. Free [^{14}C]oleic acid and [^{14}C]palmitic acid were not incorporated into phosphatidylcholine. In two experiments, the same workers showed that plasma lipoproteins were essential for the formation of the diacylglycerophospholipids. Webster [258] failed to demonstrate the enzyme lysolecithin acyltransferase and Stein and Stein [251] could find little degradation of injected lysophosphatidylcholine in plasma. So apart from its conversion to phosphatidylcholine and the reverse reaction, there appears to be no metabolism of the lysophosphatidylcholine in plasma and only the LCAT system involves phosphatidylcholine. There is, however, an exchange of both phosphatidylcholine and lysophosphatidylcholine between plasma and the red blood cells. Mulder and Van Deenen [259] demonstrated the following pathways:

SERUM		RED BLOOD CELLS
phosphatidylcholine	$\rightleftharpoons$	phosphatidylcholine
		↑ Fatty acyl CoA
lysophosphatidylcholine	$\rightleftharpoons$	lysophosphatidylcholine

Since then, this exchange mechanism has been studied in more detail. The exchange appears to be unrelated to the presence of a serum lipoprotein fraction but to be controlled by the concentration of serum phospholipid. Only a part of the red blood cell phosphatidylcholine appears to be available for exchange and of this pool the tetraenoic acid-containing phosphatidylcholine shows the greatest degree of exchange. The exchangeable pool is increased on lysis of the cell, the disrupted membranes demonstrating nearly three times the exchange capability of the intact cell [260]. The membrane of the red blood cell has 90% of its phospholipid in the form of phosphatidylcholine, phosphatidylethanolamine, phosphatidylserine and sphingomyelin [261]. The outer surface contains most of the phosphatidylcholine and sphingomyelin and one-fifth of the phosphatidylethanolamine while the inner surface contains phosphatidylserine, most of the phosphatidylethanolamine and only a small amount of phosphatidylcholine and sphingomyelin [262], see p. 24.

An exchange of phosphatidylcholine also occurs between platelets and the phosphatidylcholine of plasma HDL. Bereziat et al. [263] demonstrated that both 1-acyl-2-[1-^{14}C]arachidonyl glycerophosphocholine and 1-acyl-2-[1-^{14}C]linoleoyl glycerophosphocholine in HDL particles would exchange with platelet phosphatidylcholine, the linoleoyl species showing the greater exchange rate.

(d) Metabolism in amniotic fluid

The amniotic fluid surrounding the foetus is rich in phospholipids, particularly phosphatidylcholine and sphingomyelin. The composition of the phosphatidylcholine is almost identical to the lung surfactant from which it derives during the development of the foetus in that over 80% of the fatty acids are saturated. Pulmonary surfactant in the monkey contains 13% protein, and 85% lipid of which 75% is phosphatidylcholine [171], predominantly 1,2-dipalmitoyl glycerophosphocholine. During development in utero, the foetal lung and its surfactant develop, and much of this surfactant is expelled into the amniotic fluid. The sphingomyelin content is relatively stable but the phosphatidylcholine content rises rapidly towards term. The so-called L/S ratio (phosphatidylcholine/sphingomyelin) in amniotic fluid is used extensively to judge the age of the foetus. For example in the rhesus monkey the L/S ratio at 120 days gestation is 0.23 while at term (166 days) it has risen to 8.13 [264]. The L/S ratio is also used to detect the risk of neonatal respiratory distress syndrome [265], p. 41. The phosphatidylcholine in lung membranes is formed from choline by the cytidine pathway in preference to stepwise methylation of phosphatidylethanolamine [173]. Thus, in the foetal rhesus monkey of 120-day gestation the ratio of the rate of the synthesis by the cytidine pathway to that by the methylation pathway is approx. 100:4, in the 24-h neonate 217:4 and in the adult 137:4. The dipalmitoyl phoshatidylcholine, the chief constituent of the surfactant, may be formed by a deacylation followed by re-acylation with palmitate or possibly by the combination of two saturated lysophosphatidylcholines. There is also the possibility of a stepwise methylation [266]. The recent advances in lung surfactant biochemistry have been discussed earlier (p. 18).

8. The effects of drugs and other agents on metabolism

Until recently there has been little indication that the metabolism of choline-, ethanolamine- and serine-containing phospholipids can be modified by the action of drugs, hormones or other agents. Many of the effects reported in the older literature were indirect. It is also reasonable to state that, since there is little evidence for metabolic diseases or disorders involving these particular phospholipids, there has been little stimulus to investigate drug action. In recent years, however, there have been some findings which may prove to be of great significance in relation to drug action.

(a) Some effects on biosynthesis

One or two studies have demonstrated inhibition or stimulation of biosynthetic mechanisms. Possmayer et al. [267] showed that administration of oestradiol-17β to pregnant rabbits resulted in a significant rise in the capacity of the foetal lung to incorporate [^{14}C]choline into choline glycerophospholipids. It was concluded that the oestradiol-17β administered to the mother increased the activity of cholinephosphate cytidylyltransferase (but not the total amount of enzyme) in the foetal lung; ethanolaminephosphate cytidylyltransferase was unaffected. It is interesting, therefore, that in another study it has been shown that a potent tumour-promoting agent, 12-*O*-tetradecanoyl-phorbol-13-acetate, accelerates synthesis by stimulating this enzyme in HeLa cells [268]. Other studies [269] have shown that certain agents which are well-known inducers of the microsomal drug-metabolising systems of liver have interesting effects on the cytidine pathway. After only two daily injections, phenobarbitone decreased the pool size of phosphocholine but 3-methylcholanthrene increased it. This could be explained by the finding that phenobarbitone decreased the activity of choline kinase whereas 3-methylcholanthrene increased it [270]; ethanolamine kinase was not affected. In addition the activity of the microsomal, but not the cytosolic, cytidylyltransferase, probably the rate-limiting step in phosphatidylcholine synthesis, was depressed by phenobarbitone. Microsomal cholinephosphotransferase activity was decreased after 3-methylcholanthrene and increased after pentobarbitone. These different effects may be related to the fact that the two compounds function differently as inducers of drug-metabolising enzymes because polychlorinated biphenyls, which share the differing inducing capacities of both phenobarbitone and 3-methylcholanthrene, had effects on the enzymes intermediate between those of the other compounds [270]. The final effects of phosphatidylcholine metabolism in the liver were found to be rather complex. For example polychlorinated biphenyls, in particular, caused a considerable decrease in the synthesis of phosphatidylcholine in the liver microsomal fraction and reduced the labelling of phosphatidylcholine secreted into the plasma, though the effect was also seen to a lesser extent with the other two compounds.

In the experiments mentioned above the agent presumably either prevented or stimulated the synthesis of choline kinase rather than inhibited its action. In fact, few inhibitors of this enzyme are known. Hemicholinium-3 ($\alpha\alpha'$-dimethylethanolamino-4,4'-bisacetophenone) has been shown to be a largely uncompetitive inhibitor of the brain enzyme which can be demonstrated both in vivo [271] and in vitro [272]. In vivo the incorporation of [*Me*-^{14}C]choline into phosphatidylcholine was, however, enhanced, which may have been due to indirect stimulation of the base-exchange reaction (p. 16) paralleling the inhibition of the kinase.

(b) The modulation of methylation and decarboxylation by drugs and neurotransmitters

One of the most significant areas of research in relation to the effects of hormones, drugs and neurotransmitters in the last 20 years has been concerned with the manner in which the final biological response of the cell to these agents occurs. In some instances the response is rapid e.g. the opening of the sodium channels in response to acetylcholine when it binds to the nicotinic receptor. In others the response is relatively slow as in the response to acetylcholine at the muscarinic receptor. The discovery that the levels of cyclic AMP can be raised inside cells as a result of stimulation by a whole variety of agents was important in that it showed that stimulation of relevant receptors could cause intracellular responses which could not only amplify the effect but modulate intracellular events.

A more recently discovered metabolic response is the hydrolysis of phosphatidylinositol (Chapter 7) which occurs in response to a number of agents which have one thing in common – they do not affect levels of cyclic AMP.

An early response which may be of wide biological significance since it occurs within the membrane, has been discovered by Hirata and Axelrod [234]. Essentially they have shown that the small amount of methylation of phosphatidylethanolamine that occurs in a wide variety of cells is a consequence of the action of two methyltransferases (p. 14) and can be modulated by a wide variety of agents. An important observation was that the synthesis of phosphatidylmonomethylaminoethanol in erythrocyte ghosts was accompanied by a change of microviscosity from 1.62P to 1.09P [273]. With ghosts of immature erythrocytes (reticulocytes) which contain β-adrenergic receptors coupled to adenylate cyclase (EC 4.6.1.1) it was shown that L-isoproterenol, a β-agonist, stimulated methylation and the translocation of the methylated phospholipid and stimulated adenylcyclase activity simultaneously. The two activities increased in parallel and both were blocked by a β-antagonist (propranolol). Fig. 5 shows the proposed mechanism [274]. The idea is that the increased fluidity of the reticulocyte membrane enhances the mobility of the β-receptor, facilitating the coupling of this receptor with the adenylcyclase on the cytoplasmic surface of the membrane.

In further experiments [275] on reticulocytes it was shown that when methylation proceeded to completion with the increased formation of phosphatidylcholine then the number of β-adrenergic binding sites increased. It appeared that synthesis of phosphatidylcholine exposed binding sites not available in the absence of the methylation. It has also been observed that benzodiazepines stimulated phospholipid methylation in C_6 astrocytoma cells in a dose-dependent fashion and the more potent the benzodiazepine was in displacing [^{3}H]diazepam from the receptor the more effective it was in stimulating methylation. These receptors are "peripheral-type" benzodiazepine receptors. In the same cells β-adrenergic agonists were also effective in stimulating methylation but clearly the domains of methylation responding to stimulation of β-receptors were different from those responding to benzodiazepines since the two drugs produced additive methylation [276].

One of the most intriguing studies in this series of experiments is that on the release of histamine from mast cells, an important pathological response in allergic reactions. It is generally agreed that the release of histamine is initiated when a divalent or multivalent antigen reacts with two adjacent immunoglobulin E (IgE) molecules attached to the external surface of the mast cell. It has also been known

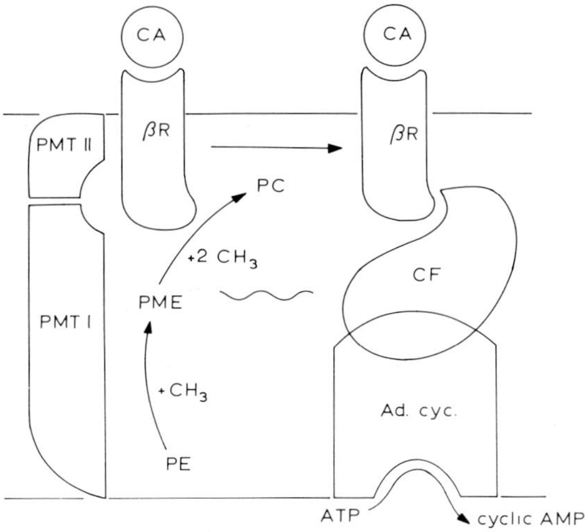

Fig. 5. A proposed mechanism for the coupling of phospholipid methylation to the β-adrenergic receptor. When a catecholamine (CA) binds to a β-adrenergic receptor (R), it stimulates phospholipid methyltransferase I (PMT I) and phospholipid methyltransferase II (PMT II). This increases the methylation of phosphatidylethanolamine (PE) to phosphatidyl-*N*-monomethylethanolamine (PME) and to phosphatidylcholine (PC). As the phospholipids are methylated they 'flip-flop' and increase fluidity (wavy line). This facilitates the lateral mobility of the β-adrenergic receptor to interact with the guanylnucleotide coupling factor (CF) and adenylate cyclase (Ad. cyc.) to generate cyclic AMP. Reproduced from [234] by kind permission of the authors and the American Association for the Advancement of Science (copyright, 1980, by the American Association for the Advancement of Science).

for some time that external calcium is necessary and that the cross-linking of the IgE receptors with an anti-receptor antibody causes an influx of Ca^{2+} and the release of histamine [277]. An influx of Ca^{2+} into the mast cell facilitated by the ionophore A23187 also caused histamine release [278]. There have also been a number of reports that phosphatidylserine enhances dextran- or antigen-induced histamine release in the presence of Ca^{2+} and in a dose-dependent manner (for refs. see [279]). It has now been suggested that cross-linking of the receptors is linked to the opening of calcium channels by a sequence of events involving the decarboxylation of phosphatidylserine, methylation of the phosphatidylethanolamine so produced and a subsequent deacylation to lysophosphatidylcholine.

Hirata et al. [280] showed that concanavalin A could cross-link adjacent IgE molecules, including release of histamine and also methylation of phospholipids. Blocking the binding of concanavalin A inhibited methylation and release; inhibition of methylation with *S*-adenosylhomocysteine also inhibited release. Thus the three processes appear to be linked [280]. Further experiments confirmed that phosphatidylserine was an essential component of this process and that added phosphatidyl [^{14}C]serine could be incorporated into the mast cell membrane, be decarboxylated and the resulting phosphatidylethanolamine methylated as a continuous sequence resulting in phosphatidylcholine formation. The finding of labelled lysophosphatidylcholine indicated yet a further metabolic step related to histamine release, namely the action of phospholipase A_2.

Because Ca^{2+} ions are necessary for histamine release, Ishizaka et al. [281] performed further experiments on the sequence of events occurring in the mast cell when exposed to divalent or monovalent fragments of the antibodies to IgE receptors of rat basophil leukaemia cells. In Fig. 6 it can be seen that the divalent fragments caused a transient increase in methylation followed by an influx of calcium ions which paralleled histamine release. Such a sequence of events could not be produced by monovalent fragments which are incapable of bridging. Inhibition of methylation by prior incubation with the methyltransferase inhibitor 5′-deoxyisobutylthio-3-deaza-adenosine inhibited the influx of Ca^{2+} and the release of histamine from challenged mast cells. These results suggested that methylation is a requirement for Ca^{2+} entry, though how this occurs is unknown.

The formation of lysophosphatidylcholine was shown in basophils (rather than mast cells) [282] which also release histamine. When these were preincubated with IgE and methyl-labelled methionine and then challenged with the antigen ovalbumin there was an increase in methylation of phospholipid followed by a slow decline of methylation which paralleled the release of histamine. The appearance of labelled lysophosphatidylcholine indicated phospholipase A_2 activity, an observation sup-

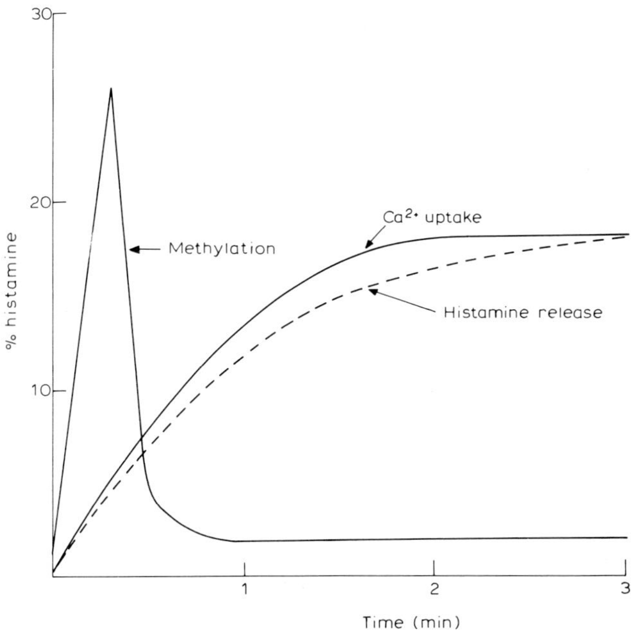

Fig. 6. The time sequence of some events in the release of histamine from rat mast cells following their incubation with $F(ab')_2$ fragments of antibodies of rat basophil leukaemia cells. Note the maximum methylation of phospholipids is reached very quickly whereas maximum influx of Ca^{2+} and the release of histamine occur somewhat later (after Ishizaka et al. [281]). For further details see text.

ported by the release of arachidonic acid which also paralleled histamine release. The inhibition of phospholipase A_2 by mepacrine also inhibited the release of histamine. Phospholipase A_2 activity in most tissues is dependent on Ca^{2+} whose entry into the mast cells appears in these and other experiments to be dependent on stepwise methylation so that a complex series of metabolic events occurs. These events are indicated in Fig. 7.

How exactly these events allow calcium to enter the mast cell or basophil promoting the exocytosis of histamine is, as yet, unclear. There are several other conceptions which require further study. For example if added phosphatidylserine enters the mast cell outer membrane it presumably has to pass to the inner layer before decarboxylation. The reader is referred to the excellent summary of these studies together with speculative comments by Hirata and Axelrod [234].

The functional significance of the methylation reaction (operating at about

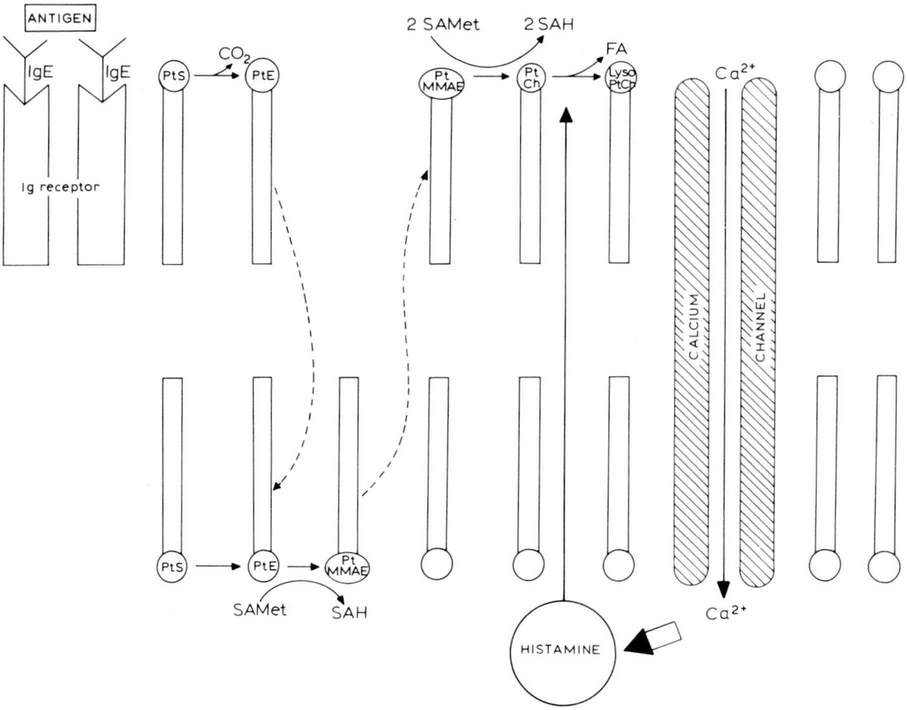

Fig. 7. A diagram to indicate a possible sequence of events linking the binding of an antigen to two IgE molecules attached to the receptor on the mast cell surface. Phosphatidylserine (PtS) is decarboxylated to phosphatidylethanolamine (PtE) either on the outer or inner leaflet and the PtE so formed methylated first in the inner leaflet to yield phosphatidylmonomethylaminoethanolamine (PtMMAE) which is then converted to phosphatidylcholine (PtCh) in the outer leaflet. PtCh is deacylated by a phospholipase A_2 which is believed to be responsible for opening the calcium channel. Ca^{2+} then enters the mast cell to induce exocytosis of the histamine granule. This diagram is adapted from those of Hirata and Axelrod [234] and Foreman [312].

one-thousandth the rate found in liver microsomes) has been challenged by Vance and De Kruijff [236] who argue that a reduction in microviscosity in the erythrocyte membrane from 1.62P to 1.09P could not occur as a result of the methylation of 0.0012% of the phosphatidylethanolamine present. However, Axelrod and Hirata [234] note that very few molecules of prostaglandin, for example, are necessary to induce a change in viscosity and the large range of receptor-activated methylations they have observed certainly strongly indicate that their observations were not artifactual. It will be interesting to see if they are confirmed in other laboratories.

(c) Phosphatidylcholine and acetylcholine synthesis in the brain

An interesting off-shoot of the studies discussed in the previous section is the finding by three laboratories that a small amount of methylation by enzymes of the methyltransferase types I and II occurs in mammalian nervous tissue [283–285]. In view of the observations discussed above it is likely that this is related to the action of central transmitters and studies on these responses may be rewarding. However, it is possible that this methylation might provide choline for acetylcholine synthesis. Until recently it seemed that the brain was incapable of synthesising choline and received all its choline supply largely in the form of lipid-bound choline (p. 31) [286]. Though it is clear that free choline in the blood can enter the brain and be incorporated into a lipid-bound form and acetylcholine there is no evidence that this is a major source [287,288]. There is considerable evidence for an enzymic mechanism for the release of choline from a lipid-bound form because the level of free choline rises rapidly post mortem [289]. According to Crews et al. [284] a brain synaptosomal fraction can synthesise 2.6 pmol phosphatidylcholine/mg protein/h and this rate is increased ten-fold if phosphatidylmonomethylaminoethanol is added. The formation of the latter is likely to be a rate-limiting step. Much lower rates in the presence of phosphatidylethanolamine have been obtained by Blusztajn and Wurtman [287] with a similar preparation but they also show that free choline is liberated from the newly synthesised phosphatidylcholine and that the phosphatidylcholine synthesised in this way represents a pool which turns over very rapidly. The mechanism of the release of choline is not known but numerous possibilities have been discussed [288,290]. It is by no means certain that it would all derive from the newly synthesised pool. In any event the answer is complex in that free choline leaves the brain in significant amounts [291] and then has to be replaced.

It would of course be most interesting to know whether the pool of choline synthesised by the methylation pathway is that used for the synthesis of acetylcholine. The turnover rate of acetylcholine in the rat occipital cortex and striatum is 150 and 1300 nmol/g/h, respectively, according to Racagni et al. [292] and 1560 and 3240 nmol/g/h in cortex and striatum [293]. On this basis free choline is acetylated in mammalian brain at a rate varying from 15 pmol to 300 pmol/mg protein/h. The rates are much higher than the rates of choline synthesis so far obtained in vitro but it must be realised that the choline resulting from hydrolysis of acetylcholine (this is the same as the synthesis at a steady rate of turnover) can be re-taken up into the pre-synaptic terminal and re-acetylated. Loss of choline from the brain amounts to only about 7 nmol/g/min [291,294] or about 42 pmol/mg protein/h. On this basis the methylation of phosphatidylethanolamine might make only a small contribution to the supply of choline for acetylcholine synthesis.

There are some neurological conditions e.g. tardive dyskinesia and Alzheimer's disease, where a cholinergic deficit occurs and attempts have been made to alleviate the defect by raising choline levels by the oral administration of choline or phosphatidylcholine. Some success has been achieved in the treatment of tardive dyskinesia by Davis's group using choline chloride [295,296] and Growdon's group using lecithin [297]. For a more recent review of tardive dyskinesia see Ansell [298]. There has not, however, been a significant response in Alzheimer's disease as was discussed at a recent meeting on this disease and related disorders [299].

(d) Roles of phosphatidylserine

Some of the more intriguing studies of recent years on the relationships of phospholipids to drug action have concerned phosphatidylserine. Reference has already been made to its possible role in the release of histamine (p. 35). There is now some evidence that this phospholipid may be a constituent of the opiate pharmacophore. Abood and Hoss [300] noted that phosphatidylserine was capable of binding morphine in a stereospecific manner. It was subsequently shown [301] that the binding of dihydromorphine to synaptic and microsomal brain membranes was enhanced by phosphatidylserine but inhibited by phosphatidylethanolamine and phosphatidylcholine; other acidic phospholipids were without effect. These indications of a role for phosphatidylserine at the opiate receptor have been supported by further studies [302] which showed that opiate binding to synaptic membranes, previously reduced by pre-treatment with lipolytic enzymes, may be restored by the addition of phosphatidylserine. Removal of the C22:6 fatty acid from the 2-position caused a precipitate fall in opiate binding; decarboxylation by phosphatidylserine decarboxylase had a lesser effect. Curiously, though the binding of the specific opiate antagonist naloxone was similarly affected, that of another antagonist naltrexone was not. These experiments are strongly suggestive of a role for phosphatidylserine in the opiate pharmacophore. Whether the phospholipid has a more general role at receptors is unknown but Aronstam et al. [303] have produced some evidence for its participation in the binding of antagonists at muscarinic receptors in the brain.

Some very unusual properties of phosphatidylserine have been observed. It is unique among phospholipids in its ability to depress brain energy metabolism in vivo, as measured by a rise in brain glucose [304]. Sonicated (liposomal) phosphatidylserine (20 μmol/kg body weight) injected intravenously into mice produced a significant rise but the effect was greater if the phospholipid was sonicated with *rat* serum. This was shown to be due to the production of 1-acyl-glycerophosphoserine because, when phosphatidylserine was previously incubated with phospholipase A_2 the glucose level in the brain was raised from 2 to 6 μmol/g brain. Phosphatidylserine and to a lesser extent phosphatidylethanolamine are capable of stimulating the output of acetylcholine from the brain when injected intravenously [305]. The significance of this remains to be explained.

Phosphatidylserine appears to be essential for the activity of galactosylceramidase [306] and human placental glucocerebrosidase [307]. It has been postulated that the purified glucocerebrosidase prepared in the presence of phosphatidylserine may be used in replacement therapy in Gaucher's disease. Until now attempts to purify the enzyme without added phosphatidylserine have resulted in loss of activity. Another brain enzyme, adenylate cyclase, when solubilised and rendered inactive by treatment with deoxycholate can have its activity restored by the addition of phosphatidylcholine and the lysolipid and to a lesser extent by phosphatidylethanolamine [308].

Phospholipids, particularly phosphatidylcholine and phosphatidylserine, are used extensively in the formation of liposomes, the potential of which for drug therapy promises to be enormous and the reader is referred to the excellent account by Ryman and Tyrrell [309] for a summary. Two recent developments are of specific interest. Morley et al. [310] have developed the use of dipalmitoylphosphatidylcholine coupled with phosphatidylglycerol and prepared as a dried powder for the relief of respiratory distress syndrome in premature infants. The powder is blown down an endotracheal tube, thus replacing the deficient surfactant in the infant's lung. Other workers have had considerable success in the use of erythrocytes as carriers of enzymes in deficiency diseases. The use of erythrocytes as carriers instead of artificially produced liposomes may be of great value since they evoke no immune response [311].

9. Conclusion

The aim of this chapter has been to give a concise and up-to-date survey, with due deference to fundamental findings in the past, of the metabolism in animal tissues of glycerophospholipids containing choline, ethanolamine and serine. Analogues containing ether groups (plasmalogens) are considered in detail in Chapter 2. In the foregoing account more attention has been paid to mechanisms of synthesis than to catabolic mechanisms because the phospholipases are described in detail in Chapter 9. With the exception of a detailed description of the peculiarities of phospholipid metabolism in the lung, metabolic variations occurring in organs such as the brain, eye and kidney have not been considered for reasons of space. However, the attention of the reader has been drawn to new work which indicates that enzymes metabolising these particular phospholipids may be involved in the response of cells to pharmacological agents and transmitters. This is an area which we believe is of increasing significance for the understanding of how cells respond to external stimuli.

References

1 Gobley, M. (1850) J. Pharm. Chim. (Paris) 17, 401–417.
2 Diakanow, C. (1867) Med. Chem. Untersuch. 2, 221–227.
3 Diakanow, C. (1868) Zbl. Med. Wiss. 2, 434–435.
4 Strecker, A. (1868) Ann. Chem. Pharm. 148, 77–90.
5 Strecker, A. (1862) Ann. Chem. Pharm. 123, 353–360.
6 MacLean, H. and MacLean, I.S. (1927) Lecithin and Allied Substances, Longmans, Green, London.
7 Wittcoff, H. (1951) The Phosphatides, Reinhold, New York.
8 Ansell, G.B. and Hawthorne, J.N. (1964) The Phospholipids–Chemistry, Metabolism and Function, Elsevier, Amsterdam.
9 Baer, E. and Kates, M. (1950) J. Am. Chem. Soc. 72, 942–949.
10 Strickland, K.P. (1973) in Form and Function of Phospholipids (Ansell, G.B., Dawson, R.M.C. and Hawthorne, J.N., eds.), pp. 9–42, Elsevier, Amsterdam.
11 The Nomenclature of Lipids, IUPAC–IUB Commission on Biochemical Nomenclature (CBN Recommendations–1976) (1977) Eur. J. Biochem. 79, 11–21.
12 Robinson, N. (1961) J. Pharm. Pharmacol. 13, 321–354.
13 Van den Bosch, H. and Van Deenen, L.L.M. (1965) Biochim. Biophys. Acta 106, 326–337.
14 Webster, G.R. and Thompson, R.H.S. (1962) Biochim. Biophys. Acta 63, 38–45.
15 Gjone, E., Berry, J.F. and Turner, D.A. (1959) J. Lipid Res. 1, 66–71.
16 Lüdecke, K. (1905) Dissertation, University of München, cited by Wittcoff, H. (1951) The Phosphatides, p. 103, Reinhold, New York.
17 Thudichum, J.L.W. (1884) A Treatise on the Chemical Constitution of the Brain, Ballière, Tindall and Cox, London.
18 Renall, M.H. (1913) Biochem. Z. 55, 296–300.
19 Baumann, A. (1913) Biochem. Z. 54, 30–39.
20 Rudy, H. and Page, I.H. (1930) Z. Physiol. Chem. 193, 251–268.
21 Spanner, S. (1973) in Form and Function of Phospholipids (Ansell, G.B., Dawson, R.M.C. and Hawthorne, J.N., eds.), pp. 43–65, Elsevier, Amsterdam.
22 Lea, C.H., Rhodes, D.N. and Stoll, R.D. (1955) Biochem. J. 60, 353–363.
23 Klenk, E. and Dohmen, H. (1955) Z. Physiol. Chem. 299, 248–252.
24 Baer, E., Maurukas, J. and Russell, M. (1952) J. Am. Chem. Soc. 74, 152–157.
25 MacArthur, C.G. (1914) J. Am. Chem. Soc. 36, 2397–2401.
26 Folch, J. and Schneider, H.A. (1941) J. Biol. Chem. 137, 51–62.
27 Folch, J. (1941) J. Biol. Chem. 139, 973–974.
28 Folch, J. (1948) J. Biol. Chem. 174, 439–450.
28a Baer, E. and Maurukas, J. (1955) J. Biol. Chem. 212, 25–38.
29 White, D.A. (1973) in Form and Function of Phospholipids (Ansell, G.B., Dawson, R.M.C. and Hawthorne, J.N., eds.), pp. 441–482, Elsevier, Amsterdam.
30 Folch, J., Lees, M. and Sloane-Stanley, G.H. (1957) J. Biol. Chem. 226, 497–509.
31 Nelson, G.J. (1975) in Analysis of Lipids and Lipoproteins (Perkins, E.G., ed.), pp. 70–89, American Oil Chemist's Society Publication.
32 Kawamoto, T., Okano, G. and Akino, T. (1980) Biochim. Biophys. Acta 619, 20–34.
33 Simon, G. and Rouser, G. (1969) Lipids 4, 607–614.
34 Rouser, G., Simon, G. and Kritchevsky, G. (1969) Lipids 4, 599–606.
35 Renooij, W., Van Golde, L.M.G., Zwaal, R.F.A. and Van Deenen, L.L.M. (1976) Eur. J. Biochem. 61, 53–58.
36 Boon, J., Broekhuyse, R.M., Van Munster, P. and Schretlen, E. (1969) Clin. Chim. Acta 23, 453–456.
37 Peeters, H. (1976) in Phosphatidylcholine (Peeters, H., ed.), pp. 10–33, Springer-Verlag, Berlin.
38 Nelson, G.J. (1967) Lipids 2, 323–328.
39 Rooney, S.A., Canaran, P.M. and Motoyama, E.K. (1974) Biochim. Biophys. Acta 360, 56–67.

40 Das, S.K. and Foster, H.W. (1980) Am. J. Obstet. Gynec. 136, 211–215.
41 Koski, C.L., Jungalwala, F.B. and Kolodny, E.H. (1978) Clin. Chim. Acta 85, 295–298.
42 Kobayashi, T., Mawatain, S. and Kuroiwa, Y. (1978) Clin. Chim. Acta 85, 259–266.
43 Nelson, G.J. (1967) Biochim. Biophys. Acta 144, 221–232.
44 Malhotra, S. and Kritchevsky, D. (1978) Mech. Aging Develop. 8, 445–452.
45 Singh, E.J. and Swartwont, J.R. (1972) Lipids 7, 26–29.
46 Godinez, R.I., Sanders, R.L. and Longmore, W.J. (1975) Biochemistry 14, 830–834.
47 Kawamoto, T., Akino, T., Nakamura, M. and Mori, M. (1980) Biochim. Biophys. Acta 619, 35–47.
48 Owen, J.S., Hutton, R.A., Day, R.C., Bruckdorfen, K.R. and McIntyre, N. (1981) J. Lipid Res. 22, 423–430.
49 Soodsma, J.F., Mims, L.C. and Harlow, R.D. (1976) Biochim. Biophys. Acta 424, 159–167.
50 Montfoort, A., Van Golde, L.M.G. and Van Deenen, L.L.M. (1971) Biochim. Biophys. Acta 231, 335–342.
51 Holub, B. and Kuksis, A. (1978) Adv. Lipid Res. 16, 1–125.
52 Kuksis, A., Marai, L., Breckenridge, W.C., Gornall, D.A. and Stachnyk, O. (1968) Canad. J. Physiol. Pharmacol. 46, 511–534.
53 Okano, G. and Akino, T. (1978) Biochim. Biophys. Acta 528, 373–384.
54 Wassef, M.K., Lim, Y.N. and Horowitz, M.I. (1979) Biochim. Biophys. Acta 573, 222–226.
55 Marai, L. and Kuksis, A. (1973) Canad. J. Biochem. 51, 1365–1379.
56 Yeung, S.K.F. and Kuksis, A. (1974) Canad. J. Biochem. 52, 830–837.
57 Marai, L. and Kuksis, A. (1973) Canad. J. Biochem. 51, 1248–1261.
58 Hübscher, G., Dils, R.R. and Pover, W.R.F. (1959) Biochim. Biophys. Acta 36, 518–528.
59 Hübscher, G. (1962) Biochim. Biophys. Acta 56, 555–561.
60 Barańska, J. (1980) Biochim. Biophys. Acta 619, 258–266.
61 Porcellati, G., Arienti, G., Pirotta, M. and Giorgini, D. (1971) J. Neurochem. 18, 1395–1417.
62 Goracci, G., Blomstrand, C., Arienti, G., Hamberger, A. and Porcellati, G. (1973) J. Neurochem. 20, 1167–1180.
63 Gaiti, A., de Medio, G.E., Brunetti, M., Amaducci, L. and Porcellati, G. (1974) J. Neurochem. 23, 1153–1159.
64 Kanfer, J.N. (1972) J. Lipid Res. 13, 468–476.
65 Taki, T. and Kanfer, J.N. (1978) Biochim. Biophys. Acta 528, 309–317.
66 Yavin, E. and Zeigler, B.P. (1977) J. Biol. Chem. 252, 260–267.
67 Dils, R.R. and Hübscher, G. (1961) Biochim. Biophys. Acta 46, 505–513.
68 Kanfer, J.N. (1980) Canad. J. Biochem. 58, 1370–1380.
69 Kanfer, J. and Kennedy, E.P. (1964) J. Biol. Chem. 239, 1720–1726.
70 Patterson, P.H. and Lennarz, W.J. (1971) J. Biol. Chem. 246, 1062–1072.
71 Marshall, H.O. and Kates, M. (1974) Canad. J. Biochem. 52, 469–482.
72 Bygrave, F.L. and Bücher, Th. (1968) Eur. J. Biochem. 6, 256–263.
73 McMurray, W.C. (1973) in Form and Function of Phospholipids (Ansell, G.B., Dawson, R.M.C. and Hawthorne, J.N., eds.), pp. 205–251, Elsevier, Amsterdam.
73a Van Golde, L.M.G., Raben, J., Batenberg, J.J., Fleischer, B., Zambrano, F. and Fleischer, S. (1974) Biochim. Biophys. Acta 360, 179–192.
74 Kiss, Z. (1976) Eur. J. Biochem. 67, 557–561.
75 Pullarkat, R.K., Sbaschnig-Agler, M. and Reha, H. (1981) Biochim. Biophys. Acta 663, 117–123.
76 James, O.A., MacDonald, G. and Thompson, W. (1979) J. Neurochem. 33, 1061–1066.
77 Arnstein, H.R.V. (1951) Biochem. J. 48, 27–32.
78 Bremer, J., Figard, P.H. and Greenberg, D.M. (1960) Biochim. Biophys. Acta 43, 477–488.
79 Wilson, J.D., Gibson, K.D. and Udenfriend, S. (1960) J. Biol. Chem. 235, 3539–3543.
80 Borkenhagen, L.F., Kennedy, E.P. and Fielding, L. (1961) J. Biol. Chem. 236, PC 28–30.
81 Dennis, E.A. and Kennedy, E.P. (1972) J. Lipid Res. 13, 263–267.
82 Suda, T. and Matsuda, M. (1974) Biochim. Biophys. Acta 369, 331–337.
83 Stoffel, W., Sticht, G. and LeKim, D. (1968) Z. Physiol. Chem. 349, 1745–1748.

84 Kennedy, E.P. and Weiss, S.B. (1956) J. Biol. Chem. 222, 193–214.
85 Weinhold, P.A. and Rethy, V.B. (1974) Biochemistry 13, 5135–5141.
86 Brophy, P.J. and Vance, D.E. (1976) FEBS Lett. 62, 123–125.
87 Ulane, R.E., Stephenson, L.L. and Farrell, P.M. (1978) Biochim. Biophys. Acta 531, 295–300.
88 Infante, J.P. and Kinsella, J.E. (1976) Lipids 11, 727–735.
89 Spanner, S. and Ansell, G.B. (1979) Biochem. J. 178, 753–760.
90 Ansell, G.B. (1973) in Form and Function of Phospholipids (Ansell, G.B., Dawson, R.M.C. and Hawthorne, J.N., eds.), pp. 327–422, Elsevier, Amsterdam.
91 Korniat, E.K. and Beeler, D.A. (1975) Anal. Biochem. 69, 300–305.
92 Stoffel, W., Sticht, G. and LeKim, D. (1969) Z. Physiol. Chem. 350, 63–68.
93 Henning, R. and Stoffel, W. (1969) Z. Physiol. Chem. 350, 827–835.
94 Stoffel, W., LeKim, D. and Sticht, G. (1969) Z. Physiol. Chem. 350, 1233–1242.
95 Offenbartl, K., Wennerberg, J., Sundler, R., Åkesson, B. and Nilsson, Å. (1973) Biochim. Biophys. Acta 306, 460–465.
96 Åkesson, B. (1979) Biochim. Biophys. Acta 573, 481–488.
97 Borkenhagen, L.F. and Kennedy, E.P. (1957) J. Biol. Chem. 227, 951–962.
98 Ansell, G.B. and Chojnacki, T. (1969) in Methods in Enzymology (Lowenstein, J.M., ed.), Vol. 14, pp. 121–125, Academic Press, New York.
99 Chojnacki, T., Radomińska-Pyrek, A. and Korzybski, T. (1967) Acta Biochim. Polon. 14, 383–388.
100 Radomińska-Pyrek, A. (1969) Acta Biochim. Polon. 16, 17–23.
101 Sundler, R. (1975) J. Biol. Chem. 250, 8585–8590.
102 Sundler, R. (1973) Biochim. Biophys. Acta 306, 218–226.
103 Sundler, R. and Åkesson, B. (1975) J. Biol. Chem. 250, 3359–3367.
103a Plantavid, M., Maget-Dana, R. and Douste-Blazy, L. (1976) FEBS Lett. 72, 169–172.
104 Ansell, G.B. and Metcalfe, R.F. (1971) J. Neurochem. 18, 647–665.
105 Snyder, F., Blank, M.L. and Malone, B. (1970) J. Biol. Chem. 245, 4016–4018.
106 Radomińska-Pyrek, A. and Horrocks, L.A. (1972) J. Lipid Res. 13, 580–587.
107 Radomińska-Pyrek, A., Strosznajder, J., Dabrowiecki, Z., Chojnacki, T. and Horrocks, L.A. (1976) J. Lipid Res. 17, 657–662.
108 Roberti, R., Binaglia, L., Francescangeli, E., Goracci, G. and Porcellati, G. (1975) Lipids 10, 121–127.
109 Radomińska-Pyrek, A., Strosznajder, J., Dabrowiecki, Z., Goracci, G., Chojnacki, T. and Horrocks, L.A. (1977) J. Lipid Res. 18, 53–58.
110 Radomińska-Pyrek, A., Pilarska, M. and Zimniak, P. (1978) Biochem. Biophys. Res. Commun. 85, 1074–1081.
111 Bell, R.M. and Coleman, R.A. (1980) Annu. Rev. Biochem. 49, 459–487.
111a Coleman, R. and Bell, R.M. (1977) J. Biol. Chem. 252, 3050–3056.
112 Kanoh, H. and Ohno, K. (1973) Biochim. Biophys. Acta 306, 203–217.
113 Kanoh, H. and Ohno, K. (1975) Biochim. Biophys. Acta 380, 199–207.
114 Bjerve, K.S. (1973) Biochim. Biophys. Acta 306, 396–402.
115 Bjerve, K.S. (1973) Biochim. Biophys. Acta 296, 549–562.
115a Gaiti, A., Goracci, G., de Medio, G.E. and Porcellati, G. (1972) FEBS Lett. 27, 116–120.
116 De Medio, G.E., Gaiti, A., Goracci, G. and Porcellati, G. (1973) Biochem. Soc. Trans. 1, 348–352.
117 Merkl, I. and Lands, W.E.M. (1963) J. Biol. Chem. 238, 905–906.
118 Lands, W.E.M. and Hart, P. (1965) J. Biol. Chem. 240, 1905–1911.
119 Van Golde, L.M.G., Scherphof, G.L. and Van Deenen, L.L.M. (1969) Biochim. Biophys. Acta 176, 635–637.
120 Åkesson, B. (1970) Biochim. Biophys. Acta 218, 57–70.
121 Sundler, R. and Åkesson, B. (1975) Biochem. J. 146, 309–315.
122 Åkesson, B., Elovson, J. and Arvidson, G. (1970) Biochim. Biophys. Acta 210, 15–27.
123 Bjørnstad, P. and Bremer, J. (1966) J. Lipid Res. 7, 38–45.
124 Du Vigneaud, V., Cohn, M., Chandler, J.P., Schenck, J.R. and Simmonds, S. (1941) J. Biol. Chem. 140, 625–641.

125 Stekol, J.A., Anderson, E.I. and Weiss, S. (1958) J. Biol. Chem. 233, 425–429.
126 Bremer, J. and Greenberg, D.M. (1961) Biochim. Biophys. Acta 46, 205–216.
127 Wilson, J.D., Gibson, K.D. and Udenfriend, S. (1960) J. Biol. Chem. 235, 3213–3217.
128 Gibson, K.D., Wilson, J.D. and Udenfriend, S. (1961) J. Biol. Chem. 236, 673–679.
129 LeKim, D., Betzing, H. and Stoffel, W. (1973) Z. Physiol. Chem. 354, 437–444.
130 Rehbinder, D. and Greenberg, D.M. (1965) Arch. Biochem. Biophys. 109, 110–115.
131 Hirata, F., Viveros, O.H., Diliberto Jr., E.J. and Axelrod, J. (1978) Proc. Natl. Acad. Sci. USA 75, 1718–1721.
132 Skurdal, D.N. and Cornatzer, W.E. (1975) Int. J. Biochem. 6, 579–583.
133 Arvidson, G.A.E. (1968) Eur. J. Biochem. 4, 478–486.
134 Morgan, T.E., Finley, T.N. and Fialkow, H. (1965) Biochim. Biophys. Acta 106, 403–413.
135 Morgan, T.E. (1969) Biochim. Biophys. Acta 178, 21–34.
136 Weinhold, P.A. (1968) J. Lipid Res. 9, 262–266.
137 Wolfe, B.M.J., Anhalt, B., Beck, J.C. and Rubenstein, D. (1970) Canad. J. Biochem. 48, 170–177.
138 Vereyken, J.M., Montfoort, A. and Van Golde, L.M.G. (1972) Biochim. Biophys. Acta 260, 70–81.
139 Wittenberg, J. and Kornberg, A. (1953) J. Biol. Chem. 202, 431–444.
140 McCaman, R.E. (1962) J. Biol. Chem. 237, 672–676.
141 McCaman, R.E. and Cook, K. (1966) J. Biol. Chem. 241, 3390–3394.
142 Oldenborg, V. and Van Golde, L.M.G. (1977) Biochim. Biophys. Acta 489, 454–465.
143 Infante, J.P. and Kinsella, J.E. (1976) Int. J. Biochem. 7, 483–496.
144 Choy, P.C., Lim, P.H. and Vance, D.E. (1977) J. Biol. Chem. 252, 7673–7677.
145 Sundler, R., Arvidson, G. and Åkesson, B. (1972) Biochim. Biophys. Acta 280, 559–568.
146 Domschke, W., Keppler, D., Bischoff, E. and Decker, K. (1971) Z. Physiol. Chem. 352, 275–279.
146a Infante, J.P. (1977) Biochem. J. 167, 847–849.
147 Wilgram, G.F. and Kennedy, E.P. (1963) J. Biol. Chem. 238, 2615–2619.
148 Choy, P.C., Farren, S.B. and Vance, D.E. (1979) Canad. J. Biochem. 57, 605–612.
149 Fiscus, W.G. and Schneider, W.C. (1966) J. Biol. Chem. 241, 3324–3330.
150 Choy, P.C. and Vance, D.E. (1978) J. Biol. Chem. 253, 5163–5167.
151 Vance, D.E. and Choy, P.C. (1979) Trends Biochem. Sci. 4, 145–148.
151a Feldman, D.A., Dietrich, J.W. and Weinhold, P.A. (1980) Biochim. Biophys. Acta 620, 603–611.
152 Strickland, K.P., Subrahmanyam, D., Pritchard, E.T., Thompson, W. and Rossiter, R.J. (1963) Biochem. J. 87, 128–136.
153 Holub, B.J. (1978) J. Biol. Chem. 253, 691–696.
153a Kanoh, H. and Ohno, K. (1973) Biochim. Biophys. Acta 326, 17–25.
153b Ewing, R.D. and Finamore, F.J. (1970) Biochim. Biophys. Acta 218, 474–481.
154 Dils, R.R. and Hübscher, G. (1959) Biochim. Biophys. Acta 32, 293–294.
155 Saito, M., Bourque, E. and Kanfer, J. (1975) Arch. Biochem. Biophys. 169, 304–317.
156 Miura, T. and Kanfer, J. (1976) Arch. Biochem. Biophys. 175, 654–660.
157 Treble, D.H., Frumkin, S., Balint, J.A. and Beeler, D.A. (1970) Biochim. Biophys. Acta 202, 163–171.
158 Salerno, D.M. and Beeler, D.A. (1973) Biochim. Biophys. Acta 326, 325–338.
159 Bjerve, K. (1971) FEBS Lett. 17, 14–16.
160 Arvidson, G.A.E. (1968) Eur. J. Biochem. 5, 415–421.
161 Lands, W.E.M. (1960) J. Biol. Chem. 235, 2233–2237.
162 Lands, W.E.M. (1965) Annu. Rev. Biochem. 34, 313–344.
163 Thompson Jr., G.A. (1970) in Comprehensive Biochemistry, Vol. 18, Lipid Metabolism (Florkin, M. and Stotz, E.H., eds.), pp. 157–199, Elsevier, Amsterdam.
164 Lands, W.E.M. and Hart, P. (1965) J. Biol. Chem. 240, 1905–1911.
165 Lands, W.E.M. and Merkl, I. (1963) J. Biol. Chem. 238, 898–904.
166 Yamashita, S., Hosaka, K. and Numa, S. (1973) Eur. J. Biochem. 38, 25–31.
166a Holub, B.J., Macnaughton, J.A. and Piekarski, J. (1979) Biochim. Biophys. Acta 572, 413–422.
166b Åkesson, B., Elovson, J. and Arvidson, G. (1970) Biochim. Biophys. Acta 218, 44–56.
167 Holub, B.J., Breckenridge, W.C. and Kuksis, A. (1971) Lipids 6, 307–313.

168 Erbland, J.F. and Marinetti, G.V. (1965) Biochim. Biophys. Acta 106, 128–138.
169 Van den Bosch, H., Bonte, H.A. and Van Deenen, L.L.M. (1965) Biochim. Biophys. Acta 98, 648–651.
169a Klaus, M.H., Clements, J.A. and Havel, R.J. (1961) Proc. Natl. Acad. Sci. USA 47, 1858–1859.
170 Macklin, C.C. (1954) Lancet i, 1099–1104.
171 Farnell, P.M. and Avery, M.L. (1975) Am. Rev. Resp. Dis. 111, 657–688.
172 Van Golde, L.M.G. (1976) Am. Rev. Resp. Dis. 114, 977–1000.
173 Epstein, M.F. and Farrell, P.M. (1975) Pediat. Res. 9, 658–665.
174 Ishidate, K. and Weinhold, P.A. (1981) Biochim. Biophys. Acta 664, 133–147.
175 Van Heusden, G.P.H., Noteborn, H.P.J.M. and Van den Bosch, H. (1981) Biochim. Biophys. Acta 664, 49–60.
175a Longmore, W.J., Oldenborg, V. and Van Golde, L.M.G. (1979) Biochim. Biophys. Acta 572, 452–460.
176 Gatt, S., Barenholz, Y. and Roitman, A. (1966) Biochem. Biophys. Res. Commun. 24, 169–172.
177 Gatt, S. (1968) Biochim. Biophys. Acta 159, 304–316.
178 Cooper, M.F. and Webster, G.R. (1970) J. Neurochem. 17, 1543–1554.
179 Woelk, H., Fürniss, H. and Debuch, H. (1972) Z. Physiol. Chem. 353, 1111–1119.
180 Waite, M. and Van Deenen, L.L.M. (1967) Biochim. Biophys. Acta 137, 498–517.
181 Woelk, H. and Porcellati, G. (1973) Z. Physiol. Chem. 354, 90–100.
182 Waite, M., Scherphof, G.L., Boshouwers, F.M.G. and Van Deenen, L.L.M. (1969) J. Lipid Res. 10, 411–420.
183 Newkirk, J.D. and Waite, M. (1971) Biochim. Biophys. Acta 225, 224–233.
184 Nachbaur, J., Colbeau, A. and Vignais, P.M. (1972) Biochim. Biophys. Acta 274, 426–446.
185 Waite, M. and Sisson, P. (1976) Biochim. Biophys. Acta 450, 301–310.
186 Woelk, H., Goracci, G., Gaiti, A. and Porcellati, G. (1973) Z. Physiol. Chem. 354, 729–736.
187 Waite, M. and Sisson, P. (1971) Biochemistry 12, 2377–2383.
188 Lumb, R.H. and Allen, K.F. (1976) Biochim. Biophys. Acta 450, 175–184.
189 McMurray, W.C. and Magee, W.L. (1972) Annu. Rev. Biochem. 41, 129–160.
190 Leibovitz-BenGershon, Z., Kobiler, I. and Gatt, S. (1972) J. Biol. Chem. 247, 6840–6847.
191 Van den Bosch, H., Aarsman, A.J., Slotboom, A.J. and Van Deenen, L.L.M. (1968) Biochim. Biophys. Acta 164, 215–225.
192 Leibovitz-BenGershon, Z. and Gatt, S. (1974) J. Biol. Chem. 249, 1525–1529.
193 Shapiro, B. (1953) Biochem. J. 53, 663–666.
194 De Jong, J.G.N., Van den Bosch, H., Rijken, D. and Van Deenen, L.L.M. (1974) Biochim. Biophys. Acta 369, 50–63.
195 Williams, D.J., Spanner, S. and Ansell, G.B. (1973) Biochem. Soc. Trans. 1, 466–467.
196 Matsuzawa, Y. and Hostetler, K.Y. (1980) J. Biol. Chem. 255, 646–652.
197 Hostetler, K.Y. and Hall, L.B. (1980) Biochem. Biophys. Res. Commun. 96, 388–393.
198 Saito, M. and Kanfer, J.N. (1973) Biochem. Biophys. Res. Commun. 53, 391–398.
199 Saito, M. and Kanfer, J.N. (1975) Arch. Biochem. Biophys. 169, 318–323.
200 Taki, T. and Kanfer, J.N. (1979) J. Biol. Chem. 254, 9761–9765.
201 Wykle, R.L. and Schremmer, J.M. (1974) J. Biol. Chem. 249, 1742–1746.
202 Wykle, R.L., Kraemer, W.F. and Schremmer, J.M. (1977) Arch. Biochem. Biophys. 184, 149–155.
203 Heller, M. (1978) Adv. Lipid Res. 16, 267–326.
203a Hayaishi, O. and Kornberg, A. (1954) J. Biol. Chem. 206, 647–663.
204 Dawson, R.M.C. (1956) Biochem. J. 62, 689–693.
205 Baldwin, J.J. and Cornatzer, W.E. (1968) Biochim. Biophys. Acta 164, 195–204.
206 Webster, G.R., Marples, E.A. and Thompson, R.H.S. (1957) Biochem. J. 65, 374–377.
207 Baldwin, J.J., Lanes, P. and Cornatzer, W.E. (1969) Arch. Biochem. Biophys. 133, 224–232.
208 Lloyd-Davis, K.A., Michell, R.H. and Coleman, R. (1972) Biochem. J. 127, 357–368.
209 Mann, S.P. (1975) Experientia 31, 1256–1257.
210 Ansell, G.B. and Spanner, S. (1981) in Cholinergic Mechanisms (Pepeu, G. and Ladinsky, H., eds.), pp. 393–404, Plenum, New York.

211 Dawson, R.M.C. (1955) Biochem. J. 59, 5–8.
212 Rothman, J.E. and Lenard, J. (1977) Science 195, 743–753.
213 Op den Kamp, J.A.F. (1979) Annu. Rev. Biochem. 48, 47–71.
214 Op den Kamp, J.A.F. (1981) in New Comprehensive Biochemistry (Neuberger, A. and Van Deenen, L.L.M., eds.), Vol. 1 (Finean, J.B. and Michell, R.H., eds.), pp. 83–126, Elsevier, Amsterdam.
215 Bretscher, M.S. (1972) J. Mol. Biol. 71, 523–528.
216 Bretscher, M.S. (1972) Nature New Biol. 236, 11–12.
217 Zwaal, R.F.A., Roelofsen, B. and Colley, C.M. (1973) Biochim. Biophys. Acta 300, 159–182.
218 Van Meer, G., Porthuis, B.J.H.M., Wirtz, K.W.A., Op den Kamp, J.A.F. and Van Deenen, L.L.M. (1980) Eur. J. Biochem. 103, 283–288.
219 Kornberg, R.D. and McConnell, H.M. (1971) Biochemistry, 10, 1111–1120.
220 Wirtz, K.W.A. and Zilversmit, D.B. (1968) J. Biol. Chem. 243, 3596–3602.
221 Higgins, J.A. and Dawson, R.M.C. (1977) Biochim. Biophys. Acta 470, 342–356.
222 Sundler, R., Sarcione, S.L., Alberts, A.W. and Vagelos, P.R. (1977) Proc. Natl. Acad. Sci. USA 74, 3350–3354.
223 Nilsson, O.S. and Dallner, G. (1977) Biochim. Biophys. Acta 464, 453–458.
224 Nilsson, O.S. and Dallner, G. (1977) J. Cell Biol. 72, 568–583.
225 Higgins, J.A. (1979) Biochim. Biophys. Acta 558, 48–57.
226 Higgins, J.A. (1981) Biochim. Biophys. Acta 640, 1–15.
227 Zilversmit, D.B. and Hughes, M.E. (1977) Biochim. Biophys. Acta 469, 99–110.
228 Van den Besselaar, A.M.H.P., De Kruijff, B., Van den Bosch, H. and Van Deenen, L.L.M. (1978) Biochim. Biophys. Acta 510, 242–255.
229 Zilversmit, D.B. (1978) Ann. N.Y. Acad. Sci. 308, 149–163.
230 Coleman, R. and Bell, R.M. (1978) J. Cell Biol. 76, 245–253.
231 Vance, D.E., Choy, P.C., Farren, S.B., Lim, P.H. and Schneider, W.J. (1977) Nature 270, 268–269.
232 Ballas, L.M. and Bell, R.M. (1980) Biochim. Biophys. Acta 602, 578–590.
233 Bell, R.M., Ballas, L.M. and Coleman, R.A. (1981) J. Lipid Res. 22, 391–403.
234 Hirata, F. and Axelrod, J. (1980) Science 209, 1082–1090.
235 Hirata, F. and Axelrod, J. (1978) Proc. Natl. Acad. Sci. 75, 2348–2352.
236 Vance, D.E. and De Kruijff, B. (1980) Nature 288, 277–278.
237 Ansell, G.B. and Spanner, S. (1972) in Current Trends in the Biochemistry of Lipids (Ganguly, J. and Smellie, R.M.S., eds.), pp. 151–159, Academic Press, New York.
238 Coleman, R. (1973) in Form and Function of Phospholipids (Ansell, G.B., Dawson, R.M.C. and Hawthorne, J.N., eds.), pp. 345–375, Elsevier, Amsterdam.
239 Arnesjo, B., Nilsson, Å., Barrowman, J. and Borgström, B. (1969) Scand. J. Gastroent. 4, 653–665.
240 Boucrot, P. (1972) Lipids 7, 282–288.
241 Nilsson, Å. and Zilversmit, D.B. (1971) Biochim. Biophys. Acta 248, 137–142.
242 Van Tol, A. and Van Berkel, T.J.C. (1980) Biochim. Biophys. Acta 619, 156–166.
243 Eisenberg, S. (1979) Prog. Biochem. Pharmacol. 15, 139–165.
244 Smith, L.C., Pownall, H.J. and Gotto Jr., A.M. (1978) Annu. Rev. Biochem. 47, 751–777.
245 Assman, G., Schmitz, G., Donorth, N. and LeKim, D. (1978) Scand. J. Clin. Lab. Invest. 38 Suppl. 150, 16–20.
246 St. Clair, R.W. (1976) Atherosclerosis Rev. 1, 61–117.
247 Portman, O.W. and Alexander, M. (1976) Biochim. Biophys. Acta 450, 322–334.
248 Young, D.L. and Hanson, K.C. (1972) J. Lipid Res. 13, 244–252.
249 Swell, L., Bell Jr., C.C. and Entenman, C. (1968) Biochim. Biophys. Acta 164, 278–284.
250 Jackson, R.L., Westerman, J. and Wirtz, K.W.A. (1978) FEBS Lett. 94, 38–42.
251 Stein, Y. and Stein, O. (1966) Biochim. Biophys. Acta 116, 95–107.
252 Illingworth, D.R. and Portman, O.W. (1972) Biochem. J. 130, 557–567.
253 Ansell, G.B. and Spanner, S. (1971) Biochem. J. 122, 741–750.
254 Spanner, S., Hall, R.C. and Ansell, G.B. (1976) Biochem. J. 154, 133–140.
255 Eisenberg, S., Stein, Y. and Stein, O. (1967) Biochim. Biophys. Acta 137, 115–120.

256 Coêlho, L.C.B.B. and Gillett, M.P.T. (1979) Biochem. Soc. Trans. 7, 988–990.
257 Subbaiah, P.V. and Bagclade, J.D. (1978) Life Sci. 22, 1971–1978.
258 Webster, G.R. (1965) Biochim. Biophys. Acta 98, 512–519.
259 Mulder, E. and Van Deenen, L.L.M. (1965) Biochim. Biophys. Acta 106, 348–356.
260 Smith, N. and Rubinstein, D. (1974) Canad. J. Biochem. 52, 706–717.
261 Kahlenberg, A., Walker, C. and Rohrlick, R. (1974) Can. J. Biochem. 52, 803–806.
262 Verkleij, A.J., Zwaal, R.F.A., Roelofsen, B., Comfurius, P., Kastelijn, D. and Van Deenen, L.L.M. (1973) Biochim. Biophys. Acta 323, 178–193.
263 Béréziat, G., Chambaz, J., Trugnan, G., Pépin, D. and Polonovski, J. (1978) J. Lipid Res. 19, 495–500.
264 Curbelo, V., Gail, D.B. and Farrell, P.M. (1978) Am. J. Obstet. Gynecol. 131, 764–769.
265 Gluck, L., Kulovich, M.V., Borer Jr., R.C., Brenner, P.H., Anderson, G.G. and Spallacy, W.N. (1971) Am. J. Obstet. Gynecol. 109, 440–445.
266 Farrell, P.M. and Morgan, T.E. (1977) in the Development of the Lung (Hodson, W.A., ed.), pp. 309–347, Marcel Dekker, New York.
267 Possmayer, F., Casola, P.G., Chan, F., MacDonald, P., Ormseth, M.A., Wong, T., Harding, P.G.R. and Tokmakjian, S. (1981) Biochim. Biophys. Acta 664, 10–21.
268 Paddon, H.B. and Vance, D.E. (1980) Biochim. Biophys. Acta 620, 636–640.
269 Ishidate, K., Yoshida, M. and Nakazawa, Y. (1978) Biochem. Pharmacol. 27, 2595–2603.
270 Ishidate, K., Tsuruoka, M. and Nakazawa, Y. (1980) Biochim. Biophys. Acta 620, 49–58.
271 Ansell, G.B. and Spanner, S. (1975) Biochem. Pharmacol. 24, 1719–1723.
272 Ansell, G.B. and Spanner, S. (1974) J. Neurochem. 22, 1153–1155.
273 Hirata, F. and Axelrod, J. (1978) Nature 275, 219–220.
274 Hirata, F., Axelrod, J. and Crews, F.T. (1979) Proc. Natl. Acad. Sci. USA 76, 4813–4816.
275 Strittmatter, W.J., Hirata, F. and Axelrod, J. (1979) Science 204, 1205–1207.
276 Strittmatter, W.J., Hirata, F., Axelrod, J., Mallorga, P., Tallman, J.T. and Henneberry, R.C. (1979) Nature 282, 857–859.
277 Ishizaka, T., Foreman, J.C., Sterk, A.R. and Ishizaka, K. (1979) Proc. Natl. Acad. Sci. USA 76, 5858–5862.
278 Foreman, J.C., Mongar, J.L. and Gomperts, B.D. (1973) Nature 245, 249–251.
279 Goth, A. and Johnson, A.R. (1975) Life Sci. 16, 1201–1214.
280 Hirata, F., Strittmatter, W.J. and Axelrod, J. (1979) Proc. Natl. Acad. Sci. USA 76, 368–372.
281 Ishizaka, T., Hirata, F., Ishizaka, K. and Axelrod, J. (1980) Proc. Natl. Acad. Sci. USA 77, 1903–1906.
282 Crews, F.T., Morita, Y., Hirata, F., Axelrod, J. and Siraganian, R.P. (1980) Biochem. Biophys. Res. Commun. 93, 42–49.
283 Mozzi, R. and Porcellati, G. (1979) FEBS Lett. 100, 363–366.
284 Crews, F.T., Hirata, F. and Axelrod, J. (1980) J. Neurochem. 34, 1491–1498.
285 Blusztajn, J.K., Zeisel, S.H. and Wurtman, R.J. (1979) Brain Res. 179, 319–327.
286 Ansell, G.B. and Spanner, S. (1975) in Cholinergic Mechanisms (Waser, P.G., ed.), pp. 117–129, Raven, New York.
287 Blusztajn, J.K. and Wurtman, R.J. (1981) Nature 290, 417–418.
288 Ansell, G.B. and Spanner, S. (1977) in Cholinergic Mechanisms and Psychopharmacology (Jenden, D.J., ed.), pp. 431–445, Plenum, New York.
289 Freeman, J.J. and Jenden, D.J. (1976) Life Sci. 19, 949–962.
290 Ansell, G.B. and Spanner, S. (1979) in Nutrition and the Brain, Vol. 5; Choline and Lecithin in Brain Disorders (Barbeau, A., Growdon, J.H. and Wurtman, R.J., eds.), pp. 35–46, Raven, New York.
291 Dross, K. and Kewitz, H. (1972) Arch. Pharmacol. 274, 91–106.
292 Racagni, G., Cheney, D.L., Trabucchi, M., Wang, C. and Costa, E. (1974) Life Sci. 15, 1961–1975.
293 Nordberg, A. (1978) J. Neurochem. 30, 383–389.
294 Choi, R.L., Freeman, J.J. and Jenden, D.J. (1975) J. Neurochem. 24, 735–741.
295 Davis, K.L., Hollister, L.E., Barchas, J.D. and Berger, P.A. (1976) Life Sci. 19, 1507–1516.

296 Davis, K.L. and Berger, P.A. (1978) Biol. Psychiat. 13, 23–49.
297 Growdon, J.H., Hirsch, M.J., Wurtman, R.J. and Wiener, W. (1977) New. Eng. J. Med. 297, 524–527.
298 Ansell, G.B. (1981) Neuropharmacology, 20, 311–317.
299 Kolata, G.B. (1981) Science 211, 1032–1033.
300 Abood, L.G. and Hoss, W. (1975) Eur. J. Pharmacol. 32, 66–75.
301 Abood, L.G. and Takeda, F. (1976) Eur. J. Pharmacol. 39, 71–77.
302 Abood, L.G., Salem, N., Macneil, M. and Butler, M. (1978) Biochim. Biophys. Acta 530, 35–46.
303 Aronstam, R.R., Abood, L.G. and Baumgold, J. (1977) Biochem. Pharmacol. 26, 1689–1695.
304 Bigon, E., Boarato, E., Bruni, A., Leon, A. and Toffano, G. (1979) Br. J. Pharmacol. 67, 611–616.
305 Mantovani, P., Pepeu, G. and Amaducci, L. (1976) in Advances in Experimental Medicine, Vol. 72 (Porcellati, G., Amaducci, L. and Galli, C., eds.), pp. 285–292, Plenum, New York.
306 Hanada, E. and Suzuki, K. (1979) Biochim. Biophys. Acta 575, 410–420.
307 Dale, G.L., Villacorte, D.G. and Beutler, E. (1976) Biochem. Biophys. Res. Commun. 71, 1048–1053.
308 Hebdon, G.M., Levine, H., Sakyoun, N.E., Schmitges, C.J. and Cuatrecasas, P. (1981) Proc. Natl. Acad. Sci. USA 78, 120–123.
309 Ryman, B.E. and Tyrrell, D.A. (1980) in Essays in Biochemistry, Vol. 16 (Campbell, P.N. and Marshall, R.D., eds.), pp. 49–98, Academic Press, New York.
310 Morley, C.J., Miller, J., Bangham, A.D. and Davis, J.A. (1981) Lancet i, 64–68.
311 Anon (1981) New Sci. 90, 162.
312 Foreman, J. (1980) Trends Pharmacol. Sci. 1, 460–462.
313 Prottey, C. and Hawthorne, J.N. (1966) Biochem. J. 101, 191–196.
314 Cohen, P. and Derksen, A. (1969) Br. J. Haematol. 17, 359–371.
315 Nelson, G.J. (1967) Biochim. Biophys. Acta 144, 221–232.

CHAPTER 2

Plasmalogens and *O*-alkyl glycerophospholipids

LLOYD A. HORROCKS and MUKUT SHARMA

Department of Physiological Chemistry,
Ohio State University, 1645 Neil Avenue,
Columbus, OH 43210, U.S.A.

1. Introduction

Glycerophospholipids with ether linkages are found in nearly all animal and bacterial cells with the exception of most aerobic bacteria. In contrast to the diacyl types of glycerophospholipids, the plasmalogens contain a hydrocarbon chain attached to glycerol through a dehydrated hemiacetal (vinyl ether) linkage and the alkylacyl glycerophospholipids contain a hydrocarbon chain attached through an ether linkage. The oxidation levels of the ether-linked side-chains are aldehyde and alcohol for plasmalogens and alkylacyl glycerophospholipids, respectively.

Snyder [1] edited a treatise *Ether Lipids: Chemistry and Biology* which was published in 1972. This book contains chapters on all types of organisms, chemical synthesis, and enzymic pathways. In this chapter we have tried to bring the information on glycerophospholipids up to date with references to the most recent papers. A shorter review published by Mangold [2] in 1979 on chemical synthesis and biosynthesis of ether lipids includes coverage of non-polar lipids.

The distinctive functions and biological roles of the ether lipids are generally not known. They are often overlooked by biochemists for that reason and because their content is relatively low in liver and blood plasma. However, plasmalogens account for about 18.7% of the phospholipids in adult man. The very potent effects of the platelet activating factor, 1-alkyl-2-acetyl-*sn*-glycero-3-phosphocholine, may stimulate further interest in the area of ether-containing glycerophospholipids.

2. Nomenclature

The systematic nomenclature for glycerophospholipids [3] is illustrated in Fig. 1. The terms "alk-1′-enyl" and "alkyl" refer to the presence or absence of unsaturation at the 1 and 2 carbons of the side-chain. Either type of side-chain may have double bonds at other positions. For example, selachyl alcohol, 1-octadec-9′-enyl-*sn*-glycerol, is classified as an alkylglycerol because it does not have a double bond adjacent to the ether linkage.

Hawthorne/Ansell (eds.) Phospholipids

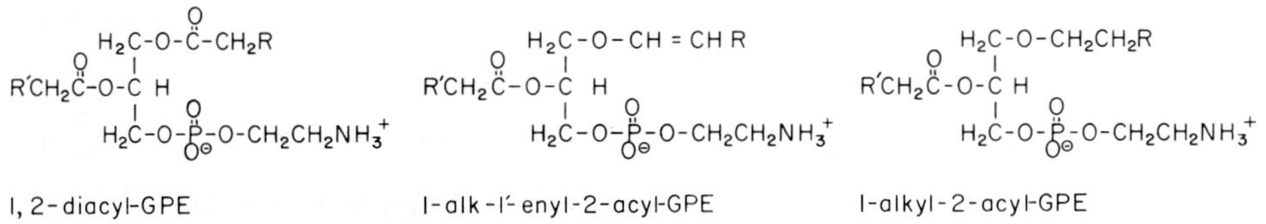

Fig. 1. Structure of ethanolamine glycerophospholipids. In this chapter, GPE is used as an abbreviation for *sn*-glycero-3-phosphoethanolamine. The common name for diacyl-GPE is phosphatidylethanolamine and for alkenylacyl-GPE is ethanolamine plasmalogen.

At present, the common nomenclature for glycerophospholipids is rather confused. The term "phosphatidyl" is supposed to be used as an abbreviation for the 1,2-diacyl-*sn*-glycero-3-phospho radical, thus phosphatidylethanolamine should refer only to those ethanolamine glycerophospholipids that have two acyl groups. However, in practice, "phosphatidyl" is often used for the mixture of diacyl, alk-1-enylacyl, and alkylacyl compounds that are isolated together by chromatography. The best term for the mixture is "ethanolamine glycerophospholipid" (EGP). The alternative term "ethanolamine phosphoglyceride" (EPG) is no longer recommended by the nomenclature commission [3].

Plasmalogens are glycerophospholipids containing a potential aldehyde that were first discovered in the plasma of cells. Thus the name was derived from plasm + al + ogen. A correct common name for 1-alk-1′-enyl-2-acyl-*sn*-glycero-3-phosphoethanolamine is "ethanolamine plasmalogen". Recently, the terms "plasmenyl ethanolamine" for alkenylacyl-GPE and "plasmanyl ethanolamine" for alkylacyl-GPE have been proposed by the nomenclature commission [3]. These terms have not been popular, perhaps because they are too much alike. Another common name for alk-1′-enylacyl-GPE is "phosphatidal ethanolamine" [4]. This term has had some usage but has been criticized because of the close resemblance to phosphatidyl ethanolamine which is "confusing for the ear and eye" [5]. An acceptable common name for alkylacyl-GPE does not exist. It was formerly called "kephalin B".

3. *Discovery and structure*

The history of ether lipids up to 1960 was described in considerable detail by Debuch and Seng [6], where references for this section can be found. Although fairly pure batyl alcohol had been prepared earlier, Tsujimoto and Toyama were the first to isolate and partially characterize batyl and selachyl alcohols. They recognized that the liver oils of elasmobranch fish contained these alcohols in the form of diesters with fatty acids. Toyama suggested an ether linkage and reported the correct molecular formula for batyl alcohol in 1924. He was also the first to isolate chimyl alcohol. The ether structure was proved by cleavage with hydriodic acid by Heilbron and Owens. Later investigators synthesized batyl alcohol and proved the location of the ether group on C-1. The natural diester form was partially purified by classical techniques but was not completely purified until the advent of thin-layer chromatography.

Bound aldehydes in the plasma of cells (plasmalogens) were discovered by an accidental, serendipitous observation by Feulgen and Voit in 1924. By a histochemical method, plasmalogens were found in a wide range of tissues in the animal kingdom. Quantitative methods for the assay of plasmalogens were developed with fuchsin and sulfurous acid, *p*-nitrophenylhydrazine, and iodine uptake. The first attempt at isolation of a plasmalogen gave the 1,2-cyclic acetal of *sn*-glycero-3-phosphoethanolamine after saponification of bovine muscle lipids. The groups of Klenk and Rapport eventually obtained the correct structure for choline and ethanolamine plasmalogens during the 1950s. Key steps included the recognition that two long-chains were present and the proof that an acyl group was present.

In the meantime, Brante reported evidence for the presence of 1-alkyl-GPE in brain in 1949. After saponification of egg-yolk lipids, Carter et al. isolated 1-octadecyl-GPE. This compound and the corresponding choline compound were found in several mammalian tissues by Svennerholm and Thorin. Hydrogenation and saponification of plasmalogen-rich phospholipids gave 1-alkyl-*sn*-glycero-3-phosphates for which the relationship to 1-alkyl-*sn*-glycerols from marine and mammalian sources was recognized. The enol ether form of the bound aldehydes was proved by three different approaches. Due to confusion about the site of action of phospholipase A_2 in the 1950s, the location of the enol ether was placed mainly at C-2. Debuch proved the exclusive location of the ether at C-1.

4. Methods and chemical properties

In 1972, Kates [7] published a book on the techniques of lipidology. Included are assay methods specific for plasmalogens such as long-chain aldehydes by *p*-nitrophenylhydrazone formation and vinyl ethers by iodine addition, and methods for acetolysis and acid hydrolysis of alkyl lipids. An excellent review of the application of chromatographic methods to the study of ether lipids was published in 1975 by Schmid et al. [8].

Purified preparations of plasmalogens have been made by removal of the diacyl glycerophospholipids by mild alkaline hydrolysis, hydrolysis by phospholipase A_2 from snake venoms and by hydrolysis by phospholipase D from cabbage [8,9]. All of these methods depend on the lesser reactivity of plasmalogens [10,11]. Much better yields of 80–90% of the starting plasmalogens can be obtained by removal of diacyl glycerophospholipids by a lipase specific for primary ester groups. The lipase from *Rhizopus delemar* has been used for the preparation of ethanolamine plasmalogens from bovine brain and heart and also for purification of alkylacyl-GPC from rat brain [12]. Choline and ethanolamine plasmalogens have been prepared with the lipase from *Rhizopus arrhizus* by Paltauf [13]. Previously, Woelk et al. had prepared plasmalogens with pancreatic lipase, but this lipase must be highly purified in order to avoid interference from other lipases [14]. The products with *Rhizopus* lipases have plasmalogen contents of 93–97%. Most of the remaining lipids are the corresponding alkylacyl compounds.

Intact plasmalogens have not been separated from alkylacyl or diacyl glycerophospholipids. Separations are possible after removal of the polar head group or modification by methylation of the phosphate [8,15]. Removal of the polar head group may be done by acetolysis or by treatment with phospholipase C. Acetolysis may cause some loss of plasmalogens. Both methods are applicable to alkylacyl glycerophospholipids. A recent paper by Waku and Nakazawa [16] illustrates the use of phospholipase C followed by acetylation of the diradylglycerols and separation of the classes by thin-layer chromatography (TLC). Each class was then separated into fractions differing in degree of unsaturation by argentation TLC. The diradylglycerol classes have also been separated by Curstedt [17] by column chromatography on Lipidex-5000 followed by the separation of 10–20 molecular species by reversed-phase column chromatography of the trimethylsilyl ether derivatives. Presumably even better separations could be made with current high-pressure liquid chromatography (HPLC) methodology.

Hydrogenolysis with lithium aluminum hydride or Vitride gives alkylglycerols, alk-1′-enylglycerols, and alcohols from glycerophospholipids [8]. The acetates of these compounds can be formed by decomposition of the metal salts with acetic anhydride or by acetylation of the extracted products. If the latter is done with radioactive acetic anhydride, the quantities of each product can be assayed after separation by TLC [18].

The vinyl ether linkages of plasmalogens are easily cleaved by mercuric ion and hydrogen ion to release aldehydes [8]. The toxic action of methyl mercuric salts may be due to their ability to liberate aldehydes from plasmalogens [19]. Released aldehydes may be quantitated by reaction with fuchsin-sulfurous acid, dinitrophenylhydrazine, or *p*-nitrophenylhydrazine [7,8]. A method for the assay of both free and bound aldehydes with the latter reagent [20] gives high results for free aldehydes because the hydrogen ion concentration is too high and bound aldehydes are released from plasmalogens [7]. The iodine addition method is very good for the spectrophotometric measurement of plasmalogens.

Other methods for the assay of plasmalogens involve differential hydrolysis and separation of products [8,21]. Hydrolysis with mild alkali gives water-soluble products such as glycerophosphoethanolamine from diacyl glycerophospholipids. Subsequent hydrolysis of the chloroform-soluble products with mild acid releases water-soluble products from plasmalogens. The remaining chloroform-soluble products include the alkyl glycerophospholipids. Two-dimensional TLC with hydrolysis of alkenyl groups with HCl vapors between dimensions is an effective method for the separation of plasmalogens from acid-stable glycerophospholipids [8]. The latter can be further hydrolyzed with mild alkali for separation of products from diacyl and alkylacyl glycerophospholipids [22]. Further improvements of the two-dimensional TLC procedure have been described [23,24].

Alkylglycerols can be quantitated by several spectrophotometric or GLC methods [8]. It is often desirable to measure the content of alkyl and alk-1-enyl groups in the same phospholipid sample. In addition to the sequential hydrolysis methods, with or without TLC, described above, other methods are available for mixtures of alkylglycerols and alk-1-enylglycerols. These include the assay of aldehydes with fuchsin reagent before and after periodate oxidation and GLC assays of isopropylidene and 1,3-dioxane derivatives with internal standards. Another method involving hydrogenolysis with lithium aluminum hydride, separation of alkylglycerols and alk-1′-enylglycerols by TLC and quantitation by densitometry after charring, seems to give high results for alkyl groups and quite low results for alk-1-enyl groups.

Several methods for the preparation of derivatives of alk-1-enyl and alkyl groups are available [8]. Recommended are isopropylidine and bistrimethylsilyl derivatives for alkylglycerols and 1,3-dioxanes for alk-1-enyl groups. The latter which are formed with 1,3-propanediol, were originally named cyclic acetals. The 1,3-dioxane derivative can be made directly from small amounts of glycerophospholipids [22]. Dimethylacetal derivatives are not recommended because they are sensitive to decomposition during GLC. Several derivatives of ether lipids can be used for structural proof by gas chromatography-mass spectrometry [25–29].

The marked susceptibility of alkylglycerols and derivatives to peroxidation is often not appreciated. Like other aliphatic ethers, they react with oxygen to form hydroperoxides at the 1-position of the glycerol and alkyl moieties because the C–H bonds adjacent to ether bonds are labile [30].

5. *Chemical synthesis*

A variety of methods are available for the synthesis of *O*-alkylglycerols [2,31]. The method used most often is the condensation of isopropylideneglycerol with an alkylmethanesulfonate in boiling benzene in the presence of potassium or in xylene in the presence of potassium hydroxide [32]. The isopropylidene group is then cleaved with hydrochloric acid. Optically active alkylglycerols are obtained from optically active isopropylideneglycerol [33]. Conventional acylation methods will give alkyldiacylglycerols which can be treated with a lipase to obtain alkylacylglycerols [2,31].

The synthesis of *O*-alk-1-enylglycerols was also reviewed by Paltauf [31] and Mangold [2]. The double bond is obtained by removal of HCl from the carbonate derivative of a 1-chloroalkylglycerol. Improvements in this synthesis were described recently [34]. *Cis-trans* mixtures are produced which can be separated on silica gel-silver nitrate.

The synthesis of alkylacyl glycerophospholipids from alkylacylglycerols was reviewed recently by Eibl [35]. Several approaches are possible for the synthesis of alkylacylglycerophosphate, the ether analogue of phosphatidic acid. Of these, direct phosphorylation with phosphorus oxychloride and triethylamine is rapid and gives product yields of greater than 90%. This synthesis is also suitable for alk-1-enylacylglycerophosphate. A very similar synthesis of dihexadecylglycerophosphate was reported by Kertscher et al. [36]. The intermediate in this phosphorylation, the alkylacylglycerophosphoric acid dichloride, can also be reacted with ethanolamine in the presence of triethylamine to give a cyclic intermediate that is hydrolyzed with formic or boric acid to give alkylacylglycerophosphoethanolamine in yields of 90–95% based on the starting diradylglycerol. The synthesis of ethanolamine plasmalogens by this method is feasible but has not been reported. Paltauf has described the synthesis of 1-*O*-[9′,10′-^{3}H]octadecyl-2-octadecenoyl-*sn*-glycero-3-phosphoethanolamine for use in studies of plasmalogen biosynthesis [37].

Dialkylglycerols have been converted to dialkylglycerophosphocholines in good yield by Brockerhoff and Ayengar [38]. Alternatively, Eibl has methylated ethanolamine glycerophospholipids with methyliodide to give the corresponding choline glycerophospholipids [35]. This reaction can be used with [^{14}C]methyliodide. Eibl's group has also produced dialkyl analogues of phosphatidylserine and phosphatidylglycerol. The latter synthesis is not suitable for plasmalogens. Phosphatidyl serine analogues were also prepared by Doerr et al. [39].

The synthesis of double-labeled *S*-alkylglycerols and their derivatives was described by Ferrell et al. [40]. Several *O*-alkylglycerols with F or Cl substituted for a hydroxyl group were synthesized as potential cytostatic agents [41]. The most recent synthesis of *O*-alkylcholesterols was reported by Halperin and Gatt [42]. Both *S*-alkylglycerols and *O*-alkylcholesterols have been detected in bovine cardiac muscle.

6. *Content and composition*

(a) Bacteria

(i) Phytanyl ethers

Phytanyl ether lipids constitute nearly all of the membranous lipids in three groups of bacteria, the anaerobic methane-producing bacteria, the extremely halophilic bacteria, and the thermoacidophilic groups *Thermoplasma* (*Caldariella*) and *Sulpholobus* [43]. Distinctive lipids and rRNA sequences are evidence that these bacteria are ancestral life-forms, designated as archaebacteria [44,45]. The phytanyl ether lipids have the hydrophobic moieties shown in Fig. 2, a and c. The halophiles contain diethers, the thermoacidophiles contain tetraethers, and the methanogens may contain a mixture of diethers and tetraethers.

Structural characterization of the diether compounds in extreme halophiles was done mainly by Kates and his colleagues [46,47]. By very thorough chemical characterization and by chemical synthesis, they established the 2,3-diphytanyl-*sn*-glycerol structure [48] (Fig. 2a). In *Halobacterium cutirubrum*, the diether structures are found in analogues of phosphatidylglycerophosphate [49] (Fig. 2b) 64%, glycolipid sulfate [50] 25%, phosphatidylglycerol [51] 4%, phosphatidylglycerol sulfate [52] 4%, and minor glycolipids [52a]. The glycolipid sulfate contains glucose, mannose, and galactose-3-sulfate [50]. The same phospholipids are present in *Halobacterium marismortui* [52b].

The tetraether structures (Fig. 2c) are equally unique [53–55]. Two molecules of

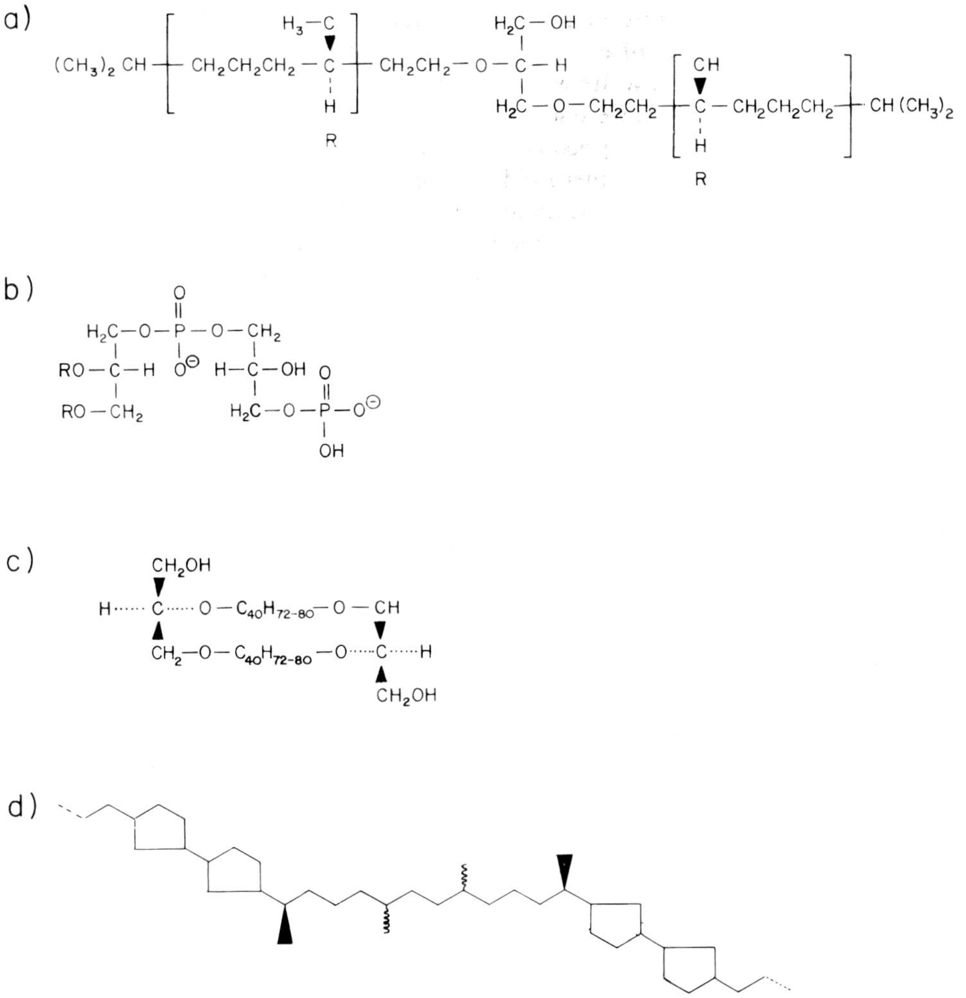

Fig. 2. Structures of representative phytanyl ether lipids. (a) 2,3-Diphytanyl-*sn*-glycerol; (b) 2,3-diphytanyl-*sn*-glycero-1-phospho-3′-*sn*-glycero-1′-phosphate; (c) di(biphytanyl)diglycerol tetraethers; (d) a C_{40}-isoprenoid with 4 cyclopentane rings.

glycerol are bridged through *sn*-2,3 ether linkages by two biphytanyl chains [43,56]. The phytanyl chains are connected by a C—C bond at the 16, 16′ positions. In the thermoacidophiles, but not in the methanogens, some of the biphytanyl chains are modified by formation of 1–4 cyclopentane rings [43,53,54,56] (Fig. 2d). The extent of cyclization is a function of environmental temperature [57]. In some of the tetraethers, one glycerol is replaced by calditol, a 9-carbon polyol [55,58]. Lipids from the extreme thermophiles are glycolipids or acidic lipids [59,60]. Acidic compounds are either sulfate or phosphoinositol derivatives of the glycolipids or phosphoinositol derivatives of the tetraether. Up to 30% of the lipid from methanogens was found in lipids with the properties of phosphonolipids [43,61]. In a recent study, the methanogen lipids were identified as glycolipids, phosphoglycolipids, and 2,3-di-*O*-phytanyl-*sn*-glycero-1-phospho-*sn*-1′-glycerol [62,62a].

Membranes formed of tetraether lipids are bipolar monolayers. With only a single hydrocarbon chain to span the membrane, considerable added stiffness and rigidity is expected. In both diethers and tetraethers, the presence of phytanyl ethers gives stability to peroxidation, to chemical hydrolysis over a wide pH range, and to enzymic hydrolysis. The stereochemical configuration is opposite to that of nearly all glycerolipids in all other organisms in which the lipids have been characterized. The only other examples of the *sn*-glycero-1-phospholipids are the lysobisphosphatidates that are found in lysosomes [63]. The unusual stereochemistry also provides resistance to lipases. The presence of tetraethers and diethers in organisms living in a wide variety of environments is taken as evidence of evolutionary relationships within the archaebacteria and not of a specifically adaptive response [43,57]. Other modifications of the molecule may be adaptive. For tetraether compounds, the degree of cyclization determines the transition temperature and is dependent on growth conditions [57,60]. Calditol substitution may also be adaptive [60]. The membranes of all three groups have a relatively high negative charge. The predominant compound in *H. cutirubrum*, the diphytanyl analogue of phosphatidylglycerophosphate, should be highly selective for K^+ as opposed to Na^+ [64] and bind Mg^{2+} strongly [65].

The phytanyl groups are probably formed through the mevalonate pathway and geranylgeranyl pyrophosphate [49,66]. The formation of the ether linkage differs from that in mammals because ^3H is not lost from the 1 or 3 position of glycerol in *H. cutirubrum*. Since ^3H is completely lost from the 2 position, the ether linkage might be formed from a reaction of geranylgeranyl pyrophosphate with dihydroxyacetone or dihydroxyacetone phosphate. The unusual stereochemistry also requires the formation of phospholipids by a pathway different from those in other organisms [46]. Before the complete reduction of the digeranylgeranyl glycerol to diphytanyl glycerol, other reactions are possible [56]. The cyclization reaction to form cyclopentane rings must compete with the reduction steps. Tetraether formation must take place by a 16,16′ coupling reaction before the last reduction. Further metabolic information, particularly on the reaction for formation of the ether linkage, is necessary. Nothing is known about the asymmetry of the hydrophilic moieties of these glycolipids and phospholipids in the membranes.

(ii) Plasmalogens

Bacterial plasmalogens are found almost exclusively in obligate anaerobes. This area has been reviewed by Goldfine and Hagen [67], Goldfine [68] and Goldfine and Johnston [69]. Bacteria were surveyed for plasmalogens by Kamio et al. [70]. They were not detected in microaerophilic, aerobic, and facultative anaerobic organisms. In anaerobes, molar ratios of alkenyl groups to lipid phosphorus ranged up to 1.04.

Most of the plasmalogens are ethanolamine plasmalogens. In *Clostridium butyricum*, 40% of the phospholipids are plasmalogens with ethanolamine, *N*-methylethanolamine, and glycerol as the polar head groups [71]. Serine glycerophospholipids usually do not accumulate in bacteria, but they account for 37% of the phospholipids in *Megasphaera elsdenii*. The remainder are ethanolamine glycerophospholipids. The plasmalogen contents of these classes are 72% and 87%, respectively [72]. Cardiolipin as well as ethanolamine and glycerol glycerophospholipids contain plasmalogens in *Sphaerophorus ridiculosis* [73]. The principal plasmalogen in *Anaeroplasma abactoelasticum* accounting for one-third of the phospholipids is glycerol plasmalogen (plasmenyl glycerol) [74]. The glycerol plasmalogen in *Butyrivibrio* accounts for all of the glycerol glycerophospholipids [75]. Choline plasmalogen has been detected in the anaerobic spirochetes, *Treponema phagedenis* [77] and *T. hyodysenteriae* [76]. The latter also contains plasmalogen forms of cardiolipin, glycerol glycerophospholipids, monogalactosyl diglycerides, and an unidentified galactolipid. Generally, the anaerobes with substantial amounts of plasmalogens include Gram-positive clostridia, Gram-negative bacteroides and cocci, and some spirochetes.

The compositions of the alkenyl groups vary widely with species but are qualitatively similar to the compositions of the acyl groups. Depending on species, they may include branched and straight chains, odd and even numbers of carbons, and chains that are saturated, monounsaturated or have cyclopropane groups.

Alkyl groups are often present in these organisms but in low amounts not exceeding 4% of the total phospholipids [67,78]. Compositions of the alkyl groups are quite different from those of the acyl and alkenyl groups.

Unusual ether lipid structures, the glycerol acetal of the ethanolamine and *N*-methylethanolamine plasmalogens, are found in 20–30% of the phospholipids of *C. butyricum* [79,80]. The systematic name for the ethanolamine compound is 1-(1′-glyceroalkyl)-2-acyl-*sn*-glycero-3-phosphoethanolamine. This lipid has an unusually high degree of hysteresis in heating–cooling curves [69].

Bacterial plasmalogens have a role in regulation of membrane fluidity. In *C. butyricum*, the adaptations to changes in growth temperature include changes in unsaturation of acyl groups and in proportions of plasmalogens and their glycerol acetals. In *Veillonella parvula*, changes in the unsaturation of the alkenyl groups were found [69,81,82]. Plasmalogens have a phase transition temperature 4°C less than that of the diacyl analogue when nearly all the side-chains of both are 18:1 *trans* acyl groups [69].

Two reports of plasmalogens in aerobic bacteria have appeared. In the phospholipids of *Proteus*, a relative of *Escherichia coli*, plasmalogens account for 2.5% and alkyl glycerophospholipids account for 1.3% [83]. More than 30% of the side-chains are alkenyl groups in the bisphosphatidate of the Gram-negative marine bacterium MB-45. That lipid had not been reported previously [84] in bacteria. Both analyses were done by generally reliable methods but alkenyl group derivatives were not isolated for a positive identification.

(b) Protozoa, fungi and plants

A polymorphic fungus, the yeast *Pullularia pullulans* is the only fungus in which plasmalogens have been detected [85]. The choline glycerophospholipids contain 12% plasmalogen and the ethanolamine glycerophospholipids contain 32% plasmalogen. The common yeasts such as *Saccharomyces cerevisiae* contain little or no plasmalogens. The slime mold, *Physarum polycephalum*, is a protozoon with some characteristics of fungi. In this organism, plasmalogens account for 21–24% and alkylglycerols are found in 12% of the phospholipids [86]. More than half of the ethanolamine glycerophospholipids are plasmalogens but very little plasmalogen is in the choline glycerophospholipids. Alkylglycerols are found in both phospholipid classes. The cellular slime mold, *Dictyostelium discoideum*, contains *N*-acylethanolamine glycerophospholipids during early development [87]. The *N*-acylethanolamine plasmalogen accounts for 5% of the total phospholipid.

Lipids of protozoa from the genus *Leishmania* were examined by Beach et al. [88]. In eight different species, phospholipids are generally about 70% of the total lipids. Of the phospholipids, 6% are ethanolamine plasmalogens, 4% alkylacyl-GPE, 1% inositol plasmalogens, and 1% alkylacyl-GPI. Compositions of side-chains were also reported. Phospholipids from rumen protozoa also contain both alkenyl and alkyl groups [89].

Ciliated protozoa are rich in alkylacyl glycerophospholipids but do not contain plasmalogens. In *Tetrahymena pyriformis* NT-I, 29% of the phospholipids contain alkylglycerols, including 66% of the choline glycerophospholipids and 20% of the 2-aminoethylphosphonolipids [90]. Strain W of *T. pyriformis* contains the same proportion of alkylglycerols in the phospholipids but 75% of the 2-aminoethylphosphonolipids have alkyl groups [91]. All of the alkyl groups in the phospholipids are 16:0 [91]. Earlier studies on *Tetrahymena* were reviewed by Thompson [92]. In *Paramecium tetraurelia*, alkylglycerols are found in 80% of the choline glycerophospholipids, 15% of the ethanolamine glycerophospholipids, and 90–95% of the 2-aminoethylphosphonolipids [93]. Acyl groups at the 2-position are mostly polyunsaturated. In the 2-aminoethylphosphonolipids, 83% contain arachidonate in a mature culture. The acyl group and phospholipid compositions change during development of the culture [94].

The existence of ethanolamine and choline plasmalogens in a plant tissue was unequivocably shown by Kaufman et al. [95]. Alkylacyl glycerophospholipids are probably present also in peas and seedlings. The alkenyl groups are predominantly 16:0, 18:1, and 18:2. The substantial amount (15%) of 18:2 alkenyl groups is very unusual. The plasmalogen contents of pea and bean phospholipids range from 0.1 to 1% [95,96]. Other plant sources have not been examined.

(c) Invertebrates

Small amounts of alk-1′-enyl and alkyl groups are present in the phospholipids of the cockroach, *Periplaneta americana*, the boll weevil, *Anthonumus grandis*, and in the tobacco budworm *Heliothis virescens* [97]. In the tobacco budworm, most of the ether lipids are ethanolamine glycerophospholipids [98,99]. Choline and ethanolamine plasmalogens each account for 1–6% of the glycerophospholipids in the tobacco hornworm, *Manduka sexta* [100]. About half of the ethanolamine glycerophospholipids are plasmalogens in a ganglion from the polychaetus annelid, *Nereis virens* [101]. Chitwood and Krusberg [102,103] found substantial proportions of ethanolamine plasmalogens, some alkylacyl glycerophosphoethanolamines, and smaller amounts of ether-containing choline glycerophospholipids in phospholipids of the plant parasitic nematode, *Meloidogyne javanica*, and the free-living nematode, *Turbatrix aceti*. At least 88% of the alkyl and alk-1′-enyl groups are 18:0 and 18:1 is predominant at the 2-position in these nematodes. Dembitsky and Vaskovsky [104–107] also found substantial proportions of ethanolamine plasmalogens and smaller amounts of choline plasmalogens in the phospholipids from a large number of marine invertebrates (Table 1). Molluscs also contain serine plasmalogens. In one echinoderm, *Ophiura sarsi*, the ethanolamine glycerophospholipids are 99.5% ethanolamine plasmalogens and 0.5% alkylacyl type [107]. The ethanolamine glycerophospholipids in a sponge, *Haliochondria panicea*, are 83% alkylacyl and 17% alkenylacyl types and contain high proportions of 26:2 and 26:3 acyl groups at the 2-position [108]. The nerve tissue of the horseshoe crab has nearly the same proportions of ethanolamine and choline plasmalogens [109] as the other *Crustacea* in Table 1. Additional analyses were reviewed by Thompson [110]. Alkyl groups are present in both phospholipid and non-polar lipid fractions [110].

(d) Fish

Although many studies have been done on the 1-alkyl-2,3-diacyl-*sn*-glycerols from elasmobranch fish [2] very little is known about the ether-linked phospholipids. Alkylacyl and alk-1-enylacyl types of choline and ethanolamine glycerophospholipids are present in the livers of two of these cartilaginous fish, *Hariotta raleighana* and *Rhinochimera atlantica* [111]. Kreps [112] has summarized analyses of phospholipids from brains of 15 species of cartilaginous fish (Elasmobranchs), 22 species of bony fish (Teleosts), and 6 species of mammals (Table 2). Only the teleost fish had significant amounts of choline plasmalogens (2–4% of total phospholipids), but these fish also had the lowest proportions of ethanolamine plasmalogens. In garfish olfactory nerve, an unmyelinated nerve, the ethanolamine glycerophospholipids are 58% plasmalogen whereas the choline and serine glycerophospholipids contain only 1.4 and 1.9% plasmalogen, respectively [113]. Alkylacyl forms of ethanolamine and choline glycerophospholipids are also present. The plasmalogen content of goldfish, *Carassius auratus*, and trout, *Salvelinus fontinalis*, brains varies as a function of water temperature [114–117].

TABLE 1

Proportions of plasmalogens and alkylglycerols in lipids from marine invertebrates [104–107]

	Plasmalogen % of class			Total PL	Alkylglycerols % of total lipid
	EGP	CGP	SGP		
Porifera (Spongia)					
Demospongiae	19.1	3.9	–	6.9	2.0
Coelenterata					
Seyphozoa	82.4	7.2	–	25.8	0.7
Anthozoa	79.5	12.8	–	31.9	9.8
Annelida					
Polychaeta	63.0	2.5	–	15.0	2.9
Echiuridea					2.2
Sipunculidea	72.0	5.2		20.4	2.8
Arthropoda					
Crustacea	44.7	7.0		15.3	2.8
Tentaculata					
Brachiopoda					4.4
Mollusca					
Loricata	77.4	6.2	45.3	29.5	8.4
Gastropoda	74.7	13.6	56.2	39.5	2.8
Bivalvia	66.8	6.9	46.0	30.1	3.9
Cephalopoda	29.3	1.2	–	10.8	4.7
Echinodermata					
Holothuroidea	72.5	8.2	–	19.9	13.7
Echinoidea	65.2	4.9	–	21.5	3.5
Asteroidea	87.0	12.0	–	35.2	3.0
Ophiuroidea	99.2	10.1	–	35.3	4.0
Chordata, Tunicata					
Ascidiacea	67.5	14.3	–	27.4	4.0

(e) Mammals and birds

Yolks of chicken eggs contain small proportions of alkyl groups in the ethanolamine and choline glycerophospholipids and traces of ethanolamine plasmalogens [118,119]. Small amounts of choline and ethanolamine plasmalogens are present in the salt glands of the herring gull and eider duck [120]. Other reported values for ether-linked phospholipids in brain, heart, skeletal muscle, kidney, liver and blood plasma of birds are in the review by Horrocks [119] and in following portions of this chapter.

The best values were estimated for the contents of plasmalogens and phospholipids in various human tissues and used to calculate the proportion of plasmalogens in the phospholipids of the entire human body (Table 3). About one-fifth of the

TABLE 2

Plasmalogens in ethanolamine glycerophospholipids from brains of elasmobranchs, teleosts, and mammals [112]

	Diacyl-GPE	Alkenylacyl-GPE
	% of total phospholipid	
Elasmobranch fish	13.8	29.0
Teleost fish	18.7	15.8
Mammals	12.2	21.0

phospholipids are plasmalogens. The proportion of plasmalogens is highest in heart, striated muscle, and nervous tissue with about two-thirds of the total content of plasmalogens found in muscles. Nearly all values for content, concentration, and composition of ether-linked phospholipids in mammalian tissues in the literature through 1971 were collected and reviewed [119].

(i) Heart and skeletal muscle

The proportions of plasmalogens in the glycerophospholipids from heart and skeletal muscles display pronounced species differences (Table 4). Plasmalogens account for more than 40% in bovine heart, 32% in human hearts, but only 8% in rat hearts. Within species, different types of muscles have different proportions of plasmalogens within their phospholipids. Rabbit heart muscle has a high proportion of plasmalogens but the iris muscle in the eye has only 4% choline plasmalogen and 11% ethanolamine plasmalogen [128a]. A small proportion ($< 1\%$) of the phospholipids in bovine heart are dialkylglycerophosphocholines [125]. The infarct plasmalogen of canine heart is the *N*-acyl derivative of the ethanolamine glycerophospholipids [129]. More than 40% of the *N*-acylethanolamine glycerophospholipids are plasmalogens. Molecular species of ether-linked choline and ethanolamine glycerophospholipids of sarcotubular membranes are different in rabbit, rat, chicken, and man [130]. Molecular species of ether-linked choline glycerophospholipids from bovine heart were separated [124].

(ii) Nervous system

A large proportion of the ethanolamine glycerophospholipids and a substantial portion of the total phospholipids are accounted for by the ethanolamine plasmalogens (Table 5). The ethanolamine plasmalogens are particularly enriched in the myelin where they account for 32 to 43% of the total phospholipid in the central nervous system (Table 6). The proportion is lower, 27 to 30%, in peripheral nerve myelin. Values for the ethanolamine plasmalogens in various species and regions of the central nervous system depend on the degree of myelination and proportion of white matter. This explains the relatively low values for retina and for isolated cells (Table 5). Microsomal fractions from white matter or spinal cord have higher values

TABLE 3

Plasmalogens in tissues of adult man (calculated from values in [119] and [121] and in this review)

Tissue	Tissue weight (kg)	Plasmalogen content (mmol)	Phospholipid content (mmol)	Plasmalogen (% of PL)
Liver	2.3	1.6	201	0.8
Striated muscle	40.0	122	488	25
Kidneys	0.5	1.6	13.7	12
Skin	6.0	6.6	46.8	14
Alimentary tract	1.9	3.4	21.3	16
Heart	0.5	2.6	8.2	3.2
Nervous tissue	3.0	38.7	169	23
Skeleton	17.6	–	–	–
Adipose tissue	11.4	8.0	37.6	21
Total	83.2	184	986	19

for ethanolamine plasmalogens than the corresponding fractions from grey matter (Table 6). Rabbit brain fractions have relatively high values.

The content of ethanolamine plasmalogens increases considerably during development [119]. The plasmalogen content is less than 10% in rat cerebrum [133] and chicken brain [151] before myelination. During development, this proportion dou-

TABLE 4

Ether-linked glycerophospholipids in mammalian heart and skeletal muscle

	Alkenylacyl-GPC	Alkylacyl-GPC	Alkenylacyl-GPE	Alkylacyl-GPE
	% choline GPL		% ethanolamine GPL	
Human heart [122]	36	5.2	59	3.6
Human rectus abdominis muscle [123]	17	–	60	–
Human gastrocnemius muscle [123]	27	–	60	–
Bovine heart [124]	46	2.1	–	–
Bovine heart [125]	39	3	–	–
Rabbit heart, male [126]	39	2.2	43	2.2
Rabbit heart, female [126]	41	2.6	48	2.7
Rat heart [127]	–	–	11.2	4.0
Rat heart [128]	1.7	0.6	12.4	5.9
Rat diaphragm [128]	2.4	0.9	24.6	2.4
Rat soleus muscle [128]	2.4	1.0	17.7	5.2
Rat rectus femoris muscle [128]	3.0	0.9	32.5	2.1

GPC, glycero-3-phosphocholine; GPE, glycero-3-phosphoethanolamine; GPL, glycerophospholipid.

TABLE 5

Ethanolamine plasmalogens in the nervous system.

Mole ratio alk-1′-enyl groups: lipid P	% ethanolamine-GPL	Animal, tissue, ref.
0.21	64	Rat, brain [131]
0.22	56	Rat, brain [132]
0.19	43	Rat, cerebrum [133]
0.29	60	Rat, spinal cord [133]
0.23	55	Rabbit, brain [134]
0.19	47	Mouse, brain [135]
0.19	57	Calf, grey matter [136]
0.27	68	Calf, white matter [136]
0.32	71	Man, white matter [137,138]
0.26	56	Chicken, brain [136]
0.12	23	Frog, retina [136a]
0.11	35	Chicken, retina [136]
0.18	44	Duck, retina [136a]
0.08	26	Calf, retina [136]
0.15	47	Rabbit, astroglia [139]
0.19	51	Rabbit, neurons [139]
0.17	49	Calf, oligodendroglia [140]
0.12	46	Rat, endothelial cells [141]
0.21	75	Rat, sciatic nerve [142]

GPL, glycerophospholipid.

bles while the phospholipid content is increasing many-fold. Similar increases in content and proportion of ethanolamine plasmalogens and of alkylacyl glycerophosphoethanolamines take place in human brain [152] and cerebellum [153]

TABLE 6

The content of ethanolamine plasmalogens in subcellular fractions from the nervous system

Mole ratio, alk-1′-enyl groups: lipid P.

Myelin	Microsomes	Synaptosomes	Animal, tissue, ref.
0.32	0.19	–	Mouse brain [143]
0.43	0.19	–	Rat brain [144]
0.42	0.27	–	Rat spinal cord [144]
0.34	0.16	–	Hamster brain [145]
0.42	–	0.27	Rabbit brain [134,146]
0.30	–	–	Rabbit PNS [147]
0.27	–	–	Squirrel monkey PNS [148]
0.38	0.16	–	Rhesus monkey brain [149]
0.39	0.27	–	Rhesus monkey spinal cord [149]
0.32	0.28	–	Man, white matter [150]
–	0.17	0.18	Man, grey matter [150]

TABLE 7

The content of alkylacyl glycerophosphoethanolamines in the nervous system

% lipid class	Animal, tissue, ref.
7.6	Chicken, retina [136]
8.9	Calf, retina [136]
4.2	Chicken, brain [136]
4.7	Calf grey matter [136]
6.6	Calf white matter [136]
3.6	Mouse brain [119]
3.9	Rat, brain [157]
3.1	Man, brain [122]
7	Calf, cerebral myelin [140]
16	Calf, oligodendroglia [140]

during development. Changes in the content and proportion of ethanolamine plasmalogens in myelin during maturation and in the nervous system during ageing have been reviewed [154–156].

The content of alkylacyl glycerophospholipids is relatively low in the nervous system (Tables 7, 8). In the brain, the ethanolamine glycerophospholipids and the choline glycerophospholipids contain 3 to 7% of the alkylacyl type. For retina, the value for alkylacyl glycerophosphocholines is in the same range but the value for alkylacyl glycerophosphoethanolamines is higher, 7 to 9%. A very high value, 16%, was reported recently for calf oligodendroglia [140].

The choline glycerophospholipids also include 1 to 8% choline plasmalogens [119] (Table 8). The choline plasmalogens are often overlooked but are nearly always found in careful analyses. Serine plasmalogens were reported in relatively high proportion some years ago but this was due to an analytical artifact [119]. Although

TABLE 8

The content of alkylacyl and alk-1′-enylacyl glycerophosphocholines in the nervous system

% lipid class		Animal, tissue, ref.
Alkylacyl-GPC	Alk-1-enylacyl GPC	
4.9	2.7	Rat brain [157]
5.0	3.8	Chicken brain [136]
2.7	3.9	Calf grey matter [136]
6.3	6.1	Calf white matter [136]
4.0	2.7	Chicken retina [136]
4.0	1.0	Calf retina [136]

TABLE 9

Composition of ethanolamine plasmalogens from the nervous system

	Human myelin [154]	Human white matter [127]	Rhesus monkey [159]	Hamster myelin [158]	Mouse myelin [160]	Human grey matter [137]	Hamster microsomes [158]	Mouse microsomes [160]	Rat capillary [141]
Alkenyl groups									
16:0	26	26				17			
18:0	16	16				55			
18:1	57	55				24			
Acyl groups									
16:0	3	3	3	2	1	2	4	2	3
18:0	3	1	1	1	1	1	2	2	2
18:1	45	52	33	30	42	17	12	10	24
20:1	8	9	10	21	23	2	5	5	6
20:4	10	8	11	17	10	22	26	12	40
22:4	20	16	25	10	10	19	10	9	5
22:6	2	2	3	7	10	29	37	58	16
24:4	2	2	6	–	–	–	–	–	–

serine plasmalogens seem to exist, at least in some species, no reports were found on their content in nervous tissue for the period 1972–80. Plasmalogen analogues of phosphatidic acid and phosphatidylinositol have also been reported [119].

The alk-1′-enyl groups of ethanolamine plasmalogens from the nervous system are primarily 16:0, 18:0, and 18:1 (Table 9). In white matter, 18:1 predominates whilst in grey matter, 18:0 predominates. The 18:1 alk-1′-enyl groups are mainly (n-7), particularly in white matter, and thus differ from the 18:1 acyl groups which are primarily the (n-9) isomer, oleate [122,137]. The acyl groups from the 2-position also have a high content of 18:1 in white matter and myelin, but not in grey matter or in microsomal fractions (Table 9). Another monoene, 20:1, accounts for more than 20% of the acyl groups in the 2-position ethanolamine plasmalogens from hamster and mouse myelin. Arachidonate, 20:4 (n-6), is highest in capillary endothelium. Grey matter has lower proportions and myelin has only 10 to 17% of arachidonate. Myelin and grey matter from primates has 19 to 25% of adrenate, 22:4 (n-6). Docosahexaenoate, 22:6 (n-3), is very low in primate myelin, but quite high in grey matter and microsomal fractions. In human myelin, more than 80% of the side-chains are unsaturated. Thus, the ethanolamine plasmalogen should contribute to the fluidity of this membrane.

Alkyl galactolipids are also found in the brain at levels of about 0.2% of the total lipid [119]. This class of compounds, 1-alkyl-2-acyl-3-β-D-galactopyranosyl-*sn*-glycerol, has been partially separated into molecular species by Yahara and Kishimoto [161]. Two-thirds of the alkyl groups and 40% of the acyl groups are 16:0. A similar composition of alkyl and acyl groups was reported for the analogue with a sulphate group attached at the 3-position of the galactose [162].

(iii) Other organs

Adrenal glands of the guinea pig contain quite low proportions of ethanolamine and choline plasmalogens in the microsomes, mitochondria, and chromaffin granules [163]. The proportions are higher in these fractions of bovine adrenal glands (Table 10). In both species, the ethanolamine plasmalogens are rich in arachidonic acid which comprises more than 50% of the total acyl groups in guinea pig adrenal, more than 60% in bovine adrenal cortex, and more than 70% in bovine adrenal medulla.

Human kidney (Table 10) contains serine plasmalogens at a level of 3.2% of the total phospholipid and 28% of the serine glycerophospholipids [164]. These values are similar to those for porcine and ovine kidney [119]. The liver has very low levels of plasmalogens with ethanolamine plasmalogens accounting for about 1% of the total phospholipids [119]. According to Curstedt [124], rat liver phospholipids include 0.1% of choline plasmalogens and 0.5% of alkylacylglycerophosphocholines. The predominant molecular species of the latter are 1-hexadecyl-2-hexadecanoyl, 1-octadecenyl-2-hexadecanoyl, and 1-hexadecyl-2-eicosatetraenoyl. The high proportion of palmitoyl groups at the 2-position is unusual. Phospholipids in plasma reflect the liver composition by also having a very low level of plasmalogens [119].

The testis resembles kidney and spleen in plasmalogen content and concentration [119]. The alk-1′-enyl group composition of choline plasmalogens from porcine testis

TABLE 10

Choline and ethanolamine plasmalogens in mammalian tissues

Animal, tissue, ref.	Alkenylacyl-GPC		Alkenylacyl-GPE	
	% PL	% CGP	% PL	% EGP
Guinea pig, adrenal gland microsomes [163]	2.4	4	4.0	14
Ox, adrenal cortex microsomes [163]	6.0	22	23.9	55
Ox, adrenal medulla, microsomes [163]	3.2	8	21.1	62
Man, kidney [164]	2.3	7	6.6	22
Sheep, testis [165]	3.9	7	8.8	38
Pig, testis [165]	1.8	3	4.9	18
Rat, testis [166]	2.6	4	11	34
Rat, lung [167]	–	–	8.2	28
Rat, adipose [168]	4.2	8	11.4	33
Ox, adipose [168]	3.1	7	20.0	53
Pig, adipose [168]	3.7	8	18.0	58
Rat, intestinal smooth muscle [169]	2.5	7	8.2	40
Rat, intestinal mucosa [169]	1.1	3	2.9	14
Rat, mammary gland, pregnant [170]	2.9	8	13.4	49
Guinea pig, platelet [171]	0	0	7.4	25
Guinea pig, megakaryocyte [171]	0	0	5.9	20

GPC, glycero-3-phosphocholine; GPE, glycero-3-phosphoethanolamine; PL, phospholipid, CGP, choline glycerophospholipid; EGP, ethanolamine glycerophospholipid.

is quite unusual in its 19% of 18:2 groups [164]. Usually, 16:0, 18:0, and 18:1 account for nearly all of the alk-1′-enyl groups. Both 1-alkyl-2-acyl-3-β-D-galactopyranosyl-*sn*-glycerol and 1-alkyl-2-acyl-3-β-D-(3′-sulfo)-galactopyranosyl-*sn*-glycerol are major glycolipids of the testes and spermatozoa of many mammals [172]. In these two glycolipid classes, the alkyl groups are primarily hexadecyl and the acyl groups are primarily hexadecanoyl [172,173]. The sulfogalactosylalkylacylglycerol, the major glycolipid, is greatly enriched in a plasma membrane fraction prepared from rat testis homogenates [174]. Seminolipid is a synonym for this glycolipid.

The proportions of ether-linked lipids in spermatozoa are quite variable between species (Table 11). Ruminants have high levels (35–41%) of choline plasmalogens except Indian buffalo which have much lower levels. Very low levels (2–8%) of choline plasmalogens are found in spermatozoa from man, rhesus monkey, dog, and chicken. Boar spermatozoa have 27% alkylacyl type of choline glycerophospholipids in addition to 12% choline plasmalogen. Very few other assays of alkylacyl glycerophospholipids in spermatozoa are available. High proportions of polyunsaturated acyl groups are found in the ether-linked glycerophospholipids of spermatozoa. The ether-linked choline glycerophospholipids of bovine spermatozoa [173] have 72–75% 22:6 (n-3) and 20–23% 22:5 (n-6) at the 2-position and 96–98% 16:0 in the alk-1′-enyl or alkyl group at the 1-position. The ether-linked ethanolamine

TABLE 11

Ether-linked choline and ethanolamine glycerophospholipids in spermatozoa

Animal, ref.	Alkenylacyl-GPC	Alkylacyl-GPC	Alkenylacyl-GPE	Alkylacyl-GPE
	% of total phospholipid			
Pig [175]	11.6	27.4	11.3	13.8
Buffalo [176]	19.4	–	3.4	–
Ox [176]	36.8	–	9.0	–
Ox [173]	34.8	9.5	9.4	4.7
Sheep [177]	40.8	–	5.9	–
Man [177]	2.3	–	9.3	–
Rhesus monkey [178]	6.9	–	16.1	–
Dog [179]	3.6	–	15.3	–
Chicken [179]	7.8	–	5.2	–

glycerophospholipids have 60–62% 22:6 (n-3), 20–23% 22:5 (n-6), and smaller proportions of 20:4 (n-6) and 22:5 (n-3) at the 2-position. In other species, 22:6 (n-3) is also the predominant acyl group in the plasmalogens [177,178]. The function of the ether-linked glycerophospholipids is not known but in contrast to diacyl glycerophospholipids they are not metabolized when spermatozoa are incubated with labeled glycerol and dihydroxyacetone.

A relatively low proportion of ethanolamine plasmalogens is found in the lung [119] (Table 10). The molecular species from rat lung include 56% tetraenes and 32% polyenes with 16:2 and 18:0 alk-1′-enyl groups [167]. The very low concentration of phospholipids in adipose tissue includes 11–20% of ethanolamine plasmalogens [168]. Intestinal smooth muscle of the rat is relatively low in plasmalogen concentration, but alkylacyl compounds also account for 8 and 12% respectively of choline and ethanolamine glycerophospholipids [169]. The mucosal layer choline glycerophospholipids are 9% alkylacyl type [169]. Analyses of the ethanolamine glycerophospholipids have given values of 12% and 28% for the alkylacyl type [169,180]. Rat embryos contain substantial proportions of ethanolamine plasmalogens [181]. During pregnancy, proportions of plasmalogens are also substantial in rat mammary glands. However, the proportion of ethanolamine plasmalogens decreases markedly during lactation (6%) and early post-lactation (3%) with a return to 16% of total phospholipid at 10 days post-lactation [170]. Of the total phospholipid in bovine milk, 0.25% by weight is alkylglycerol [182].

Human red blood cell and platelet phospholipids include about 15% ethanolamine plasmalogens [119]. Their topology is discussed in the section on function and biological role of plasmalogens. Guinea pig platelets and megakaryocytes [171] and hen erythrocytes [183] have lower proportions of plasmalogens (Tables 10, 12).

The alkylacyl type of glycerophospholipid is nearly always overlooked in analyses, but sometimes the alkylacyl glycerophospholipids are major components. About 75%

TABLE 12

Ether-linked choline and ethanolamine glycerophospholipids in lymphocytes and erythrocytes

Animal, tissue, ref.	Alkenylacyl-GPC	Alkylacyl-GPC	Alkenylacyl-GPE	Alkylacyl-GPE
	% of phospholipid			
Pig lymphocytes, mesenteric lymph node [184]	1.1	11.6	12.0	1.9
Hen erythrocyte, nuclear membrane [183]	3.0	12.2	6.7	0.8
Hen erythrocyte, plasma membrane [184]	1.9	5.2	9.0	1.3

of the ethanolamine glycerophospholipids in bovine erythrocytes are the alkylacyl type [185]. An additional 5% is 1-alkyl-2-acyl-*sn*-glycero-3-phospho-*N*-acylethanolamine [186]. This type as well as the corresponding plasmalogen account for most of the ethanolamine glycerophospholipids in degenerating baby hamster kidney (BHK) cells [187]. Significant quantities of 1-alkyl-2-acyl-*sn*-glycero-3-phosphocholine are found in other blood cells (Table 12). Phospholipids from Harderian glands of rabbits contain 4.2% alkylacyl-GPC in the white portion and 5.5% alkylacyl-GPC together with 7.2% alkylacyl-GPE in the pink portion [188].

(f) Neoplasms

Several excellent reviews on the content and metabolism of ether-linked glycerolipids in cancer cells have been published by Snyder and co-workers [190–192]. The metabolism of these lipids has been studied with great interest since the discovery that many neoplastic tissues have markedly higher levels of 1-alkyl-2,3-diacyl-*sn*-glycerols [193]. This lipid has been proposed as a tumor marker but recently lower levels of alkyldiacylglycerols were reported in a carcinogen-induced hepatocellular carcinoma [194] and elevated levels are induced by staphylococcus infection [195]. Thus elevations of alkyldiacyl glycerols do not correlate completely with the presence of tumors. The remainder of this section is about the concentration of ether-linked glycerophospholipids in neoplastic tissue and cultured tumor cells.

Human brain tumors were examined for phospholipid and acyl group composition by Albert and Anderson [196,197]. As a percentage of total phospholipid, the proportions of ethanolamine plasmalogens were rather low but varied from 0.8% to 11.8%. The proportion of alkylacyl-GPC was elevated in 3 of 5 tumor types. The choline plasmalogen value was only 0.1% in the control but was elevated to values of 0.8 to 3.4% in the tumors. Yates et al. also found markedly elevated values of 2.0 to 4.2% for choline plasmalogens in tumors and in cells cultured from the tumors [198]. In phospholipids of a Yoshida ascites hepatoma, the proportions of choline plas-

TABLE 13

Choline and ethanolamine plasmalogens in cultured cells

Source, ref.	Alkenylacyl-GPC	Alkenylacyl-GPE	Alkylacyl-GPL
	% phospholipid		
Hamster astrocytes [202]	2.7	11.5	2.2
NN astroblasts [203]	3.1	10.0	2.5
C-6 astrocytoma [204]	<0.3	11.7	2.1
C-6 astrocytoma [203]	2.6	10.5	3.5
Glioma 12–18 [205]	1.9	13.8	–
Fetal neural CH [205]	1.4	13.8	–
C1300 neuroblastoma [206]	0	12.8	–
$HSDM_1C_1$ fibrosarcoma [207]	6.9	12.0	–
Ehrlich ascites [208]	2.8	6.1	11.0
L fibroblasts [209]	8.2	4.1	–

malogens, 1.3%, and alkylacyl-GPC, 4.3%, are markedly elevated above values for normal liver [199]. The choline plasmalogen in hepatomas is concentrated in the plasma membrane [200]. Not all tumors have elevated levels of ether-linked choline glycerophospholipids since none could be detected in three types of human lung carcinoma [201].

Cultured cells, both primary and malignant cells, generally have 2–3% choline plasmalogens and 1–13% ethanolamine plasmalogens (Table 13). Robert et al. also found 0.4% or less of inositol and serine plasmalogens and alkylacyl glycerophospholipids [203]. Incubation of neuroblastoma cells with ethanolamine markedly increased the diacyl-GPE and decreased the ethanolamine plasmalogen proportions [206]. Confluent cell cultures of a glioma and a fetal neural cell line were consistently higher in ethanolamine plasmalogen than sparse cells [205]. These aspects of the regulation of ethanolamine plasmalogen metabolism are not understood.

7. Biosynthetic pathways

(a) Synthesis of long-chain alcohols

Octadecanol is incorporated more quickly than octadecanoate into the alk-1-enyl groups of plasmalogens [210]. Further metabolic studies showed that fatty acids are reduced to alcohols which form alkyl groups that are oxidized to alk-1-enyl groups [2,211,212]. Lumb and Snyder illustrated the use of [1-^{3}H]hexadecanol for the selective labelling of alkyl and alk-1′-enyl groups [213]. Any hexadecanol utilized for acyl groups must first be oxidized to hexadecanoate which does not retain ^{3}H.

Long-chain alcohols are present in small amounts in tissues [214] and are important for the growth of mammalian cells and *Clostridium butyricum* [215]. Cell-free systems for the synthesis of long-chain alcohols have been described for many tissues. Fatty acids are reduced in the presence of ATP, Mg^{2+}, coenzyme A, and NADPH by microsomes from brain [216–218]. Other mammalian tissues, birds, marine organisms, plants, and bacteria also synthesize alcohols from the corresponding fatty acids [219]. Aldehydes are intermediates. The acyl-CoA reductase has some specificity for chain-length and degree of unsaturation [218–220]. The levels of long-chain alcohol may control the rate of synthesis of ether-linked lipids because it is a substrate for the enzyme that forms the alkyl bond [221]. Liver tumors with higher ether lipid contents have higher levels of acyl-CoA reductase and lower levels of long-chain alcohol: NAD oxidoreductase than does normal liver [221].

(b) Synthesis of O*-alkyl bonds*

The reaction of acyldihydroxyacetone phosphate with a long-chain alcohol to produce alkyldihydroxyacetone phosphate and a fatty acid is catalyzed by alkyldihydroxyacetone phosphate synthase (Fig. 3) [222,223]. The alkyldihydroxyacetone phosphate is then reduced by NADPH: alkyldihydroxyacetone phosphate oxidoreductase followed by acylation by 1-alkyl-*sn*-glycero-3-phosphate:acyl-CoA acyltransferase [222,223]. The alkyldihydroxyacetone phosphate synthase is a membrane-bound enzyme which has been solubilized and partially purified [224]. The enzyme has an M_r-value of 250000–300000 [224] and in the intact microsomal fraction is not exposed to the cytoplasmic surface [225]. The enzyme exhibits some chain-length specificity for the long-chain alcohol [215].

The alkyldihydroxyacetone phosphate catalyzes the replacement of an acyl group by an alkyl group with the oxygen of the ether linkage being derived from the long-chain alcohol [226]. At the same time, the C-1 *pro*-R hydrogen is exchanged with the medium and the *pro*-S hydrogen is not affected [227,228]. The hydrogen exchange does not require the presence of the alcohol so Friedberg et al. postulated an intermediate 1-*O*-acyl endiol [229]. This deprotonated intermediate reacts with the long-chain alcohol [230].

Acyldihydroxyacetone phosphate is required for the synthesis of ether-linked lipids from long-chain alcohols. The dihydroxyacetone phosphate *O*-acyltransferase is primarily localized in peroxisomes in guinea pig and rat liver according to Jones and Hajra [231]. Like the alkyldihydroxyacetone phosphate synthase, the acyltransferase is not affected by proteases in the absence of detergent and the enzyme activity is stimulated by detergent [231,232]. The acyltransferase acting on dihydroxyacetone phosphate differs from the acyltransferase acting on glycero-3-phosphate [231]. The utilization of dihydroxyacetone phosphate for glycerolipid synthesis is selectively inhibited by clofenapate. Thus this agent should inhibit synthesis of ether-linked lipids from long-chain alcohols but not from alkylglycerols [233].

Alkylglycerols are converted to 1-alkyl-*sn*-glycero-3-phosphate by ATP:al-

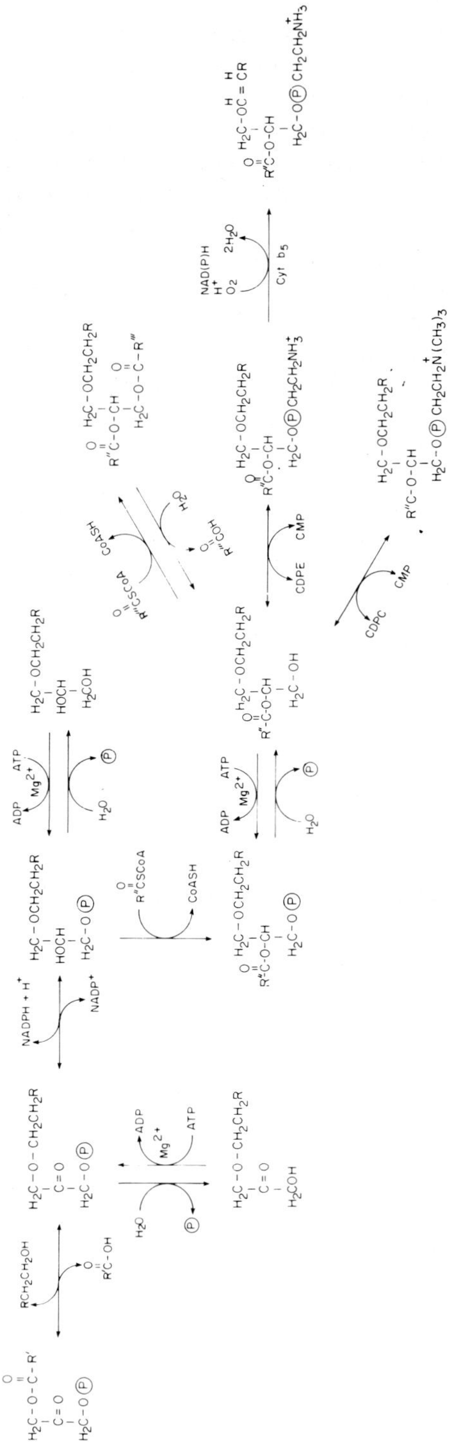

Fig. 3. Pathways involved in the synthesis of ether-linked glycerophospholipids.

kylglycerol phosphotransferase. Properties of this kinase in a cell-free system from the pink portion of the rabbit harderian gland were described by Rock and Snyder [234]. The kinase is presumed to exist in other cells because alkylglycerols can be taken up by cultured cells and incorporated into alkylacylglycerophospholipids [235]. Long-chain alcohols and alkylglycerols can be used with cultured cells to verify the pathways of incorporation [236]. In *Leishmania donovani*, it is possible that alkylglycerols are acylated directly to alkyldiacylglycerols before the alkyl side-chain appears in glycerophospholipids [236].

The specificity for incorporation of long-chain alcohols into ether-linked glycerophospholipids has been studied by Bandi and Mangold with a nutritional approach [237]. By feeding gram quantities of unlabeled alcohols and acids, they have found a limited specificity for alcohol incorporation in alkyl groups and considerable specificity for the reduction of acids to alcohols. Schmid et al. have given various alcohols to rats by intracerebral injection [238]. A number of primary alcohols can be incorporated into alkyl groups but a secondary alcohol, heptadecan-2-ol, was not incorporated into glycerophospholipids [238,239].

(c) Synthesis of plasmalogens

Structural similarities between alkylacyl and alk-1-enylacyl glycerophospholipids have suggested some form of precursor–product relationship. Proof of such a relationship from in vivo experiments requires that the specific radioactivity of the product should reach a peak after that for the precursor. If all of the precursor is converted into all of the product, then the rules of Zilversmit et al. are applicable [240]. Thus, comparisons of ratios of specific radioactivities are very useful. In the ideal case, the ratio precursor:product will exceed one before reaching the peak value for the product. After reaching the peak value for the product, the ratio will be less than one.

Thompson determined the specific radioacitivities of alkylglycerols and alk-1′-enylglycerols prepared from glycerophospholipids of the slug, *Arion ater*, after feeding of labelled alkylglycerols [241]. Over a 3-day period the ratio of specific radioactivities of alkylglycerol: alk-1′-enylglycerol decreased markedly but did not reach a ratio of one and a peak for alk-1′-enylglycerol was not found. This provided evidence but not proof for a precursor–product relationship. Horrocks and Ansell studied the ethanolamine glycerophospholipids of rat brain after labeling with intracerebral [^{14}C]ethanolamine [242]. The ratio of specific radioactivities of alkylacyl-GPE:alk-1′-enylacyl-GPE changed from 5.3 at 0.5 h to 1.1 at 48 h. These results provided additional evidence for a precursor–product relationship with alkylacyl-GPE the precursor. After incubation of [1-^{14}C, 1-^{3}H]hexadecanol with Ehrlich ascites cells in culture, Wood et al. obtained specific activity ratios for the 16:0 side-chains from ethanolamine glycerophospholipids [243]. The ratio alkyl: alk-1′-enyl changed from 1.8 at 6 h to 0.54 at 48 h. The ratio of one near 24 h was at a value for alk-1′-enyl groups that was close to the peak value. In addition the ^{3}H:^{14}C ratio for alk-1′-enyl-GPE was 37% of the ratio for alkyl-GPE at 6 h of incubation. This is consistent with a dehydrogenation of alkyl groups to produce alk-1′-enyl groups, and was the strongest evidence at the time for the precursor–product relationship of alkylacyl-

GPE and alk-1-enylacyl-GPE. Additional in vivo studies, particularly with intracerebral injections of labelled long-chain alcohols, have been reviewed [119]. Stoffel and LeKim demonstrated that the 1S and 2S hydrogens were removed during oxidation of alkyl groups to alk-1′-enyl groups [244].

The conversion in vitro of 1-alkyl-2-acyl-*sn*-glycero-3-phosphoethanolamine labelled in the alkyl group to ethanolamine plasmalogen requires molecular oxygen, NADH or NADPH, and cytochrome b_5 and is inhibited by cyanide [245–250]. Activities reported for the mixed-function oxidase include 3.5 pmol/mg protein/h for microsomes from hamster small intestine [246], 56 pmol/mg protein/h for adult rat brain microsomes [250], and 2.0 nmol/mg protein/h for pig spleen microsomes [248]. Rates for plasmalogen synthesis of 3.3 nmol/mg protein/h from 1-alkyl-GPE [251] and 6.1 nmol/mg protein/h from 1-alkyl-2-acyl GPE [251a] have been reported recently. Activities of at least this magnitude are expected in order to replace the molecules that are turned over.

Proteins from the cytosol fraction stimulate the activity of the mixed-function oxidase. Paltauf described two proteins from pig kidney with an M_r-value of 27000 that do not have enzymic activity by themselves but with microsomes they stimulate the desaturase activity by 3–10-fold [252]. Baker et al. [253] had previously isolated from rat liver cytosol a protein that stimulated the desaturase activity by 40–45%. The protein was identified as catalase. Catalase differs from the proteins isolated by Paltauf which are specific mediators between the membranous enzyme system and the lipophilic substrate [252].

The desaturase activity for alkylacyl-GPE requires the same cofactors as the fatty acid desaturases, but Lee et al. showed that the Δ9 desaturase in rat livers and tumors responds differently to a fat-free diet [254]. However, the alkylacyl-GPE desaturase inserts a double bond between atoms 5 and 6 counting from the carbonyl carbon on the 2-acyl group. Perhaps the enzyme responsible for alkylacyl-GPE desaturase activity is also a Δ6 desaturase that acts on acyl groups at the 2-position of ethanolamine glycerophospholipids.

Debuch et al. have made extensive studies on the metabolism of labeled alkyl-GPE and alkylacyl-GPE after intracerebral injection into young rats [255,256]. Deacylation of the alkylacyl-GPE takes place before formation of alk-1-enylacyl-GPE. The lyso compound, alkyl-GPE, is desaturated faster than any alkylacyl-GPE, regardless of the acyl group. They have concluded that the substrate for desaturation is the lyso compound, alkyl-GPE. Wykle and Schremmer obtained three-fold more desaturase activity with the lyso compound in vitro [257]. When the lyso compound was preincubated with microsomes but without NADH, nearly all of the alkyl-GPE was bound to the microsomes as alkylacyl-GPE but none was desaturated. Subsequent incubation with NADH gave the same desaturase activity as was obtained with direct incubation of the alkyl-GPE. They concluded that the acylation process may position the newly formed alkylacyl-GPE in the membrane so that it is more accessible to the alkyl desaturase. In earlier experiments in vitro, no desaturation was found without ATP and CoA so acylation was necessary before desaturation [247] and no alk-1′-enyl-GPE accumulated when acylation was limited to part of the substrate [245]. Our explanation of the in vivo experiments is that deacylation is necessary for transport of the alkyl-GPE into the cell. The reaction sequence may be alkylacyl-GPE → alkyl-GPE outside → alkyl-GPE inside → alkylacyl-GPE → alk-1-enylacyl-GPE. The desaturase in vivo is capable of introducing a double bond into a 1-*O*-ethyloxyhexadecyl group to form a 1-*O*-ethyleneoxyhexadecyl group [238].

The accepted pathway for ethanolamine plasmalogen synthesis includes a reaction in which alkylacylglycerols are converted to alkylacyl-GPE by the CDPethanolamine:1-alkyl-2-acyl-*sn*-glycerol phosphoethanolaminetransferase (Fig. 3). The specific activity of this enzyme is 40–70% higher in isolated oligodendroglia than in neuronal perikarya or in astroglia [258], and the total activity increases three-fold during oligodendroglial proliferation at the beginning of myelination in chicken brain [259]. This activity may be regulated differently than is the phosphoethanolaminetransferase that forms diacyl-GPE. Free fatty acids, ATP, cAMP, and biogenic amines inhibit one activity more than the other [260–262]. Ethanolamine plasmalogens may also be formed by a phosphoethanolaminetransferase activity if any alk-1-enylacylglycerol is available [263].

The main pathway for choline plasmalogen synthesis is not known. Choline plasmalogens are not formed by desaturation of alkylacylglycerophosphocholine [247,264] but they are labelled in the alk-1′-enyl group after an intracerebral injection of [^{14}C]hexadecanol [264]. When primary cultures of astrocytes were incubated with [^{32}P]phosphate, the choline plasmalogens had a specific radioactivity higher than any other glycerophospholipid including phosphatidate and phosphatidylinositol [202]. Choline plasmalogens have the most rapid turnover of any glycerophospholipid in neuronal perikarya [265]. Some of the choline plasmalogen may be formed from ethanolamine plasmalogen by base exchange [266], but this does not explain the turnover of the phosphate. In the brain, some of the choline plasmalogen is formed from ethanolamine plasmalogen by methylation [267], but this also does not explain the turnover. Choline plasmalogens may be formed by phosphocholinetransferase activity from alk-1′-enylacylglycerols and CDPcholine [268]. The source of the alk-1′-enylacylglycerols may be the reversal of the phosphoethanolaminetransferase reaction with CMP and ethanolamine plasmalogens [269]. These reactions explain the labeling of choline plasmalogens with ^{14}C but not with ^{3}H when 1-[^{3}H]alkyl-2-acyl-*sn*-glycerols are incubated with CDP [^{14}C]choline and rat brain microsomes [270]. These reactions, however, do not explain why the labeling of choline plasmalogens from CDP[^{14}C]choline is increased several-fold by addition of alkylacylglycerols but not by addition of diacylglycerols [188], nor do they explain the origin of the marked differences in alk-1′-enyl groups between choline and ethanolamine plasmalogens. A new pathway is needed for synthesis of the alk-1′-enyl groups of choline plasmalogens from long-chain alcohols.

The 2-acyl groups of ether-linked glycerophospholipids can also turn over independently of the remainder of the molecules. After hydrolysis of the 2-acyl group by a phospholipase A_2 (Fig. 4), or after the formation of 1-alkyl-*sn*-glycero-3-phosphate (Fig. 3), an acyltransferase can add an acyl group to the 2-position. The specificities of these enzyme have been studied in detail by Waku and Nakazawa [271]. A broad

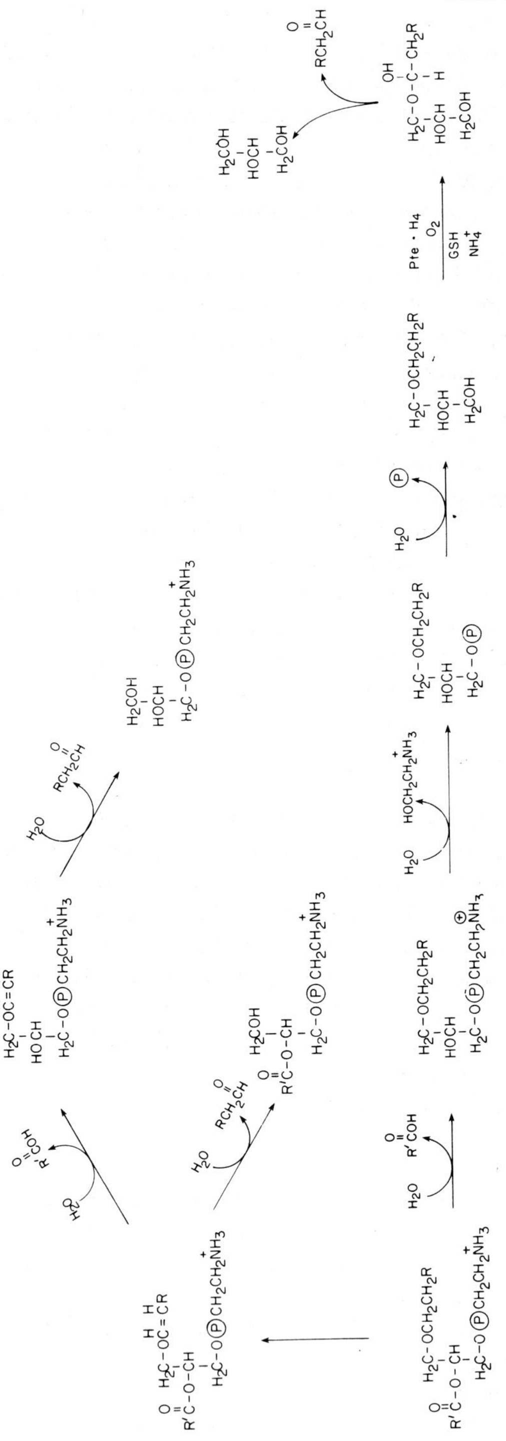

Fig. 4. Pathways involved in the cleavage of ether-linked side-chains from glycerophospholipids.

range of activities with equal rates for saturated and unsaturated acyl-CoA was found for acyl-CoA: 1-alkyl-*sn*-glycero-3-phosphate acyltransferase whereas the acyl-CoA:1-alkyl-*sn*-glycero-3-phosphocholine acyltransferase is specific for polyunsaturated acyl-CoA.

In bacteria and protozoa, plasmalogens are not formed by the oxidation of the alkyl analogues [272,273]. Also, [2-^{3}H]glycerol is incorporated into their plasmalogens with little or no loss of ^{3}H, thus ruling out dihydroxyacetone or dihydroxyacetone phosphate as intermediates [72,274–276]. Either glycerol or glycerophosphate seem to be one precursor of the alk-1′-enylglycerol. In experiments with [1-^{3}H, 1-^{14}C]hexadecanol, 15% of the ^{3}H was retained in alk-1′-enyl groups demonstrating that oxidation to the acid was not required [272]. A reduction to the alcohol before incorporation could not be excluded. A slight incorporation of [1-^{3}H, 1-^{14}C]hexadecanol into plasmalogens and alkylacyl glycerophospholipids was found by Hagen [273]. The aldehyde is more likely than the alcohol to be the precursor of the side-chain. This might take place by displacement of an acyl group [67] or by prior activation, perhaps by phosphorylation or pyrophosphorylation as was suggested for formation of saturated ether bonds in *Halobacteria* [46]. The loss of ^{3}H from [2-^{3}H]glycerol in *H. cutirubrum* may be due to the presence of glycerol dehydrogenase. The pathway for formation of choline plasmalogens in mammals, presently unknown, may be similar to the pathway in anaerobic bacteria.

8. Catabolic pathways

Ether-linked glycerophospholipids can be partially degraded by a number of lipases. Alkaline and acid phosphohydrolases hydrolyze the phosphate groups from 1-alkyldihydroxyacetone phosphate, 1-alkyl-*sn*-glycero-2-phosphate, and 1-alkyl-2-acyl-*sn*-glycero-3-phosphate [277]. Phospholipase D from cabbage has a much slower velocity with ether-linked choline glycerophospholipids than with diacyl-GPC [10]. Different snake venom phospholipases A_2 differ in their relative rate of hydrolysis of choline plasmalogens [10,278]. The rat brain mitochondrial phospholipase A_2 and the human cerebral phospholipase A_2 have a lower but significant hydrolytic activity with ether-linked choline glycerophospholipids than with diacyl-GPC [278]. Woelk et al. have also found somewhat higher activities of these phospholipases A_2 in experimental allergic encephalomyelitis, subacute sclerosing panencephalitis, and multiple sclerosis [279].

A lysophospholipase D is present in brain and liver microsomes. It hydrolyzes the amines from 1-alkyl-*sn*-glycero-3-phosphocholine and 1-alkyl-*sn*-glycero-3-phosphoethanolamine (Fig. 4) [280–281]. The product, 1-alkyl-*sn*-glycero-3-phosphate, is rapidly hydrolyzed to alkylglycerol by an endogenous phosphohydrolase. The lysophospholipase D does not hydrolyze 1-hexadecyl-2-acetyl glycerophospholipids [282]. Brain microsomes also have some phospholipase C activity with 1-alkyl-*sn*-glycero-3-phosphoethanolamine [283].

If the alkyl group in a glycerophospholipid is not oxidized to an alk-1′-enyl group to form a plasmalogen, the only known mechanism for removing the alkyl group is oxidative cleavage by alkylglycerol monooxygenase [284]. As shown in Fig. 4, prior removal of the 2-acyl-group, the amine, and the phosphate is necessary [285]. This cleavage enzyme may regulate the total alkyl ether-lipid content of cells since tissues rich in the enzyme are low in alkylglycerols and vice versa. The mammalian enzyme utilizes tetrahydropteridine, glutathione, ammonium ion, and catalase as cofactors [284,285]. In *Tetrahymena pyriformis*, NAD^+ and NADPH but not pteridines are required cofactors [286].

Alk-1′-enyl groups can be hydrolyzed by a calcium ion-requiring plasmalogenase in cabbage leaves [287]. Little use has been made of this enzyme because the alk-1′-enyl groups are easily hydrolyzed with acid. Rat liver microsomes contain alk-1′-enyl hydrolase activities for 1-alk-1′-enyl-*sn*-glycero-3-phosphocholine and 1-alk-1′-enyl-*sn*-glycero-3-phosphoethanolamine (Fig. 4) [288–290]. The enzyme acting on the lysoethanolamine plasmalogen does not require a cation and is inhibited by sodium deoxycholate and *p*-hydroxymercuribenzoate [290]. The enzyme acting on the lysocholine plasmalogen does not require a cation, is solubilized by sodium deoxycholate, and requires phospholipid for full activity [288,289]. Rat liver microsomes do not contain an alk-1′-enyl hydrolase that acts on intact plasmalogens [288,290].

Intact plasmalogens are hydrolyzed by an enzyme found in brain (Fig. 4) that seems to be responsible for much of the turnover of the plasmalogens [291]. The activity of 1-alk-1′-enyl-2-acyl-*sn*-glycero-3-phosphoethanolamine aldehydohydrolase can be assayed by measuring the amount of aldehyde released with a coupled reaction with alcohol dehydrogenase [292] or by measuring the amount of substrate remaining [291,293–295]. The cation requirement is not known. Stimulation by Mg^{2+} was reported [295] but has not been confirmed. Plasmalogenase acts on choline plasmalogens as well as ethanolamine plasmalogens [293], and also hydrolyzes lysoplasmalogens at a slower rate. Diacylglycerophospholipids inhibit plasmalogenase. Plasmalogenase activity increases during development [296], reflects the degree of myelination [297] and is much higher in oligodendroglia than in neuronal perikarya or in astroglia [298]. With a microsomal fraction from 14-day-old rat brain, some aldehydohydrolase activity was found with lysoethanolamine plasmalogen but not with intact ethanolamine plasmalogen [299]. An oxygen-dependent production of aldehydes from choline plasmalogen by rat brain cytosol was probably due to autooxidation reactions [300].

In canine white matter with early demyelinating lesions due to canine distemper virus, the plasmalogenase activity was nearly 6-fold greater than in control tissue [294]. Phospholipases acting on phosphatidylethanolamine did not seem to be involved in the demyelination. The activity of plasmalogenase is also elevated during cerebral ischemia, in white matter from subjects with multiple sclerosis, in spinal cords from monkeys with demyelination due to vitamin B_{12} deficiency, and in rat brains with demyelination due to complement-dependent anti-myelin antibodies [296]. Plasmalogens are also the glycerophospholipids most affected by anoxia of rat heart [127]. Degradation of plasmalogens may be an important factor in the pathogenesis of irreversible cellular injury.

9. Turnover of ether-linked glycerophospholipids

Previous studies on the turnover of glycerophospholipids in the brain have been reviewed [301,302]. An accurate measure of the turnover rates for rapidly turning-over pools of ethanolamine plasmalogens has been difficult because they are not pulse-labelled. This is due to the prior incorporation of the precursors into alkylacyl-GPE. After labeling with ethanolamine, apparent half-lives of 1.3 days for alkylacyl-GPE and 2.4 days for alk-1′-enylacyl-GPE were found in mouse brain [302]. Values for myelin are comparable but longer. The faster turnover of the alkylacyl-GPE is consistent with its role of precursor for the plasmalogen. The slow turnover pool of ethanolamine plasmalogens turns over slightly slower than diacyl-GPE or diacyl-GPC in myelin or in microsomes when the labeling is from glycerol [303]. These three glycerophospholipids have nearly the same half-life, 12.5 days, in microsomes from mouse brain after labeling with arachidonate [304]. In a study of modification by diet of acyl group compositions of rat brain capillaries, Selivonchick and Roots observed greater changes in ethanolamine plasmalogens than in diacyl-GPE or diacyl-GPC [305]. Sun et al. also concluded that ethanolamine plasmalogens are more active metabolically than previously supposed from results on recovery from essential fatty acid deficiency [306]. All of these observations suggest a substantial turnover of polyunsaturated fatty acids in ethanolamine plasmalogens in brain. Waku and Nakazawa reported turnover rates for various molecular species of ethanolamine plasmalogens labelled with [^{14}C]glycerol in Ehrlich ascites tumor cells [16]. Those species with six double bonds had the fastest rate of turnover. High turnover rates were also found for tetraene and saturated fractions. The monoene and diene species have much slower turnover rates. The very high apparent turnover rate of choline plasmalogens is described in section 7.

10. Platelet-activating factor

1-Octadecyl-2-acetyl-*sn*-glycero-3-phosphocholine and the 1-hexadecyl analogue, at concentrations less than 1 nM, cause platelets to change shape, aggregate, and release 5-hydroxytryptamine. The semisynthetic compound with predominantly hexadecyl groups was made from choline plasmalogens from bovine heart [307,308]. The choline glycerophospholipids, about one-half plasmalogens, were isolated, hydrogenated, and subjected to base-catalyzed methanolysis. This produced 1-alkyl-GPC which was acetylated to produce the platelet-activating factor. Natural platelet-activating factor with properties identical to the semisynthetic material was isolated and purified from the secretions of stimulated human neutrophils and monocytes and rabbit neutrophils and basophils [309]. The natural material from rabbit leukocytes is 1-octadecyl-2-acetyl-*sn*-glycero-3-phosphocholine with less than 10% of the hexadecyl analogue [310]. The octadecyl compound made by total chemical synthesis has about 2-fold greater activity than the semisynthetic material for aggregation of rabbit platelets [311].

The antihypertensive renomedullary polar lipid is also a mixture of 1-alkyl-2-acetyl-*sn*-glycero-3-phosphocholines with about two-thirds of the alkyl groups as 16:0 [312]. The semisynthetic material from choline plasmalogen has the same biological effects as the natural material when tested with one-kidney hypertensive rats with mean arterial pressures between 170 and 190 mm Hg [308]. When given as an intravenous bolus, 63 ng produces an immediate decrease of 60 mm in arterial pressure and 6 μg decreases the pressure to 50–70 mm Hg. Larger amounts are fatal. The compound is a vasodilator and prolonged vasodepressor effects are obtained with multiple intravenous doses or with oral doses [312]. In normal rabbits, spaced oral doses have no effect on platelets or leukocytes.

Intravenous injections into rabbits of semisynthetic platelet-activating factor in doses of 0.3–1.3 μg cause an acute, reversible depression of the numbers of circulating platelets and neutrophils. The release of platelet factor 4 is dose-dependent [313]. With doses of more than 1.5 μg, there is complete depletion of platelets, neutrophils, and basophils, platelet factor 4 release, severe systemic hypotension, and death within 2 min. This resembles an anaphylactic reaction. Intradermal injection, as little as 52 pg, causes wheal and flare reactions in man [314]. At concentrations of 200 nM in human and 20 nM in rabbit platelet-rich plasma, platelet-activating factor causes irreversible aggregation of platelets with maximal secretion of 5-hydroxytryptamine and platelet factor 4 [315]. Exogenous platelet-activating factor (10 nM) causes rabbit platelets to synthesize thromboxane A_2 [316]. Platelet-activating factor is also chemotactic for neutrophils and mononuclear leukocytes at 10–100 μM, much higher concentrations than are required for platelet activation [317].

In addition to secretion by leukocytes, platelet-activating factor is also formed by rabbit platelets when they are stimulated by the calcium ionophore A23187, thrombin, or collagen [318]. Platelets do not form platelet-activating factor when they are stimulated by exogenous platelet-activating factor, ADP, or arachidonate. The platelet-aggregating factor seems to be responsible for aggregation that is independent of ADP and thromboxane A_2. An involvement of phospholipase A_2 in the synthesis of platelet-activating factor was suggested [318]. Hog leukocytes release 1-alkyl-GPC simultaneously with the release of 1-alkyl-2-acetyl-GPC [319]. The lyso compound may be a precursor that escaped acetylation or a product of enzymic inactivation of platelet-activating factor.

The lyso compound, 1-alkyl-GPC, and the phosphatidylcholine, 1-acyl-2-acetyl-GPC, have little or no biological activity [307,308,320]. Substitution of maleyl, succinyl, or phthaloyl groups for the acetyl group at the 2-position causes a loss of the secretion of lysosomal enzymes from leukocytes but the chemotactic properties are retained [317]. Substitution of propionyl or butyryl groups at the 2-position increases the concentrations required for platelet aggregation and release of 5-hydroxytryptamine [307,311]. Ether groups at the 2-position reduce the activity. Platelet aggregation is reduced 28-fold and 5-hydroxytryptamine release is reduced 45-fold with a 2-ethyl group [320]. The latter activity is reduced 66-fold with a 2-methyl group [321]. No activity is found with 2-benzyl groups or with the deoxy compound [320].

The bioassay used by Hanahan's group is the release of 50% of the 5-hydroxytryptamine from rabbit platelets. The activity of 1-hexadecyl-2-acetyl-GPC is 3-fold greater than that of 1-octadecyl-2-acetyl-GPC [321]. The semisynthetic material is even more active than the hexadecyl compound. Substitution of dodecyl for hexadecyl at the 1-position gives 5-fold less activity. Surprisingly, the unnatural stereoisomer, 3-hexadecyl-2-acetyl-*sn*-glycero-1-phosphocholine, has only 11-fold less activity than the natural isomer. Activity of the racemic mixture is intermediate. Modifications of the amine also decrease biological activity [322]. Reductions of activity when compared with the phosphocholine compound are 2.5-fold for removal of one methyl group, 20-fold for removal of two methyl groups and 3800-fold for removal of all three methyl groups to produce 1-alkyl-2-acetyl-GPE. The phosphatidate derivative, 1-alkyl-2-acetyl-*sn*-glycero-3-phosphate, has 4600-fold less activity then the phosphocholine compound.

Snyder's group has investigated the metabolism of platelet-activating factor. Synthesis is possible by an acetyltransferase [323] and by phosphocholinetransferase [324]. The acetyl-CoA:1-alkyl-2-lyso-GPC acetyltransferase is a microsomal enzyme with activities up to 300 nmol/min/g tissue in spleen, lung, lymph nodes, and thymus. The enzyme differs from palmitoyltransferase [323]. The CDPcholine:1-alkyl-2-acetyl-*sn*-glycerol-3-phosphocholinetransferase is more active in spleen and lung than in liver, kidney, and heart. It is partially cross-inhibited by 1,2-diacyl-*sn*-glycerols [324]. An inactivating enzyme, 1-alkyl-2-acetyl-GPC acetylhydrolase is in the cytosol of kidney. Relatively high activities are also present in lung and brain. The enzyme is most active at pH 7.5–8.5, does not require divalent cations, and is sensitive to diisopropylphosphofluoridate [325].

11. Function and biological role

The ether-linked glycerophospholipids in mammalian cells tend to be concentrated in the plasma membrane [119,326,327]. Within the plasma membrane, like diacyl-GPE, the ethanolamine plasmalogens are localized primarily on the inner surface of the membrane [328]. Some of the plasmalogens are in clusters associated with proteins and may play a role, together with other amino phospholipids, in cation and anion leak [328]. The localization of plasmalogens and their affinity for mercuric ions also suggest a role in ion transport systems and the control of water movement across plasma membranes [119].

In the white matter of brain, the ethanolamine plasmalogens contain a large proportion of the (n-6) family of polyunsaturated acyl groups. Plasmalogens may function as a reservoir of these prostaglandin and thromboxane precursors [291]. Plasmalogens have a specific bond at the 1-position that is hydrolyzed by a specific enzyme. A lysophospholipase acts on the 2-acyl-GPE to release the polyunsaturated fatty acid [293]. In platelets, prelabeled with arachidonate, exposure to thrombin to release arachidonate causes a transfer of radioactivity from phosphatidylcholine and phosphatidylinositol to ethanolamine plasmalogen [329]. Thus, an active turnover of the 2-acyl groups of ethanolamine plasmalogens is involved in platelet aggregation and the unlabeled fatty acids released could be precursors of prostaglandins. The extent of turnover of the remainder of the molecule is not known. The platelet membrane has 84% of the ethanolamine plasmalogen in the inner layer [330]. A substantial portion of the arachidonate in both layers is present in the ethanolamine plasmalogen. With mouse fibrosarcoma cells prelabeled with arachidonate, then stimulated with bradykinin for release of arachidonate, no detectable change was found in ethanolamine plasmalogens [207].

Phospholipases A_1 are localized in membranes of neuronal and glial cells with an appreciable activity in the plasma membranes [331]. Hydrolysis of 1 μmol of diacyl-GPC is inhibited 54% by alkenylacyl-GPE, 56% by alkenylacyl-GPC, and 68% by alkylacyl-GPC when 0.5 μmol is included in the incubation. This is similar but opposite to the inhibition of plasmalogenase, another phospholipase A_1, by diacyl-GPE and diacyl-GPC [293]. A mixture of plasmalogens and diacyl glycerophospholipids may increase the stability of plasma membranes. Membrane stability may be increased further by formation of membranes with alkyl groups [92].

The proportion of ether-linked glycerophospholipids in membranes is increased by culturing LM fibroblasts or Ehrlich ascites cells in medium containing long-chain alcohols [327]. The increase includes both alkylacyl and alk-1′-enylacyl glycerophospholipids. Proportions of alkylacyl glycerophospholipids in membranes of LM cells are also increased by addition of *N,N*-dimethylethanolamine, *N*-methyl-ethanolamine, or ethanolamine in place of choline [332]. The ability to manipulate the content of ether-linked glycerophospholipids should help in understanding their function and regulation of their synthesis and hydrolysis.

Another approach for understanding the effects of ether-linked glycerophospholipids on properties of membranes is to study the phase transitions. Dialkyl glycerophospholipids have phase-transition temperatures 3–5°C higher than those for the corresponding diacyl compounds, but the corresponding alkylacyl glycerophospholipids have slightly lower phase-transition temperatures [333]. For alk-1-enylacyl glycerophospholipids, the transition temperature is 5–6°C lower than for the corresponding diacyl compound from *Clostridium butyricum* [334]. Thus ether-linked glycerophospholipids generally do not pack as closely in the membrane and increase membrane fluidity. Lipid–lipid interactions were reviewed recently by Boggs [335]. Since the ether-linked glycerophospholipids lack a carbonyl group at the 1-position, they have a lower surface potential. Brockerhoff suggested that cholesterol hydrogen-bonds to a greater extent to the carbonyl group at the 2-position and this bonding is greater with ether-linked than with diacyl glycerophospholipid because the former lack the 1-position carbonyl [336]. This should influence permeability. A glycerol acetal of ethanolamine plasmalogen exists in *C. butyricum* [337]. Perhaps other alcohols such as cholesterol can form acetals with the hydrated form of plasmalogens.

Some alkyl lysoglycerophospholipid analogues have a direct cytotoxic effect on tumor cells in vitro at concentrations of only 1–5 μg/ml [338]. Surface activity has been ruled out as the mechanism. Tumor cells are particularly susceptible due to their low levels of ether-cleavage enzyme. Activity may require substitution at the 2-position because 1-alkyl-GPC is acted on by the acyltransferase to form 1-alkyl-2-acyl-GPC. Effective compounds include 1-octadecyl-2-methyl-GPC and 1-octadecylpropanediol-3-phosphocholine. Both the 1-octadecyl-GPC and 1-octadecyl-2-methyl-GPC are effective in vivo in inhibiting development of a lung tumor [339]. Macrophage may be involved. Tumor growth is also inhibited by oral administration of 1-*O*-methoxyhexadecyl glycerol [340].

Acknowledgement

The preparation of this review was supported in part by research grant NS-08291 from the National Institutes of Health, U.S. Public Health Service.

References

1 Snyder, F. (1972) Ether Lipids: Chemistry and Biology, Academic Press, New York.
2 Mangold, H.K. (1979) Angew. Chem. Int. Ed., 18, 493–503.
3 IUPAC-IUB Commission on Biochemical Nomenclature (1978) The Nomenclature of Lipids (Recommendations 1976), J. Lipid Res. 19, 114–128.
4 Rapport, M.M. and Alonzo, N. (1955) J. Biol. Chem. 217, 199–204.
5 Gray, G.M. and Macfarlane, M.G. (1958) Biochem. J. 70, 409–425.
6 Debuch, H. and Seng, P. (1972) in Ether Lipids: Chemistry and Biology (Snyder, F., ed.), pp. 1–24, Academic Press, New York.
7 Kates, M. (1972) Techniques of Lipidology. Isolation, Analysis and Identification of Lipids, North-Holland, Amsterdam.
8 Schmid, H.H.O., Bandi, P.C. and Su, K.L. (1975) J. Chromatogr. Sci. 10, 478–486.
9 Ansell, G.B. and Spanner, S. (1971) in Methods in Neurochemistry (Fried, R., ed.), Vol. 1, pp. 32–76, Marcel Dekker, New York.
10 Waku, K. and Nakazawa, Y. (1972) J. Biochem. (Tokyo) 72, 149–155.
11 Woelk, H. and Debuch, H. (1971) Z. Physiol. Chem. 352, 1275–1281.
12 Cox, J.W. (1977) Fed. Proc. 36, 852.
13 Paltauf, F. (1978) Lipids 13, 165–166.
14 Woelk, H., Debuch, H. and Porcellati, G. (1973) Z. Physiol. Chem. 354, 1265–1270.
15 Snyder, F. (1973) J. Chromatogr. 82, 7–14.
16 Waku, K. and Nakazawa, Y. (1979) Eur. J. Biochem. 100, 317–320.
17 Curstedt, T. (1977) Biochim. Biophys. Acta 489, 79–88
18 Totani, N. and Mangold, H.K. (1981) Mikrochim. Acta 1, 73–78.
19 Segall, H.J. and Wood, J.M. (1974) Nature 248, 456–458.
20 Ferrell, W.J., Radloff, J.F. and Jackiw, A.B. (1969) Lipids 4, 278–282.
21 Viswanathan, C.V. (1974) J. Chromatog. 98, 129–155.
22 Horrocks, L.A. and Sun, G.Y. (1972) in Research Methods in Neurochemistry (Rodnight, R. and Marks, N., eds.), Vol. 1, pp. 223–231, Plenum, New York.
23 Vaskovsky, V.E. and Dembitsky, V.M. (1975) J. Chromatog. 115, 645–647.
24 Yavin, E. and Zutra, A. (1977) Analyt. Biochem. 80, 430–437.
25 Yanishlieva, N., Becker, H. and Mangold, H.K. (1977) Chem. Phys. Lipids, 18, 149–153.
26 Myher, J.J., Marai, L. and Kuksis, A. (1974) J. Lipid Res. 15, 586–592.
27 Blank, M.L., Rainey, W.T., Christie, W.H., Piantadosi, C. and Snyder, F. (1976) Chem. Phys. Lipids, 17, 201–206.

28 Satouchi, K. and Saito, K. (1976) Biomed. Mass Spectrom. 3, 122–126.
29 Satouchi, K. and Saito, K. (1977) Biomed. Mass Spectrom. 4, 107–112.
30 Phillipou, G. and Poulos, A. (1978) Chem. Phys. Lipids. 22, 51–54.
31 Paltauf, F. (1973) Chem. Phys. Lipids 11, 270–294.
32 Baumann, W.J. and Mangold, H.K. (1964) J. Org. Chem. 29, 3055–3057.
33 Baumann, W.J. and Mangold, H.K. (1966) Biochim. Biophys. Acta 116, 570–576
34 Chebyshev, A.V., Serebrennikova, G.A. and Evstigneeva, R.P. (1979) Bioorg. Khim. 5, 628–632.
35 Eibl, H. (1980) Chem. Phys. Lipids 26, 405–429.
36 Kertscher, P., Rüger, H.-J. and Nuhn, P. (1980) J. Prakt. Chem. 322, 1067–1068.
37 Paltauf, F. (1976) Chem. Phys. Lipids 17, 148–154.
38 Brockerhoff, H. and Ayengar, N.K.N. (1979) Lipids 14, 88–89.
39 Doerr, I.L., Tang, J.C., Rosenthal, A.F., Engel, R. and Tropp, B.E., (1977) Chem. Phys. Lipids 19, 185–202.
40 Ferrell, W.J., Garces, A. and Desmyter, E.A. (1976) Chem. Phys. Lipids 16, 276–284.
41 Brachwitz, H., Langen, P., Otto, A. and Schildt, J. (1979) J. Prakt. Chem. 321, 775–786.
42 Halperin, G. and Gatt, S. (1980) Steroids 35, 39–42.
43 Tornabene, T.G. and Langworthy, T.A. (1979) Science 203, 51–53.
44 Woese, C.R., Magrum, L.J. and Fox, G.E. (1978) J. Mol. Evol. 11, 245–252.
45 Bayley, S.T. (1979) Trends Biochem. Sci. 4, 223–225.
46 Kates, M. (1972) in Ether Lipids: Chemistry and Biology (Snyder, F., ed.), pp. 351–398, Academic Press, New York.
47 Kates, M. (1978) Prog. Chem. Fats other Lipids 15, 301–342.
48 Kates, M., Palameta, B. and Yengoyan, L.S. (1965) Biochemistry 4, 1595–1599.
49 Kates, M., Park, C.E., Palameta, B. and Joo, C.N. (1971) Canad. J. Biochem. 49, 275–281.
50 Kates, M. and Deroo, P.W. (1973) J. Lipid Res. 14, 438–445.
51 Joo, C.N. and Kates, M. (1969) Biochim. Biophys. Acta 176, 278–297.
52 Hancock, A.J. and Kates, M. (1973) J. Lipid Res. 14, 430–437.
52a Smallbone, B.W. and Kates, M. (1981) Biochim. Biophys. Acta 665, 551–558.
53 Langworthy, T.A. (1977) Biochim. Biophys. Acta 487, 37–50.
54 DeRosa, M., DeRosa, S., Gambacorta, A., Minale, L. and Bu'Lock, J.D. (1977) Phytochemistry 16, 1961–1965.
55 DeRosa, M., DeRosa, S., Gambacorta, A. and Bu'Lock, J.D. (1977) Chem. Commun. 514–515.
56 DeRosa, M., Gambacorta, A., Nicolaus, B., Sodano, S. and Bu'Lock, J.D. (1980) Phytochemistry 19, 833–836.
57 DeRosa, M., Esposito, E., Gambacorta, A., Nicolaus, B. and Bu'Lock, J.D. (1980) Phytochemistry 19, 827–831.
58 DeRosa, M., DeRosa, S., Gambacorta, A. and Bu'Lock, J.D. (1980) Phytochemistry 19, 249–254.
59 Langworthy, T.A. (1977) J. Bacteriol. 1326–1332.
60 DeRosa, M., Gambacorta, A., Nicolaus, B. and Bu'Lock, J.D. (1980) Phytochemistry 19, 821–826.
61 Makula, R.A. and Singer, M.E. (1978) Biochem. Biophys. Res. Commun. 82, 716–722.
62 Kushwaha, S.C., Kates, M., Sprott, G.D. and Smith, I.C.P. (1981) Science 211, 1163–1164.
62 a. Kushwaha, S.C., Kates, M., Sprott, G.D. and Smith, I.C.P. (1981) Biochim. Biophys. Acta 664, 156–173.
63 Joutti, A. and Renkonen, O. (1979) J. Lipid Res. 20, 840–847.
64 Vandenheuvel, F.A. (1965) J. Am. Oil. Chem. Soc. 42, 481–491.
65 Rayman, M.K., Gordon, R.C. and MacCleod, R.A. (1967) J. Bacteriol. 93, 1465–1466.
66 Kates, M. and Kushwaha, S.C. (1978) in Energetics and Structure of Halophilic Microorganisms (Caplan, S.R. and Ginzburg, M., eds.), pp. 461–480, Elsevier/North-Holland Biomedical Press, Amsterdam.
67 Goldfine, H. and Hagen, P.O. (1972) in Ether Lipids: Chemistry and Biology (Snyder, F., ed.), pp. 329–350, Academic Press, New York.
68 Goldfine, H., Khuller, G.K. and Lueking, D.R. (1976) in Lipids-Biochemistry (Paoletti, R., Porcellati, G. and Jacini, G., eds.), Vol. 1, pp. 11–22, Raven, New York.

69 Goldfine, H. and Johnston, N.C. (1980) in Membrane Fluidity: Biophysical Techniques and Cellular Regulation (Kates, M. and Kuksis, A., eds.), pp. 365–380, Humana Press, Clifton, NJ.
70 Kamio, Y., Kanegasaki, S. and Takahashi, H. (1969) J. Gen. Appl. Microbiol. 15, 439–451.
71 Baumann, N.A., Hagen, P-O. and Goldfine, H. (1965) J. Biol. Chem. 240, 1559–1567.
72 VanGolde, L.M.G., Prins, R.A., Franklin-Klein, W. and Akkermans-Kruyswijk (1973) Biochim. Biophys. Acta 326, 314–323.
73 Hagen, P-O. (1974) J. Bacteriol. 119, 643–645.
74 Langworthy, T.A., Mayberry, W.R., Smith, P.F. and Robinson, I.M. (1975) J. Bacteriol. 122, 785–787.
75 Clarke, N.G., Hazelwood, G.P. and Dawson, R.M.C. (1976) Chem. Phys. Lipids 17, 222–232.
76 Meyer, H. and Meyer, F. (1971) Biochim. Biophys. Acta 231, 93–106.
77 Matthews, H.M., Yang, T.K. and Jenkin, H.M. (1980) Biochim. Biophys. Acta 618, 273–281.
78 Kim, K.C., Kamio, Y. and Takahashi, H. (1970) J. Gen. Appl. Microbiol. 16, 321–325.
79 Matsumoto, M., Tamiya, K. and Koizumi, J. (1971) J. Biochem. (Tokyo) 69, 617–620.
80 Khuller, G.K. and Goldfine, H. (1975) Biochemistry 14, 3642–3647.
81 Khuller, G.K. and Goldfine, H. (1974) J. Lipid Res. 15, 500–507.
82 Goldfine, H., Khuller, G. K., Borie, R.P., Silverman, B., Selick, H., Johnston, N.C., Vanderkooi, J.M. and Horwitz, A.F. (1977) Biochim. Biophys. Acta 488, 341–352.
83 Rebel, G. (1972) Arch. Mikrobiol. 81, 333–343.
84 McAllister, D.J. and De Siervo, A.J. (1975) J. Bacteriol. 123, 302–307.
85 Goni, F.M., Dominguez, J.B. and Uruburu, F. (1978) Chem. Phys. Lipids 22, 79–81.
86 Poulos, A. and Le Stourgeon, W.M. (1971) Lipids 6, 466–469.
87 Ellingson, J.S. (1980) Biochemistry 19, 6176–6182.
88 Beach, D.H., Holz, Jr., G.G. and Anekwe, G.W. (1979) J. Parasitol. 65, 203–216.
89 Dawson, R.M.C. and Kemp, P. (1967) Biochem. J. 105, 837–842.
90 Fukushima, H., Watanabe, T. and Nozawa, Y. (1976) Biochim. Biophys. Acta 436, 249–259.
91 Berger, H., Jones, P. and Hanahan, D.J. (1972) Biochim. Biophys. Acta 260, 617–629.
92 Thompson, G.A. (1972) in Ether Lipids: Chemistry and Biology (Snyder, F., ed.), pp. 321–329, Academic Press, New York.
93 Kaneshiro, E.S. (1980) J. Lipid Res. 21, 559–570.
94 Rhoads, D.E. and Kaneshiro, E.S. (1979) J. Protozool. 26, 329–338.
95 Kaufmann, H.P., Hamza, Y. and Mangold, H.K. (1971) Chem. Phys. Lipids 6, 325–332.
96 Kuroda, N. and Negishi, T. (1975) Obihiro Chikusan Daigaku Gakujutsu Kenkyu Hokoku, Dai-1-Bu 9, 371–374, cited in Chem. Abs. 83, 176978.
97 Lambremont, F.N. and Wood, R. (1968) Lipids 3, 503–510.
98 Lambremont, E.N. (1972) Comp. Biochem. Physiol. 41B, 337–342.
99 Lambremont, E.N., Ferguson, J.R. and Greer, G.J. (1976) Insect Biochem. 6, 363–366.
100 Kulkarni, A.P., Smith, E. and Hodgson, E. (1971) Insect Biochem. 1, 348–362.
101 Marsden, J.R. (1971) Comp. Biochem. Physiol. 40B, 871–884.
102 Chitwood, D.J. and Krusberg, L.R. (1981) J. Nematol. 13, 105–111.
103 Chitwood, D.J. and Krusberg, L.R. (1981) Comp. Biochem. Physiol. 69B, 115–120.
104 Isay, S.V., Makarchenko, M.A. and Vaskovsky, V.E. (1976) Comp. Biochem. Physiol. 55B, 301–305.
105 Dembitsky, V.M. and Vaskovsky, V.E. (1976) Biol. Morya 68–72.
106 Dembitsky, V.M. (1979) Biol. Morya 86–90.
107 Dembitsky, V.M. (1980) Bioorg. Khim. 6, 426–430.
108 Dembitsky, V.M., Svetashev, V.I. and Vaskovsky, V.E. (1977) Bioorg. Khim. 3, 930–933.
109 Lee, R.F. and Gonsoulin, F. (1979) Comp. Biochem. Physiol. 64B, 375–379.
110 Thompson, G.A., Jr. (1972) in Ether Lipids: Chemistry and Biology (Snyder, F., ed.), pp. 313–320, Academic Press, New York.
111 Kaiser, H., Grosse-Oetringhaus, S. and Hudalla, B. (1978) J. Chromatog. 154, 93–98.
112 Kreps, E.M. (1981) Comp. Biochem. Physiol. 68B, 363–367.
113 Chacko, G.K., Goldman, D.E. and Pennock, B.D. (1972) Biochim. Biophys. Acta 280, 1–16.
114 Driedzic, W. and Roots, B.I. (1975) J. Thermal. Biol. 1, 7–10.

115 Driedzic, W., Selivonchick, D.P. and Roots, B.I. (1976) Comp. Biochem. Physiol. 53B, 311–314.
116 Selivonchick, D.P. and Roots, B.I. (1976) J. Thermal Biol. 1, 131–135.
117 Selivonchick, D.P., Johnston, P.V. and Roots, B.I. (1977) Neurochem. Res. 2, 379–393.
118 Do, U.H. and Ramachandran, S. (1980) J. Lipid Res. 21, 888–894.
119 Horrocks, L.A. (1972) in Ether Lipids: Chemistry and Biology (Snyder, F., ed.), pp. 177–272, Academic Press, New York.
120 Bergh, C.H., Larson, G. and Samuelsson, B.E. (1975) Lipids 10, 299–302.
121 Forbes, R.M., Cooper, A.R. and Mitchell, H.H. (1953) J. Biol. Chem. 203, 359–366.
122 Panganamala, R.V., Horrocks, L.A., Geer, J.C. and Cornwell, D.G. (1971) Chem. Phys. Lipids 6, 97–106.
123 Hughes, B.P. (1972) J. Neurol. Neurosurg. Psychiat. 35, 658–663.
124 Curstedt, T. (1977) Biochim. Biophys. Acta 489, 79–88.
125 Pugh, E.L., Kates, M. and Hanahan, D.J. (1977) J. Lipid Res. 18, 710–716.
126 Osanai, A. and Sakagami, T. (1979) J. Biochem. 85, 1453–1459.
127 Rabinowitz, J.L. and Hercker, E.S. (1976) Biochem. J. 160, 463–466.
128 Okano, G., Matsuzaka, H. and Shimojo, T. (1980) Biochim. Biophys. Acta 619, 167–175.
128a Abdel-Latif, A.A. and Smith, J.P. (1979) Exp. Eye Res. 29, 131–140.
129 Epps, D.E., Natarajan, V., Schmid, P.C. and Schmid, H.H.O. (1980) Biochim. Biophys. Acta 618, 420–430.
130 Marai, L. and Kuksis, A. (1973) Canad. J. Biochem. 51, 1365–1379.
131 Norton, W.T. and Poduslo, S.E. (1973) J. Neurochem. 21, 759–773.
132 Moscatelli, E.A. and Duff, J.A. (1978) Lipids 13, 294–296.
133 de Sousa, B.N. and Horrocks, L.A. (1979) Dev. Neurosci. 2, 122–128.
134 Taranova, N.P. and Dvorkin, V.Y. (1975) Bull. Exptl. Biol. Med. 79, 259–261.
135 Sun, G.Y. and Horrocks, L.A. (1968) Lipids 3, 79–83.
136 Dorman, R.V., Dreyfus, H., Freysz, L. and Horrocks, L.A. (1976) Biochim. Biophys. Acta 486, 55–59.
136a Urban, P.F., Edel-Harth, S. and Dreyfus, H.L. (1975) Exp. Eye Res. 20, 397–405.
137 Pullarkat, R.K. and Reha, H. (1978) J. Neurochem. 31, 707–711.
138 Winterfeld, M. and Debuch, H. (1977) J. Neurol. 215, 261–272.
139 Goracci, G., Francescangeli, E., Piccinin, G.L., Binaglia, L., Woelk, H. and Porcellati, G. (1975) J. Neurochem. 24, 1181–1186.
140 Pleasure, D., Hardy, M., Johnson, G., Lisak, R. and Silberberg, D. (1981) J. Neurochem. 37, 452–460.
141 Matheson, D.F., Oei, R. and Roots, B.I. (1980) Neurochem. Res. 5, 683–695.
142 Klein, F. and Mandel, P. (1976) Lipids 11, 506–512.
143 Sun, G.Y. and Horrocks, L.A. (1970) Lipids 5, 1013–1019.
144 Toews, A.D., Horrocks, L.A. and King, J.S. (1976) J. Neurochem. 27, 25–31.
145 Blaker, D.W. and Moscatelli, E.A. (1978) J. Neurochem. 31, 1513–1518.
146 Wender, M., Adamczewska-Goncerzewicz, Z. and Goncerzewicz, A. (1979) Exp. Path. 17, 334–339.
147 Sobue, G. and Koizumi, K. (1980) J. Neurol. Sci. 44, 229–239.
148 Horrocks, L.A. (1967) J. Lipid Res. 8. 569–576.
149 Toews, A.D., King, J.S., Yashon, D. and Horrocks, L.A. (1980) Neurol. Res. 1, 271–279.
150 Sun, G.Y. (1973) J. Lipid Res. 14, 656–663.
151 Freysz, L., Bieth, R. and Mandel, P. (1971) Biochimie 53, 399–405.
152 Paterrakis, S.P. and Kalofoutis, A.T. (1972) Experientia 28, 1416–1417.
153 Martinez, M. and Ballabriga, A. (1978) Brain Research 159, 351–362.
154 Horrocks, L.A. (1973) Prog. Brain Res. 40, 383–395.
155 Horrocks, L.A., Sun, G.Y. and D'Amato, R.A. (1975) in Neurobiology of Aging (Ordy, J.M., ed.), pp. 359–367, Plenum, New York.
156 Horrocks, L.A., Van Rollins, M. and Yates, A.J. (1981) in The Molecular Basis of Neuropathology (Thompson, R.H.S. and Davison, A.N., eds.), pp. 601–630, Edward Arnold, London.

157 Clarke, N.G. and Dawson, R.M.C. (1981) Biochem. J. 195, 301–306.
158 Blaker, W.D. and Moscatelli, E.A. (1979) Lipids 14, 1027–1031.
159 Sun, G.Y. and Samorajski, T. (1973) Biochim. Biophys. Acta 316, 19–27.
160 Sun, G.Y. and Yau, T.M. (1976) J. Neurochem. 26, 291–295.
161 Yahara, S. and Kishimoto, Y. (1981) J. Neurochem. 36, 190–194.
162 Levine, M., Kornblatt, M.J. and Murray, R.K. (1975) Can. J. Biochem. 53, 679–689.
163 Sun, G.Y. (1979) Lipids 14, 918–924.
164 Druilhet, R.E., Overturf, M.L. and Kirkendall, W.M. (1975) Int. J. Biochem. 6, 893–901.
165 Neill, A.R. and Masters, C.J. (1975) Comp. Biochem. Physiol. 51B, 99–101.
166 Blank, M.L., Wykle, R.L. and Snyder, F. (1973) Biochim. Biophys. Acta 316, 28–38.
167 Okano, G., Kawamoto, T. and Akino, T. (1978) Biochim. Biophys. Acta 528, 385–393.
168 Grigor, M.R., Moehl, A. and Snyder, F. (1972) Lipids 7, 766–768.
169 Yamada, K., Imura, K., Taniguchi, M. and Sakagami, T. (1976) J. Biochem. (Tokyo) 79, 809–817.
170 Sun, G.Y. and Leung, B.S. (1976) Lipids 11, 322–327.
171 Schick, B.P., Schick, P.K. and Chase, P.R. (1981) Biochim. Biophys. Acta 663, 239–248.
172 Murray, R.K., Narasimhan, R., Levine, M., Pinteric, L., Shirley, M., Lingwood, C. and Schachter, H. (1980) in Cell Surface Glycolipids (Sweeley, C.C., ed.), pp. 105–125, American Chemical Society, Washington, DC.
173 Selivonchick, D.P., Schmid, P.C., Natarajan, V. and Schmid, H.H.O. (1980) Biochim. Biophys. Acta 618, 242–254.
174 Shirley, M.A. and Schachter, H. (1980) Can. J. Biochem. 58, 1230–1239.
175 Evans, R.W., Weaver, D.E. and Clegg, E.D. (1980) J. Lipid Res. 21, 223–228.
176 Jain, Y.C. and Anand, S.R. (1976) J. Reprod. Fert. 47, 255–260.
177 Darin-Bennett, A., Poulos, A. and White, I.G. (1976) Andrologia 8, 37–45.
178 Darin-Bennett, A., White, I.G. and Hoskins, D.D. (1977) J. Reprod. Fert. 49, 119–122.
179 Darin-Bennett, A., Poulos, A. and White, I.G. (1974) J. Reprod. Fert. 41, 471–474.
180 Yurkowski, M. and Walker, B.L. (1971) Chem. Phys. Lipids 6, 103–108.
181 Chepenik, K.P., Blank, M., Snyder, F. and Waite, B.M. (1980) Int. J. Biochem. 11, 605–607.
182 Ahrne, L., Björck, L., Raznikiewicz, T. and Claesson, O. (1980) J. Dairy Sci. 63, 741–745.
183 Kleinig, H., Zentgraf, H., Comes, P. and Stadler, J. (1971) J. Biol. Chem. 246, 2996–3000.
184 Sugiura, T., Masuzawa, T. and Waku, K. (1980) Lipids 15, 475–478.
185 Hanahan, D.J., Ekholm, J. and Jackson, C.M. (1963) Biochemistry 2, 630–641.
186 Matsumoto, M. and Miwa, M. (1973) Biochim. Biophys. Acta 296, 350–364.
187 Somerharju, P. and Renkonen, O. (1979) Biochim. Biophys. Acta 573, 83–89.
188 Radominska-Pyrek, A., Dabrowiecki, Z. and Horrocks, L.A. (1979) Biochim. Biophys. Acta 574, 248–257.
189 Deleted
190 Snyder, F. (1972) in Ether Lipids: Chemistry and Biology (Snyder, F., ed.), pp. 273–395, Academic Press, New York.
191 Snyder, F. and Snyder, C. (1975) Prog. Biochem. Pharmacol. 10, 1–41.
192 Lee, T-C. and Snyder, F. (1976) in Lipid Metabolism of Mammals (Snyder, F., ed.), Vol. 2, pp. 293–310, Plenum, New York.
193 Snyder, F. and Wood, R. (1969) Cancer Res. 29, 251–257.
194 Lin, H.J., Ng, W.L., Tung-Ma, L. and Lee, C.L.H. (1981) Cancer Lett. 11, 231–237.
195 Finke, E. and Scheid, P. (1973) Acta Biol. Med. Germ. 30, 353–363.
196 Albert, D.H. and Anderson, C.E. (1977) Lipids 12, 188–192.
197 Albert, D.H. and Anderson, C.E. (1977) Lipids 12, 722–728.
198 Yates, A.J., Thompson, D.K., Boesel, C.P., Albrightson, C. and Hart, R.W. (1979) J. Lipid Res. 20, 428–436.
199 Ruggieri, S. and Fallani, A. (1979) Lipids 14, 323–333.
200 Selkirk, J.K., Elwood, J.C. and Morris, H.P. (1971) Cancer Res. 31, 27–31.
201 Nakamura, M., Onodera, T. and Akino, T. (1980) Lipids 15, 616–623.
202 Eichberg, J., Shein, H.M. and Hauser, G. (1976) J. Neurochem. 27, 679–685.

203 Robert, J., Rebel, G. and Mandel, P. (1976) J. Neurochem. 26, 771–777.
204 Hauser, G., Eichberg, J. and Shein, H.M. (1976) Brain Res. 109, 636–642.
205 Liepkalns, V.A., Icard, C., Yates, A.J., Thompson, D.K. and Hart, R.W. (1981) J. Neurochem. 36, 1959–1965.
206 Robert, J., Rebel, G., Mandel, P. and Yavin, E. (1978) Life Sci. 22, 211–216.
207 Schremmer, J.M., Blank, M.L. and Wykle, R.L. (1979) Prostaglandins 18, 491–505.
208 Waku, K., Nakazawa, Y. and Mori, W. (1976) J. Biochem. 79, 407–411.
209 Weinstein, D.B., Marsh, J.B., Glick, M.C. and Warren, L. (1969) J. Biol. Chem. 244, 4103–4111.
210 Keenan, R.W., Brown, J.B. and Marks, B.H. (1961) Biochim. Biophys. Acta 51, 226–229.
211 Snyder, F. (1972) in Ether Lipids: Chemistry and Biology (Snyder, F., ed.), pp. 121–156, Academic Press, New York.
212 Wykle, R.L. and Snyder, F. (1976) in Enzymes of Biological Membranes (Martinosi, A., ed.), pp. 87–117, Plenum, New York.
213 Lumb, R.H. and Snyder, F. (1971) Biochim. Biophys. Acta 244, 217–221.
214 Natarajan, V. and Schmid, H.H.O. (1977) Lipids 12, 128–130.
215 Gilbertson, J.R., Gelman, R.A., Chiu, T.H., Gilbertson, L.I. and Knauer, T.E. (1978) J. Lipid Res. 19, 757–762.
216 Natarajan, V. and Schmid, H.H.O. (1978) Arch. Biochem. Biophys. 187, 215–222.
217 Bourre, J.M. and Daudu, O. (1978) Neurosci. Lett. 7, 225–230.
218 Bishop, J.E. and Hajra, A.K. (1978) J. Neurochem. 30, 643–647.
219 Wykle, R.L., Malone, B. and Snyder, F. (1979) J. Lipid Res. 20, 890–896.
220 Reichwald, I. and Mangold, H.K. (1977) Nutrit. Metab. 21, 198–201.
221 Lee, T.C., Fitzgerald, V., Stephens, N. and Snyder, F. (1980) Biochim. Biophys. Acta 619, 420–423.
222 Hajra, A.K. (1969) Biochem. Biophys. Res. Commun. 37, 486–492.
223 Wykle, R.L., Piantadosi, C. and Snyder, F. (1972) J. Biol. Chem. 247, 2944–2948.
224 Brown, A.J. and Snyder, F. (1979) Biochem. Biophys. Res. Commun. 90, 278–284.
225 Rock, C.O., Fitzgerald, V. and Snyder, F. (1977) Arch. Biochem. Biophys. 181, 172–177.
226 Snyder, F., Rainey, Jr., W.T., Blank, M.L. and Christie, W.H. (1970) J. Biol. Chem. 245, 5853–5856.
227 Friedberg, S.J. and Alkek, R.D. (1977) Biochemistry 16, 5291–5294.
228 Davis, P.A. and Hajra, A.K. (1979) J. Biol. Chem. 254, 4760–4763
229 Friedberg, S.J., Gomillion, D.M. and Stotter, P.L. (1980) J. Biol. Chem. 255, 1074–1079.
230 Friedberg, S.J. and Gomillion, M. (1981) J. Biol. Chem. 256, 291–295.
231 Jones, C.L. and Hajra, A.K. (1980) J. Biol. Chem. 255, 8289–8295.
232 Rock, C.O., Fitzgerald, V. and Snyder, F. (1977) J. Biol. Chem. 252, 6363–6366.
233 Bowley, M. and Brindley, D.N. (1976) Int. J. Biochem. 7, 141–147.
234 Rock, C.O. and Snyder, F. (1974) J. Biol. Chem. 249, 5382–5387.
235 Cabot, M.C. and Snyder, F. (1980) Biochim. Biophys. Acta 617, 410–418.
236 Herrmann, H. and Gercken, G. (1980) Lipids 15, 179–185.
237 Bandi, Z.L. and Mangold, H.K. (1978) Nutrit. Metab. 22, 190–199.
238 Schmid, H.H.O., Bandi, P.C., Madson, T.H. and Baumann, W.J. (1977) Biochim. Biophys. Acta 488, 172–178.
239 Chang, N.C., Muramatsu, T. and Schmid, H.H.O. (1973) Biochim. Biophys, Acta 306, 437–445.
240 Zilversmit, D.B., Entenman, C. and Fishler, M.C. (1943) J. Gen. Physiol. 26, 325–331.
241 Thompson, Jr., G.A. (1966) Biochemistry 5, 1290–1296.
242 Horrocks, L.A. and Ansell, G.B. (1967) Lipids 2, 329–333.
243 Wood, R., Walton, M., Healy, J. and Cumming, R.B. (1970) J. Biol. Chem. 245, 4276–4285.
244 Stoffel, W. and LeKim, D. (1971) Z. Physiol. Chem. 352, 501–511.
245 Blank, M.L., Wykle, R.L. and Snyder, F. (1972) Biochem. Biophys. Res. Commun. 47, 1203–1208.
246 Paltauf, F. (1972) FEBS Lett. 20, 79–82.
247 Paltauf, F. and Holasek, A. (1973) J. Biol. Chem. 248, 1609–1615.
248 Paltauf, F., Prough, R.A., Masters, B.S. and Johnston, J.M. (1974) J. Biol. Chem. 249, 2661–2662.
249 Wykle, R.L., Blank, M.L., Malone, B. and Snyder, F. (1972) J. Biol. Chem. 247, 5442–5447.

250 Wykle, R.L. and Schremmer Lockmiller, J.M. (1975) Biochim. Biophys. Acta 380, 291–298.
251 Strosznajder, J. and Dabrowiecki, Z. (1977) Bull. Acad. Pol. Sci. Ser. Sci. Biol. 25, 133–139.
251a Woelk, H. and Peiler-Ichikawa, K. (1978) Arzneim. Forsch. 28, 1752–1756.
252 Paltauf, F. (1978) Eur. J. Biochem. 85, 263–270.
253 Baker, R.C., Wykle, R.L., Lockmiller, J.S. and Snyder, F. (1976) Arch. Biochem. Biophys. 177, 299–306.
254 Lee, T.C., Wykle, R.L., Blank, M.L. and Snyder, F. (1973) Biochem. Biophys. Res. Commun. 55, 574–579.
255 Tjiong, H.B., Gunawan, J. and Debuch, H. (1976) Z. Physiol. Chem. 357, 707–712.
256 Gunawan, J. and Debuch, H. (1977) Z. Physiol Chem. 358, 537–543.
257 Wykle, R.L. and Schremmer, J.M. (1979) Biochemistry 18, 3512–3517.
258 Freysz, L. and Horrocks, L.A. (1980) in Neurological Mutations Affecting Myelination (Baumann, N.A., ed.), pp. 223–230, Elsevier/North-Holland Biomedical Press, Amsterdam.
259 Freysz, L., Horrocks, L.A. and Mandel, P. (1980) J. Neurochem. 34, 963–969.
260 Freysz, L., Horrocks, L.A. and Mandel, P. (1978) Adv. Exp. Med. Biol. 101, 253–268.
261 Strosznajder, J., Radomińska-Pyrek, A., Lazarewicz, J. and Horrocks, L.A. (1979) Bull. Acad. Pol. Sci. Ser. Sci. Biol. 27, 693–700.
262 Strosznajder, J., Radomińska-Pyrek, A. and Horrocks, L.A. (1979) Biochim. Biophys. Acta 574, 48–56.
263 Binaglia, L., Roberti, R., Goracci, G., Francescangeli, E. and Porcellati, G. (1974) Lipids 10, 738–747.
264 Goracci, G., Francescangeli, E. and Horrocks, L.A. (1977) Ital. J. Biochem. 26, 262–263.
265 Freysz, L., Bieth, R. and Mandel, P. (1969) J. Neurochem. 16, 1417–1424.
266 Brunetti, M., Gaiti, A. and Porcellati, G. (1979) Lipids 14, 925–931.
267 Mozzi, R., Andreoli, V. and Porcellati, G. (1980) in Natural Sulfur Compounds. Novel Biochemical and Structural Aspects (Cavallini, D., Gaull, G.E. and Zappia, V., eds.), pp. 41–54, Plenum, New York.
268 Poulos, A., Hughes, B.P. and Cumings, J.N. (1968) Biochim. Biophys. Acta 152, 629–632.
269 Goracci, G., Francescangeli, E., Horrocks, L.A. and Porcellati, G. (1981) Biochim. Biophys. Acta 664, 373–379.
270 Goracci, G., Horrocks, L.A. and Porcellati, G. (1978) Adv. Exp. Med. Biol. 101, 269–278.
271 Waku, K. and Nakazawa, Y. (1977) J. Biochem. (Tokyo) 82, 1779–1784.
272 Hagen, P-O. and Goldfine, H. (1967) J. Biol. Chem. 242, 5700–5708.
273 Hagen, P-O. (1970) unpublished results cited in [67].
274 Hill, E.E. and Lands, W.E.M. (1970) Biochim. Biophys. Acta 202, 209–211.
275 Prins, R.A., Akkermans-Kruyswijk, J., Franklin-Klein, W., Lankhorst, A. and Van Golde, L.M.G. (1974) Biochim. Biophys. Acta 348, 361–369.
276 Prins, R.A. and Van Golde, L.M.G. (1976) FEBS Lett. 63, 107–111.
277 Blank, M.L. and Snyder, F. (1970) Biochemistry 9, 5034–5036.
278 Woelk, H., Goracci, G. and Porcellati, G. (1974) Z. Physiol. Chem. 355, 75–81.
279 Woelk, H., Jakumeit-Morgott, V. and Schenck, K. (1976) J. Neurochem. 26, 275–279.
280 Wykle, R.L. and Schremmer, J.M. (1974) J. Biol. Chem. 249, 1742–1746.
281 Wykle, R.L., Kraemer, W.F. and Schremmer, J.M. (1977) Arch. Biochem. Biophys. 184, 149–155.
282 Wykle, R.L., Kraemer, W.F. and Schremmer, J.M. (1980) Biochim. Biophys. Acta 619, 58–67.
283 Vierbuchen, M., Gunawan, J. and Debuch, H. (1979) Z. Physiol. Chem. 360, 1091–1097.
284 Soodsma, J.F., Piantadosi, C. and Snyder, F. (1972) J. Biol. Chem. 247, 3923–3929.
285 Rock, C.O., Baker, R.C., Fitzgerald, V. and Snyder, F. (1976) Biochim. Biophys. Acta 450, 469–473.
286 Kapoulas, V.M., Thompson Jr., G.A. and Hanahan, D.J. (1969) Biochim. Biophys. Acta 176, 250–264.
287 Matsumoto, M., Suzuki, Y. and Tamiya, K. (1967) Jap. J. Exp. Med. 37, 355–358.
288 Warner, H.R. and Lands, W.E.M. (1961) J. Biol. Chem. 236, 2404–2409.
289 Ellingson, J.S. and Lands, W.E.M. (1968) Lipids 3, 111–120.
290 Gunawan, J. and Debuch, H. (1981) Z. Physiol. Chem. 362, 445–452.

291 Horrocks, L.A. and Fu, S.C. (1978) Adv. Exp. Med. Biol. 101, 397–406.
292 Freeman, N.M. and Carey, E.M. (1980) Biochem. Soc. Trans. 8, 612–613.
293 D'Amato, R.A., Horrocks, L.A. and Richardson, K.E. (1975) J. Neurochem. 24, 1251–1255.
294 Fu, S.C., Mozzi, R., Krakowka, S., Higgins, R.J. and Horrocks, L.A. (1980) Acta Neuropathol. 49, 13–18.
295 Ansell, G.B. and Spanner, S. (1965) Biochem. J. 94, 252–258.
296 Horrocks, L.A., Spanner, S., Mozzi, R., Fu, S.C., D'Amato, R.A. and Krakowka, S. (1978) Adv. Exp. Med. Biol. 100, 423–438.
297 Dorman, R.V., Freysz, L., Horrocks, L.A. and Mandel, P. (1978) J. Neurochem. 30, 157–159.
298 Dorman, R.V., Toews, A.D. and Horrocks, L.A. (1977) J. Lipid Res. 18, 115–117.
299 Gunawan, J., Vierbuchen, M. and Debuch, H. (1979) Z. Physiol. Chem. 360, 971–978.
300 Yavin, E. and Gatt, S. (1972) Eur. J. Biochem. 25, 437–446.
301 Hennacy, D.M. and Horrocks, L.A. (1978) Bull. Mol. Biol. Med. 3, 207–221.
302 Horrocks, L.A., Toews, A.D., Thompson, D.K. and Chin, J.Y. (1976) Adv. Exp. Med. Biol. 72, 37–54.
303 Miller, S.L., Benjamins, J.A. and Morell, P. (1977) J. Biol. Chem. 252, 4025–4037.
304 Sun, G.Y. and Su, K.I. (1979) J. Neurochem. 32, 1053–1059.
305 Selivonchick, D.P. and Roots, B.I. (1979) Lipids 14, 66–69.
306 Sun, G.Y., Winniczek, H., Go, J. and Sheng, S.L. (1975) Lipids 10, 365–373.
307 Demopoulos, C.A., Pinckard, R.N. and Hanahan, D.J. (1979) J. Biol. Chem. 254, 9355–9358.
308 Blank, M.L., Snyder, F., Byers, L.W., Brooks, B. and Muirhead, E.E. (1979) Biochem. Biophys. Res. Comm. 90, 1194–1200.
309 Clark, P.O., Hanahan, D.J. and Pinckard, R.N. (1980) Biochim. Biophys. Acta 628, 69–75.
310 Hanahan, D.J., Demopoulos, C.A., Liehr, J. and Pinckard, R.N. (1980) J. Biol. Chem. 255, 5514–5516.
311 Godfroid, J.J., Heymans, F., Michel, E., Redeuilh, C., Steiner, E. and Benveniste, J. (1980) FEBS Lett. 116, 161–164.
312 Muirhead, E.E., Byers, L.W., Desiderio, D.M., Brooks, B. and Brosius, W.M. (1981) Fed. Proc. 40, 2285–2290.
313 McManus, L.M., Hanahan, D.J., Demopoulos, C.A. and Pinckard, R.N. (1980) J. Immunol. 124, 2919–2924.
314 Pinckard, R.N., McManus, L.M., Demopoulos, C.A., Halonen, M., Clark, P.O., Shaw, J.O., Kniker, W.T. and Hanahan, D.J. (1980) J. Reticul. Soc., 28, 95S–103S.
315 McManus, L.M., Hanahan, D.J. and Pinckard, R.N. (1981) J. Clin. Invest. 67, 903–906.
316 Shaw, J.O., Hanahan, D.J. and Klusick, S.J. (1981) Biochim. Biophys. Acta 663, 222–229.
317 Goetzl, E.J., Derian, C.K., Tauber, A.I. and Valone, F.H. (1980) Biochem. Biophys. Res. Commun. 94, 881–888.
318 Chignard, M., LeCouedic, J.P., Vargaftig, B.B. and Benveniste, J. (1980) Br. J. Haematol. 46, 455–464.
319 Polonsky, J., Tence, M., Varenne, P., Das, B.C., Lunel, J. and Benveniste, J. (1980) Proc. Natl. Acad. Sci. USA 77, 7019–7023.
320 O'Flaherty, J.T., Wykle, R.L., Miller, C.H., Lewis, J.C., Waite, M., Bass, D.A., McCall, C.E. and DeChatelet, L.R. (1981) Am. J. Pathol. 103, 70–79.
321 Hanahan, D.J., Munder, P.G., Satouchi, K., McManus, L. and Pinckard, R.N. (1981) Biochem. Biophys. Res. Commun. 99, 183–188.
322 Satouchi, K., Pinckard, R.N., McManus, L.M. and Hanahan, D.J. (1981) J. Biol. Chem. 256, 4425–4432.
323 Wykle, R.L., Malone, B. and Snyder, F. (1980) J. Biol. Chem. 255, 10256–10260.
324 Renooij, W. and Snyder, F. (1981) Biochim. Biophys. Acta 663, 545–556.
325 Blank, M.L., Lee, T.C., Fitzgerald, V. and Snyder, F. (1981) J. Biol. Chem. 256, 175–178.
326 Friedberg, S.J. and Halpert, M. (1978) J. Lipid Res. 19, 57–64.
327 Cabot, M.C. and Snyder, F. (1980) Biochim. Biophys. Acta 617, 410–418.

328 Marinetti, G.V. and Crain, R.C. (1978) J. Supramol. Struct. 8, 191–213.

329 Rittenhouse-Simmons, S., Russell, R.A. and Deykin, D. (1976) Biochem. Biophys. Res. Commun. 70, 295–301.

330 Schick, P.K., Schick, B.P., Brandeis, G. and Mills, D.C.B. (1981) Biochim. Biophys. Acta 643, 659–662.

331 Woelk, H., Rubly, N., Arienti, G., Gaiti, A. and Porcellati, G. (1981) J. Neurochem. 36, 875–880.

332 Schroeder, F. and Vagelos, P.R. (1976) Biochim. Biophys. Acta 441, 239–254.

333 Lee, T.C. and Fitzgerald, V. (1980) Biochim. Biophys. Acta 598, 189–192.

334 Goldfine, H., Johnston, N.C. and Phillips, M.C. (1981) Biochemistry 20, 2908–2916.

335 Boggs, J.M. (1980) Can. J. Biochem. 58, 755–770.

336 Brockerhoff, H. (1974) Lipids 9, 645–650.

337 Goldfine, H. and Johnston, N.C. (1980) in Membrane Fluidity–Biophysical Techniques and Cellular Regulation (Kates, M. and Kuksis, A., eds.), pp. 365–380, Humana Press, Clifton, NJ.

338 Andreesen, R., Modolell, M., Weltzien, H.U., Eibl, H., Common, H.H., Löhr, G.W. and Munder, P.G. (1978) Cancer Res. 38, 3894–3899.

339 Berdel, W.G., Bausert, W.R., Weltzien, H.U., Modolell, M.L., Widmann, K.H. and Munder, P.G. (1980) Eur. J. Cancer 16, 1199–1204.

340 Boeryd, B. and Hallgren, B. (1980) Acta Pathol. Microbiol. Scand. Sect. A. 88, 11–18.

CHAPTER 3

Phosphonolipids

TARO HORI and YOSHINORI NOZAWA *

Department of Chemistry, Shiga University
Otsu, Shiga, 520, Japan, and
** Department of Biochemistry, Gifu University School of Medicine, Tsukasamachi-40, Gifu, 500, Japan*

1. Historical introduction and classification

The isolation of 2-aminoethylphosphonic acid (ciliatine, AEPn) in 1959 by Horiguchi and Kandatsu [1] from ciliated protozoa of the rumen initiated extensive research into the biochemistry of compounds containing a C—P bond. AEPn and other aminophosphonic acids structurally related to AEPn have since been identified in various species [2–10]. Additional phosphonic acids, namely 2-amino-3-phosphonopropionic acid and three *N*-methyl derivatives of AEPn, have been discovered in coelenterates [11,12]. AEPn and *N*-methyl AEPn (MAEPn) have been implicated as components of certain complex lipids, for which Baer and Stanacev [13] proposed the generic name of phosphonolipids.

The first evidence for the occurrence of phosphonolipids was obtained in 1963 by Rouser et al. [14] during an analysis of the sea anemone, *Anthropleura elegantissima*. This new sphingolipid was tentatively identified as *N*-acyl-sphingosyl-1-(2-aminoethyl)phosphonate (ceramide aminoethylphosphonate, CAEPn). In the following year Hori and colleagues [15] independently proved the existence of CAEPn in the fresh-water bivalve, *Corbicula sandai*. Subsequent work showed that CAEPn was widely distributed in a variety of species of fresh-water and marine bivalves [16], snail [17], cephalopods [17], and gametes of the fresh-water bivalve, *Hyriopsis schlegelii* [18,19]. The same compound was shown to be present in oysters [20,21], abalone [22], snail [23], gastropods [24], sea anemone [25–27], ox [28], protozoa [29–36], *Bdellovibrio bacteriovorus* [37] and *Pythium prolatum* [38]. However, it was absent from the lobster, *Pannlirus interruptus*, the sea urchin, *Stronglylocentrotus franciscamus* [27], and from various crustaceans [17].

An *N*-methyl derivative of CAEPn (CMAEPn) was found in a number of snail species in 1969 by Hayashi et al. [39,40] and Hori et al. [41]. The same compound has also been isolated from the protozoon, *Tetrahymena pyriformis* by Viswanathan and Rosenberg [33] and Sugita et al. [36].

The occurrence of glycerophosphonolipids (GPnL), glycerolipids containing AEPn, was originally demonstrated in the protozoon, *T. pyriformis* by Liang and Rosenberg [42]. Thompson [43] presented a more detailed analysis showing that this

Hawthorne/Ansell (eds.) Phospholipids

I Glycerophosphonolipids (GPnL)

(a)

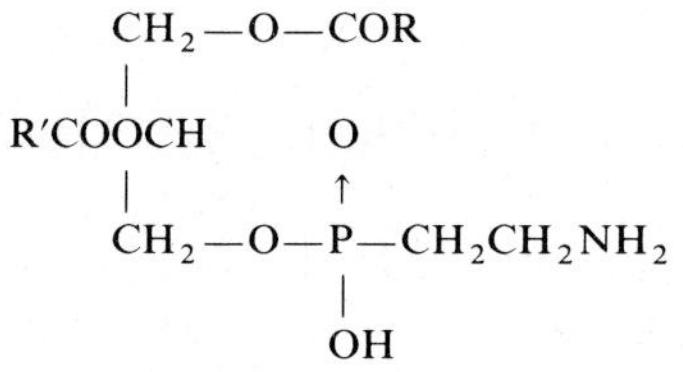

(b)

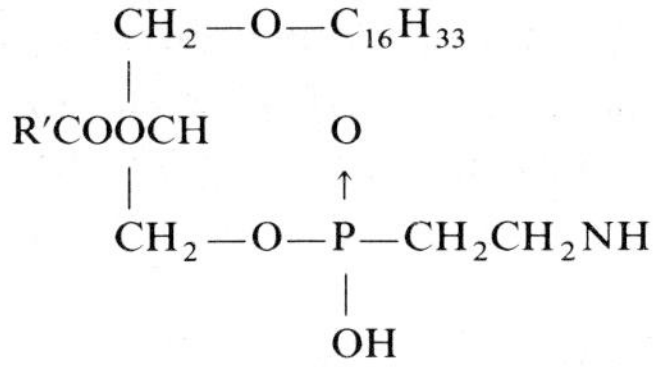

(a) (3-*sn*-phosphatidyl)-ethylamine *
(1,2-diacyl-*sn*-glycero(3)-2-aminoethylphosphonate)
(b) 1-hexadecyl-2-acyl-*sn*-glycero(3)-2-phosphonoethylamine *
* Recommended by the IUPAC-IUB Commission (1976).

II Sphingophosphonolipids (SPnL)

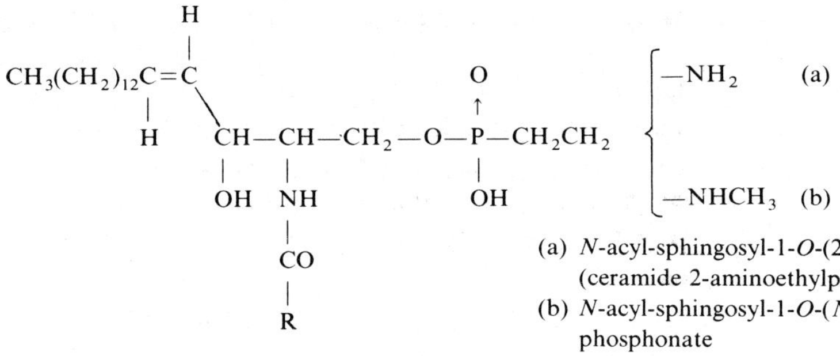

(a) *N*-acyl-sphingosyl-1-*O*-(2-aminoethyl)phosphonate (ceramide 2-aminoethylphosphonate)
(b) *N*-acyl-sphingosyl-1-*O*-(*N*-methyl-2-aminoethyl)-phosphonate

III Sphingophosphonoglycolipids (SPnGL)

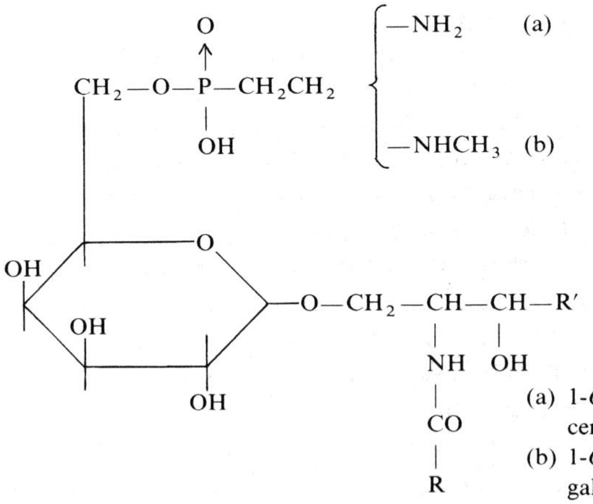

(a) 1-*O*-[6-*O*-(aminoethylphosphono)galactopyranosyl] ceramide
(b) 1-*O*-[6-*O*-(*N*-methylaminoethylphosphono)-galactopyranosyl] ceramide

Fig. 1. Chemical structures of some phosphonolipids.

GPnL was a derivative of chimyl alcohol, 1-*O*-hexadecyl-2-acyl-glycero(3)-2-aminoethylphosphonate. Kennedy and Thompson [44] subsequently demonstrated that large amounts of the GPnL were contained in the ciliary membrane of *Tetrahymena*. Tamari et al. [28] reported the presence of diacyl-GPnL in bovine gall bladder bile, and Sugita and Hori [36] also reported isolation of a diacyl-GPnL from *Tetrahymena*, although its yield was very low.

Hayashi and colleagues [45] described new classes of phosphonolipids containing a glycolipid from the marine snail, *Turbo cornutus*. In 1977, detailed structural analysis of these snail sphingophosphonoglycolipids (SPnGL) showed them to be 1-*O*-[6-*O*-(*N*-methyl-2-aminoethylphosphono)galactopyranosyl]ceramide [46–48]. Recently, Araki et al. [49] isolated another SPnGL from the skin of the marine gastropod, *Aplysia krodai*. This lipid was shown to have 2 mol of AEPn and an oligosaccharide chain.

The structures of three types of well-known phosphonolipids mentioned above are shown in Fig. 1. Since the discovery of AEPn a little over two decades have passed, and the existence of phosphonolipids is well established only in the molluscs, coelenterates and protozoa, where they occur in significant amounts. Despite considerable interest in phosphonolipids, their specific physiological role in living cells remains unclear.

Several reviews have already dealt with the natural distribution, chemistry, analysis and metabolism of phosphonolipids, as well as their possible biochemical function [50–52]. The present chapter is an effort to summarize these and to incorporate more recent information. In the following discussion, we have used the IUPAC-IUB Commission on Biochemical Nomenclature (1976) and other proposed new rules for lipid nomenclature.

2. *Methods of isolation and characterization*

(a) Isolation and purification

The combined use of DEAE-cellulose and silicic acid column chromatography described by Rouser et al. [14] has been effective in the isolation of phosphonolipids. A mixture of phosphonolipids and phosphatidylethanolamine can be recovered from a DEAE-cellulose column by elution with chloroform–methanol (7:3). Their further separation can be achieved by silicic acid column chromatography. For the isolation of SPnL from the tissues of molluscs [16,17,41] and coelenterates [27], the lipid extracts obtained with chloroform–methanol were subjected to mild alkaline hydrolysis to remove glycerolipids, and the alkali-stable lipid fraction was fractionated by silicic acid column chromatography, using increasing proportions of methanol in chloroform as solvent. The individual phosphonolipids were finally purified by preparative thin-layer chromatography [41,53,54]. Several attempts [31,55–57] at the preparative isolation of GPnL and their further subfraction into ether-GPnL and diacyl-GPnL have been made, but it has not been possible to completely separate these two types of GPnL by any simple column chromatographic procedure. Viswanathan and Nagabhushanam [58] described a procedure which yielded substantial amounts of ether-GPnL and diacyl-GPnL by ascending dry column chromatography with chloroform–acetic acid–methanol–water (75:25:5:1.5) of total *Tetrahymena* lipids.

Quantitative analysis of phospholipid mixtures containing phosphonolipids by two-dimensional thin-layer chromatography has been described by Simon and Rouser [59] and Liang and Strickland [23].

Itasaka and Hori [60] analyzed CAEPn and CMAEPn by high performance liquid chromatography of the alkali-stable lipid fraction from a number of shellfish, using a Zorbax SIL column with *n*-hexane–*iso*-propanol–water (30:40:7) as the solvent. Detection was with an ultraviolet spectrometer at 207 nm.

(b) Characterization

(i) Infrared spectrometry of intact phosphonolipids

GPnL and SPnL can be readily distinguished by infrared spectroscopy. The most characteristic difference between the two spectra is the absence of the ester carbonyl absorption around 1735/cm and the presence of typical amide I and amide II absorptions around 1640/cm and 1550/cm, respectively, in the spectra of the sphingolipids. All infrared spectra of phosphonolipids have a typical phosphonate absorption around 1200/cm [16,39,54,58,61]. This observation demonstrates that the stretching vibrational frequency due to P—O around 1230/cm in phosphate ester-containing lipid (P—O—C) is shifted to a lower frequency of around 1200/cm in phosphonate lipids (C—P). Thomas [62] noted an increase in the frequency of the P—O stretching vibration when an electronegative substituent was linked on the phosphorus, but a decrease in frequency when the phosphorus was directly linked to either carbon or hydrogen.

Baer and colleagues have presented an almost complete set of spectra for synthetic phosphonolipids including diacyl- [13] and diether- [63] GPnL, and SPnL (*erythro-N*-palmitoyl-D-sphingosyl-1-(2-aminoethyl)phosphonate) [64]. The spectra for phosphonolipids isolated from natural sources closely resemble those of the synthetic compounds.

(ii) Gas-liquid chromatography and mass spectrometry

Trimethylsilyl derivatives are satisfactory for the characterization of the aminoalkylphosphonic acids, and both AEPn and MAEPn can easily be estimated by gas-liquid chromatography of their derivatives [65]. Using this method, Karlsson [25,26] confirmed that SPnL of the sea anemone, *Metridium senile* contains predominantly CAEPn, with small amounts of CMAEPn. Matsubara [66] has analyzed the trimethylsilyl derivatives of intact SPnL from the oyster, *Ostrea gigas* by direct injection into a gas chromatograph-mass spectrometer. The main molecular species in the adductor muscle of the oyster was hexadecanoyl-hexadeca-4-sphingenyl- and in the viscera was hexadecanoyl-octadeca-4,8-sphingadienyl-2-aminoethylphosphonate. Molecular species of SPnL from a number of marine shellfish have been studied in detail by Hayashi and colleagues [67–69], the characterization of SPnL from the protozoon, *T. pyriformis* WH-14 has been reported by Sugita et al. [36], and the sea anemone, *M. senile* by Karlsson and Samuelsson [27].

SPnGL can be partially hydrolyzed with aqueous HCl and the product, an aminoethylphosphonate-containing carbohydrate can be converted to a trimethylsilyl derivative and identified by its retention times on gas-liquid chromatography and mass spectra [46,47]. No identification of intact GPnL by mass spectrometry has been reported although mass spectral studies of the aminoethylphosphonic acids have been reported by several workers [70–72].

(iii) Nuclear magnetic resonance spectroscopy

NMR should provide the ideal method for direct detection and determination of phosphonolipids. At the present time, however, only limited evidence is available.

The absorption band of phosphonolipids is well-separated from the band of phospholipids in the NMR spectra of intact lipids by [^{31}P]NMR. Glonek et al. [73] found that the spectra of lipid fractions of two sea anemones, *Bunadosoma* sp. and *Metridium* sp., and the protozoon, *T. pyriformis* show two major adsorption bands. The phosphonolipids come into resonance within the range of -18 to -24 ppm, while the resonance band of the phospholipids occurs at -3 to $+3$ ppm. The amounts of the phosphonolipids, determined from the respective areas of the absorption bands in each spectrum were identical to those determined by the classical colorimetric procedure for the quantitation of phosphonate phosphorus [74].

Proton spectra of aminoethylphosphonic acid and its related compounds are described in several publications [12,75].

(3) Occurrence and distribution

(a) Qualitative and quantitative distribution of phosphonolipids

From the results of quite a number of studies, it is obvious that the distribution of phosphonolipids in nature is primarily limited to lower animals such as the molluscs, coelenterates and protozoa, although these lipids can also be detected in minute quantities in mammalian tissues [28,54]. Phosphonolipids have been detected in bacteria [37], but the levels found were generally very low. Among the three types of structures of phosphonolipids (Fig. 1), GPnL occurs in high concentration in *Tetrahymena* and is the major phospholipid of the ciliary membranes of protozoa. In determining the phospholipid composition of various membrane fractions from *T. pyriformis* WH-14 cells, Nozawa and Thompson [76] and Kennedy and Thompson

TABLE 1

Relative quantitative distribution of phospholipids in animals (as % of total phospholipid)

	GPnL	SPnL	PC	PE	PS	PI	PA	LP	CA	SP	Others [c]	Ref.
Protozoa												
Tetrahymena pyriformis WH-14 [a]												
Whole cells	23		33	37				2	5			
Cilia	47		28	11				10	1		3	
Pellicles	30		25	34				8	2		1	
Mitochondria	18		35	35				2	10			
Microsomes	23		35	34				4	1		3	
T. pyriformis NT-1												82
Whole cells	23		26	39				3	5		4	
T. pyriformis GL												
Whole cells	26		24	38				4	4		4	
T. pyriformis W												
Whole cells	29		27	35				3	4		2	
T. pyriformis E												
Nuclear	23		31	26				12	3		5	86
Entodinium caudatum	20 [b]	6	26	20		4	3			5	16	29
Paramecium tetraurelia												
Cilia	32	14	13	24	7	2		3		5		35

Mollusca											
Abalone											
Haliotis midae	5	44	37	5	4			4	1		22
Pink abalone											
Haliotis corrugata	9	40	27	10	4	0.3	0.7			9	59
Water snail											
Lymnaea stagnalis	8	50	31	6	4					1	23
Land snail											
Cepaea nemoralis	7	47	21	8	6	0.6	1			6	83
Scallop											
Hinnifes giganteum	17	35	26	12	5		1			4	59
Gastropoda											
Aplysia kurodai											
Ganglion	11	51	28	10					tr		} 24
Fiber	13	45	30	12					tr		
Coelenterata											
Arthropleura, elegantissima	20	22	20	14	3	0.3		1		19.7	84
Metridium senile	11	34	21	15	3					16	85

[a] As mol %.

[b] Includes plasmenyl lipids.

[c] Includes unidentified lipids.

Glycerophosphonolipid (GPnL), sphingophosphonolipid (SPnL), 3-*sn*-phosphatidylcholine (PC), 3-*sn*-phosphatidylethanolamine (PE), 3-*sn*-phosphatidylserine (PS), 3-*sn*-phosphatidylinositol (PI), cardiolipin (CA), 2-lysophospholipid (LP), 1,2-diacyl-*sn*-glycero-3-phosphate (PA), sphingomyelin (SP).

[44], indicated that the concentration of GPnL in the ciliary membrane is about 60% and in mitochondria and microsomes about 20–30% of the total phospholipids. The concentration of the *Tetrahymena* GPnL changes rapidly when the physiological condition of this organism is varied. These details will be mentioned later in this chapter. Examinations of shellfish and sea anemones did not reveal detectable amounts of GPnL.

On the other hand, high concentrations of CAEPn are found in molluscs and coelenterates, while their concentration is low in *Tetrahymena* (Tables 1 and 2). In a distribution study of oyster tissues, Matsubara [66] found that adductor muscle contained the highest concentration of CAEPn, 45% of total sphingolipids, while the equivalent percentage in gills was 22%, in mantle, 21%, and viscera, 19%. A very similar value was reported by Swift [77] for the adductor muscle of the oyster, *Crassoftrea virginica*, namely, 40.4% of total phospholipids. Swift also reported that the phosphonolipid content increases in the starved oyster and more still post-spawning, while the percentage of other phospholipids decreases correspondingly (Table 3). As the glycogen deposits are exhausted, the phosphonolipids are conserved at the expense of ester phospholipids.

Komai et al. [24] provided a quantitative analysis of phospholipids found in the nervous system of a marine gastropod, *A. kurodai*, which has a simpler anatomical structure than that of vertebrates. CAEPn accounted for approx. 11% of the phospholipids, but neither gangliosides, cerebrosides nor sulphatides, which occur in vertebrate nerve cell membranes, were detected. Based on this fact Komai et al. speculated that CAEPn may be indispensible in the shellfish for neuronal function.

CAEPn occurs in high concentration in some shellfish such as *T. cornutus, Monodonta labio* and *Tegula lischkei* while the concentration is lower in other shellfish, such as *Liolophira japonica* and *Celluna eucosmia*, although they contain measurable amounts of CMAEPn [69]. The concentration ratios of CAEPn to CMAEPn in some shellfish, as determined by high performance liquid chromatography, were 89:11 in *Sinotaia histrica*, 68:32 in *Semisulcospira bensoni* 69:31 in *M. labio* and 85:15 in the land snail, *Helix pomatia* [60].

Sphingophosphonoglycolipids (SPnGL), the third type of phosphonolipid, have been found in the viscera of *T. cornutus* and *M. labio*. The backbone of SPnGL is a cerebroside to which AEPn or MAEPn are linked as the phosphorus-containing component [48]. Although sphingolipids containing both carbohydrate and phosphorus have been isolated from a few kinds of plants [78,79] and termed "phytoglycolipids", they consist of *N*-acyl phytosphingosyl phosphoinositol linked glycosidically to an oligosaccharide moiety. Recently, another SPnGL has been isolated from the skin of a marine gastropod, *A. kurodai* [49]. Its structure is tentatively characterized as ceramide bis(2-aminoethylphosphono)-pentaoside having an oligosaccharide chain consisting of 1 mol each of glucose, 3-*O*-methylgalactose, and galactosamine and 2 mol of galactose. These SPnGL represent a few percent or less of the total complex lipids.

The relative quantitative distribution of phosphonolipids to other phospholipids varies among animals. The available data are summarized in Table 1 and the content of SPnL and quantitative distribution of CAEPn and CMAEPn in Table 2.

TABLE 2

Content of sphingophosphonolipids in Protozoa, Mollusca and Coelentrata (as % of phospholipid)

	SPnL	CAEPn	CMAEPn	Ref.
Protozoa				
T. pyriformis W	11 [a]	+	++	33
T. pyriformis WH-14	5 [a]	++	+	36
Mollusca				
Liolophira japonica	37.4	++	+	
Trubo cornutus				
muscle	10.6	−	+++	
viscera	8.2	−	+++	
Monodonta labio	16.2	−	+++	
Tegula lischkei	33.1	+	++	
Conomurex luhuanus	20.7	+++	−	
Celluna eucosmia	14.5	+++	−	69
Ostrea gigas				
adductor	45.2	+++	−	
gills	22.2	+++	−	
mantle	21.0	+++	−	
viscera	19.2	+++	−	
Mytilus edulis	25.6	+++	−	
Hyriopsis schlegelii				
ova	16 [a]	++	+	19
spermatozoa	7 [a]	++	+	18, 60
Scallop				
H. giganteum	17	+++	−	59
Pink abalone				
H. corrugata	9	++	?	59
Coelenterata				
A. elegantissima	19.9 [b]	+++	−	84
M. senile	10 [b]	++	+	27

[a] As % of alkali-stable lipid fraction.
[b] As % of total lipid.

(b) Fatty acid and sphingosine base compositions

The fatty acid composition of phosphonolipids is being investigated. The C-2 position of the GPnL from *Tetrahymena* contains almost exclusively 6,9-octadecadienoic acid (linoleic acid) and 6,9,12-octadecatrienoic acid (γ-linolenic acid) [31,80]. However, detailed information on the fatty acids at the 1- and 2-positions of diacyl-GPnL is not available. The GPnL of paramecium cells and cilia contains up to 93% polyunsaturated fatty acids such as 5,8,11,14-eicosatetraenoic and 5,8,11,14,17-eicosapentaenoic acids, although these change with the culture age [35].

SPnL of *T. pyriformis* WH-14 contain six major fatty acids, identified principally

TABLE 3

Distribution of lipid phosphorus in oysters in three physiological conditions

Tissues	Condition	Total lipid phosphorus (μg P/g wet weight tissue)	Percent phosphorus	
			Phosphonate	Ester phosphorus
Total oyster	Fresh	237	23.2	76.8
	Starved	313	33.9	66.1
	Post spawning	212	36.3	63.7
Mantle and gills	Fresh	267	24.0	76.0
	Starved	281	38.3	61.7
	Post spawning	209	28.7	71.3
Adductor muscle	Fresh	228	40.4	59.6
	Starved	289	38.8	61.2
	Post spawning	156	26.3	73.7
Digestive diverticula	Fresh	242	12.8	87.2
	Starved	404	27.2	72.8
	Post spawning	297	29.0	71.0

Taken from [77].

as 2-hydroxy palmitic, 2-hydroxy *iso*-heptadecanoic, and 2-hydroxy stearic acids [36]. Ferguson et al. [32] found that the culture medium and conditions markedly influenced the fatty acid biosynthetic pattern in *T. pyriformis*. 2-Hydroxy fatty acids could be found in this organism when it was grown in a proteose-peptone medium, but when grown in a chemically defined medium the *Tetrahymena* failed to synthesize the hydroxy fatty acid. When maintained in proteose-peptone *T. pyriformis* WH-14 has, as the major SPnL fatty acids, 2-hydroxy acids with carbon numbers of 16 to 19, which are almost exclusively the *iso*-types [36]. The fatty acid composition of SPnL from the sea anemone, *M. senile*, contained one-third branched chains [27]. Thus, the predominant fatty acid composition of phosphonolipids is complex in *Tetrahymena* and in sea anemone, while that of shellfish is simpler. In general, SPnL of shellfish tissues contains high amounts of palmitic and 2-hydroxy palmitic acids except in *M. labio* which contains the unusual *trans*-2-octadecanoic acid, but in very small amounts [69].

The sphingosine base composition is complex in *Tetrahymena* and shellfish, but simple in sea anemone. In *T. pyriformis* WH-14 sphingosine bases were predominantly C_{16}, C_{17}, C_{18} and C_{19}-*iso*-4-sphingenine homologues [36]. In contrast, the components of the marine snails were mainly dihydroxy bases with 16–22 carbons and one or two double bonds. These species of dienoic bases were identified as octadeca-4,8-sphingadienine, eicosa-4,11-sphingadienine, and docosa-4,15-sphingadienine [68]. 4-Hydroxydocosa-15-sphingenine was also identified in the viscera of *T. cornutus* [69]. The ova phosphonolipids from the fresh-water bivalve, *H. schlegelii* were found to contain *iso*-sphingosine, sphingosine (60%), *iso*-nonadeca-4-sphing-

TABLE 4

Fatty acid composition (%) of glycerophosphonolipids (GPnL) sphingophosphonolipids (SPnL) and sphingophosphonoglycolipids (SPnGL) in animals and fungi

	Class	14:0+ 14:unsat.	16:0	16:1	17:0	18:0	18:1	18:2	18:3	20:4	20:5	Others	Ref.
Fungi													
Pythium protatum	SPnL	0.3	19.6	0.3		1.6	21.2	21.6	4.7	5.9		24.8	38
Protozoa													
T. pyriformis WH-14													
Whole cells	GPnL	0.7	2.3	3.2		0.5	5.5	33.1	50.4			4.3	86
Mitochondria	GPnL						3.2	24.9	67.9			4.0	86
T. pyriformis WH-14	SPnL	2.2	17.9		22.9 (*iso*)	38.5 (*iso*)						4.1	36
		1.8 (*iso*)			6.5 (*iso*, h)	4.6 (h)							
		1.5 (h)											
T. pyriformis NT-1	GPnL	4.5	8.2	6.7		2.0	11.6	20.5	38.5			8.0	87
P. tetraurelia 51S	GPnL	tr	9.8	0.7		3.9	4.6	3.8	2.8	61.7	12.7	4.1	35
Mollusca													
Corbicula sandai	SPnL		39.9		6.7	54.7						1.6	16
H. midae	SPnL		53	4	4	15	1			5		18	22
L. japonica	SPnL		52.1		3.8	7.7							69
			36.4 (h)										
T. cornutus													
muscle	SPnL		86.7		8.4	4.4						0.5	69
viscera	SPnL		68.7		5.5	4.7							69
			21.1 (h)			3.9						12.6	46
T. cornutus	SPnGL		53.3		8.1	3.7 (h)							
			14.6 (h)		1.8 (br)								
					6.1 (h)								
					3.4 (h, br)	4.5	5.9 (Δ^2)						69

TABLE 4 (continued)

	Class	14:0+ 14:unsat.	16:0	16:1	17:0	18:0	18:1	18:2	18:3	20:4	20:5	Others	Ref.
M. labio	SPnL		78.7		3.3								
			1.2 (h)		6.5 (br)	6.3	5.7					9.5	48
M. labio	SPnGL	0.4	49.7		3.3	1.1 (br)							
			12.2 (h)		3.7 (br)	3.4 (h)							
					4.7 (h)	5.3							69
T. iischkei	SPnL		71.3		6.2	6.7							69
			5.8 (h)		11.4 (br)								
C. luhuanus	SPnL		80.1		0.8	4.4							69
			12.4 (h)										
C. eucosmia	SPnL		73.0		4.4								
			18.2 (h)			3.3							66
O. gigas						4.7						0.1	66
adductor	SPnL		89.7		7.0								
gills	SPnL		77.2		4.8	3.2						0.5	66
			13.2 (h)										
mantle	SPnL		76.2		5.0	4.0							66
			15.1 (h)										
viscera	SPnL		77.2		5.1								
H. schlegelii			13.7 (h)										
Spermatozoa	SPnL		4.0			96.0							19
M. edulis	SPnL		76.5		3.0	2.5							69
			5.7 (h)		12.1 (br)								
Coelenterata						1	6					4	27
M. senile	SPnL	1	52		2	10 (br)							
			4 (br)		20 (br)								
A. elegantissima	SPnL	0.5	82.4	2.3	7.3 (br)	0.6	3.8	0.5				2.6	84

br and h in parentheses mean branched-chain and 2-hydroxy fatty acids.
Others include unidentified fatty acids.

enine and *anteiso*-nonadeca-4-sphingenine (22%) as the bases [81]. These data are presented in Tables 4 and 5, respectively.

4. Metabolism

(a) Biosynthesis

(i) 2-Aminoethylphosphonic acid (AEPn)

There are two major metabolic steps in the biosynthesis of phosphonolipids; formation of AEPn and its incorporation into lipids. Although several biosynthetic pathways of AEPn have been proposed, they were largely, if not all, consistent with an intramolecular rearrangement of phosphonoenolpyruvate (PnEP) which was postulated independently by Warren [88] and Liang and Rosenberg [89]. This pathway proceeds via amination and subsequent decarboxylation to result in the AEPn formation. Later Horiguchi [90] demonstrated that the decarboxylation occurred prior to amination. Therefore, the intermediate was thought to be 2-phosphonoacetaldehyde (PnAA) in the latter route, whereas it would be 2-amino-3-phosphonopropionic acid (APnP) in the former route. More detailed information regarding the origin of C—P bond was given in a useful review [52].

(ii) Glycerophosphonolipids (GPnL)

Liang and Rosenberg [42] have first demonstrated by radioisotope labelling experiments in *Tetrahymena* that radioactive AEPn was incorporated into a nucleotide-bound substance and that this reaction has a specific requirement for cytidine nucleotide, producing cytidine monophosphate-aminoethylphosphonate. The formation of CMP-AEPn was also confirmed in vitro using CTP and AEPn as the substrates. Then attempts were made to demonstrate synthesis of AEPn-containing phospholipid by incorporation of labelled CMP-AEPn into GPnL. When cell-free preparations of *Tetrahymena* were incubated with CMP-[^{32}P]AEPn in the presence of dipalmitin, the highest activity of ^{32}P incorporation into lipids was found in the "mitochondrial" fraction. The requirement of dipalmitin was evident from the fact that omission of the diacylglycerol from the reaction mixture resulted in a great reduction in the activity to incorporate AEPn into lipids. This may imply that the reaction to form an AEPn-containing glycerolipid would be catalysed by CMP-AEPn:diacylglycerol AEPn transferase. Thus there exists a pathway of the GPnL biosynthesis comparable to the CDP-ethanolamine pathway for phosphatidylethanolamine formation. The former reaction utilizes phosphonate rather than ester phosphate.

$$\text{Ethanolamine-P} + \text{CTP} \rightarrow \text{CDP-ethanolamine} + \text{pyrophosphate} \qquad (1)$$

TABLE 5

Sphingosine base composition (%) of sphingophosphonolipids in animals

	d16:1	d17:1	d18:1	d18:2
Protozoa				
T. pyriformis WH-14	5.6	3.5	9.2	
	2.1 (*iso*)	58.9 (*iso*)	15.3 (*iso*)	
Mollusca				
L. japonica	18.9	17.1	18.8	
T. cornutus				
muscle	5.4	11.0	37.2	18.0
viscera	10.3	8.2	41.5	11.7
T. cornutus [a]	2.7	5.7	18.0	
			7 (br)	
M. labio	15.0	7.2	37.6	13.5
M. labio [a]	9.0	3.7	26.0	12.4
			9.2 (*iso*)	
T. iischkei	4.4	5.3	39.4	29.9
C. luhuanus	13.0	12.1	32.6	30.1
C. eucosmia	13.9	6.9	10.2	
O. gigas				
adductor	75.9		4.7	19.3
gills	18.4	4.1	25.3	30.7
mantle	20.4	5.4	23.0	29.0
viscera	10.8	4.2	17.9	52.3
M. edulis	17.7	4.0	40.0	33.2
H. schlegelii				
ova			59.7	
			12.0 (*iso*)	
Coelenterata				
M. senile			5	95

In the shorthand formulae, d means dihydroxy and t trihydroxy: the figure before the colon means carbon chain length and the figure after the colon number of double bonds.

$$\text{CDP-ethanolamine} + \text{diacylglycerol} \rightarrow \text{phosphatidylethanolamine} + \text{CMP} \quad (2)$$

$$\text{CTP} + \text{AEPn} \rightarrow \text{CMP-AEPn} + \text{pyrophosphate} \quad (3)$$

$$\text{CMP-AEPn} + \text{diacylglycerol} \rightarrow \text{GPnL} + \text{CMP} \quad (4)$$

However, since no evidence was presented that free AEPn can be a precursor of phosphonolipid, it was concluded that the above classical pathway did not play a dominant role in GPnL biosynthesis in *Tetrahymena*, serving as a salvage mechanism for free AEPn derived from the breakdown of lipids and other components

d19:1	d20:1	d20:2	d22:0	t22:1	Others [b]	Ref.
5.2 (*iso*)					0.2	36
10.6	4.9	29.8				69
2.7			25.3		0.4	69
3.8			8.2	16.4		69
1.3			23.5		38.7	46
3.1 (br)						
19.1			5.2		2.4	69
1.9			9.6		3.4	48
6.5 (*iso*)						
18.3 (*anteiso*)						
18.4		2.6			1.3	69
10.9						69
21.9		47.1				69
					0.1	69
12.9		7.6			1.0	69
13.1		8.5			0.6	69
6.4		8.1			0.3	69
5.3						69
2.1 (*iso*)					4.6	81
21.6 (*anteiso*)						
						27

[a] Sphingophosphonoglycolipid.
[b] Includes unidentified sphingosine bases.

containing AEPn [42]. Recently, Smith and O'Malley [91] have demonstrated that AEPn supplementation in the growth medium caused a marked increase in GPnL content in *Tetrahymena* cells (Fig. 2). This elevation of GPnL level was accompanied by a decrease in the phosphatidylethanolamine as well as by small decreases in the minor phospholipid components, lysophosphatidylethanolamine and lysophosphonolipid. Although it has not been determined whether reactions 2 and 4 are catalysed by the same or different enzymes, they suggested that the increase in GPnL in the AEPn-supplemented cells reflected a competition between substrates for either the same enzymes or for the diacylglycerol. An alternate pathway of phosphonolipid biosynthesis, called the 3-phosphonoalanine decarboxylase cycle, has been proposed, and it does not involve the free AEPn step of the CMP-AEPn pathway (reaction 3). The establishment of a unified mechanism for phosphonolipid formation awaits future extensive studies.

On the other hand, it was shown in mammalian tissues that the phosphonolipids are made mainly by the "Kennedy pathway" involving a CMP derivative [92–94]. Bjerve [92] has compared the incorporation into rat phospholipids of phosphocholine and its phosphonate analogue, *N,N,N*-trimethyl-2-aminoethylphosphonic acid (TM-AEPn), and demonstrated that the phosphonate analogue was incorporated into phosphonolecithin by the same mechanism as phosphocholine into lecithin. The CMP derivative was found to be intermediate in the pathway and actually a partially purified cytidylyltransferase catalysed an active incorporation of TM-AEPn into the CMP derivative. Phosphocholine and CDP-choline were strong competitive inhibitors of the incorporation, indicating that CMP-TM-AEPn is a probable intermediate in the phosphonolecithin synthesis.

The time-course studies of the incorporation of radioactive AEPn into the tissue lipids of rats have been performed by Curley-Joseph and Henderson [94], revealing that maximum incorporation into liver lipids is seen within 12 to 30 h of [^{3}H]AEPn injection, as compared to a few hours for the phosphoethanolamine incorporation. The highest radioactivity was observed in diacyl-GPnL, but methylation of this lipid to phosphonolecithin did not occur. The much faster incorporation of phosphorylethanolamine than AEPn into liver lipids suggested that the enzymes involved in phospholipid metabolism have higher rates of reaction than the corresponding

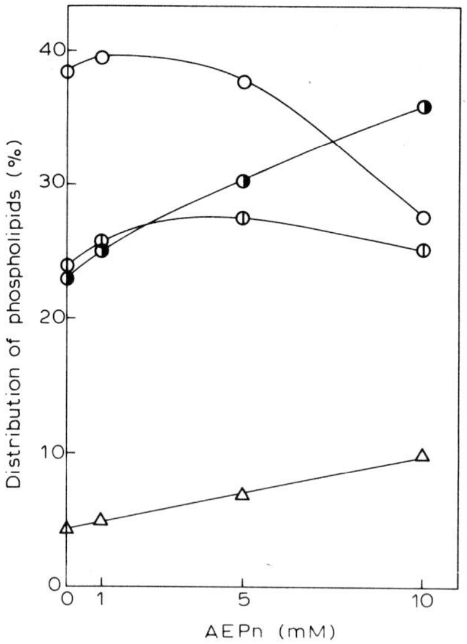

Fig. 2. Effects of addition of 2-aminoethylphosphonic acid (AEPn) upon phospholipid composition in *T. pyriformis*. ◑, glycerophosphonolipid; ○, phosphatidylethanolamine; ⦶, phosphatidylcholine; △, cardiolipin. Taken from [91].

enzymes for GPnL synthesis. Another possibility seemed more likely: that the same enzyme acts in both phospholipid and phosphonolipid biosynthesis, but has a greater affinity for phosphoethanolamine than AEPn.

(iii) Sphingophosphonolipids (SPnL)

Little information is available regarding the biosynthesis of SPnL. Itasaka et al. [95] showed that the mussel, *Hyriopsis schlegelii* can incorporate ^{32}P into CAEPn. As seen in Table 6, the highest incorporation occurred in lecithin regardless of incubation period, and the radioactivity was found to a considerable extent in CAEPn but not in ceramide phosphoethanolamine. The results of ^{32}P incorporation into CAEPn of various organs of the mussel indicated that the specific radioactivity was maximum in the liver lipid, being approx. 2–4 times greater than that in other organs, adductor, mantle and gill. The formation of CAEPn was confirmed also by an in vitro system in which a liver homogenate was incubated in the presence of $^{32}P_i$. In rat liver, a small but significant incorporation of [^{32}P]AEPn into CAEPn [93] was observed.

TABLE 6

Distribution of radioactivity in individual phospholipids from the mussel

Phospholipids	Total radioactivity			
	24 h after injection of $^{32}P_i$		48 h after injection of $^{32}P_i$	
	Counts/min	%	Counts/min	%
CAEPn	3709	7.1	3495	5.0
Phosphatidylethanolamine	13228	25.5	12989	18.9
Phosphatidylcholine	25524	48.9	22813	33.2
Phosphatidylserine + phosphatidylinositol	2391	4.6	3515	5.1

Taken from [95].

(b) Degradation

It has generally been known that because of the C—P bond the phosphonolipids are extremely resistant to chemical and enzymatic hydrolysis. At the present time there is no evidence to indicate the presence of enzyme(s) degrading the C—P bond itself in organisms which contain the phosphonolipids, such as *Tetrahymena*, rat, sea anemone, housefly and mollusc. However, an enzyme capable of cleaving the C—P bond was isolated by La Nauze [96] from *Bacillus cereus*, and was trivially called "phosphonatase". This enzyme (2-phosphonoacetaldehyde hydrolase EC 3.11.1.1) catalyses the following reaction (II), and its

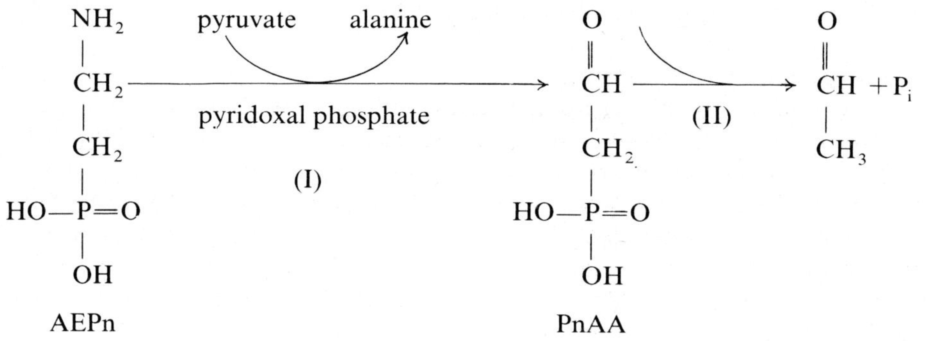

active form with an M_r-value of 75000 is composed of two similar subunits. Mg^{2+} is required at high concentration to maintain the enzyme in its dimeric active state rather than to affect directly the active site, and could be replaced by Mn^{2+} but not by Zn^{2+} or Ca^{2+}.

The degradation of the lipid moiety in the phosphonolipid has been demonstrated in several cell lines. *Tetrahymena* can break down GPnL to release the AEPn base, but at a slower rate than the phospholipid degradation [56]. CAEPn was broken down by phospholipase C from *Clostridium perfringens* [97,98], yielding ceramide and AEPn.

5. *Phosphonolipids and membranes of* Tetrahymena

(a) Intracellular distribution

The *Tetrahymena* cell has most of the typical subcellular organelles found in higher eukaryotic cells and also contains some other specific membranes, cilia, oral apparatus, and mucocysts. Nozawa and Thompson [76] have established a procedure for isolating various organelles including ciliary and pellicular membranes. The lipid composition of these isolated membrane fractions is shown in Table 7, which indicates variations among the different membranes. The highest proportional content of AEPnL is found in cilia, followed by pellicles, while the internal membranes, mitochondria, microsomes and nuclear envelopes contain lower amounts, being rich in phosphatidylethanolamine (PE) and phosphatidylcholine (PC). Tetrahymanol, a principal component of neutral lipids is also present abundantly in the surface membrane fractions, cilia and pellicles, and may interact preferentially with phosphonolipids. Recently, Andrews and Nelson [99] have analysed the phospholipid composition of cilia and deciliated bodies of *Paramecium tetraurelia*, and demonstrated that the cilia contained ether-GPnL (24%) and CAEPn (3%). Furthermore, the phospholipid constituents of mutants of three distinct phenotypes (pawn,

TABLE 7

Phospholipid composition of various membrane fractions from *T. pyriformis* WH-14 cells

Membrane fractions	Phosphonate (% of total lipid phosphorus)	Mol % of total phospholipids						Tetra-hymanol (mol/mol lipid phosphorus)	Glyceryl ether (mol/100 mol lipid phosphorus)
		GPnL	PE	PC	Lyso PC	Lyso GPnL and Lyso PE	CL		
Whole cells	29	23	37	33	2	0	5	0.057	29.7
Cilia	67	47	11	28	1	9	1	0.30	52.6
Ciliary supernatant	44	35	16	19	8	13	1	0.16	23.1
Pellicles	42	30	34	25	5	3	2	0.084	32.8
Mitochondria	26	18	35	35	2	0	10	0.048	24.7
Nuclear envelopes	–	23	26	31	6	6	3	0.036	–
Microsomes	33	23	34	35	1	3	1	0.041	18.3
Post-microsomal supernatant	26	22	30	34	5	4	2	0.016	27.4

Taken from [76] and [100]. For abbreviations see Table 1.

TABLE 8

Fatty acid composition of major phospholipids from whole cells of *Tetrahymena* and *Paramecium*

Fatty acids	*Tetrahymena pyriformis* NT-I [a]			*Paramecium tetraurelia* [b]		
	GPnL	PE	PC	AEPnL	PE	PC
14:0	4.0	13.2	10.0	tr [c]	tr	tr
16:0	7.8	12.1	11.2	9.8	26.0	20.6
16:1	7.8	17.5	14.1	tr	2.0	tr
18:0	3.2	2.6	3.3	3.9	1.7	1.1
18:1	6.3	4.9	6.2	4.6	14.2	8.0
18:2 [d]	6.8	1.3	3.2	—	—	—
18:2	12.5	12.1	12.0	3.8	23.8	28.5
18:3	41.4	14.4	24.1	2.8	26.7	23.8
20:4	–	–	–	61.7	3.1	12.8
20:5	–	–	–	12.7	0.4	2.7
Unsaturated fatty acids	74.8	50.2	59.6	81.0	54.6	69.1

Values are expressed as mol % of total fatty acids.

[a] Taken from Nozawa, Kasai and Sekiya [unpublished data].

[b] Taken from [35].

[c] Trace.

[d] Cilienic acid, $18:2\Delta^{6,11}$.

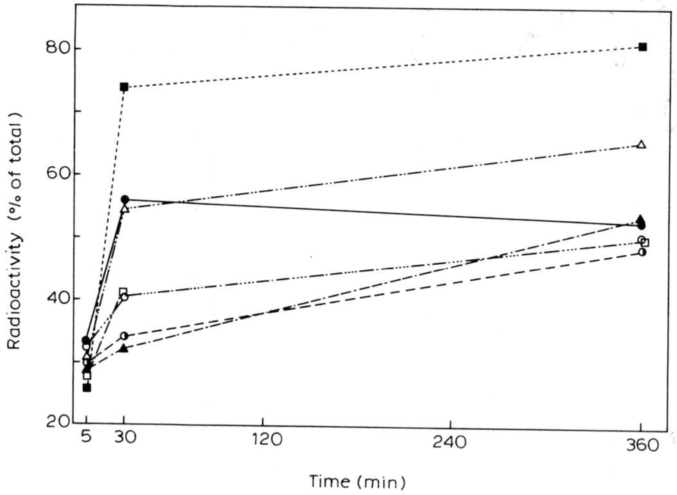

Fig. 3. Incorporation of [8,9-^{3}H]hexadecyl glycerol into glycerophosphonolipids (GPnL) of various membranes in *T. pyriformis* WH-14. ■, cilia; △, pellicles; ●, ciliary supernatant; □, microsomes; ○, mitochondria; ◑, postmicrosomal supernatant; ▲, whole cells. Taken from [100].

paranoic and fast) were also examined, and the fast mutant was observed to contain diacyl-GPnL.

It should be noted that the phosphonolipids have much higher amounts of polyunsaturated fatty acids, compared to the other two major phospholipids, PE and PC. This trend, first found for *Tetrahymena* is also seen in *Paramecium* [35], summarised in Table 8. The *Tetrahymena* GPnL is rich in γ-linolenic ($18:3\Delta^{6,9,12}$), linoleic ($18:2\Delta^{9,12}$) and palmitoleic acids ($16:1\Delta^{9}$), whereas the fatty acid composition of the *Paramecium* GPnL is dominated by arachidonic acid ($20:4\Delta^{5,8,11,14}$) and eicosapentaenoic acid ($20:5\Delta^{5,8,11,14,17}$).

(b) Mechanism for enrichment of GPnL in the surface membranes

Although, as discussed in section 4 a-ii, the biosynthetic pathway of phosphonolipids remains to be clarified, one would expect that enzyme(s) for the biosynthesis of the specific lipids are localized in the endoplasmic reticulum, together with other lipid-synthesizing enzymes. Therefore, the preferential accumulation of GPnL in the surface membranes is presumed to occur by some mechanism after the biosynthetic step. Two possible mechanisms might operate; (1) specificity in a system transporting newly made lipids to their destinations (carrier-protein theory), and (2) specificity at the acceptor membranes in selecting certain lipids to accumulate (reshuffling theory). To test these possible mechanisms, the rates of incorporation of [^{3}H]chimyl alcohol into GPnL of various membrane fractions were examined [100]. Fig. 3 depicts a typical labelling profile. At 5 min of labelling time, the radioactive precursor of GPnL appears to be incorporated to a similar extent into all cell fractions (cilia, ciliary supernatant, pellicles, mitochondria, microsomes, and post-microsomal supernatant). This suggests that a selective transport system is not involved in the mobilization of synthesized lipids. However, 30 min after the administration of the radioisotope great differences in the incorporation rate occur among various membranes. For example, the ciliary membrane shows the highest level of radioactivity, followed by pellicles and ciliary supernatant, whereas internal membrane fractions such as microsomes and mitochondria are much less labelled. These observations indicate that an assortment of lipids made in endoplasmic reticulum may arrive at other cell compartments without selection during transport and then it may be subject to a reshuffling at the acceptor membrane. The phospholipases are considered to participate in the reshuffling. It seems therefore reasonable that phosphonolipids which are highly resistant to the endogenous phospholipases are allowed to reside but certain phospholipids are rejected. There is some indication that GPnL appears to be more stable to clostridial phospholipase C than other phospholipids in *Tetrahymena* (Watanabe and Nozawa, unpublished data).

(c) Roles in membrane lipid adaptation

There are several lines of evidence that a variety of microorganisms adjust their membrane lipid composition in response to changes of environmental conditions,

TABLE 9

Lipid composition of various membranes from *T. pyriformis* NT-I cells grown at different temperatures

Phospholipids	Cilia			Pellicles			Microsomes		
	15°C	24°C	39.5°C	15°C	24°C	39.5°C	15°C	24°C	39.5°C
Glycerophosphonolipid	40.0	37.0	36.4	30.8	32.9	19.1	22.9	20.8	14.7
Phosphatidylethanolamine	20.8	21.4	20.0	33.2	36.2	49.2	34.1	35.5	43.9
Phosphatidylcholine	15.8	18.1	30.0	18.9	21.4	23.5	30.0	27.6	32.4

Values are expressed as mol % of total fatty acids. Taken from [103].

temperature, nutrition, metabolic inhibitors, etc. It is furthermore accepted that this adaptive modification of membrane lipids is required to maintain the physical state of membranes within the optimum range. *Tetrahymena* is one of the most typical cells which undergo marked lipid modification depending upon altered growth conditions [101].

(i) Temperature

It is well known that poikilothermic organisms modify the fatty acid composition of their membrane phospholipids in response to changes in growth temperature [102]. The general trend is an increase in the proportion of unsaturated fatty acids at low growth temperatures. On the other hand, the phospholipid composition has been observed usually to remain rather constant. However, *Tetrahymena* cells were found to undergo marked alteration in phospholipid polar head composition in accordance with temperature changes [103,104]. As indicated in Table 9, there was an increase in the GPnL content and a corresponding decrease of PE in whole cells as the growth temperature decreased. The relative concentration of PC did not change. These alterations were also reflected to various extents in three major membrane fractions. Both the pellicular and microsomal membranes followed the trend observed in the whole cell lipids, whereas the ciliary membrane containing the largest amount of GPnL differed in having a profound change in PC content. The increase in GPnL at low temperature was small and the PE level was unchanged. These findings imply that GPnL which is richest in polyunsaturated fatty acids may play some important role in thermal adaptations of *Tetrahymena* membranes. Since, as shown in Fig. 1, two types of GPnL—ether-GPnL and diacyl-GPnL—are present in this cell, the relative content of the two lipids was compared between the cells grown at 15°C and 39.5°C [105]. In the 15°C cells, there was a small but significant increase (24.6 → 29.0%) in diacyl-GPnL whereas ether-GPnL was decreased (75.4 → 71.0%). The analysis of fatty acid composition demonstrated a striking difference in overall profile between the two types of cells (Table 10). The diacyl-GPnL from 15°C cells was much richer in linoleic and γ-linolenic acids than that from 39.5°C cells. The positional distribution in diacyl-GPnL showed that regardless of growth temperature polyunsaturated fatty acids were preferentially located at the 2-position whereas saturated fatty acids favoured the 1-position. The increase in polyunsaturated fatty acids in the cold cells was principally due to the pronounced enhancement of linoleic acid at the 2-position. The ether-GPnL has its sole acyl chain at the 2-position, mostly γ-linolenic acid. It should be noted that this phospholipid from the cold cells contains a very large amount of an unusual fatty acid, cilienic acid ($18{:}2\Delta^{6,11}$) [34]. Since little or no cilienic acid was present at the 2-position of diacyl-GPnL, the ether-GPnL appears to be a favourite acceptor for the cilienic acid. The findings that in the cold cells the proportion of GPnL became greater and that its fatty acid profile became more unsaturated, may suggest that GPnL serves as a principal regulatory lipid for cold acclimatization in *Tetrahymena*.

It has been shown in our earlier studies that an increase in the content of unsaturated fatty acids leads to a decrease in the microviscosity of the membrane

TABLE 10

Fatty acid composition of 1,2-diacyl- and 1-*O*-alkyl-2-acyl-glycero(3)-2-aminoethylphosphonate in *T. pyriformis* NT-1 cells grown at 39.5°C or 15°C

Fatty acids	1,2-Diacyl-glycerophosphonolipid						1-*O*-alkyl-2-acyl-glycerophosphonolipid Overall composition (2-position)	
	Overall composition		1-position		2-position			
	39.5°C	15°C	39.5°C	15°C	39.5°C	15°C	39.5°C	15°C
14:0	15.2	10.4	26.0	20.1	4.5	1.1	0.4	0.3
16:0	19.3	10.9	39.1	23.4	0.5	–	1.3	0.9
16:1	19.8	12.8	7.8	14.6	31.8	11.0	5.9	1.0
16:2 [a]	8.0	4.5	3.2	3.7	12.8	5.2	3.6	0.7
18:1	8.3	9.3	0.8	3.3	15.8	15.4	4.1	3.3
18:2 [b]	tr	3.8	0.7	3.8	–	3.9	10.9	39.5
18:2	9.1	24.5	0.8	5.9	17.4	43.1	9.2	5.5
18:3	8.0	18.4	3.7	18.6	12.3	18.3	58.0	47.6
Unsaturation index [c]	71	135	23	93	121	177	239	237
ΣU/ΣS	0.9	2.5	0.2	0.9	3.9	13.5	16.8	50.0

Values are expressed as mol % of total fatty acids.

[a] Contains also small amounts of 17:0.

[b] Cilienic acid, $18:2\Delta^{6,11}$.

[c] Denotes the number of double bonds per fatty acyl chain and calculated from [(number of double bonds per fatty acid)×(percentage of the fatty acid)].

Taken from [105].

lipid bilayer [106]. However, there has been no information about involvement of the C—P bond of GPnL in the membrane's physical state. Recently we have examined the membrane fluidity of dipalmitoyl GPnL using electron spin resonance. Liposomes of this lipid and dipalmitoyl phosphatidylethanolamine (DPPE) were labelled with a spin probe, TEMPO (2,2,6,6-tetramethylpiperidine-1-oxyl), and a spectral parameter, f was measured as a function of temperature. The parameter, f is calculated from the equation, $f = H/H + P$ [107], where H is the proportion of spin label dissolved in the membrane bilayer and P is the proportion of the label dissolved in the aqueous region. In Fig. 4, Arrhenius plots of TEMPO spectral parameter (f) of dipalmitoyl GPnL, which is approximately the fraction of spin label dissolved in the membrane bilayer, exhibit an abrupt decrease in magnitude in the range 58.3°C to 64.4°C [Kameyama, Ohki and Nozawa, unpublished data]. This profile of TEMPO titration for DP-GPnL was found to be identical with that for DPPE, which has a transition temperature of 61.4°C, defined as the midpoint of the transition curve. This experimental fact that no difference in the spectral parameter curve was present between DP-GPnL and DPPE, implies that the C—P bond in GPnL may not exert any significant effect upon the physical behaviour of the membrane lipid bilayer. Therefore it is likely that GPnL would contribute to the cold adaptation of membrane lipids as an efficient acceptor for polyunsaturated fatty acids (cilienic and γ-linolenic acid) and not because of specific polar head structure.

An increased concentration of GPnL was also demonstrated during cold acclimatization [104,108]. As shown in Table 11, there was little significant modification of phospholipid class composition up to 10 h after the temperature shift, where no cell division occurred. However, a marked change was seen in the proportions of individual phospholipids thereafter, and GPnL increased markedly at the expense of PE, whereas the level of PC was fairly constant. Furthermore, the fatty acyl chain composition was altered in the increased GPnL (Table 12). The relative content of palmitic acid diminished gradually until 10 h after the shift, and then increased up to 24 h. In contrast, γ-linolenic and cilienic acids showed a marked elevation for the first 10 h and then decreased. After 10 h incubation at 15°C, 16:0 content was lowest whereas $18{:}2\Delta^{6,11}$ and $18{:}3\Delta^{6,9,12}$ content was highest in the three major phospholipids including PE and PC. Such modified fatty acid composition of GPnL

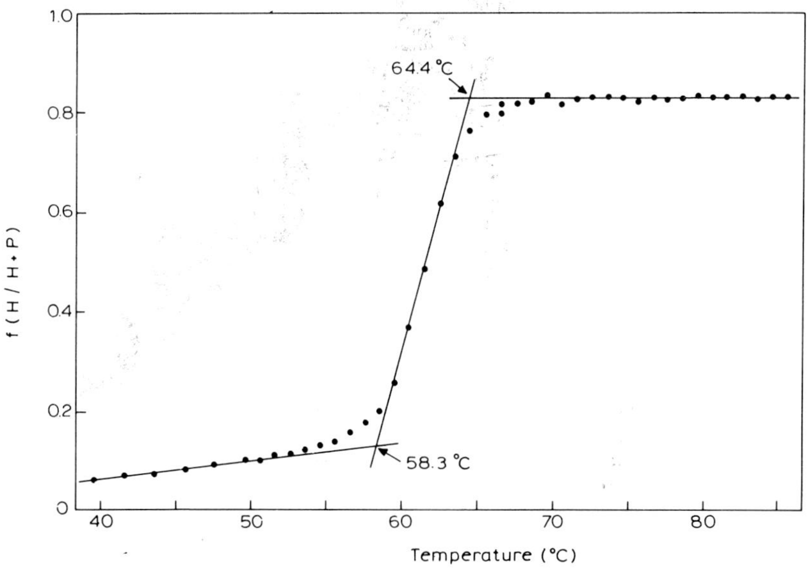

Fig. 4. Tempo spectral parameter (*f*) vs. temperature for dipalmitoyl glycerophosphonolipid. *H*, proportion of the label dissolved in the lipid bilayer; *P*, proportion of the label dissolved in the aqueous region; TEMPO, 2,2,6,6-tetramethylpiperidine-1-oxyl. Taken from Kameyama, Ohki and Nozawa [unpublished data].

was well reflected in the fluidity of its liposomes as measured by ESR. Table 13 demonstrates time-dependent changes after adaptation to 15°C in the order parameter, *S*, determined with a 5-ketostearate spin probe [109]. There was no change in *S* within the first 4 h, and the maximum decrease was observed at 10 h after adaptation when the GPnL liposome was in the most fluid state. These results suggest that the content of both γ-linolenic and cilienic acids plays an important role in regulating thermal adaptation process.

TABLE 11

Alteration in phospholipid composition due to temperature shift in *T. pyriformis* NT-1

Phospholipids	Isothermal 39.5°C	Time after shift-down to 15°C				Isothermal 15°C
		4 h	10 h	24 h	48 h	
Glycerophosphonolipid	17.4	14.9	16.5	26.3	30.2	29.0
Phosphatidylethanolamine	44.9	46.5	46.7	36.0	32.4	25.6
Phosphatidylcholine	30.7	32.5	28.7	29.2	28.0	27.2

Values are expressed as % of total phospholipids. Taken from [104].

TABLE 12

Alteration in fatty acid composition of glycerophosphonolipid from *T. pyriformis* NT-1 cells during adaptation to 15°C

Fatty acids	Time after temperature shift from 39.5°C to 15°C				
	0 h	2 h	4 h	10 h	24 h
14:0	6.8	3.3	3.1	1.5	4.7
16:0	11.2	10.8	8.5	3.3	6.3
16:1	12.7	9.2	12.0	7.3	11.8
16:2 [a]	5.6	4.4	5.6	5.4	6.1
18:0	5.4	5.7	4.5	3.5	4.4
18:1	5.2	11.0	6.5	4.9	6.1
18:2 [b]	5.3	6.7	8.8	12.7	10.5
18:2	9.7	9.5	9.5	8.2	9.6
18:3	30.7	35.7	38.5	50.7	36.0
Unsaturated fatty acids	63.6	72.1	74.1	85.2	74.0

Values are expressed as mol % of total fatty acids.

[a] Contains also small amounts of 17:0.

[b] Cilienic acid, $18:2\Delta^{6,11}$. Taken from [109].

(ii) Nutrition

Glyceryl ether supplementation. When *Tetrahymena* cells were grown in medium supplemented with 1-*O*-hexadecyl glycerol (HDG, chimyl alcohol), a GPnL precursor, membrane phospholipids became richer in GPnL with an accompanying reduction in PE (Table 14). This higher content of GPnL would be expected to result from the increase in ether-GPnL rather than diacyl-GPnL, because there was an increased concentration of glyceryl ether in membrane phospholipids from HDG-supplemented cells [87]. ESR studies indicated that the pellicular and microsomal membranes from HDG-cells had greater fluidities than the membranes from the control, unsupplemented cells [110,111].

Starvation. Tetrahymena cells can survive for a rather long time after transfer to a

TABLE 13

Changes in order parameter of glycerophosphonolipid from *T. pyriformis* NT-1 during adaptation to 15°C

Measured at (°C)	Time after temperature shift from 39.5°C to 15°C			
	2	4	10	24
15	0.676	0.680	0.656	0.670
24	0.621	0.620	0.603	0.610
39.5	0.546	0.540	0.532	0.539

Taken from [109].

TABLE 14

Alteration induced by hexadecyl glycerol-feeding in glyceryl ether content and phospholipid composition in pellicles and microsomes from *T. pyriformis* NT-1

Membrane fractions	Glyceryl ether (mol/100 mol lipid phosphorus)	Glycero-phosphono-lipid	Phosphatidyl-ethanolamine	Phosphatidyl-choline
Pellicles				
Native	32.7	21.9	46.8	25.0
Hexadecyl glycerol-fed	40.7	30.5	37.1	25.5
Microsomes				
Native	33.1	20.9	40.9	33.2
Hexadecyl glycerol-fed	39.3	22.6	36.1	35.5

Values for individual phospholipids are expressed as % of total phospholipids. Taken from [87].

non-nutrient inorganic medium [112]. Shortly after the nutritional shift-down, cells gradually shrank to approximately half their original size. The total number of cilia covering the whole body was unchanged even in 24-h-starved cells. The cell motility

TABLE 15

Modification during starvation of acyl chain composition of GPnL in whole cells of *T. pyriformis* NT-1

Fatty acids	Control	Starved for			
		4 h	6 h	12 h	24 h
14:0	4.0	4.3	3.3	2.4	4.2
16:0	7.8	6.3	6.2	5.3	6.6
16:1	7.8	6.9	5.8	3.5	2.6
16:2 [a]	3.4	3.1	2.8	2.0	1.3
18:1	6.3	9.6	9.0	8.2	6.7
18:2 [b]	6.8	7.9	8.8	10.4	10.9
18:2	12.5	13.6	13.4	11.3	10.0
18:3	41.4	42.5	45.4	51.6	49.2
Unsaturated fatty acids	74.8	80.5	82.4	85.0	79.4
Unsaturation index [c]	176.9	187.0	195.4	209.9	198.7

Values are expressed as mol % of total fatty acids.

[a] Contains also small amounts of 17:0.

[b] Cilienic acid, $18:2\Delta^{6,11}$.

[c] See Table 10.

also was not different from that of control growing cells. The ultrastructure of intracellular membrane systems was strikingly altered in the starved cell. The most obvious change was the marked increase in degenerating mitochondria which were

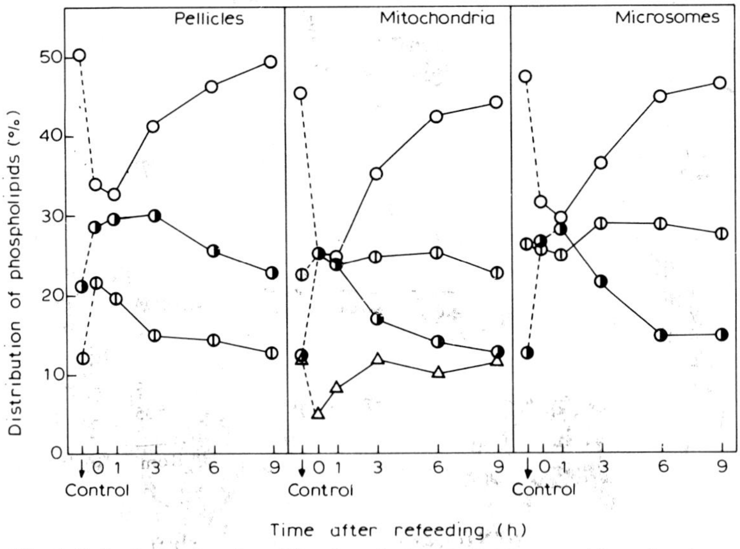

Fig. 5. Refeeding-induced modification of phospholipid composition of various membrane fractions of *T. pyriformis* NT-1 cells. The 24-h starved cells were transferred into proteose-peptone medium enriched with glucose and yeast extract. ◑, glycerophosphonolipid; ○, phosphatidylethanolamine; ⦶, phosphatidylcholine, △, cardiolipin. Taken from Nozawa, Kasai and Sekiya [unpublished data].

rather round, electron-dense, and often sequestered into autophagic vacuoles for digestion. At the later stages the numbers of mitochondria and endoplasmic reticulum membranes were greatly decreased. However, these changes in size and ultrastructure of the starved cells could be reversed by transferring them back to the nutrient-rich medium.

A reversible modification in phospholipid composition accompanied the nutritional changes [113]. During the 24-h starvation period there was a marked increase in GPnL content with a corresponding decrease in PE. The PC level was unchanged. But after the 24-h-starved cells were transferred back to the proteose-peptone medium enriched with glucose and yeast extract, reversal of the lipid changes began. The processes of amelioration of various membranes are shown in Fig. 5. At 9 h after the shift-up, the pellicular, mitochondrial and microsomal membranes were all restored in terms of their phospholipid composition to the normal profile. The fatty acyl chain composition in GPnL was also modified in a reversible manner by starvation and refeeding. Table 15 represents the fatty acid composition of GPnL at different intervals during starvation, demonstrating an increase in γ-linolenic acid accompanied with a decrease in palmitoleic acid. This trend was more marked in PC and PE. The refeeding-induced recovery of the fatty acid profile of GPnL from the pellicular membranes is summarized in Table 16, which indicates the occurrence of great reduction in γ-linolenic acid and marked elevation in palmitoleic acid. This variation occurred in both mitochondrial and microsomal membranes.

Phosphonolipid changes in response to starvation in the American oyster [77] have already been mentioned (p. 102 and Table 3).

TABLE 16

Modification following nutritional shift-up of acyl chain composition of GPnL in pellicular membrane of *T. pyriformis* NT-1

Fatty acids	Time after transfer to nutrient-rich medium					Control
	0 h	1 h	3 h	6 h	9 h	
14:0	0.6	6.7	7.9	6.4	6.1	5.2
16:0	3.8	9.7	9.5	8.0	7.3	8.2
16:1	1.2	3.6	5.9	10.0	1.4	9.9
16:2 [a]	0.9	1.9	2.5	3.8	4.7	4.7
18:1	4.1	6.7	7.1	7.1	8.2	8.2
18:2 [b]	11.6	8.9	8.5	10.0	9.5	6.8
18:2	11.4	13.0	10.3	10.3	12.1	13.6
18:3	63.8	42.1	41.4	37.5	35.1	35.8
Unsaturated fatty acids	92.1	74.3	73.2	74.9	76.3	74.3
Unsaturation [c] index	242.7	180.4	174.8	170.2	168.1	166.3

Values are expressed as mole % of total fatty acids.

[a] Contains also small amounts of 17:0.

[b] Cilienic acid, 18:2$\Delta^{6,11}$.

[c] See Table 10. Taken from Nozawa, Kasai and Sekiya [unpublished data].

(iii) Alcohols

Addition of phenethyl alcohol (2-phenyl ethanol, $C_6H_5CH_2CH_2OH$) caused a drastic perturbation in the phospholipid composition in *Tetrahymena* membranes [114]. Compared with membranes (pellicles, mitochondria, microsomes) of the control cells, the membranes from phenethyl alcohol-treated cells were found to contain a higher level of PC content with compensating decrease in PE, while GPnL showed a small decrease in these membranes. Nandini-Kishore et al. [115] have demonstrated that the cells grown in the present of ethanol have a lower proportion of GPnL with corresponding decrease in PE.

(iv) Aging

The membrane lipid composition of *Tetrahymena* was observed to change in a manner markedly dependent on the age of the culture [116]. Although the phospholipid composition varied somewhat among membrane fractions (pellicles, mitochondria, microsomes), the general trend with age was a profound decrease in the content of PE and a slight increase in PC. The GPnL level showed an increase in microsomes and pellicles from the late stationary phase cells, but was unchanged in mitochondria. As for fatty acid composition, the most notable variation occurred in unsaturated fatty acids, i.e. a great increase in oleic and linoleic acids and a compensatory decrease in palmitoleic acid.

In contrast to small changes in GPnL content in *Tetrahymena* associated with culture age, it should be noted that another ciliated protozoon, *Paramecium*, underwent a great increase in relative percentage of GPnL [35]. A compensating decrease occurred in PE. The level of PC, PS and CAEPn was kept constant. These changes in the ciliary membrane of *P. tetraurelia* are demonstrated in Table 17. There was no significant modification of fatty acyl chains with age of culture in *Paramecium* GPnL. Taken together with the enrichment in the unusual fatty acid (cilienic acid), and γ-linolenic acid in *Tetrahymena*, it is of great interest to notice that arachidonic acid is the major fatty acid of this phospholipid. This finding

TABLE 17

Age-dependent alteration in membrane phospholipid composition of *Paramecium tetraurelia*

Phospholipids	Age of culture		
	3 days	5 days	7 days
Glycerophosphonolipid	32.1	38.6	42.3
Phosphatidylethanolamine	23.8	11.0	8.8
Sphingophosphonolipid	13.9	14.5	15.5
Phosphatidylcholine	13.4	13.9	11.9
Phosphatidylserine + phosphatidylinositol	8.7	9.2	8.1
Sphingolipid	5.3	8.1	8.6

Values are expressed as mol % of total fatty acids. Taken from [35].

supports the hypothesis proposed for *Tetrahymena* that GPnL may serve as a potential acceptor for polyunsaturated acyl chains.

6. Other possible physiological functions

There has been no clear-cut evidence to explain the precise function of phosphonolipids. Since this specific phospholipid is highly resistant to chemical as well as enzymatic attack, and is localized predominantly in the surface membranes, it might serve as a protecting mediator in naked, free-living cells such as *Tetrahymena* and *Paramecium*, which are devoid of the protective coating seen in bacterial and fungal cells. Indeed, Rosenthal et al. [117,118] have demonstrated that various phosphonolipid analogues were not degraded by phospholipase A from the water moccasin (*A. piscivorus*) and phospholipase C from *C. perfringens*, but that they acted as potential inhibitors of these phospholipases. However, this hypothesis cannot be directly applied to other organisms [119], since, for example, phosphonolipids exist in mycobacteria which have a strong wall outside the cytoplasmic membrane.

References

1 Horiguchi, M. and Kandatsu, M. (1959) Nature 184, 901–902.
2 Kandatsu, M. and Horiguchi, M. (1960) Agr. Biol. Chem. 24, 565–570.
3 Kandatsu, M. and Horiguchi, M. (1962) Agr. Biol. Chem. 26, 721–722.
4 Kittredge, J.S., Roberts, E. and Simonsen, D.G. (1962) Biochemistry 1, 624–628.
5 Quin, L.D. (1965) Biochemistry 4, 324–329.
6 Kittredge, J.S., Horiguchi, M. and Williams, P.M. (1969) Comp. Biochem. Physiol. 29, 859–863.
7 Shimizu, H., Kakimoto, Y., Nakajima, T., Kanazawa, A. and Sano, I. (1965) Nature 207, 1197–1198.
8 Tamari, M. (1971) Agr. Biol. Chem. 35, 1799–1802.
9 La Nauze, J.M. and Rosenberg, H. (1968) Biochim. Biophys. Acta 165, 438–447.
10 Alhadeff, J.A. and Daves, G.D. (1971) Biochim. Biophys. Acta 244, 211–213.
11 Kittredge, J.S. and Hughes, R.R. (1964) Biochemistry 3, 991–996.
12 Kittredge, J.S., Isbell, A.F. and Hughes, R.R. (1967) Biochemistry 6, 289–295.
13 Baer, E. and Stanacev, N.Z. (1964) J. Biol. Chem. 239, 3209–3214.
14 Rouser, G., Kritchevsky, G., Heller, D. and Lieber, E. (1963) J. Am. Oil Chem. Soc. 40, 425–454.
15 Hori, T., Itasaka, O., Inoue, H. and Yamaka, K. (1964) J. Biochem. 56, 477–479.
16 Hori, T., Itasaka, O. and Inoue, H. (1966) J. Biochem. 59, 570–573.
17 Hori, T., Arakawa, I. and Sugita, M. (1967) J. Biochem. 62, 67–70.
18 Higashi, S. and Hori, T. (1968) Biochim. Biophys. Acta 152, 568–575.
19 Hori, T., Sugita, M., Kanbayashi, J. and Itasaka, O. (1975) Yukagaku 24, 181–184.
20 Hayashi, A., Matsubara, T. and Mishima, Y. (1967) J. Fac. Sci. Tech. Kinki Univ. 2, 39–51.
21 Tamari, M. and Kametaka, M. (1972) Agr. Biol. Chem. 36, 1147–1152.
22 De Koning, A.J. (1966) J. Sci. Fd. Agric. 17, 460–464.
23 Liang, C.R. and Strickland, K.P. (1969) Can. J. Biochem. 47, 85–89.
24 Komai, Y., Matsukawa, S. and Satake, M. (1973) Biochim. Biophys. Acta 316, 271–281.
25 Karlsson, K.-A. (1970) Chem. Phys. Lipids 5, 6–43.
26 Karlsson, K.-A. (1970) Biochem. Biophys. Res. Commun. 39, 847–851.
27 Karlsson, K.-A. and Samuelsson, B.E. (1974) Biochim. Biophys. Acta 337, 204–213.
28 Tamari, M., Ogawa, M., Hasegawa, S. and Kametaka, M. (1976) Agr. Biol. Chem. 40, 2057–2062.

29 Dawson, R.M.C. and Kemp, P. (1967) Biochem. J. 105, 837–842.
30 Carter, H.E. and Gaver, R.C. (1967) Biochem. Biophys. Res. Commun. 26, 886–891.
31 Berger, H., Jones, P. and Hanahan, D.J. (1972) Biochim. Biophys. Acta 260, 617–629.
32 Ferguson, K.A., Conner, R.L., Mallory, F.B. and Mallory, C.W. (1972) Biochim. Biophys. Acta 270, 111–116.
33 Viswanathan, C.V. and Rosenberg, H. (1973) J. Lipid Res. 14, 327–330.
34 Ferguson, K.A., Davis, F.G., Conner, R.L., Landrey, J.R. and Mallory, F.B. (1975) J. Biol. Chem. 250, 6998–7005.
35 Rhoads, D.E. and Kaneshiro, E.S. (1979) J. Protozool. 26, 329–338.
36 Sugita, M., Fukunaga, Y., Ohkawa, K., Nozawa, Y. and Hori, T. (1979) J. Biochem. 86, 281–288.
37 Steiner, S., Conti, S.F. and Lester, R.L. (1973) J. Bacteriol. 116, 1199–1211.
38 Wassef, M.K. and Hendrix, J.W. (1977) Biochim. Biophys. Acta 486, 172–178.
39 Hayashi, A., Matsuura, F. and Matsubara, T. (1969) Biochim. Biophys. Acta 176, 208–210.
40 Hayashi, A., Matsubara, T. and Matsuura, F. (1969) Yukagaku 18, 118–123.
41 Hori,T. Sugita, M. and Itasaka, O. (1969) J. Biochem. 65, 451–457.
42 Liang, C.R. and Rosenberg, H. (1966) Biochim. Biophys. Acta 125, 548–562.
43 Thompson, G.A. (1967) Biochemistry 6, 2015–2022.
44 Kennedy, K.E. and Thompson, G.A. (1970) Science 168, 989–991.
45 Hayashi, A. and Matsuura, F. (1971) Biochim. Biophys. Acta 248, 133–136.
46 Matsuura, F. (1977) Chem. Phys. Lipids 19, 223–242.
47 Hayashi, A. and Matsuura, F. (1978) Chem. Phys. Lipids 22, 9–23.
48 Matsuura, F. (1979) J. Biochem. 85, 433–441.
49 Araki, S., Komai, Y. and Satake, M. (1980) J. Biochem. 87, 503–510.
50 Quin, L.D. (1967) in Topics in Phosphorus Chemistry (Grayson, M. and Griffith, E.j., eds.), pp. 23–48, Interscience, New York.
51 Horiguchi, M. (1972) in Analytical Chemistry of Phosphorus Compounds (Halmann, M., ed.), Vol. 4, pp. 703–724, Wiley-Interscience, New York.
52 Rosenberg, H. (1973) in Form and Function of Phospholipids (Ansell, G.B., Dawson, R.M.C., and Hawthorne, J.N., eds.), BBA Library Vol. 3, pp. 333–344, Elsevier.
53 Matsuura, F., Matsubara, T. and Hayashi, A. (1973) J. Biochem. 74, 49–57.
54 Hasegawa, S., Tamari, M. and Kametaka, M. (1976) J. Biochem. 80, 531–535.
55 Kapoulas, V.M. (1969) Biochim. Biophys. Acta 176, 324–329.
56 Thompson, G.A. (1969) Biochim. Biophys. Acta 176, 330–338.
57 Viswanathan, C.V. (1973) J. Chromatogr. 75, 141–145.
58 Viswanathan, C.V. and Nagabhushanam, A. (1973) J. Chromatogr. 75, 227–233.
59 Simon, G. and Rouser, G. (1969) Lipids 4, 607–614.
60 Itasaka, O. and Hori, T. (1979) Proc. Jap. Conf. Biochem. Lipids 21, 62–65.
61 Sugita, M. and Hori, T. (1971) J. Biochem. 69, 1149–1150.
62 Thomas, L.C. (1974) Interpretation of the Infrared Spectra of Organophosphorus Compounds, pp. 18, 97, Heyden, London.
63 Baer, E. and Stanacev, N.Z. (1965) J. Biol. Chem. 240, 44–48.
64 Baer, E. and Sarma, G.R. (1969) Can. J. Biochem. 47, 603–610.
65 Harvey, D.J. and Horning, M.G. (1973) J. Chromatogr. 79, 65–74.
66 Matsubara, T. (1975) Biochim. Biophys. Acta 388, 353–360.
67 Harashi, A. and Matsuura, F. (1973) Chem. Phys. Lipids 10, 51–65.
68 Matsubara, T. and Hayashi, A. (1973) Biochim. Biophys. Acta 296, 171–178.
69 Matsubara, T. (1975) Chem. Phys. Lipids 14, 247–259.
70 Harvey, D.J. and Horning, M.G. (1974) Org. Mass Spectrom. 9, 111–124.
71 Harvey, D.J. and Horning, M.G. (1974) Org. Mass Spectrom. 9, 955–969.
72 Rueppel, M.L., Suba, L.A. and Marvel, J.T. (1976) Biomed. Mass Spectrom. 3, 28–31.
73 Glonek, T., Henderson, T.O., Hiderbrand, R.L. and Myers, T.C. (1970) Science 169, 192–194.
74 Kirkpatrick, D.S. and Bishop, S.H. (1971) Anal. Chem. 43, 1707–1709.

75 Benezra, C., Pavanaram, S.K. and Baer, E. (1970) Can. J. Biochem. 48, 991–993.
76 Nozawa, Y. and Thompson, G.A. (1971) J. Cell Biol. 49, 712–721.
77 Swift, M.L. (1976) Lipids 12, 449–451.
78 Carter, H.E., Stroback, D.R. and Hawthorne, J.N. (1969) Biochemistry 8, 383–388.
79 Hsieh, T.C.-Y., Kaul, K., Laine, R.A. and Lester, R.L. (1978) Biochemistry 17, 3575–3581.
80 Nozawa Y., (1975) Methods Cell Biol. 10, 105–133.
81 Hori, T., Sugita, M., Nishimori, C., Itasaka, O. (1975) Yukagaku 24, 611–614.
82 Fukushima, H., Kasai, R., Akimori, N. and Nozawa, Y. (1978) Japan J. Exp. Med. 48, 373–380.
83 Van der Horst, D.J., Kingma, F.J. and Oudejans, R.C.H.M. (1973) Lipids 8, 759–765.
84 Simon, G. and Rouser, G. (1966) Lipids 2, 55–59.
85 Mason, W.T. (1970) Biochim. Biophys. Acta 280, 538–544.
86 Nozawa, Y., Fukushima, H. and Iida, H. (1973) Biochim. Biophys. Acta 318, 335–344.
87 Fukushima, H., Watanabe, T. and Nozawa, Y. (1976) Biochim. Biophys. Acta 436, 249–259.
88 Warren, W.A. (1968) Biochim. Biophys. Acta 156, 340–346.
89 Liang, C.R. and Rosenberg, H. (1968) Biochim. Biophys. Acta 156, 437–439.
90 Horiguchi, M. (1972) Biochim. Biophys. Acta 261, 102–113.
91 Smith, J.D. and O'Malley, M.A. (1978) Biochim. Biophys. Acta 528, 394–398.
92 Bjerve, K.S. (1972) Biochim. Biophys. Acta 270, 348–363.
93 Maget-Dana, R., Tamari, M., Marmauyet, J. and Douste-Blazy, L. (1974) Eur. J. Biochem. 42, 129–134.
94 Curley-Joseph, J. and Henderson, T.O. (1977) Lipids 12, 75–84.
95 Itasaka, O., Hori, T. and Sugita, M. (1969) Biochim. Biophys. Acta 176, 783–788.
96 La Nauze, J.M., Rosenberg, H. and Shaw, D.C. (1970) Biochim. Biophys. Acta 212, 332–350.
97 Hori, T., Arakawa, I., Sugita, M. and Itasaka, O. (1968) J. Biochem. 64, 533–536.
98 Sato, K. and Mukoyama, K. (1968) Biochim. Biophys. Acta 164, 596–598.
99 Andrews, D. and Nelson, D.L. (1979) Biochim. Biophys. Acta 550, 174–187.
100 Thompson, G.A., Bambery, R.J. and Nozawa, Y. (1971) Biochemistry 10, 4441–4447.
101 Nozawa, Y. and Thompson, G.A. in Biochemistry and Physiology of Protozoa (Levandowsky M. and Hutner, S.H., eds.), Vol. 2, pp. 275–338, Academic Press, New York.
102 Fulco, A.J. (1974) Annu. Rev. Biochem. 43, 215–241.
103 Fukushima, H., Martin, C.F., Iida, H., Kitajima, Y., Thompson, G.A. and Nozawa, Y. (1976) Biochim. Biophys. Acta 431, 165–179.
104 Nozawa, Y. and Kasai, R. (1978) Biochim. Biophys. Acta 529, 54–66.
105 Watanabe, T., Fukushima, H. and Nozawa, Y. (1980) Biochim. Biophys. Acta 620, 133–141.
106 Nozawa, Y., Iida, H., Fukushima, H., Ohki, K. and Ohnishi, S. (1974) Biochim. Biophys. Acta 367, 134–147.
107 Shimshick, E.J. and McConnell, H.M. (1973) Biochemistry 12, 2351–2368.
108 Martin, C.E., Hiramitsu, K., Kitajima, Y., Nozawa, Y., Skriver, L. and Thompson, G.A. (1976) Biochemistry 15, 5218–5227.
109 Ohki, K., Kasai, R. and Nozawa, Y. (1979) Biochim. Biophys. Acta 558, 273–281.
110 Shimonaka, H., Fukushima, H., Kawai, K., Nagao, S., Okano, Y. and Nozawa, Y. (1978) Experientia 34, 586–587.
111 Nozawa, Y. (1980) in Membrane Fluidity (Kates, M. and Kuksis, A., eds.), pp. 399–418, Humana Press, Clifton, NJ.
112 Thompson, G.A., Bambery, R.J. and Nozawa, Y. (1972) Biochim. Biophys. Acta 260, 630–638.
113 Nozawa, Y., Kasai, R. and Sekiya, T. (1980) Biochim. Biophys. Acta 603, 347–365.
114 Nozawa, Y., Kasai, R. and Sekiya, T. (1979) Biochim. Biophys. Acta 552, 38–52.
115 Nandini-Kishore, S.G., Mattox, S., Martin, C.E. and Thompson, G.A. (1979) Biochim. Biophys. Acta 551, 315–327.

116 Nozawa, Y., Kasai, R., Kameyama, Y. and Ohki, K. (1980) Biochim. Biophys. Acta 599, 232–245.
117 Rosenthal, A.F. and Pousada, M. (1968) Biochim. Biophys. Acta 164, 226–237.
118 Rosenthal, A.F. and Ham, S.C.-H. (1970) Biochim. Biophys. Acta 218, 213–220.
119 Sarma, G.R., Chandramouli, V. and Venkitasubramanian, T.A. (1970) Biochim. Biophys. Acta 218, 561–563.

CHAPTER 4

Sphingomyelin: metabolism, chemical synthesis, chemical and physical properties

YECHEZKEL BARENHOLZ and SHIMON GATT

Laboratory of Neurochemistry, Hebrew University-Hadassah Medical School, Jerusalem, Israel

1. Introduction

Sphingomyelin * (SPM; *N*-acyl sphingosine-1-phosphocholine; ceramide-1-phosphocholine, see Scheme 1), a major lipid constituent of animal tissues, was first described by Thudichum in his book *The Chemical Constitution of the Brain* [2], as a compound whose hydrolytic products are sphingosine, fatty acid and choline. Its general structure, as *N*-acyl sphingosine-1-phosphocholine, was provided by Pick and Bielschowsky [3]. In 1930 it was shown that sphingomyelin actually is a mixture of molecules, each containing a different fatty acid [4]. But it was not until 1962 that a considerable research effort defined the detailed structure and configuration of the naturally occurring sphingosine as *trans*-D-*erythro*-2-amino-4-octadecene-1,3-diol or, (2S,3R,4E)-2-amino-4-octadecene-1,3-diol. Further details can be found in the review by Shapiro [5] and in *Chemistry and Physics of Lipids* Vol. 5, No. 1 (1970), a volume dedicated to H.E. Carter.

(a) Sphingomyelin composition

When isolated from various natural sources, sphingomyelin varies in two of its components: sphingosine (long-chain base, LCB) and the fatty acyl residues. With the development of chromatographic and analytical procedures, several LCBs were characterized [6–9]. The compositional analysis of SPM requires its complete hydrolysis to intact LCB and fatty acid, but most of the procedures used for the degradation of sphingomyelin lead to incomplete dephosphorylation of the base [10]. This is so in case of hydrolysis by HCl in anhydrous methanol or in methanol–water mixtures [7,11]. Hydrolysis by HCl also results in the formation of several derivatives of sphingosine [8], whereas alkaline hydrolysis results in a low yield of sphingosine bases [12]. These problems can be overcome by initial dephosphoryla-

* Sphingomyelin structure, physical properties and its presence in and contribution to the properties of lipid bilayers, biological membranes and lipoproteins are also described in a recent review by Barenholz and Thompson [1]. (List of abbreviations on p. 177.)

Hawthorne/Ansell (eds.) Phospholipids

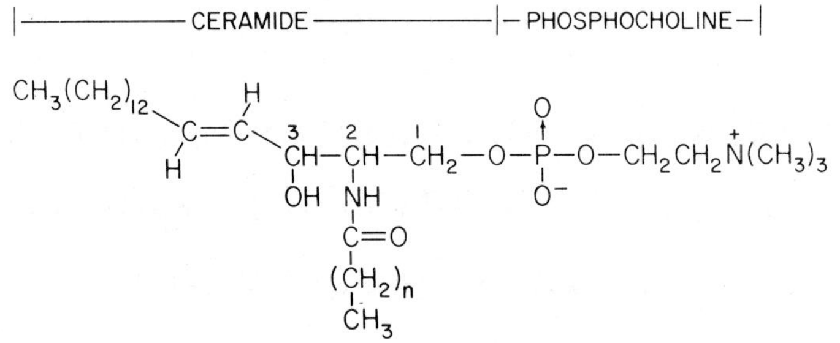

Scheme 1. Sphingomyelin.

tion of SPM using either phospholipase C of *Clostridium perfringens* [13–16] or by hydrofluoric acid [17]; neither treatment alters the stereochemical configuration of the sphingosine base. These procedures yield ceramide (*N*-acylsphingosine) which is then completely hydrolyzed by alkali to LCB and fatty acids [13,15]. There are three main structures of LCBs and all have been found in SPM: sphinganine, or in its trivial name dihydrosphingosine; 4E-sphingenine (sphingosine) and 4D-hydroxy sphinganine (phytosphingosine). Sphingosine (*trans*-D-*erythro* 1,3-dihydroxy-2-amino-4-octadecene) and dihydrosphingosine (*trans*-D-*erythro*-1,3-dihydroxy-2-aminooctadecane) are present in SPM of most tissues. Phytosphingosine ((2S,3S,4R)-2-amino-1,3,4-octadecanetriol) was also found in SPM of bovine kidney [18]. Traces of other LCBs have also been reported [18–20]. SPM of most mammalian sources is characterized by a relatively high content of very long chain, saturated or monounsaturated fatty acids and a very low content of polyunsaturated fatty acids [21,22]. In most mammalian tissues, palmitic acid (C16:0) is the prevalent fatty acid, followed, in decreasing abundance by nervonic acid (24:1), ligonoceric acid (24:0) and behenic acid (C22:0) [21,23]. In the nervous system stearic (C18:0) rather than palmitic acid is the main component [22,24], followed by C24:1 and C24:0; the fatty acid composition varies with the exact location and animal age [25]. In other tissues there is very little effect of aging although changes in fatty acid composition of the SPM can be induced by diet [26–28]. Compositional changes were found also in SPM of the red blood cell membrane [29]. For further references on SPM composition, see Rouser et al. [22] and Table 27 of White [21].

2. *Total and partial chemical synthesis of sphingomyelin*

Sphingomyelin can be prepared by complete or partial chemical synthesis, for the latter, sphingosylphosphocholine (SPC) is used. The complete chemical synthesis permits labelling of SPM with radioactive isotopes such as ^{3}H, ^{14}C or heavy isotopes such as ^{2}H, ^{15}N, or ^{13}C. This, therefore, provides a procedure of great importance for biochemical and, even more so, for biophysical studies on the properties of sphingomyelin in membranes and lipoproteins (for more details see section 4). Using the complete synthesis, one can label specifically the phosphocholine head group, the acyl chain (at any desired position in the molecule) as well as carbons 1–3 of the sphingosine base, which form the interfacial region of the sphingomyelin molecules in bilayers and lipoproteins (see sections 4a, 4b). The complete chemical synthesis also permits studying the effect on physicochemical parameters, of the chain length of the sphingosine base as well as the difference in chain length between the LCB and the fatty acyl residue. The partial chemical synthesis is much simpler but also less versatile since it permits only changes in the acyl chain. The detailed procedures of sphingomyelin synthesis are out of the scope of this review but the origin of the various molecular regions is described.

(a) Complete chemical synthesis of sphingomyelin

The synthesis of sphingomyelin was described by Shapiro and Flowers [30] and later reviewed by Shapiro in his monograph *Chemistry of Sphingolipids* [5].

(i) Synthesis of LCB

There are two synthetic procedures for sphingosine. Grob and Gadient [31] used 2-hexadecanal-1 whereas the procedure described by Shapiro and his coworkers [5,32–35] varied the chain length of the aldehyde which subsequently determines the chain length of the sphingosine base. Thus, when myristaldehyde was used as starting material, C18 sphingosine was the LCB produced. Carbon atoms 3 and 4 of sphingosine were derived from malonic acid and carbons 1 and 2 from ethyl sodium acetoacetate. Benzediazonium (as the chloride) was used as the source of the nitrogen of the primary amino group at carbon atom 2. The products of the synthesis were the racemates, namely DL-*erythro* and DL-*threo* sphingosine. Separation of the *erythro* and *threo* stereoisomers could be obtained in good yield while the separation between D and L enantiomers was less efficient [5]. Recently, Shoyama et al. [36] modified this procedure to obtain either the D- or L-isomers of the *erythro* or *threo* isomers of sphingosine or ceramide. Several procedures for the synthesis of dihydrosphingosine have been described. Because of absence of the double bond, synthesis of this compound is simpler, requires fewer steps compared to that of sphingosine [5], and is therefore available commercially. Such procedures were developed by Shapiro [5], Grob et al. [37], Egerton et al. [38], Carter and Shapiro [39] and Jenny and Grob [40,41]. A stereospecific synthesis of phytosphingosine was described by Gigg and coworkers [42,43].

*(ii) Synthesis of ceramide (*N*-acyl sphingosine)*

The appropriate 3-benzyol ceramide was prepared to prevent an *N* to *O* acyl migration in ethyl *erythro*-2-amino-3-hydroxy-4-octadecenoate. The primary amino group at C2 was then acylated by one of the following procedures: (a) via the desired acyl chloride [5]; (b) using a fatty acid and dicyclohexyl carbodi-imide (Dagan, Cohen, Gatt and Barenholz, in preparation) or its ethyl-analog [44], or (c) using a *p*-nitrophenyl ester of the fatty acid [45].

(iii) Synthesis of sphingomyelin

Introduction of the phosphocholine group was described by Shapiro et al. [5,30], using chloroethyl phosphoryldichloride for the phosphorylation of the primary hydroxy group at position 1. Trimethylamine was used as the precursor of the choline group. For more details see Shapiro [5]. Using this procedure mostly the racemic (DL-*erythro* or DL-*threo*) SPMs were prepared. The same procedures were also used to prepare the specific stereoisomers (enantiomers) but the yields were low. Alpes [46] suggested a simpler modification to the last steps (ceramide to SPM) in the synthesis of DL-*erythro* sphingomyelins. The specific stereoisomers could be obtained in better yield by adopting the procedure of Shoyama et al. [36] to prepare stereospecific isomers of ceramides.

(b) Partial chemical synthesis of sphingomyelin

The initial step involves the preparation of SPC from SPM, using aqueous-methanolic HCl [7,11], or butanolic HCl [47]. The yield of the SPC using either procedure is less than 30%. The SPC is subsequently acylated to sphingomyelin using the desired acyl chloride [5], the *p*-nitrophenyl ester of the desired fatty acid [45] or a fatty acid in the presence of carbonyldi-imidazole, dicyclohexyl carbodi-imide or its ethyl derivative [48] (also Dagan, Cohen, Gatt and Barenholz, in preparation). The SPM thus obtained is purified by standard procedures. The stereospecificity is unaffected by the above procedures; the stereoconfiguration of SPM will therefore be determined by the SPM used as starting material.

(c) Determination of SPM stereospecificity

The stereochemical characterization of the SPMs is based on either its optical rotation [5] or on its circular dichroism [49].

As summarized in Table 1, there are big differences in $[\alpha]_D^{25}$ between SPM and dihydrosphingomyelin but the acyl chain length has practically no effect.

TABLE 1

Stereochemical characterization of sphingomyelins [a]

Sphingosine base	Fatty acyl residue	$[\alpha]_D^{25}$ in chloroform-methanol
D-*erythro*	palmitoyl	+6.1
L-*erythro*	palmitoyl	−6.5
D-*erythro*	stearoyl	+6.1
D-*erythro*	lignoceryl	+6.5
D-*erythro*, dihydro	palmitoyl	+22.5
D-*erythro*, dihydro	stearoyl	+20.5

[a] Data are taken from [5].

3. Metabolic pathways of biosynthesis and degradation

(a) Biosynthesis of sphingomyelin

The SPM molecule consists of three components; sphingosine, fatty acid, and phosphocholine. It therefore could be constructed biosynthetically by first condensing sphingosine with acyl coenzyme A [50–53] or with free fatty acid [54,55] to produce ceramide. The latter could then bind phosphocholine either by condensation with cytidine diphosphocholine (CDPC) or by transfer from PC. An alternative pathway would involve condensation of sphingosine with CDPC to form SPC, which would further condense with acyl coenzyme A to form SPM. Indeed, all these possibilities have been proposed and supported experimentally. Sribney and Kennedy [56] suggested the condensation of ceramide and CDPC. It is of interest that ceramides containing the unnatural *threo*-configuration in the sphingosine residue were preferred over ceramides having the *erythro*-configuration [56]. Sribney [57] subsequently showed utilization of the D-*erythro* as well as D-*threo* derivatives. These observations were supported by these of Fujino and his co-workers [58,59]. The condensation of SPC with fatty acyl coenzyme A was suggested by Brady et al. [60]. This was supported by Fujino and his coworkers [61,61a]. Transfer of the phosphocholine groups from PC to ceramide was shown by Ullman and Radin [62] and by Anderer, Diringer, Koch, and Magraff [63–66]. Further experimentation is required to ascertain if the three pathways indeed occur in intact cells and if so, which one predominates. Kanfer and Spielvogel [67] showed that phospholipase-C catalyzed the formation of [^{14}C]sphingomyelin from PC and [^{14}C]oleoyl-sphingosine (ceramide).

A considerable number of papers dealing with the biosynthesis and metabolism of SPM in organs, cells and subcellular organelles have been published. These cannot be discussed in detail and the reader is referred to refs. [68–95].

Recently, Stoffel and Melzner [96] re-evalued the proposed pathways of SPM biosynthesis. Using synthetic stereo- and radiochemically pure substrates of high specific radioactivity they confirmed the sequence of acylation of sphingosine bases via acyl coenzyme A and transfer of phosphocholine group from CDP-choline to ceramide to give sphingomyelin. In their system the free sphingosine base did not serve as the phosphocholine acceptor nor did phosphatidylcholine serve as a phosphocholine donor. Sphingosine was a better acceptor than dihydrosphingosine and the unnatural L-*threo* enantiomer was a better acceptor than the natural D-*erythro* enantiomer.

The interrelations between pathways leading to SPM biosynthesis and cholesterol biosynthesis are worth noting. Glucocorticoid treatment of HeLa-65 cells reduced the level of cholesterol while SPM biosynthesis almost doubled [97]. Gatt and Bierman [98] showed that uptake of SPM by human skin fibroblasts resulted in a marked increase in acetate incorporation into sterols, suggesting regulation of intracellular cholesterol balance by SPM. This is of special interest because of the interaction between cholesterol and SPM in normal and abnormal biological membranes [1,99].

(b) Enzymatic degradation of sphingomyelin

The SPM molecule might be catabolized by hydrolysis to ceramide and phosphocholine or, alternatively to SPC and fatty acid. When working on the enzymatic hydrolysis of ceramide, Gatt et al. [54,100] could find no evidence for the latter pathway in extracts of rat brain, but did describe an enzyme which cleaves SPM to phosphocholine and ceramide [100]. The latter product can then be further degraded to sphingosine and fatty acid by ceramidase [54,55]. Three other papers described a similar enzyme in extraneuronal tissues [101–103]. After the description of this enzyme, which was universally termed sphingomyelinase, workers in two laboratories showed its deficiency in subjects with Niemann-Pick disease [103,104]. Fowler [105] further characterized the enzyme as having a lysosomal origin, though Weinreb et al. [106] found some activity in other subcellular fractions. Similar to most hydrolytic, lysosomal enzymes, sphingomyelinase exhibits an acidic pH optimum and requires no cofactor. The lysosomal sphingomyelinase is stereospecific, in that the D-*erythro* SPM is a preferred substrate over the L-*erythro* or the DL-*threo* SPM [100]. It is usually assayed in the presence of detergents (non-ionic, such as Triton X-100 or anionic such as sodium cholate or taurocholate), which are used to disperse the substrate [100,101]. Gatt et al. [107] showed that a solubilized preparation of this enzyme from rat brain hydrolyzed liposomal dispersions of SPM in the absence of a detergent.

Schneider and Kennedy [103] mentioned that spleens of patients with Niemann-Pick disease have sphingomyelinase activity which has an optimal pH at 7.4, requires magnesium ions and probably is not lysosomal. Such an enzymatic activity was indeed found in brain tissue [108–112]. Its activity in brain exceeds many-fold that observed in other tissues [108,113]. Furthermore, in brain, the magnesium-dependent enzyme has an interesting developmental pattern, having high activity in young rats or humans and decreasing in activity with age [108]. It is of interest that the specific activity of this sphingomyelinase in brain of young rats or humans considerably exceeds that of the lysosomal enzyme. The magnesium-dependent enzyme is not lysosomal, but enriched in microsomes [108] and perhaps associated with myelin [114]. Gatt et al. [115] have shown that brain tissue from a patient with Niemann-Pick disease, which was deficient in sphingomyelinase activity when assayed at pH 5 in the absence of added divalent cations or in the presence of EDTA, had near-normal activity of the magnesium-dependent enzyme. This again emphasises that the two sphingomyelinases in brain are indeed two separate enzyme entities. The higher specific activity of the alkaline enzyme in infantile brain suggests that it might have an important role in regulating SPM levels in developing animals or humans.

Attempts were made to extract and purify the lysosomal sphingomyelinase. Gatt [54,100] used sodium cholate to extract the enzyme; this detergent could subsequently be removed by dialysis or filtration through Sephadex beads. Subsequently, Gatt and Gottesdiner [117] showed that, if a lysosomal preparation of rat brain was extracted overnight with buffer or an isotonic sucrose solution, most of the acidic enzyme was solubilized and did not sediment at $100000 \times g$, although gel filtration studies suggested that the extract is still a multi-protein aggregate of about M_r 300000. The solubilized enzyme could be further purified on an affinity column having sphingomyelin bound to Sepharose. Sphingomyelinase was bound to this column and could not be eluted by non-ionic detergents but was displaced by a mixed dispersion of SPM in Triton X-100 (Gatt, unpublished). Several laboratories have purified the lysosomal sphingomyelinase from placenta [118,119] or brain [120,121]. Isoelectric focussing of the enzyme was attempted by Callahan and his colleagues [122,123] and Harzer et al. [124].

The enzymatic activity of sphingomyelinase is usually determined using a dispersion of SPM, labelled with a radioactive tracer in the sphingosine, fatty acid or choline moiety [100,102]. This requires dispersing the lipid substrate in a detergent. Yedgar, Barenholz and Gatt studied the effect of detergents on the utilization of SPM by brain sphingomyelinase [125–127]. Their studies demonstrated the importance of the type of detergent used and analyzed the mixed dispersions of detergent and SPM which are utilized by sphingomyelinase.

Several authors attempted modification of the standard procedures used for determining sphingomyelinase activity [128–133]. Others have used substrates other than SPM. Fensom et al. [134] used *bis*(4-methylumbelliferyl)pyrophosphate diester and Gal et al. [135,136] introduced a chromogenic substrate, 2-hexadecanoylamino-4-nitrophenyl-phosphocholine. Gatt and his co-workers synthesized a coloured derivative of SPM by condensing trinitrophenylaminolauric acid with SPC to yield the corresponding derivative of SPM having a trinitrophenylamino group on the terminal position of the fatty acid moiety [137]. They subsequently synthesized several fluorescent derivatives of SPM, using a similar procedure [48,138]. Both the coloured and fluorescent analogues were readily hydrolyzed by sphingomyelinase of mammalian or bacterial origin. The coloured or fluorescent ceramides formed in the reaction were isolated by solvent extraction and estimated, respectively, in a spectrophotometer or spectrofluorimeter [138]. Because of the considerable sensitivity of the fluorescent probes used (e.g., anthracene), the fluorescent SPM was not used in its pure form, but was mixed with authentic SPM isolated from brain tissue. Sphingomyelinase hydrolyzed the fluorescent and natural SPM molecules at equal rates [48]. Several other derivatives of SPM with well-defined fatty acids have been synthesized by Dagan, Cohen, Gatt and Barenholz (in preparation).

Of interest is the effect of the apparent microviscosity of the membrane on degradation of SPM by calf brain synaptosomal plasma membrane sphingomyelinase [139]. Halothane stimulated the degradation of SPM by reducing the apparent microviscosity.

Hirschfield and Loyter [140] showed that, following brief lysis in a hypotonic medium, chicken erythrocyte ghosts exhibited sphingomyelinase activity. This suggested that chicken erythrocytes have this enzyme in a form which is inoperative ("latent") in the intact cell but which is expressed, enzymatically, after lysis. Record et al. [141] showed that, following lysis, this enzyme can hydrolyze the entire SPM of the chicken erythrocyte ghosts and, when mixed with human erythrocyte ghosts up to 50–60% of the SPM content of the latter. Sphingomyelinase activity has been studied in several tissues [142–151]. Sphingomyelinase with phospholipase-C activity is also present in some bacterial strains and has been investigated and purified [152,153]. Among them the best studied is the sphingomyelinase of *Staphylococcus aureus* [154–160]. Sphingomyelinase with phospholipase-D activity (i.e., splitting off the choline moiety only) has also been described [161–163].

(c) Niemann-Pick disease

Niemann-Pick (or Niemann-Pick's) disease is a name given to a group of disorders, which are diverse in clinical symptoms as well as enzymatic deficiency but have one common feature, an increased accumulation of sphingomyelin. The name relates to Niemann, a pediatrician who in 1914 described the first patient [164] and Pick, who provided histological evidence that this disease is entirely different from Gaucher disease [3,116,165]. SPM was subsequently identified in visceral organs by Klenk [166]. A considerable literature exists on Niemann-Pick disease and for further information the reader is referred to several reviews published [167–179,305]. Seven types of Niemann-Pick disease have been described. Crocker [180] introduced an initial classification of 5 types. Type A is an acute neuropathic form which shows storage of SPM in visceral organs as well as brain; death from this type usually occurs by the 3rd year of life. Type B is a chronic non-neuropathic form; onset is later than type A; visceral organs are affected but not brain. This form prevails in Ashkenazi Jewish families. Type C is a chronic, subacute form with initial onset after the 2nd year of life and gradual development, including some neurological involvement. Type D ("Nova Scotia") resembles Type C, but occurs in Catholic families from Nova Scotia. Type E appears in adults and few clinical symptoms are seen, though SPM is somewhat increased in liver and spleen. Recently, Schneider et al. [181,359] added Type F, characterized by a temperature-labile sphingomyelinase. Other, unclassified cases are found in the literature [150,182]. While Types A and B show a very marked reduction in sphingomyelinase activity [103,104,104a,284,285], Types C–E have normal enzyme activity. Types A and B can therefore be easily diagnosed, pre- and post-natally, by determining the specific activity of the lysosomal sphingomyelinase in leukocytes or skin fibroblasts [168,183–188,382], hepatic cells [188a], and even hair roots [132]. Gatt et al. [109] have shown that brain tissue of Type A has a near-normal activity of the non-lysosomal, magnesium-dependent sphingomyelinase. Since this enzyme is most active in brain, these authors suggested that its presence might be the reason that Type A patients have little neurological involvement and a relatively low SPM level in brain. In contrast, the enzymatic defect in Types C–E is still being debated. Callahan and his coworkers [122,123,178,184] used isoelectric focussing and suggested a deficiency of one isoenzyme in Type C. Besley [189] used these findings for diagnosis of Type C in skin fibroblasts. Besley et al. [190], using complementation studies, suggested that the genes coding for Types C and A are located on separate chromosomes. This is supported by recent findings of Christomanou [191] who suggests the deficiency of an activator which stimulates SPM and glucocerebroside degradation as the cause of

Type C. For diagnosis of the enzymatic defect, a variety of substrates were used. These included unlabelled SPM [123] radioactively labelled SPM [104,192,193], coloured or fluorescent derivatives [48,137,138], 4-methylumbelliferyl-pyrophosphate diester or bis(4-methylumbelliferyl)phosphate [134] and the coloured analogue of Gal et al. [135] (see also [194]). The disease and heterozygotic state could also be diagnosed by determining the SPM content [195,196]. No animal model exists yet for Niemann-Pick disease Type A, but certain strains of mice show increased SPM [197,198,388] and Aubert-Tulkens et al. [199] studied an experimentally induced model of Niemann-Pick disease. An attempt to treat Niemann-Pick disease has been made by liver transplantation into a patient with Type A disease [200].

4. *Physical properties of sphingomyelin*

SPM and PC are both choline phospholipids which seem to replace each other in biological membranes [1]. Both are part of the group of lipids classified as non-soluble swelling amphipaths [201] which serve as the matrix of the biological membrane and the envelope of lipoproteins. In this section, the properties of SPM and PC will be compared.

(a) Atom numbering

The method of atom numbering of sphingophospholipids is described by Sundaralingam [202] and compared to those used for the glycerophospholipids. The author uses a similar method to that which numbers the atoms of the sphingomyelin molecule. In glycerophospholipids the atom numbering is based on the glycerol skeleton and therefore characterizes 3 separate chains: the acyl chains in position 1 and 2 and the phosphocholine in position 3 of the glycerol. But this is not the case for SPM where the long-chain amino alcohol (sphingosine) serves as the SPM skeleton. Sundaralingam [202] defines part of the sphingosine as chain 3, the amide bond and the acyl residue as chain 2 and carbon atom No. 1 of the sphingosine, including the phosphocholine groups as chain 1. In this review, the atom numbering suggested by Pascher [203] will be used as shown in Scheme 2.

(b) Molecular structure of sphingomyelin (a comparison with phosphatidylcholine)

All phospholipids have a general common structure comprising three regions: the hydrophobic region, an interfacial region and the ionogenic group (Scheme 2). SPM and PC are defined as "insoluble, swelling amphipaths" which in excess water will form a lamellar structure [201,204]. This is composed of multiple bilayers, in which the polar regions are directed towards the interspacing water while their apolar paraffinic chains comprise the core of the bilayer. This structure prevents the unfavourable contact of this region with the aqueous environment.

It is clear from Scheme 2 that the two choline phospholipids resemble each other

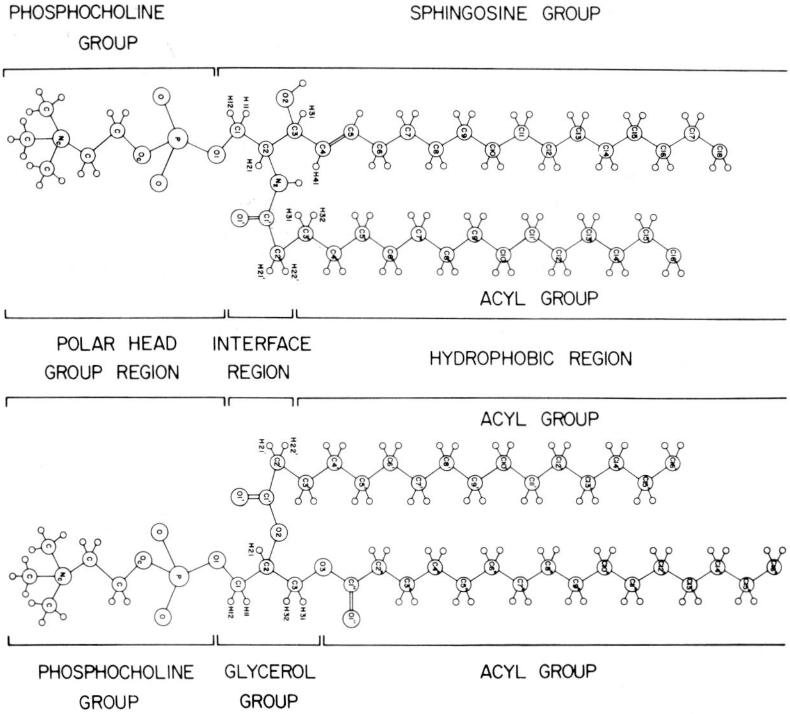

Scheme 2. Structures of *N*-palmitoyl sphingosylphosphocholine (top) and *sn*-dipalmitoyl phosphatidylcholine (bottom).

at the macroscopic level. But a more careful examination reveals that, although phosphocholine is the polar head group common to both, the other regions of each respective molecule have distinctly different structural features. The hydrophobic region of PC is composed of two acyl chains which are esterified to the glycerol and which are almost identical in length. In most cases, one of these, in the 1 (or 2A) position is saturated while the other, in the 2-position, is unsaturated, containing from one to six double bonds, all in the *cis*-configuration. In SPM, this region is composed of one acyl chain, bound in amide linkage to the primary amino group at C2 of the sphingosine. Over 60% of the fatty acids in SPM are saturated with a chain length of 16–24 carbon atoms. The unsaturated acyl chains are *cis*-monoenoic, mostly nervonic acid (24:1). More than half of the acyl chains are longer than 20 carbon atoms. The second chain is the paraffinic residue of the sphingosine base which contributes only 13–15 carbon atoms to the apolar region. The result is that more than 50% of the SPM molecules are very asymmetric (see Scheme 3); this might have a considerable effect on the bilayer organization. The average number of *cis* double bonds in the molecule of naturally occurring lecithin is 1.1–1.5 while for the SPM molecule it is only 0.1–0.35 [21,22,205].

The differences in the interfacial region are even more striking. This region comprises the interface between the hydrophobic region and the ionogenic group which protrudes into the aqueous solution. In PC, this region includes the carbon atoms 1, 2, and 3 of the glycerol backbone as well as the components of the two ester bonds (see Scheme 2).

In SPM (Scheme 2) the interfacial region contains the components of the amide linkage (between the acyl chain and the primary amino group at position C2 of the sphingosine) as well as the free hydroxyl group attached to C3 and probably the *trans* double bond between C4 and C5. This structure of the interfacial region of SPM provides possibilities for hydrogen bonding. In this respect it might be a donor as well as an acceptor of hydrogen. In comparison, PC can act only as a hydrogen acceptor. Since the hydrogen bond is electrostatic in nature it is affected by the dielectric constant of the medium; the lower the dielectric constant the greater the hydrogen bond strength [206,207]. In SPM, the region which is a hydrogen donor resides in the interfacial portion of the molecule and has a relatively low dielectric constant. This makes the hydrogen bonds of SPM (with other phospholipids, cholesterol, or with protein) stronger than the corresponding bonds of glycerophospholipids, such as phosphatidylethanolamine (PE) or phosphatidylserine. In the latter, the groups which donate the hydrogen protrude into the aqueous phase, where the dielectric constant is considerably greater. The *trans* double bond between C4 and C5 of the sphingosine moiety has the ability to induce dipoles in the interfacial region. (review by Barenholz and Thompson [1]). It may also aid in chain-stacking and close packing of the lipid molecules [208].

The above differences in the interfacial regions of SPM and PC are of considerable significance since they affect the hydrophobic regions, mainly the packing of the paraffinic chains, though they also affect the orientation of the PC polar head group. The presence of the hydroxyl groups, amide bond and the *trans* double bond in the interfacial region of SPM increases the polarity of this region which thereby enhances its interaction with water. The presence of these groups provides a series of hydrogen bonds within the lipid bilayer which results in increased stability of the membrane.

The number of physical studies done with systems containing SPM is considerably smaller than those available for PC [1,209]. Information on the detailed molecular structure, derived from single crystal X-ray diffraction was recently described for dimyristoyl-PC [210] (DMPC); similar data are yet not available for SPM. However, some information might be derived from data available for ceramide [202,203,211].

Most of the physical studies on SPMs used liposomal dispersions which served as models for biological membranes. The various types of liposomes were reviewed by Sozoka and Papahadjopoulos [212]. These comprise two main groups: the large multilamellar liposomes [213] (MLV) and the small unilamellar vesicles (SUV). The latter group is further divided into many sub-groups based on size and preparative procedure [212]. The small, unilamellar vesicles which are prepared by ultrasonic irradiation are the most common [214]. It is worth noting that the properties of the liposome depend on its composition as well as size. The effect of curvature on the artificial membrane properties was clearly demonstrated for PCs [215–217]. This emphasizes the importance of using well-defined systems for physical measurement, but, because of difficulties in preparing synthetic SPM, many studies were done using the naturally occurring lipid (for a recent review see [1]). This is not the case for the glycerophospholipids where most studies have been carried out with synthetic lipids. The physical properties of the SPMs might vary with changing composition of the acyl chains. This is true for such properties as thermotropic behaviour [218,219], osmotic properties of liposomes [220], liposomal size [221], interaction with Triton X-100 (Yedgar and Barenholz, unpublished results). Thus, specifying the origin of natural SPM may not suffice, since the composition differs with different preparations [219]. This might stem from a dietary effect [27,28,222] or from variations in the purification procedures (Barenholz, unpublished results).

(c) Studies on monomolecular films at air–water interfaces

Considerable work has been done using monomolecular layers of phospholipid as an air–water interface, but some reservations have been voiced against the information obtained by this technique. Many experiments were performed using supercompressed films when working above the equilibrium spreading pressure [223]. Also, it was suggested that the phase change from the liquid-expanded to the liquid-condensed phases in insoluble monolayers is not a true equilibrium phenomenon [224]. However, it is possible that if the compression-decompression time scales are short compared with the long-term transition of the film to the bulk phase, thermodynamic treatment of this transition would be acceptable [225]. The values of the limiting surface area per molecule, obtained from monolayer studies, are in good agreement with the corresponding values obtained from X-ray diffraction studies of lipid bilayers. The following average values of limiting surface area were reported (reviewed by Jain and Wanger [226]): dipalmitoyl phosphatidylcholine (DPPC)—44.5 $Å^2$; DOPC—72 $Å^2$; Egg PC—62 $Å^2$; bovine brain SPM—42–50 $Å^2$ (the exact value depends on the fatty acid composition). The limiting surface area of synthetic SPM (racemic, DL-*erythro*) is similar to those of synthetic lecithins: *N*-palmitoyl sphingosylphosphocholine (C16 SPM)—40.5 $Å^2$; *N*-stearoyl (C18) SPM—39 $Å^2$; *N*-lignoceryl (C24) SPM—38 $Å^2$ (Yedgar et al. [127]). When SPM, isolated from bovine brain, was enriched with C18 SPM, the liquid condensed film had a minimal molecular surface area of only 40 $Å^2$, while a fraction enriched in *N*-nervonyl (C24:1) SPM had a corresponding value of 56 $Å^2$, and a film of the liquid-extended type. The difference in the surface properties of these two fractions is most probably related to the *cis* double bond between carbon atoms 15 and 16 in the nervonyl residue, but not to the *trans* double bond between carbon atoms 4 and 5 of the sphingosyl residue. The latter conclusion was derived from the fact that SPM, having stearic or lignoceric acid residues in which the *trans* double bond was reduced by hydrogenation have each a limiting area of 40 $Å^2$ per molecule, a value identical with that obtained prior to the hydrogenation [227,228]. It is worth noting that this *trans* double bond has a considerable effect on the force-area curves of ceramides. Thus, *N*-octadecanoyl-D-dihydrosphingosine which has a limiting area similar to that of *N*-octadecanoyl-D-sphingosine (42 Å^2), is much more expanded at lower pressure. This is due to facilitation of chain-stacking and close-packing of the lipid molecule, thereby imposing a more condensed organization of the lipid [208]. The reason for the different behavior of SPM and ceramide is not yet clear. Another possible role of the *trans* double bond of the sphingosine base was derived from measurement of the surface potential of films of dipalmitoyl phosphatidylcholine and bovine brain sphingomyelin. Although measurements were made under similar conditions, the values were markedly different. The relatively larger surface potential of sphingomyelin monolayers was reduced by hydrogenation, suggesting a contribution of the *trans* double bond to the surface potential [227]. Also, the binding affinity of calcium ions to monolayers of these two lipids, as evident from surface potential measurements, was less towards sphingomyelin than phosphatidylcholine. Similar results were also obtained with other multivalent cations [229]. These differences in binding affinity may be due to an ion-dipole interaction, or alternatively, hydrogen bond formation between the hydroxyl group and the phosphate oxygens of sphingomyelin, which cannot occur in phosphatidylcholine [227,230]. It has, however, been suggested that the observed differences in surface potential of these two phospholipids may be due to impurities present in both preparations [231].

(d) Solubility in organic solvents

Considerable differences in solubility in organic solvents were found between SPM and PC. Thus, in a non-polar solvent (such as decane which is used for preparation of planar lipid bilayers), SPM forms a gel [46]. Naturally occurring SPM and most synthetic SPMs are practically insoluble in chloroform, even at high temperatures, whereas most PC species form micellar solutions in this solvent [232–234]. Solubilization of SPM as micelles in chloroform-based solvents can be increased by addition of a polar solvent such as methanol. Even then, considerable differences are observed in the solubility of SPM and PC (Barenholz, unpublished). Alpes [46] suggested that the solubility of SPM can be further increased and made comparable to that of PC by adding water. It is plausible that the differences in solubility in organic solvents are caused mostly by differences in the interfacial region of the molecule (see section 2(b)).

(e) Thermotropic behaviour

The melting point of natural SPM (of unspecified source) to an isotropic liquid is 210°C [204]. The melting points of synthetic SPMs are summarized in Table 2. It is clear that for molecules having saturated fatty acids there is almost no effect of acyl chain length. This is true also for PC and PE [204]. Unsaturated fatty acids reduce the melting point; this is related to the location of double bonds and possibly to *cis* or *trans* isomerism. The double bond in the sphingosine also affects this property, since dihydrosphingomyelin has a higher melting point than SPM. As expected, there is almost no difference in melting point between the various stereoisomers (enantiomers) of the same SPM.

As with other phospholipids, phase changes other than the melting of SPM to an isotropic liquid occur at low temperature [1,204,209]. In a pioneering study of the thermotropic and lyotropic behaviour of lipids, using low angle X-ray diffraction, Reiss-Husson [235] found that an aqueous dispersion of bovine brain SPM forms a gel phase at 25°C but a liquid-crystalline lamellar phase at 40°C. At 40°C the lamellar phase incorporates a maximum of 40% water (by weight) with additional water forming a bulk phase. Reiss-Husson also found that the surface area per molecule, as well as the interlamellar spacing, increase as the SPM is progressively hydrated. At the maximum hydration of 40%, the area per molecule was found to be 54 $Å^2$, the lipid bilayer and water layer thicknesses being 48 and 30 Å, respectively. In many cases SPM, which is described as "anhydrous", exists in hydrated form, as was concluded from infrared spectroscopy [5].

The phase behaviour of aqueous dispersions of SPM having a well-defined fatty acid composition was studied by Shipley and coworkers [236] using polarized light microscopy, differential scanning calorimetry and X-ray diffraction. Fig. 1 shows the temperature-composition phase-diagram of the bovine brain SPM/water system as

TABLE 2

Melting point of sphingomyelins

Sphingosine base	Fatty-acyl residue	M.P. (°C)
DL-*erythro*	palmitoyl	209–211 [a] 209–210 [b]
DL-*erythro*	stearoyl	209–210 [a]
DL-*erythro*	lignoceroyl	213–216 [a]
DL-*erythro*	oleoyl	188–190 [b]
DL-*erythro*	2 DL-hydroxypalmitoyl	219–221 [b]
DL-*erythro* dihydro	palmitoyl	224–225 [b]
DL-*erythro* dihydro	oleoyl	208–210 [b]
DL-*erythro* dihydro	2 DL-hydroxypalmitoyl	228–230 [b]
DL-*erythro* dihydro	*trans* hexadecenoyl	205–206 [b]
D-*erythro*	palmitoyl	215–217 [a]
D-*erythro*	stearoyl	213–214 [a]
D-*erythro*	lignoceroyl	213–214 [a]
L-*erythro*	palmitoyl	216–217 [a]
D-*erythro* dihydro	palmitoyl	222–223 [a]
D-*erythro* dihydro	stearoyl	221–222 [a]
DL-*threo* dihydro	palmitoyl	224–225 [a]

[a] Data from [5].
[b] Data from [46].

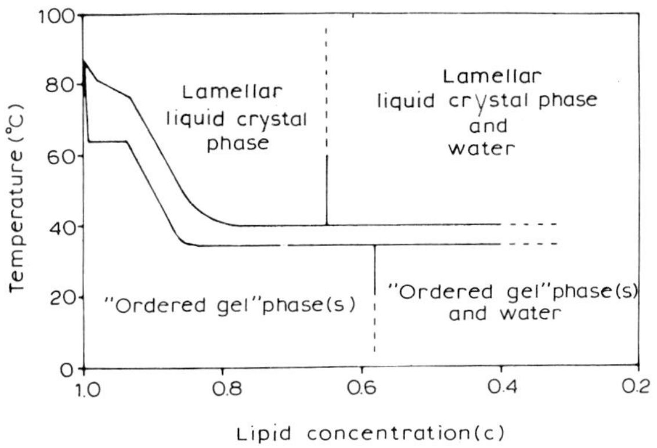

Fig. 1. Phase diagram relating the effect of temperature and composition of bovine brain SPM in water [236].

constructed by these authors. Lamellar phases, in which water is intercalated between sheaths of lipid molecules arranged as bilayers, were found to be present over much of the phase diagram. An order–disorder transition separates the high-temperature liquid-crystalline lamellar phase from a more ordered lamellar phase at low temperature. The thermotropic behaviour in the absence of water proved to be similar to that exhibited by various PCs. At 87°C, a transition occurs from a crystalline phase, in which SPM is organized in bilayers, to a liquid crystal-like phase with a mobile lamellar structure. Formation of a viscous isotropic phase occurs at 144°C which at 170°C is transformed to give a hexagonal-type structure with the lipid head groups forming a core of parallel rods packed in a regular two-dimensional hexagonal lattice. Increasing the water content to about 10% causes the gel–liquid crystalline transition temperature to decrease progressively to a value of about 40°C, which is then independent of water content above 10%. At 47°C, this SPM preparation shows a maximum water uptake of 35% (w/w). Above this water content, the maximally swollen lamellar lipid phase coexists with an excess bulk water phase. At this limiting hydration, the area per molecule was found to be 60 $Å^2$ and the lipid bilayer and water layer thickness 38 and 22.2 Å, respectively. The differences between these values and those reported earlier by Reiss-Husson [235] may be due to differences in fatty acid composition of the SPM but are more probably due to differences in the temperature of the measurements [236]. Shipley and coworkers have pointed out that the maximum hydration of 35%, observed with bovine brain SPM at 47°C, is similar to that of egg PC at 5°C. Since either system, at these respective temperatures is just above the gel–liquid crystalline phase transition, these authors suggested that the phosphocholine head group, which is common to both, is the principal factor controlling the swelling behaviour of the two phospholipids [236].

Below the transition temperature (25°C) this SPM exists in a bilayered structure which exhibits a maximum hydration of 42%. At the transition temperature, maximally hydrated SPM is found to have an area per molecule of 57.6 $Å^2$ and a bilayer thickness of 42.5 Å. The anhydrous SPM has an area per molecule of 36.1 $Å^2$ and a bilayer thickness of 63.5 Å at the same temperature. These data led the authors to assume a β-type structure in which the hydrocarbon chains are packed in a pseudohexagonal lattice with rotational disorder [237,238], similar to PC with heterogeneous acyl chains. They also suggested that the changes in the molecular packing which occur with increasing hydration may be due to a progressive tilt of the hydrocarbon chain axis in a β-type structure; this structure is known to occur in synthetic diacyl PCs [239]. The polymorphic phase behaviour of bovine brain SPM was also confirmed using ^{31}P-NMR studies [240]. Yeagle et al. [241], using such studies, proposed the presence of non-lamellar structure in dispersions of bovine brain SPM. Later, the same group [242] modified its conclusions and suggested that only a lamellar phase is present over the entire temperature range, as was previously suggested by the elaborate study of Cullis and Hope [240].

Most SPMs isolated from natural sources in excess water, exhibit a complex thermotropic behaviour—having a distinct gel–liquid crystalline phase transition. In most cases, the region of the phase transition is in the physiological temperature range [16,219,236,243] (see also Fig. 3). It is possible that the complex thermotropic behaviour is related to fatty acid composition of the SPM. Therefore, the thermotropic behaviour of synthetic SPMs is herewith described.

The thermotropic behaviour of aqueous liposomal dispersions of four synthetic SPMs of the DL-*erythro* configuration, prepared by the method of Shapiro [5], is described by Barenholz et al. [218]. It is worth emphasizing that Calhoun and Shipley [244] have shown that the thermotropic behaviour of D-*erythro*-*N*-palmitoyl sphingomyelin is very similar to that of the DL-*erythro* mixture [218].

The single sharp transition exhibited by each of the above four synthetic SPMs (C16 SPM, C18 SPM and *N*-palmitoyl dihydrosphingosylphosphocholine) is reminiscent of the gel–liquid-crystalline phase transition of synthetic PCs [245,246]. However, the simple linear increase in the transition enthalpy change and transition temperature with increasing acyl chain length, which holds for saturated diacyl phosphatidylcholines, is not obtained for these synthetic SPMs. This is described in Table 3.

When ΔH is plotted as a function of T_m, straight lines are obtained for both DL-*erythro* SPM and L-PC though the parameters which define the curves (i.e., the slope and intercept) differ considerably for the two lipid species [1]. Thus, for SPM, in spite of the linear relationship between ΔH and T_m, the ordering of the data does not conform with increasing acyl chain length. Furthermore, bilayers of C18 SPM exist below the transition in a gel form which exhibits an unusually high degree of order [247]. The transition of the gel phase of bilayers of this material to the liquid-crystalline state is associated with the large ΔH and high T_m (Table 3). When MLV of this SPM are prepared above the transition temperature, brought quickly to 20°C and then examined in the differential scanning calorimeter, a transition at about 45°C is observed which has an enthalpy of 7 kcal/mol. The gel phase of this quenched material provides an X-ray diffraction pattern exhibiting the degree of order usually associated with bilayered systems of glycerophospholipids [247]. The transition in the gel phase from the lesser to a more ordered system has a half-time of several hours at 20°C. As a result, composite patterns showing both low- and high-temperature transitions are frequently obtained. Exotherms in these patterns have also been occasionally observed. Traces of impurities such as the fluorescent probe, 1,6-diphenylhexatriene (DPH) or cholesterol cause the system to remain in the less-ordered form.

These two different gel phases of the C18 SPM were also monitored using Raman spectroscopy [248]. The thermotropic behaviour of the L-*erythro* C16 SPM is almost identical to that of the DL-*erythro* C16 SPM. The L-enantiomer, synthesized by Prof. D. Shapiro, was obtained from the DL-*erythro* C16 SPM. Both of these contained less than 2% of dihydrosphingosine which is of importance since the presence of

TABLE 3

Thermodynamic parameters for the gel-to-liquid crystalline transition of sphingomyelin-containing liposomes

Stereo configuration	Sphingomyelin species	MLV		SUV	Source
		T_m (°C)	ΔH kcal/mol	T_m (°C)	
DL-*erythro*	C16:0 dihydro	47.8	9.4	–	a
DL-*erythro*	C16:0	41.3	6.8	40.7	a
DL-*erythro*	C18:0	57.0	20.0	–	a,b,c
L-*erythro*	C18:0	45.0	7.0	43.6	a,b,c
DL-*erythro*	C24:0	48.6 (42.6)	15.3 (1.9)	46.3	a,b,k
L-*erythro*	C16:0	40.8	6.1	–	h
D-*erythro*	C16:0	37.5	6.7	g	d,j
D-*erythro*	C16:0	40.5	5.8	–	e
D-*erythro*	C18:0	44.5	–	g	i
D-*erythro*	C22:0	44.5	6.5	g	d
D-*erythro*	C24:0	43.8 (35.5)	7.3 (4.6)	–	d,k
D-*erythro*	C24:1	27.5	very broad	27.0	d

[a] Barenholz and Thompson [1].
[b] Barenholz et al. [218].
[c] C18 SPM has a complex thermotropic behaviour exhibiting a phase transition from a stable gel to the liquid-crystalline state at 57°C; or from a metastable gel to liquid-crystalline state at 45°C [286].
[d] All SPMs having the *erythro* configuration were prepared by partial synthesis from SPC.
[e] Calhoun and Shipley [244].
[f] Based on results obtained using DPH [249].
[g] D-*erythro* sphingomyelins having saturated acyl chains do not form SUV; D-*erythro* C24:1 SPM forms SUV which aggregate and/or fuse to larger but leaky aggregates.
[h] Prepared by Prof. D. Shapiro by complete synthesis. The thermotropic behaviour was determined as described by Dagan et al. [249].
[i] Determined by using DPH and *trans* parinaric acid (Cohen and Barenholz, unpublished).
[j] Neuringer et al. [392].
[k] The numbers in parentheses are for a second endotherm (a pretransition-like endotherm which may represent interdigitation of the paraffinic chains (section 4(f)).

dihydrosphingosine may decrease the values of T_m as well as ΔH [247]. The D-*erythro* SPMs (Table 3) were prepared by partial synthesis from SPC and the desired fatty acid; SPC was prepared from bovine brain SPM and therefore had a high content of dihydrosphingosine, approx. 20% [249]. This may explain the lower T_m and ΔH values obtained for the D-*erythro* SPMs, when compared with those of the DL-*erythro* SPMs. SUV of sphingomyelin have a lower T_m than MLV [249] and in this they resemble those prepared from synthetic PC [250,250a]. The above differences between the thermal behaviour of saturated diacyl PCs and synthetic SPMs are probably related to the fact that, while in diacyl PCs the two methylene chains of each molecular species are always of equal length, in SPM the two methylene chains are of roughly equal length only when the acyl chain is 16 carbon atoms long (Scheme 3). In SPMs, the methylene chain, contributed by the sphingosine base, is of constant length in all molecular species. Increasing the acyl chain length increases the length disparity between the two methylene chains of SPMs. Thus, it is possible that when there is no disparity between the methylene chain lengths the thermal properties and degree of order of SPM and the corresponding PC are indeed similar

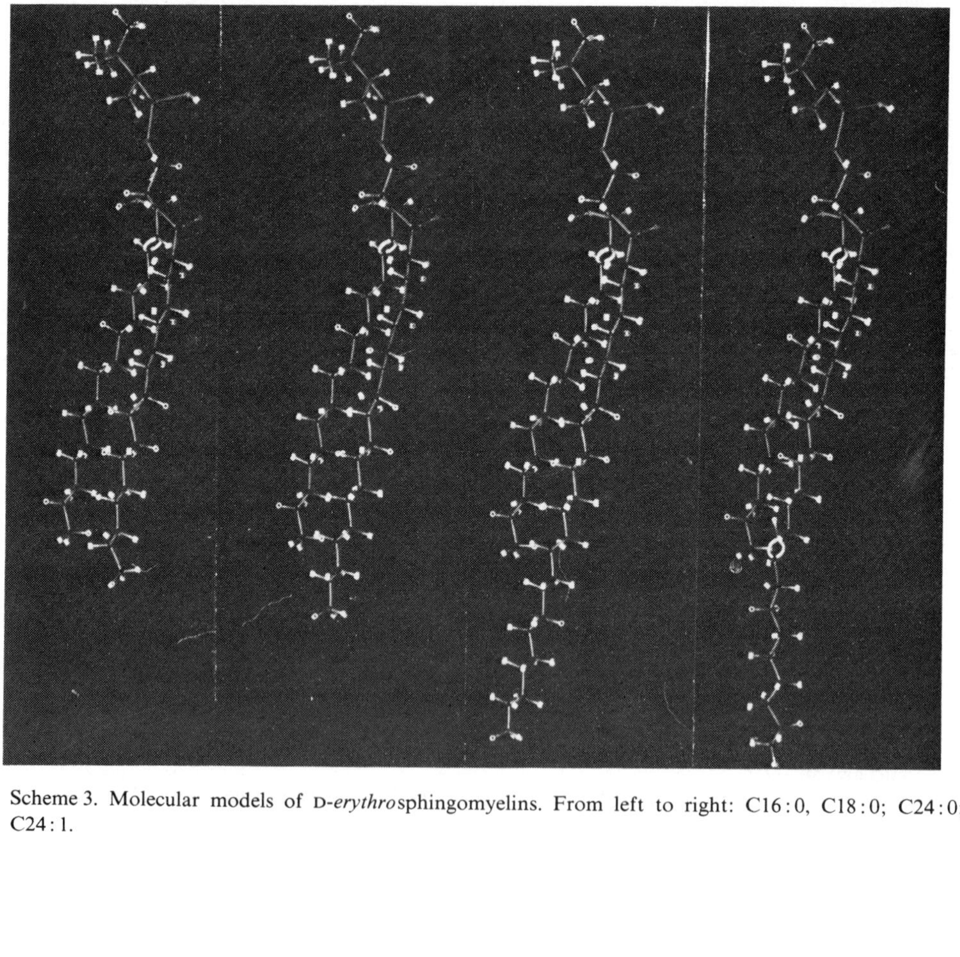

Scheme 3. Molecular models of D-*erythro*sphingomyelins. From left to right: C16:0, C18:0; C24:0; C24:1.

[1]. A larger degree of chain-length disparity may force the molecules, in the gel phase, to eventually assume a higher degree of order and display larger values of T_m and ΔH. But a still larger disparity in chain length may cause the molecules to be trapped in a more disordered state, with a resultant lowering of both ΔH and T_m [1].

Mechanical coupling between the two monolayers comprising the bilayer of SUV, prepared from synthetic C24 SPM [251] has been demonstrated with the aid of ^{1}H-NMR. The experimental approach utilized two properties of trivalent, paramagnetic lanthanide ions: the well known fact that they can be used to separate resonances due to molecules on the external part of small single-walled vesicles from those of molecules inside, and that the binding of these ions to PC head groups increases the thermotropic transition temperature. The transition was monitored using the line widths of the methyl group. These resonances provide good outside/inside resolution, are of high and constant intensity through the phase transition, and show a sharp break at a temperature which corresponds well with the onset of the thermotropic transition of small single-walled vesicles, as monitored by calorimetry. Data obtained using C24 SPM clearly show that the onset temperature of the internal monolayer is affected by increasing that of the external monolayer. No evidence for monolayer coupling was found in experiments on C18 SPM or dipalmitoyl phosphatidylcholine (DPPC), lipids which contain methylene chains of nearly equal length. Thus, it is possible that such coupling may result from interdigitation of the methylene chains of the two monolayers; this is favoured by the marked difference in the respective lengths of the two methylene chains comprising the molecule of C24 SPM. This was also supported by recent Raman studies [248]. The two endotherms observed for C24 SPM (see Table 3) may be the result of two populations of molecules, of which one includes the molecules involved in interdigitation between the paraffinic chains of two opposing monolayers of the same bilayer.

TABLE 4

Thermodynamic parameters for the gel-to-liquid crystalline phase transition and phase separation of mixtures of synthetic SPMs and 1-palmitoyl, 2-oleoylphosphatidylcholine (POPC)

Preparation (MLV)	T_m (°C)	ΔH kcal/mol	Source
C16 SPM–C18 SPM–C24 SPM, 1 : 1 : 1	39.5	6.8	a
C16 SPM–C24 : 1 SPM, 1 : 1	30	–	b
C18 SPM–POPC, 2 : 1	47.6	9.8	a
C18 SPM–POPC, 1 : 1	35 (24;15)		c

[a] Barenholz et al. [218]; all SPMs are of the DL-*erythro* configuration.
[b] Cohen and Barenholz (unpublished); data were obtained by fluorescence depolarization of DPH using D-*erythro* SPMs [249].
[c] The numbers in parentheses represent the two peaks of the shoulder of the endotherm. The curve is very broad and asymmetric.

The incorporation of a *trans* double bond between carbon atoms 4 and 5 of the sphingosine moiety appears to have little effect on the character of the phase transition. Thus, the difference between the T_m values for C16 SPM and C16 dihydrosphingomyelin (DHSPM) is only 6.5°C, a value considerably smaller than that observed as the effect of introducing a *cis* double bond between carbons 9 and 10 of the unsaturated acyl chain of PC. This apparent anomaly is most likely a consequence of the fact that the *trans* double bond of SPM is located mostly in the interfacial region rather than deeper, in the apolar moiety. Barton and Gunstone [252], have described similar effects resulting from variation in the position of a *cis* double bond in the acyl chains of phosphatidylcholine. In comparison, the presence of dihydrosphingosine in SPM induces a reduction of both T_m and ΔH. This is also the case when C16 SPM, C18 SPM and C24 SPM are mixed. The thermotropic behaviour of MLV formed from a 1:1:1, mole-ratio of these 3 species exhibits a single, sharp transition at a temperature below the T_m values for the individual species (Table 4). In comparison, using glycerophospholipids, such mixtures exhibit either a distinct transition with a T_m value lying between the respective values of the

TABLE 5

Fatty acid composition and thermodynamic parameters of gel-to-liquid crystalline phase transition of MLV of natural SPMs

Species	Tissue	TM_1 (°C)	TM_2 (°C)	TM_3 (°C)	ΔH kcal/mol	Fatty acid composition	Source
Sheep	brain	31.1	37.1	–	7.3	1	a
Bovine	brain	30.4	32.4	37.7	6.8	2	b
Bovine	brain	42	52		8.5	3	c1
Bovine	brain	42	52		9.1	4	c2
Bovine	brain	36	39		6.2	5	c3
Bovine	brain	22	26	32	5.5	6	c4
Egg	yolk	39			5.9	7	c5
Sheep	erythrocytes	21	31		5.3	8	c6
Human	erythrocytes	32			–	9	d
Bovine	erythrocytes	30				10	d

[a] Hertz and Barenholz [220].
[b] Barenholz et al. [218].
[c] Calhoun and Shipley [219]. The number describes the source of SPM: (1) research product; (2) United States Biochemicals; (3) prepared by Calhoun and Shipley [219]; (4 & 5) Avanti Biochemicals; (6) Supelco, Inc.
[d] Van Dijck et al. [243].

(1) 18:0—65.5%; 24:0—10.9%; 22:0—7.9%; 24:1—5.9%; 20:0—5.5%. (2) 18:0—35.9%; 24:1—30.8%; 24:0—9.7%; 25:0—4.7%; 22:0—4.4%. (3) 18:0—50.1%; 16:0—18.1%; 24:1—12.9%; 24:0—8.2%; 22:0—7.4%. (4) 18:0—42.5%; 16:0—16%; 24:1—11.5%; 22:6—13.1%; 24:0—9.5%. (5) 18:0—40.6%; 24:1—29.9%; 24:0—10.4%; 22:6—10.3%. (6) 24:1—73.4%; 16:0—19.8%; 24:0—3.5%; 22:0—3.3%. (7) 16:0—86.2%; 18:0—6%; 24:1—5.9%. (8) 24:1—55.2%; 22:0—18.7%; 24:0—10.9%; 16:0—9.5%. (9) 24:1—30.5%; 16:0—30.0%; 24:0—15.7%; 18:0—7.8%; 22:0—7.4%. (10) 24:0—42.2%; 16:0—17.1%; 22:0—10.7%; 24:1—10.5%.

individual components or, alternatively, show evidence of lateral phase separation [209,253,254]. In this particular case it appears that phase separation does not occur and that the thermotropic behaviour is a manifestation of the unique character of the mixed, but homogeneous, bilayer phase. This behaviour is similar to the thermotropic behaviour of bovine or ovine brain SPMs, which show a transition range below the transition temperatures of the principal components of the mixture. But, mixtures of the latter compounds show multiple maxima in the curve which describes the heat capacity as a function of temperature [218,236]. Recently, Calhoun and Shipley [219] Barenholz et al. [218] and Van Dijck et al. [243] have studied the thermotropic behaviour of bovine brain, egg yolk and ovine erythrocyte SPMs. The differences in acyl chain composition are reflected in the thermal behaviour of these preparations. The complexity of these systems is demonstrated in Table 5.

Clearly, considerably more work is needed to really understand the thermotropic behaviour of complex mixtures and interpret them in terms of thermal properties and interactions of the component SPMs.

(f) Molecular motion of sphingomyelin in bilayers

Information about the motion of the entire molecule or its components can be obtained using various physical techniques; each of them is related to a different time scale and therefore can give different information. Techniques include (1) NMR, using various nuclei including ^{1}H, ^{2}H, ^{13}C and ^{31}P [222,255–257]; (2) ESR [258–262]; (3) fluorescence [263,264]; (4) Raman spectroscopy [265,266]. Relatively little work of this sort has been carried out on SPM (see Fig. 3) as compared with the glycerophospholipids. These aspects were covered by a recent review [1].

5. Interactions of SPM with other lipids

Most biological membranes contain phospholipids, glycolipids and sterols (mainly cholesterol) whose distribution in the bilayer is not random, but asymmetric, as related to the two faces of the bilayer (see section 8b in this chapter). The possibility of non-random distribution in the plane of the membrane also exists. Gaining information on the relationship between membrane composition and structure or properties requires the understanding of intramembrane lipid–lipid as well as lipid–protein interaction.

The only detailed study of SPM-containing model membranes involved the interaction of SPM with PC. These two lipids, in combination with cholesterol, form the matrix of one of the monolayers which form the bilayered lipid membrane (see section 8b). Existence of non-lamellar structure in the membrane is a function of the composition. In the absence of cholesterol, bovine brain SPM resembles PC in its ability to stabilize the lamellar structure when mixed with soyabean PE [267].

(a) Interaction of sphingomyelin with phosphatidylcholine

The properties of these two choline-containing phospholipids and the interaction between them is a function of their molecular structure (see section 4b).

Untract and Shipley [268], studied mixtures of egg PC and bovine brain SPM of well-characterized acyl chain composition. A complete ternary phase diagram was constructed in the temperature range 10–44°C using data acquired by X-ray diffraction, differential scanning calorimetry and polarized light microscopy. The phase diagram shows that, at 37°C and in the presence of excess water, PC and SPM are completely miscible and are ordered in a liquid-crystalline, bilayer phase; at 20°C, lateral phase separation does occur in systems having more than 33 mol% of SPM. In these conditions, the system is composed of a lamellar gel phase of SPM and a liquid-crystalline phase containing SPM as well as PC. Below 20°C, the SPM gel phase coexists with an ordered bilayer phase composed of a stoichiometric phase containing 2 mol of PC per mol of SPM. When SPM content is less than 33%, the latter phase can be separated at temperatures below 20°C, from a liquid-crystalline bilayered phase composed largely of PC [268].

The phase behaviour of mixtures of egg PC with SPM of bovine spinal cord [269] or brain (Barenholz and Thompson, unpublished) has been studied with the aid of fluorescence polarization of DPH as probe. Systems of either SUV or MLV exhibit a broad phase transition in the range from 20–40°C when the fraction of SPM exceeds about 45 mol%. The apparent microviscosity, examined at temperatures between 0°C and 60°C, increases with increasing content of SPM. The use of *trans*-parinaric acid revealed a phase separation with concentrations as low as 25 mol% SPM (Barenholz and Thompson, unpublished). These data are compatible with the phase diagram for the closely similar system constructed by Untracht and Shipley [268]. Similar results were obtained for the lateral phase separation in liposomes made of PC/SPM mixtures isolated from LM cell plasma membrane [270].

The primary cause for the phase separation seems to be the well-defined difference in the acyl chain composition. Limited calorimetric studies were carried out on multilamellar liposomes formed from mixtures of synthetic C18 SPM and 1-palmitoyl-2-oleyl phosphatidylcholine (POPC) [218]. The heat capacity function of a mixture of these two components in which SPM comprises 66 mol%, exhibits a single, asymmetric transition with a temperature maximum at 47.6°C and ΔH of 9.8 kcal/mol. The width of the transition curve at half-height is about 5.5°C (Table 4). Comparison of these thermal characteristics with those of the pure SPM (shown in Table 3) suggests that below the transition region, a gel phase composed essentially of SPM coexists with a liquid crystalline phase of PC [218]. The function relating heat capacity to temperature for a 1 : 1 mole ratio mixture of these two compounds is considerably more complex, showing three poorly resolved maxima in the broad transition range of 10 to 40°C [218]. The thermal diagrams obtained for these two synthetic SPM-PC mixtures correspond well with the phase diagram for mixtures of egg PC and bovine brain SPM. This is expected, since the two synthetic compounds are major components of the two phospholipids prepared from natural sources [268].

The phase behaviour of 3-*sn*-DMPC mixed with DL-*erythro* C16 SPM was investigated, in MLV as well as SUV [271]. Temperature-induced changes in the membrane were studied using steady-state fluorescence polarization of DPH as well as freeze-fracture electron microscopy. The phase diagram was almost independent of the type of the vesicle used and resembled the DMPC-DPPC system, except for the immiscibility, in the ordered gel phases, below the "pre-transition" of pure C16 SPM. Also, the anisotropy of DPH fluorescence was found to be almost invariant with C16 SPM content at temperatures just above or below the gel to liquid-crystalline phase transition in SUV. This implicates the acyl chain compositions of the main structural parameter which determine the phase separation in systems containing SPM and PC and confirms the conclusions reached by Calhoun and Shipley [244] for MLV made of a mixture of D-*erythro* C16 SPM and DMPC. The proton NMR resonance of the *N*-methyl group of bovine brain sphingomyelin in an equimolar mixture with *N*-$(C^2H_3)_3$ egg PC, dispersed as SUV was investigated by Schmidt et al. [272]. Using vesicles of pure SPM the signal due to the choline methyl proton consists of two distinct, but overlapping peaks arising from molecules on the internal and external surfaces of the vesicle bilayer. At 100 MHz and 52°C, the splitting is 3.5 ppm and increases with decreasing temperature [272]. A similar splitting was observed using equimolar mixtures of SPM and PC. It is of interest that using the mixture at 29°C, the line width is markedly broader than it is in pure SPM at temperatures which induce a similar microviscosity. This observation suggests that the motion of the *N*-methyl system in the mixture is restricted. This may be the result of interactions between the two types of phospholipids in the mixed dispersion [272]. On the other hand, evidence exists for intramolecular hydrogen bonding between the phosphorus of SPM and the amide or hydroxyl hydrogens. This is derived from the observation that the T_1 values for the *N*-methyl protons of SPM are but little affected by the addition of *N*-$(C^2H_3)_3$-egg PC at concentrations as high as 67 mol% [272].

The fact that the intermolecular interactions between similar molecules differ from those observed using dissimilar molecules in mixed dispersions of SPM and PC is illustrated by the asymmetric distribution of these lipids in the bilayer of small vesicles. ^{31}P-NMR studies of equimolar mixtures of SPM of bovine brain and dipalmitoyl PC suggested that the sphingomyelin is more abundant in the outer surface and the PC in the inner surface [272–275].

(b) Interaction of sphingomyelin with cholesterol

With the exception of several microorganisms, cholesterol or similar sterols are components of all biological membranes [276,277]. Interaction of cholesterol with glycerophospholipids in bilayered membranes requires the β-hydroxyl group of the sterol and the carbonyl group of the phospholipid [207]. This topic was reviewed by Demel and De Kruijff [278] and by Huang [207,279]. Much less attention has been devoted to the examination of the interactions between cholesterol and sphingolipids. Vandenheuvel [280] suggested, on theoretical grounds, that molecular configuration and Van der Waals interactions should lead to a stable complex between cholesterol and SPM. Indeed, it has been known for some time that there is a significant relationship between the content of cholesterol and SPM in the membranes of many mammalian cells [99].

Early ^{1}H-NMR and ESR probe studies [281,282] showed that, in multilamellar liposomes formed from cholesterol and bovine brain SPM, the effect of the added cholesterol was to fluidize the bilayer below the phase transition while making it less fluid above the transition temperature (Fig. 3d). Thus, at low mole fractions, the effect of cholesterol on the apparent viscosity of SPM bilayers parallels that observed in glycerophospholipid systems [283]. It has also been known for some time that gel–liquid crystalline phase transition is not present in SPM-containing dispersions having more than about 40 mol% cholesterol (Fig. 3) [282]. This is similar to the behaviour of cholesterol-containing bilayers formed from saturated PCs [246,286]. More recently, ^{1}H-NMR studies on SUV composed of cholesterol and SPM have provided similar results [272]. Thus, in systems containing 40 mol% cholesterol at 52°C, the line widths of the methyl and methylene protons of the SPM are increased by 50% and the T_1 values for these protons are markedly reduced.

Recently, the thermotropic behaviour of aqueous dispersions of MLV formed from mixtures of cholesterol or synthetic C16 or C24 SPM were examined in detail by differential scanning calorimetry [287]. When the cholesterol content was less than 25 mol%, the curves describing heat capacity vs. temperature exhibited overlapping, sharp as well as broad components. The enthalpy related to the sharp component decreased with increasing cholesterol reaching zero between 25 and 30 mol%. The broad component enthalpy maximized at 3 and 4 kcal/mol, between 10 and 20 mol% cholesterol and decreased as the cholesterol content deviated from this range [287]. These data were interpreted as evidence that these mixtures undergo phase separation, with the sharper endotherm corresponding to a gel–liquid-crystalline transition of a phase enriched in SPM and the broader component associated in some manner with a cholesterol-enriched phase. Possibly, the broad component arises from a transition in the boundary region between the two phases [287]. This interpretation is based on the similarities between these systems and those formed from mixtures of cholesterol and DPPC [286]. Thus, there is evidence that a phase composed of a stoichiometric complex of cholesterol and SPM exists in the systems containing synthetic SPM similar to that forming in cholesterol/DPPC bilayers [285,286].

Further evidence supporting the existence of strong interactions between cholesterol and SPM was presented by Demel and coworkers [288] and by Van Dijck [289]. On the basis of calorimetric studies, these investigators suggested that in ternary systems of brain SPM, cholesterol and PC or PE, a phase separation of the phospholipid occurred perhaps by a complex of these two compounds in the liquid-crystalline phase. This conclusion was challenged by Calhoun and Shipley [244], who have carried out a similar calorimetric study in the ternary system: cholesterol/DMPC/C16 SPM. The two phospholipid components in this system are completely miscible in both the gel and liquid-crystalline phases, and thus no lateral phase separation occurs. In this system there was no evidence for the preferential interaction of cholesterol with either component [244]. This discrepancy can be resolved by assuming that, although cholesterol does appear to interact preferentially with SPM in a laterally phase-separated gel phase of SPM, it does not seem to show a preferential interaction in a molecularly mixed system, such as the DMPC/C16 SPM in either gel or liquid-crystalline configurations. This result suggests that cholesterol does not interact preferentially with SPM in the liquid-crystalline states of the systems studied by Demel and coworkers [288].

Rintoul et al. [270] suggested that sterols modulate interactions in membranes between different phospholipids, citing as an example the interaction of sterols with SPM and PC in LM cell plasma membrane.

Additional support for the preferential interaction between cholesterol and SPM is derived from the fact that extraction of cholesterol from cells is more rapid and complete when delipidated serum is reconstituted with SPM (rather than other phospholipids) and used as the extraction medium [290]. This was recently confirmed, for the depletion of vesicular stomatitis virus cholesterol by serum lipoproteins enriched with various phospholipids [291].

Another aspect of the SPM–cholesterol interaction is the stabilization of lamellar structures of membranes. In the presence of cholesterol, a mixture containing 30 mol% of bovine brain SPM and 70 mol% soya PE displayed a considerably greater affinity for the lamellar configuration than did similar PC/soya PE systems [240].

6. Interaction of sphingomyelin with detergents

The interaction of detergents with phospholipids is useful for studying physical properties of phospholipids and biological membranes [292–294]. Also, it is well known that detergents are required for interaction of many enzymes with their lipid substrates [295]. Interaction of detergents with SPM has been studied using various physical and enzymic methods.

(a) Interaction with Triton X-100

Solubilization of SPM-MLV by Triton X-100 occurs only above the critical micellar concentration (CMC) of the detergent. In contrast, when SPM with Triton X-100 are mixed in organic solvent and the latter is removed prior to dispersion in aqueous media, Triton X-100 is incorporated into the mixed aggregates with SPM even below its CMC [297].

Hydrodynamic characterization of the aggregates using the latter procedure [221,296], showed that as long as the mole fraction of the detergent was between 0.32 and 0.79 only one population of mixed micelles was present. The properties of these depended on the composition of the mixture. Thus, the M_r-value decreases steadily with increasing molar fraction of Triton X-100. But, the "aggregation number" of the detergent remains nearly constant at a level of about 196 molecules per micelle, so that the decreasing M_r-value is due to the reduction of SPM from 442 to 50 molecules per micelle when the mole fraction of the detergent increases from 0.32 to 0.79. It was suggested that the shape of the mixed micelles and their internal organization depends on the ratio of the two components, changing from an oblate ellipsoid at low Triton to a spherical structure at increased levels of Triton X-100. Also, it was suggested that the molecules of Triton X-100 are not homogeneously distributed in the micelles, so that relatively more detergent is present in the regions of high curvature, while the SPM is concentrated in the regions of low curvature of the micelle. This is shown in Fig. 2 [296] and was supported by proton NMR studies of the mixed micelles [298].

The rate of hydrolysis of SPM by lysosomal SPM increases dramatically in the presence of detergents such as Triton X-100 [100,107]. The kinetic parameters of the enzymic reaction can also be related to the physical properties of mixed micelles of SPM and Triton X-100 [125].

Hertz and Barenholz [299] studied interaction of Triton X-100 with MLV composed of mixtures of spinal cord SPM and egg yolk lecithin, supplemented with 10% diacetylphosphate. Incorporation of Triton X-100 and MLV solubilization were time-dependent, much slower than the formation of simple micelles, and governed by mol ratios of SPM and PC, or Triton to phospholipid, as well as the absolute concentration of Triton X-100. Titration with increasing Triton concentration showed that at a low Triton to phospholipid molar ratio, the leakage of glucose entrapped inside MLV occurs without reducing the turbidity. At greater molar ratios of Triton to phospholipid, mixed micelles of Triton and phospholipid occur, followed by a drastic decline in turbidity. The above effects could be related to the molar ratio of SPM to PC. The greater the mole fraction of SPM in the membrane, the less Triton is required to reach the concentration affecting a complete solubilization. These effects are due to tighter packing and stronger phospholipid-phospholipid interactions induced by SPM and expressed as greater apparent microviscosity following increase of the mole fraction of SPM in the bilayer [299].

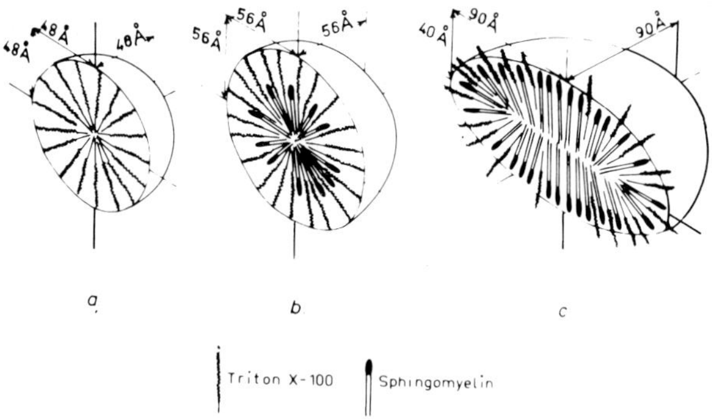

Fig. 2. Cutaway representation of mixed micelles of Triton X-100 and SPM with 3 different mol fractions of the detergent: (a) 1.0; (b) 0.79; (c) 0.3 [221,296].

(b) Interaction of sphingomyelin with bile salts

Solubilization of SPM by sodium taurodeoxycholate occurs when the molar ratio between the two compounds does not exceed 3.3. Analytical ultracentrifugation studies suggested that the size of the mixed aggregates was a function of the molar ratio when using sodium taurodeoxycholate but of the absolute concentration of the bile salt when sodium taurocholate was employed [126]. The difference between the two detergents is one extra hydroxyl group in the sodium taurocholate. This is also the reason for its greater CMC [300].

The rate of hydrolysis of SPM by the lysosomal sphingomyelinase of rat brain could be related to the physical properties of the mixed aggregates of bile salt and SPM [126].

7. *Interaction of sphingomyelin with proteins*

Although the phosphocholine moieties of SPM and PC are chemically identical, enzymes which hydrolyze this group, namely, phospholipase C and sphingomyelinase, are specific for SPM or PC [101–103,108,114,115,140,149,218,302–304]. For the sphingomyelinase of *S. aureus*, the enzymic activity could be related to the thermotropic behaviour of the MLV (Fig. 3). Optimal activity appears at the phase transition region and depression of this transition by cholesterol also depresses the enzymic activity [156]. Hydrolysis of monomolecular films of synthetic and natural SPMs by the above enzyme is also optimal at the transition from the liquid-condensed to the liquid-expanded states [127]. This is in accord with the hydrolysis of PC by pancreatic phospholipase A_2 which is also enhanced by a phase separation, imposed on either monomolecular films or multilamellar liposomes by the presence of SPM (Barenholz, Cohen and Thompson, unpublished; Barenholz and Verger, unpublished).

Specific interaction of SPM with membrane proteins other than sphingomyelinase has been proposed by Kramer and coworkers [306] who showed, in reconstitution experiments, that the proteins of the sheep erythrocyte bind more strongly to SPM than the corresponding proteins of the human erythrocyte. This preferential binding correlates with the unusually high SPM content of the sheep erythrocyte membrane shown in Table 6. Widnell and co-workers [307,308] reported that the 5′-nucleotidase from the plasma membranes of rat liver cells is isolated as a complex with SPM; about 100 mol of lipid are found per mol of protein. Removal of SPM inactivates the enzyme [307,308]. In contrast, Evans and Gurd [309] purified a similar enzyme from mouse liver plasma membranes which was active in the absence of any lipid. Sandermann [310] discussed in detail the problems of establishing specific lipid requirements for membrane-bound enzymes. The $(Na^{+}+K^{+})$-ATPase isolated from rabbit kidney by Lubrol extraction binds strongly to liposomes made from SPM but not to dispersions of PC unless the positively charged amphiphile, stearylamine, is added [311]. Recently, it was shown that the haemolytic toxin isolated from the sea anemone, *Stoichactis helianthus*, binds to aqueous dispersions of SPM but not to liposomes formed from other erythrocyte lipids. The observation suggests that the site of binding of this cytolytic glycoprotein, whose M_r-value is 16000 might be SPM of the plasma membranes [312].

Perhaps the most interesting interaction between SPM and a membrane protein is that with acetylcholinesterase [313]. This molecule, when isolated from the electric

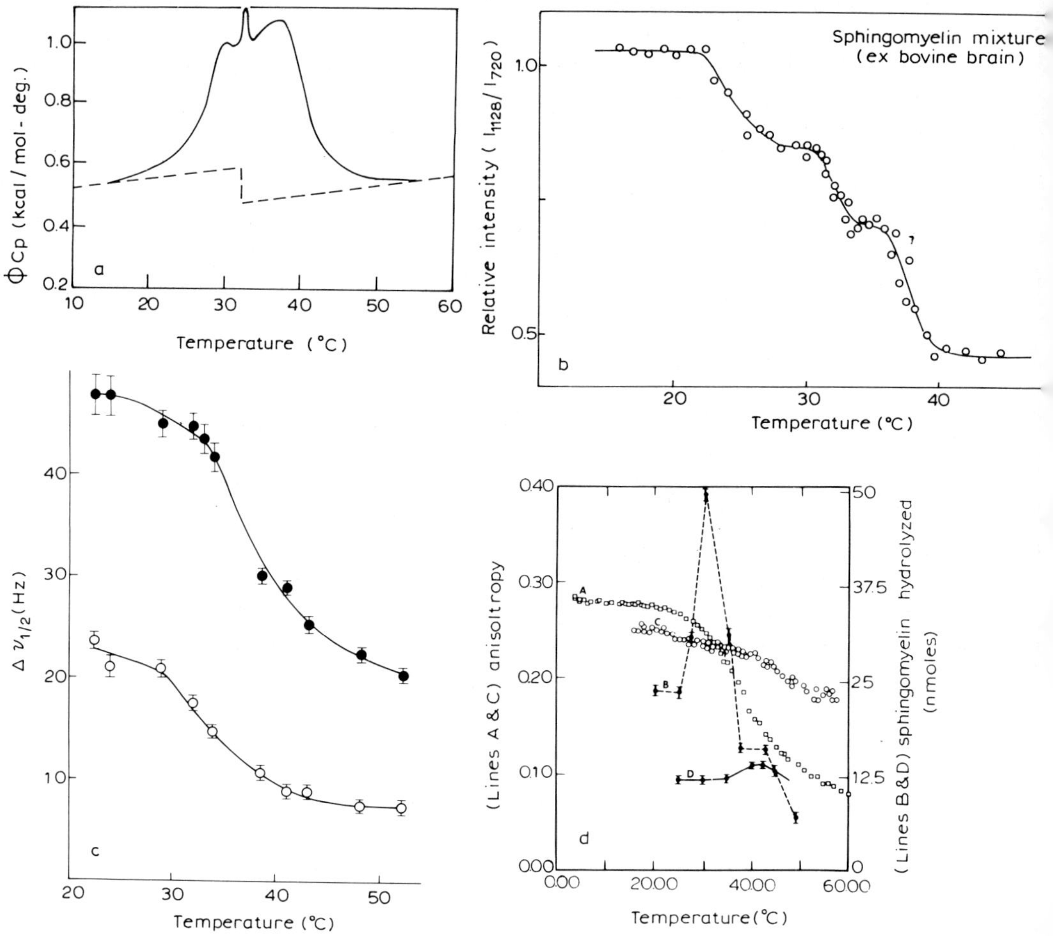

Fig. 3. Thermotropic behaviour of bovine brain SPM-(MLV) (a) A calorimetric scan relating ϕC_p and temperature; (b) "melting curve" of SPM-MLV derived from the change in the *trans* band intensity relative to the temperature invariant C-N stretch; (c) effect of temperature on the line widths of methylene (●) and choline-methyl (○) proton resonances using small unilamellar vesicles; (d) correlation between the thermotropic behaviour, as measured by fluorescence depolarization of DPH and the rate of hydrolysis of SPM by the enzyme of *S. aureus* (○, SPM; ●, SPM with 40 mol % cholesterol).

TABLE 6

PHOSPHOLIPID DISTRIBUTION IN ERYTHROCYTES FROM VARIOUS MAMMALIAN SPECIES [a]

	Dog	Rat	Guinea pig	Horse	Rabbit	Human	Cat	Pig	Sheep	Cow	Goat
PC	46.9	47.5	41.1	42.4	33.9	35.7	30.5	23.3	n.d.	n.d.	n.d.
PE	22.4	21.5	24.6	24.3	31.9	24.7	22.2	29.7	26.2	29.1	27.9
PS	15.4	10.8	16.8	18.0	12.2	13.8	13.2	17.8	14.1	19.3	20.8
PI	2.2	3.5	2.4	0.3	1.6	5.8	7.4	1.8	2.9	3.7	4.6
PA	0.5	0.3	4.2	0.3	1.6	n.r.	0.8	0.3	0.3	0.3	0.3
SPM	10.8	12.8	11.1	13.5	19.0	24.7	26.1	26.5	51.0	46.2	45.9
LPC	1.8	3.8	0.3	1.7	0.3	n.r.	0.3	0.9	n.d.	n.d.	n.d.
Others									4.8	1.7	0.8
SPM/PC	0.23	0.27	0.27	0.32	0.56	0.64	0.855	1.13	12.0	12.0	13.0

[a] Source: Rouser et al. [22]; White [21].

PC, phosphatidylcholine; PE, phosphatidylethanolamine; PS, phosphatidylserine; PI, phosphatidylinositol; PA, phosphatidic acid; SPM, sphingomyelin; LPC, lysophosphatidylcholine; X, unidentified; n.d., not determined; n.r. not recorded.

tissue of *Electrophorus electricus*, appears in electron micrographs to be composed of clusters of 4, 8 and 12 subunits attached to a 50 nm tail which is similar in amino acid composition to collagen and which may be removed from this assembly by collagenase. Treatment of intact electric tissues with this enzyme leads to solubilization of acetylcholinesterase activity. It is possible that the collagen-like tail-piece serves to anchor the enzyme assembly to the plasma membranes of the cells in the electric organ. Watkins and coworkers [313] using a flotation-type assay, showed that purified acetylcholinesterase binds strongly to liposomes composed of bovine brain SPM but not to those made from egg PC. Cohen and Barenholz (unpublished) found that the enzyme binds well also to liposomes made of synthetic DL-*erythro*-SPM. The binding to these liposomes was unaffected by the thermotropic behaviour of the lipid and did not damage the integrity of the liposomal membrane. Binding of the acetylcholinesterase to SPM liposomes was reduced by inclusion of PC in the liposome bilayer (Cohen and Barenholz, in preparation).

The observation that collagenase-treated enzyme does not bind to SPM liposomes localizes the site of interaction in the collagen-like tail of the enzyme assembly [313]. Although the molecular basis for the interaction between the collagen-like tail and the SPM bilayer system has not been established, it seems possible that the apparent specificity may rest in the ability of SPM to form hydrogen bonds with the hydroxyprolyl and hydroxylysyl residues of the tail (see section 4b).

Apart from being an important component of many biological membranes, SPM also occurs in the circulating plasma lipoproteins of higher animals. In man, the ratio of PC to SPM is about 4:1 in the total plasma lipoprotein fraction [314]. The ^{31}P-NMR resonances arising from each of these phospholipids can be resolved in both native and reconstituted human lipoprotein [315]. The chemical shift difference between the high-field signal resulting from PC and the downfield resonance from SPM is similar to that observed in vesicles formed from mixtures of these phospholipids [272,274]. The spin-lattice relaxation time for the SPM-^{31}P of 1.73 *s* is considerably smaller than the value of 2.3 *s* obtained for the PC signal. Similar results have been obtained in mixed vesicle systems [275,316]. The similarity between the environments of these two phospholipids in both reconstituted lipoproteins and small vesicles was also demonstrated by ^{13}C-NMR spectroscopy of phospholipids enriched in the *N*-methyl system [317,318]. These results suggest that the interactions between these phospholipids and the apolipoproteins do not involve the phosphocholine moiety of either molecule. In reconstitution experiments, SPM appears to bind strongly to apolipoprotein A-II but not to A-I, which interacts preferentially with PC [319]. Apolipoprotein A-II, a disulphide-linked dimer of two identical polypeptides, each of 77 amino acids, constitutes about 30% of the protein of human plasma high-density lipoprotein [314] (HDL). The molecular specificities of SPM and PC are sufficient to interact differently with antibodies. Arnon and Teitelbaum [320] showed that antibodies prepared against SPC bound to a carrier through its free amino group at C2 (see Scheme 2) interacted with membranes enriched in SPM (such as sheep red blood cell, see Table 6) but not with parallel membranes whose SPM content is low and replaced by PC, such as guinea pig red blood cells. This suggested that the interface region of the choline phospholipids is responsible for the antibody specificity [320].

8. Sphingomyelin in biological systems

(a) Distribution

SPM is a major lipid component of the cellular membranes of mammals [21,22] as well as of serum lipoproteins [321,322]. Its content varies considerably in membranes from diverse sources but, in many systems, the sum of the two choline-containing lipids, SPM and PC, constitutes about half of the total phospholipid, although the molar ratio of these two respective components varies considerably [21,22]. This is true even for membranes of different organs and tissues of a single mammalian species. The PC to SPM ratio is fairly constant in the same organ of different mammalian species (Rouser et al., 1968). This is not true for brain, however, where large variations in the ratio occur in various species [205], lipid-rich regions having higher SPM/PC ratios than lipid-poor regions. In spinal cord, this ratio is even higher than in brain [205].

There is also a very distinct pattern of SPM distribution in various parts of the eye. All the hard tissues in the bovine eye (i.e., the iris, choroid, cornea, sclera and the vitreous body) are very rich in SPM [205]. That is a characteristic of the plasma membrane of epithelial cells and is similar for oligodendroglia. This is in contrast to the very low levels of SPM (about 5% of total phospholipids) in retina, rod outer segment and neurons [205]. In the erythrocyte membrane, which has been extensively studied (for reviews see Ellory and Loew [323], Lux et al. [324]), the SPM/PC ratio varies with mammalian species from 0.25 in rats to above 12 in ruminants. Data for 11 species are summarized in Table 6. Cells which have subcellular organelles show the highest ratio in the plasma membrane and the lowest in both nuclear and mitochondrial membranes; the endoplasmic reticulum and Golgi membranes have intermediate values [205]. Based on the morphology of the generalized cell, there is thus an increasing gradient in the ratio from the cell centre to the periphery [205]. The positive correlation between the SPM and cholesterol contents of many membrane systems in mammals has already been mentioned in section 5b. This correlation does not hold for erythrocytes, where the cholesterol: total phospholipid ratio is constant [21]. Plasma lipoproteins contain two separate regions: the core which consists mostly of neutral lipids and the envelope which is composed mostly of phospholipids, cholesterol and the apolipoproteins [314,321,322]. Among the lipoproteins, the low-density lipoproteins (LDL) have the greatest content of SPM and HDL_3 the lowest (Table 7). The molar ratio of SPM to PC in serum of patients with abetalipoproteinaemia is greater than in those without. These patients do not have any detectable amounts of apoprotein B, VLDL or LDL in their plasma. Their sole lipoprotein is an HDL-like lipoprotein which is very rich in SPM, its SPM:PC molar ratio being 0.5 while HDL of normal patients has a molar ratio

TABLE 7

Lipid composition of human plasma lipoproteins

Lipoprotein	Lipid content			Ratios (molar)			Ref.
	Triglyceride	Cholesterol (total)	Phospholipids	Esterified cholesterol/ Free cholesterol	Free cholesterol/ Phospholipid	SPM/PC	
	(mg/100 mg lipoprotein lipid)						
Chylomicrons	37.7	3.0	8.8	0.88	0.35	0.17	[322]
VLDL	55.7	16.8	19.3	1.30	0.73	0.25	[322]
LDL	9.3	60.2	30.2	4.33	0.73	0.39	[328]
HDL_2	6.1	42.5	42.4	2.81	0.51	0.2	[322]
HDL_3	6.7	38.4	40.9	4.38	0.33	0.12	[322]
HDL "Abeta"	1.8	53.2	45.0	2.30	0.70	0.56	[326]
"LDL-like"	3.8	59.1	34.8	3.51	0.74	0.42	[328]

of 0.14 [314,325,326] (also see Table 7). A relationship between the lipid composition of serum lipoproteins and the composition and properties of the red blood cell membrane can be demonstrated in abetalipoproteinaemia patients [326]. Also, the cholesterol content of this abnormal HDL is greater than in normal HDL. The LDL of heterozygotes of familial hyperlipoproteinaemia is another example in which the molar ratio of SPM to PC is increased (0.52 in comparison with 0.37 in normals) as is the molar ratio of cholesterol to phospholipid [327]. This relative increase in SPM and cholesterol may be due to the longer life span of the LDL in the patients with familial hyperlipoproteinaemia when compared with normals, which may result in a more efficient and faster removal of PC compared to SPM. Incubation of VLDL with milk lipoprotein-lipase in the presence of albumin results in formation of LDL-like particles, in which the ratios of SPM to PC, and cholesterol to phospholipid increase to those in native LDL [328]. Similar changes occur using rat VLDL [301].

(b) Membrane asymmetry

Considerable evidence suggests that there is a marked asymmetric distribution of lipids between the outer and inner faces of plasma membranes [329–331]; the best documented system is the membrane of the human erythrocyte. Experiments utilizing phospholipases [154,155,332,333], phospholipid exchange proteins [334,335] or chemical labelling reagents [336–338] clearly show that essentially all of the SPM and most of the PC are located in the external surface of this membrane whereas PE and phosphatidylserine are the major lipid constituents of the cytoplasmic surface. Associated with this transmembrane lipid asymmetry is an absolute compositional asymmetry of protein components [330]. A similar asymmetric distribution of SPM has been demonstrated in rat erythrocytes [339], ovine erythrocytes [340], LM cell plasma membranes [341] and vesicular stomatitis virus [342]. In contrast to these systems, influenza A virus grown in Maden-Darby bovine kidney cells has been reported to have most of the SPM on the inner surface of the membrane [343]. In the case of viral membranes, most enveloped viruses contain relatively high levels of SPM which is a reflection of the similarity of lipid composition of the viral membrane to that of the plasma membrane of the host cell [344–347].

(c) Changes in sphingomyelin distribution associated with aging and pathological conditions

During aging of the aorta and arteries in humans, there is a striking increase in the relative contents of SPM and cholesterol in the membranes of cells comprising these tissues [22,144,348]. Rouser and Solomon [348] showed a linear correlation between the logarithm of the age (in years) and the total phospholipids. Increase of the latter was accounted for mostly by a large increase in SPM. Eisenberg et al. [144] showed a similar correlation, expressed as increased SPM : PC molar ratio with age. A similar, more pronounced increase occurs during the development of atherosclerosis

[145,348,349]. Smith and Cantab [349] showed that the principal change is in the intima where, during aging, SPM reaches 40% of the total lipid and in advanced aortic lesions can be as high as 70–80% of the total phospholipids. A striking increase in the proportion of SPM relative to other phospholipids with increasing severity of atherosclerosis, was reported for the fibrous plaques of the intima; the ratio SPM:PC increased 3- to 9-fold. This change is mainly in the "amorphous" lipid fraction. A parallel increase in the cholesterol/phospholipid ratio was also observed in this fraction [350]. These observations were described by Böttcher and Van Gent [351] in studies on aorta and coronary arteries, who also noted that the ratio between saturated and unsaturated fatty acids increases with age. Work by Smith and Cantab [349] indicates that these changes in lipid composition cannot be explained by the increase of connective tissue. Probably the increased concentration of lipids in the intima, of which 70% can be attributed to an increase in SPM, might be a consequence of changes in enzymatic activities [144,350], and by pronounced increase in the entry rate of the serum SPM into the aortic wall [350,352].

The role of the serum lipoproteins as a source for aortic wall SPM was demonstrated by Seth and Newman [352] who showed that the level of SPM in the aortic intima of rabbits increased exponentially with time, when the animals were fed a cholesterol-rich diet. This increase is a consequence of the exponential increase in the entry rate of serum SPM into the aortic wall which results in a marked change in the ratio of SPM:PC

Eisenberg and coworkers [145] showed greater incorporation of choline into the phospholipids of the aorta with increasing age; this increase parallels an increase of phospholipase A activity. But, sphingomyelinase activity remains constant or even declines. These authors showed that the SPM:PC ratio changes linearly from a value of 0.4 at birth to a value of 2.4 at the age of 90 years. Concomitantly, sphingomyelinase activity decreases by 50% relative to DNA. The authors propose that increased biosynthesis or intake of the phospholipids is followed by an increased rate of hydrolysis of PC but not SPM.

SPM accumulation was also observed in the agranular endoplasmic reticulum, the plasmalemma and smooth muscle cells, the principal cells of the intima and the inner media of the wall [353–355].

Recently it was shown that collagen binds to SPM liposomes with a 5–10-fold greater affinity than to PC liposomes (Barenholz and Cohen, 1983, in preparation). This may be related to the accumulation of collagen in the fibrous plaques [349]. There is also a significant positive correlation between the cholesterol and phospholipid content of the aortal wall. The correlation between SPM and free cholesterol is more pronounced in the non-sudanophilic portion. It has been suggested that the relation between endothelial integrity and accumulation of cholesterol during atherogenesis might be affected by the ability of the membranes of the endothelial cells to act as a barrier against excessive influx of cholesterol [356]. Bierman and coworkers [357] found that rat aortic smooth muscle cells take up "remnants" of very low-density lipoprotein (VLDL) formed by lipolysis of the VLDL by lipoprotein lipase. These are rich in cholesterol and SPM [358]. Cultured human arterial cells preferentially bind and take up the low-density lipoprotein (LDL) and VLDL fraction which includes the remnant [360]. LDL and the remnant are both rich in SPM and might be the source of the increased SPM of the aortal wall [352].

Changes in lipid composition with age occur also in the nervous system. In man, the rate of formation of new membranes and, therefore, the amount of total lipid in this tissue, is greater than the rate of loss by cell death up to the fourth decade of life. During this decade the rate of loss begins to exceed the rate of formation of new membrane [205]. Rouser et al. showed that, in human and animal brains, SPM and cerebroside gradually replace PC, and sulphatide replaces PE, but in some invertebrates SPM is replaced by ceramide phosphoethanolamine or ceramide phosphoethylamine [205]. A similar, age-dependent change in the ratio has also been noted for the lens of the eye in many species [361,362]. In man, this change is very dramatic, the SPM content of the lens rising to as high as 70% of the total phospholipid and the PC content falling to as low as 0.5% [361–363]. It is worth noting that the increase in SPM content with age does not exist in tissues or organs which have a rapid turnover of lipids such as kidney or liver [22,205] and skeletal muscle [364].

There are several pathological conditions, in addition to atherosclerosis, which are associated with a large increase in tissue SPM content. The best known of these is Niemann-Pick disease which was discussed in detail in section 3c. But changes in the SPM:PC ratio have also been noted in muscular dystrophy [365] and in some malignant diseases. Leukaemic cells appear to be deficient in both SPM and cholesterol [366,367]. In the thymus-derived leukaemia in the GR/A mouse strain, the reduction in the SPM and cholesterol content occurs by shedding the rigid plasma membranes which are enriched in SPM and cholesterol [368]. SPM deficiency was also observed in plasma membranes derived from rat thymic lymphoid cells when compared with normal thymocyte membranes. Again, this was parallelled by a lesser content of cholesterol and a smaller cholesterol: phospholipid molar ratio [369]. In other malignant types, the SPM:PC molar ratio and cholesterol content were increased. Thus, Bergelson and coworkers [370] showed that the SPM:PC ratio increased 2.3-fold in Jensen hepatomas when compared to regenerating rat liver. In a variety of hepatomas the principal increase in SPM is seen in the mitochondrial and nuclear membranes [370,371], which normally have the lowest content of this phospholipid. In at least one type of hepatoma having a marked increase in SPM the level of the SPM exchange protein is markedly elevated [372], though it is still not clear if this protein is identical with the non-specific phospholipid exchange proteins [335,373]. A similar increase in the relative content of SPM was observed in other malignant systems such as Ehrlich ascites cells in which this lipid amounts to 26% of the total phospholipids [374] (for reviews see [375–377]).

Senile cataract of the eye is another disease in which SPM level is increased above the normal increase caused by aging, while PC and PE levels are reduced. The change in lipid composition may be related to transport and membrane abnormality and/or to the insolubilization of proteins. SPM from cataract is more saturated than that isolated from normal patients [378]. The increase in SPM content is again parallelled by increased cholesterol [379].

(d) Membrane integrity and membrane properties

(i) Membrane integrity
Enzymatic hydrolysis of the lipid components of biological membranes might result in cell lysis. However, hydrolysis of 80% of the SPM in human erythrocytes by the sphingomyelinase of *S. aureus* and the resulting accumulation of ceramide in the membrane does not cause erythrocyte lysis [302]. Similar results were also obtained for ruminant erythrocytes in which SPM is the main phospholipid and the SPM : PC molar ratio might reach a value of 12 [154,333] (see Table 6 of this chapter). This shows that under iso-osmotic conditions, the conversion of SPM to ceramide, which remains in the membrane, does not lead to loss of membrane integrity. Erythrocytes treated in this fashion are, however, osmotically fragile [155] and after sphingomyelinase treatment immediate lysis of ruminant erythrocytes occurs at 4°C [157,158]. The "cold shock" is probably due to a phase separation which occurs at low temperature and results in membrane disintegration.

Conversion of erythrocyte SPM to ceramide also makes the PC of the membrane susceptible to attack by phospholipase C with resulting haemolysis [155]. Thus, haemolysis is caused by exposure of the erythrocyte to a mixture of sphingomyelinase and the phospholipase C of *B. cereus* or the non-specific phospholipase C from *Cl. perfringens* which hydrolyzes SPM as well as PC [155]. A similar effect is observed with porcine erythrocytes [380] or chicken erythrocytes depleted of ATP [333]. In comparison, in ruminant erythrocytes which contain large amounts of SPM, hydrolysis of SPM to ceramide is required for hydrolysis of PC by phospholipase C but haemolysis does not follow. In toad erythrocytes, which contain small amounts of SPM, phospholipase C is able to promote hydrolysis of glycerophospholipids without prior hydrolysis by sphingomyelinase [333]. This hydrolysis leads to lysis if the cells are depleted of ATP [333].

Haemolysis of sheep erythrocytes occurred after the synergistic action of phospholipase D of *Corynebacterium ovis* which hydrolyzed the SPM to sphingosine phosphate, and the phospholipase C of the same organism which further hydrolyzed the latter to ceramide. Lysis of the cells occurred upon chilling [381] in agreement with the results on sphingomyelinase [157,158].

Another aspect of membrane integrity is susceptibility to damage by detergents. Thus, the effect of Triton X-100 on a lipid bilayer is related to its lipid composition [220] (see section 6 of this chapter). It is of interest that the solubilization of phospholipids and cholesterol from erythrocyte membranes by Triton X-100 is differential. Of all the lipids present in this membrane, SPM and cholesterol are preferentially retained in the pellet and require much higher concentrations of detergent for their solubilization. This might indicate that in the plane of the membrane there are domains enriched in SPM and cholesterol which have a much higher apparent microviscosity [269] whereas the regions enriched in lecithin have a much lower apparent microviscosity. Because of their own low microviscosity, the Triton molecules preferentially penetrate the latter regions and only subsequently interact with SPM. Erythrocyte membranes with low SPM content are more sensitive to haemolysis by the bile salt glycocholate than erythrocytes with high SPM such as sheep erythrocytes [383]. Haemolysis of the latter is obtained when their PC content is increased, thereby reducing their SPM/PC molar ratio.

(ii) Mechanical properties and apparent microviscosity

The apparent microviscosity of biological membranes may be an expression of the degree of order and free volume in the membrane (see sections 4f and 5a). A general positive correlation between membrane SPM content and apparent microviscosity has been shown for erythrocytes from a variety of mammalian species [269,384]. The stability of erythrocytes under iso-osmotic conditions is also greater with those richer in SPM, although the resistance to osmotic shock appears to be lower [29]. A similar relationship was between SPM content and apparent microviscosity as well as osmotic fragility in MLV composed of SPM and PC [220].

The mechanical properties of the erythrocyte membranes also seem to correlate with SPM content. Cooper and coworkers [325] showed that membranes obtained from acanthocytic erythrocytes in patients with abetalipoproteinaemia are enriched in SPM and depleted in PC. The SPM:PC ratio can be as high as 1.56 in acanthocytes while the value in normal cells is about 0.86. Associated with this change is an increase in apparent microviscosity of the membrane as determined from DPH polarization studies. The acanthocytes also have a prolonged filtration time in a 0.3 μm nucleopore filter, indicative of decreased membrane deformability [325]. It was suggested that the increase of the apparent microviscosity in acanthocytes is related to the interaction of SPM with cholesterol [326]. Viral membranes are usually richer in SPM and cholesterol than the plasma membranes of their host cells [345]. It is suggested that the viral membrane is more ordered and closely packed than the host cell membrane. This was confirmed in most systems studied [218,342,344–347,385–387].

(iii) Permeability and transport in membranes

The permeability of erythrocyte membranes from a variety of mammalian species to various non-electrolytes and electrolytes correlates with their SPM content; the mole fraction of cholesterol is closely similar in the erythrocyte membranes of these species [29]. Deuticke [29] also showed that variations in fatty acid composition have only a small effect on permeability and that the permeability to molecules for which no specific transport system exists decreases with increasing SPM content. A strikingly similar correlation between SPM content and the permeability to water and glucose of multilamellar liposomes has been reported by Hertz and Barenholz [220].

The water permeability of artificial membranes of SPM differs considerably from those composed of PC. This is explained by the higher degree of saturation and the likelihood of hydrogen bonding in SPM bilayers [389]. Similar results were obtained for the permeability of the lipid bilayer to either the potassium or sodium salt of 6-carboxy-fluorescein. Liposomes of bovine brain SPM showed practically no leakage of this compound even at the gel–liquid-crystalline phase transition; this is a consequence of its high content of nervonic acid (C24:1) (Barenholz and Cohen, in preparation).

The effect of SPM content on transport is also observed in systems having an active transport. Kirk [390] showed that the level of active transport of K^+ in various mammalian erythrocytes is inversely related to the SPM:PC ratio. It is also suggested that interaction of SPM, the major lipid of sheep erythrocyte membrane, with cholesterol induces changes in the microenvironment of the Na^+ and K^+ pumps and is the reason for the discontinuity in the Arrhenius plots of K^+ flux [391].

9. Summary and Conclusions

SPM is one of the major lipids of membranes of mammalian cells. In most normal cells there is a gradient of SPM; its highest content is in the plasma membrane and its lowest in the inner mitochondrial membrane and the nuclear membrane which are almost free of the lipid, while SPM content in other subcellular membranes and organelles varies between the two above extremes.

In the plasma membrane, the two choline-containing lipids constitute more than 50% of the total phospholipids. SPM content increases with age, especially in tissues having a relatively low phospholipid turnover, i.e., nervous tissue and blood vessels. It also increases in several diseases, such as atherosclerosis, certain types of cancer (e.g. hepatoma), eye cataract and in Niemann-Pick disease. It is worth noting that there are other instances (such as leukaemia) in which the SPM content is reduced. It is of interest that, in general, there is a good positive correlation between the concentrations of SPM and cholesterol in membranes, and change in the content of one is followed by a similar change in the other. It is still not clear how cells maintain varying lipid composition in their different membranes. It is possible that the lipid composition of a membrane is established during its biosynthesis and assembly. Does membrane composition represent a thermodynamic equilibrium or is it preserved by the action of various enzymes and exchange proteins?

The pathological changes in SPM content might result from changes in its metabolism, e.g. increase in rate of biosynthesis, reduction in rate of degradation, or changes in transport in or out of the affected cells. The change occurring in a membrane might remain localized or be propagated to other membranes of the cell by transfer of the lipid.

It is clear to date that most membrane functions are closely related to the lipid composition which affects the physical properties of the membrane. A variable of all membranes is the SPM/PC molar ratio, while their sum is nearly constant, amounting to about half of membrane phospholipids. These two lipids are concentrated in one face of the membrane bilayer, namely that facing the extracellular environment. Variations in the relative contents of SPM and PC in artificial bilayers and biological membranes have a profound effect on the system properties of the bilayer. Perhaps the most striking difference between natural PC and SPM is the temperature of the gel–liquid-crystalline phase transition exhibited in bilayers. Most SPMs have their transition temperatures in the physiological temperature range, whereas almost all naturally occurring PCs are well above their transition temperature at 37°C.

The markedly different behaviour of SPMs and PCs in bilayered systems is a reflection of the differences in the molecular structures of these two classes of molecules. Although both molecular species have similar polar regions, composed of phosphocholine, and a hydrophobic region composed of two methylene chains, there are marked dissimilarities of structure elsewhere in the molecules. PCs have two methylene chains of about equal length, while SPMs have one methylene chain of constant chain-length, contributed by sphingosine and a second of varying length, contributed by the *N*-acyl group; the latter can be up to 10 carbons longer than the sphingosyl residue. This difference in the methylene chain length is in part responsible for those properties which are unique to bilayers composed of SPMs. The generally lower degree of unsaturation of SPMs relative to PCs is another contributing factor, and a third might be the difference in hydrogen bond-forming capability of the belt region, which connects the polar and apolar parts of these molecules. The amide bond and hydroxyl group in this region of SPM can act as hydrogen bond donors and acceptors whereas in PC, the carboxyl oxygens act only as hydrogen bond acceptors. Thus, bilayers formed from a mixture of naturally occurring SPMs and PCs show a distinct phase separation at 37°C.

These characteristics are further reflected in the apparent microviscosity of the mixed bilayer at 37°C which increases with increasing content of SPM. Also, the phase behaviour of bilayers composed of these two choline-containing lipids is strongly influenced by the addition of cholesterol. There is compelling evidence to suggest that the interaction between SPM and cholesterol is stronger than between this compound and PC. Thus, the microscopic phase configuration of simple bilayered systems is markedly affected by the relative concentration of SPM, PC and cholesterol. By inference, the same situation exists in the bilayers of the plasma membranes of cells. The thermotropic and phase behaviour and the permeability of membranes are related mainly to differences in the hydrophobic region of the two naturally occurring choline phospholipids. In contrast, interaction with various proteins and possibly with sterols, as well as some of the passive transport processes, seem to be related to differences in the interface region, mainly hydrogen bonding capabilities.

Even now it is possible to predict some membrane properties based on composition, but the relation between lipid composition and numerous biological properties of membranes or lipoproteins is not yet fully interpretable.

A computer search made in 1981 provided over 1000 citations of papers dealing with SPM, published in the recent decade. Many of these relate to the ratio of PC and SPM in amniotic fluid and have not been discussed at all. Furthermore, relatively less information has been added in the last few years on metabolic aspects than on physical properties related to membrane composition. This review therefore covers in greater detail the latter aspect and summarizes, in lesser detail the enzymic and metabolic aspects. It is clear that both of these aspects of the biochemistry of SPM are rapidly developing and that the coming few years will provide much new information on its intracellular metabolism as well as its function as a membrane constituent.

Acknowledgements

The work of the authors discussed in this review was supported by USPHS-NIH grants HL17576, NS 02967 and US-Israel BSF grants 1688, 2669 and 2772. We wish to thank Mrs. June Morris for her assistance in the preparation of the manuscript.

References

1 Barenholz, Y. and Thompson, T.E. (1980) Biochim. Biophys. Acta 604, 129–158.
2 Thudichum, J.L.W. (1884) A Treatise on the Chemical Constitution of the Brain, Baillière, Tindall and Cox, London.
3 Pick, L. and Bielschowsky, M. (1927) Klin. Wschr. 6, 1631–1637.
4 Merz, W.Z. (1930) Z. Physiol. Chem. 193, 59–63.
5 Shapiro, D. (1969) Chemistry of Sphingolipids, Herman, Paris.
6 Sweeley, C.C. and Moscatelli, E.A. (1959) J. Lipid Res. 1, 40–47.
7 Gaver, R.C. and Sweeley, C.C. (1965) J. Am. Oil Chem. Soc. 42, 294–303.
8 Karlsson, K.A. (1970) Chem. Phys. Lipids 5, 6–43.
9 Jungalwala, F.B., Hayssen, V., Pasquini, J.M. and McCleur, R.H. (1979) Lipid Res. 20, 579–587.
10 Michalec, C. and Kolman, Z. (1967) J. Chromatog. 31, 632–635; ibid. 636–639; ibid. 640–643.
11 Stoffel, W. and Assmann, G. (1972) Z. Physiol. Chem. 353, 65–74.
12 Carter, H.E., Rothfus, J.A. and Gigg, R. (1961) J. Lipid Res. 2, 228–234.
13 Michalec, C. (1967) J. Chromatogr. 31, 643–645.
14 Stoffel, W., Zierenberg, O., Tunggal, B. and Schreiber, E. (1974) Z. Physiol. Chem. 355, 1381–1390.
15 Shinitzky, M. and Barenholz, Y. (1974) J. Biol. Chem. 249, 2652–2657.
16 Karlsson, K.A. (1968) Acta Chem. Scand. 22, 3050–3052.
17 Reddy, P.V., Natarajan, V. and Sastry, P.S. (1976) Chem. Phys. Lipids 17, 373–377.
17a Carter, H.E., Shapiro, D. and Harrison, J.B. (1953) J. Am. Chem. Soc. 75, 1007–1013.
18 Karlsson, K.A. and Steen, G.O. (1968) Biochim. Biophys. Acta 152, 798–800.
19 Hizvisalp, E.L. and Renkonen, O. (1970) J. Lipid Res. 10, 47–55.
20 Samuelsson, B. and Samuelsson, K. (1969) J. Lipid Res. 10, 47–55.
21 White, D. (1973) in Form and Function of Phospholipids (Ansell, G.D., Dawson, R.M.C. and Hawthorne, J.N., eds.), pp. 441–482, Elsevier, Amsterdam.
22 Rouser, G., Nelson, C.J., Fleischer, S. and Simon, G. (1968) in Biological Membrane (Chapman, D., ed.), p. 5, Academic Press, New York.
23 Svennerholm, E., Stallberg-Stenhagen, S. and Svennerholm, L. (1966) Biochim. Biophys. Acta 125, 60–69.
24 O'Brien, J.S. and Rouser, G. (1964) J. Lipid Res. 5, 539–544.
25 Stallberg-Stenhagen, S. and Svennerholm, L. (1965) J. Lipid Res. 6, 146–155.
26 Leat, W.M.F. (1964) Biochem. J. 91, 444–447.
27 Van Deenen, L.L.M., De Gier, J., Houtsmuller, V.M.T., Montpoort, A. and Mulder, U. (1963) in Biochemical Problems of Lipids (Frazer, A.C., ed.), p. 404, Elsevier, Amsterdam.
28 Di Constanzo, G. and Clement, J. (1963) Bull. Soc. Chim. Biol. (Paris) 45, 137–142.
29 Deuticke, B. (1977) Rev. Physiol. Biochem. Pharm. 78, 1–97.
30 Shapiro, D. and Flowers, H.M. (1962) J. Am. Chem. Soc. 84, 1047–1050.

31 Grob, C.A. and Gadient, F. (1957) Helv. Chim. Acta 40, 1145–1156.
32 Shapiro, D. and Segal, H. (1954) J. Am. Chem. Soc. 76, 5894–5895.
33 Shapiro, D., Segal, H. and Flowers, H.M. (1958) J. Am. Chem. Soc. 80, 2170–2171.
34 Shapiro, D., Segal, H. and Flowers, H.M. (1958) J. Am. Chem. Soc. 80, 1194–1197.
35 Shapiro, D., Flowers, H.M. and Spector Slefer, S. (1958) J. Am. Chem. Soc. 81, 4360–4364.
36 Shoyama, Y., Okabe, H., Kishimoto, Y. and Costello, C. (1978) J. Lipid Res. 19, 250–259.
37 Grob, C.A., Jenny, E.F. and Utzinger, H. (1951) Helv. Chim. Acta 34, 2249–2251.
38 Egerton, M.J., Gregory, G.I. and Malkin, T. (1952) J. Chem. Soc. 74, 1952–1955.
39 Carter, H.E. and Shapiro, D. (1953) Am. Chem. Soc. 75, 5133–5138.
40 Jenny, E.F. and Grob, C.A. (1953) Helv. Chim. Acta 36, 1454–1471.
41 Jenny, E.F. and Grob, C.A. (1953) Helv. Chim. Acta 36, 1936–1952.
42 Gigg, R., Warren, C.D. and Cunningham, J. (1965) Tetrahedron Lett. 1303.
43 Gigg, J., Gigg, R. and Warren, C.D. (1966) J. Chem. Soc. 1872.
44 Hammarstrom, S.A. (1971) J. Lipid Res. 12, 760–765.
45 Shapiro, D. Rachman, Y., Robinson, Y. and Diver-Haber, A. (1967) Chem. Phys. Lipids 1, 183–191.
46 Alpes, H. (1974) Chem. Phys. Lipids 13, 109–116.
47 Kaller, H. (1961) Biochem. Z. 334, 451–456.
48 Gatt, S., Dinur, T. and Barenholz, Y. (1980) Clin. Chem. 26, 93–96.
49 Litman, B.J. and Barenholz, Y. (1975) Biochim. Biophys. Acta 394, 166–172.
50 Sribney, M. (1966) Biochim. Biophys. Acta 125, 542–547.
51 Braun, P.E., Morell, P. and Radin, N.S. (1970) J. Biol. Chem. 245, 335–341.
52 Morell, P. and Radin, N.S. (1970) J. Biol. Chem. 245, 342–350.
53 Ullman, M.D. and Radin, N.S. (1972) Arch. Biochem. Biophys. 152, 767–777.
54 Gatt, S. (1965) J. Biol. Chem. 241, 3724–3730.
55 Yavin, Y. and Gatt, S. (1969) Biochemistry 8, 1692–1698.
56 Sribney, M. and Kennedy, E.P. (1958) J. Biol. Chem. 233, 1315–1321.
57 Sribney, M. (1968) Arch. Biochem. Biophys. 126, 954–965.
58 Fujino, Y. and Negishi, T. (1968) Biochim. Biophys. Acta 152, 428–430.
59 Fujino, Y., Negishi, T. and Ito, S. (1968) Biochem. J. 109, 310–311.
60 Brady, R.O., Bradley, R.M., Young, O.M. and Kaller, H. (1965) J. Biol. Chem. 240, 3693–3694.
61 Fujino, Y., Nakano, M., Negishi, T. and Ito, S. (1968) J. Biol. Chem. 243, 4650–4651.
61a Negishi, T., Ito, S. and Fujino, Y. (1969) J. Japan. Biochem. Soc. May 41, 217–221.
62 Ullman, M.D. and Radin, N.S. (1974) J. Biol. Chem. 1506–1512.
63 Diringer, H., Marggraf, W.D., Koch, M.A. and Anderer, F.A. (1972) Biochem. Biophys. Res. Commun. 47, 1345–1352.
64 Diringer, H., Marggraf, W.D., Koch, M.A. and Anderer, F.A. (1972) Z. Naturforsch. 27b, 730–732.
65 Diringer, H. and Koch, M.A. (1973) Z. Physiol. Chem. 354, 1661–1665.
66 Marggraf, W.D. and Anderer, F.A. (1974) Z. Physiol. Chem. 355, 803–810.
67 Kanfer, J.N. and Spielvogel, C.H. (1975) Lipids 10, 391–394.
68 Crocker, A.C. and Mays, V.B. (1961) Am. J. Clin. Nutr. 9, 63–67.
69 Dvorkin, V.I. (1972) Dokl. Akad. Nauk SSSR 204, 499–502.
70 Freysz, L., Bieth, R. and Mandel, P. (1969) J. Neurochem. 16, 1417–1424.
71 Freysz, L., Bieth, R. and Mandel, P. (1971) Biochimie 53, 399–405.
72 Freysz, L., Lastennet, A. and Mandel, P. (1976) J. Neurochem. 27, 355–359.
73 Gerstl, B., Tavaststjerna, M.G., Eng, L.F. and Smith, J.K. (1972) Z. Neurol. 202, 104–120.
74 Gozlan-Devillierre, N., Baumann, N.A. and Bourre, J.M. (1976) Biochimie 58, 1129–1133.
75 Gouttler, F. (1972) Biochem. J. 128, 953–960.
76 Hanon, M. and Delaunay, A. (1971) Ann. Inst. Pasteur 120, 779–790.
77 Hirschberg, C.B., Wolf, B.A. and Robbins, P.W. (1975) J. Cell Physiol. 85, 31–39.
78 Henning, R. and Stoffel, W. (1969) Z. Physiol. Chem. 350, 827–835.

79 Kanazawa, I., Ueta, N. and Yamakawa, T. (1972) J. Neurochem. 19, 1483–1494.
80 Kiss, Z. (1977) Biochem. J. 168, 387–391.
81 Lastennet, A., Freysz, L. and Mandel, P. (1972) J. Neurochem. 191, 831–840.
82 Lastennet, A., Freysz, L. and Mandel, P. (1973) Biochim. Biophys. Acta 306, 287–297.
83 Lunt, G.G. and Lapetina, E.G. (1970) Brain Res. 17, 163–167.
84 Morin, R.J. (1969) Life Sci. 8, 613–616.
85 Nixon, R. and Kanfer, J.N. (1971) Life Sci. 10, 71–79.
86 Portman, O.W. and Alexander, M. (1970) J. Lipid Res. 11, 23–30.
87 Portman, O.W., Illingworth, D.R. and Alexander, M. (1973) J. Neurochem. 20, 1659–1667.
88 Pries, C. and Böttcher, C.J. (1968) J. Atheroscler. Res. 81, 731–734.
89 Prokhorova, M.I. (1969) Nerv. Sist. 10, 71–78.
90 Sribney, M., Duffe, M.K. and Lyman, E.M. (1973) Can. J. Biochem. 51, 1498–1504.
91 Sobotka, P. and Hinzen, D.H. (1973) Act. Nerv. Super (Praha) 15, 28.
92 Soula, G., Souillard, C., Gard, C. and Douste-Blazy, L. (1971) Eur. J. Biochem. 24, 264–270.
93 Soula, G. and Gard, C. (1972) Biochimie 54, 1213–1216.
94 Soula, G., Souillard, C. and Douste-Blazy, L. (1972) Eur. J. Biochem. 30, 93–99.
95 Sun, G.Y. and Horrocks, L.A. (1973) J. Lipid Res. 14, 206–214.
96 Stoffel, W. and Melzner, I. (1980) Z. Physiol. Chem. 361, 755–771.
97 Johnston, D., Matthews, E.R. and Melnykovych, G. (1980) Endocrinology 107, 1482–1488.
98 Gatt, S. and Bierman, E.L. (1980) J. Biol. Chem. 255, 3371–3376.
99 Patton, S. (1970) J. Theor. Biol. 29, 489–491.
100 Barenholz, Y., Roitman, A. and Gatt, S. (1966) J. Biol. Chem. 241, 3731–3737.
101 Heller, M. and Shapiro, B. (1966) Biochem. J. 98, 763–769.
102 Kanfer, J.N., Young, O.M., Shapiro, D. and Brady, R.O. (1966) J. Biol. Chem. 240, 1081–1084.
103 Schneider, P.B. and Kennedy, E.P. (1967) J. Lipid Res. 8, 202–209.
104 Brady, R.O., Kanfer, J.N., Mock, M.B. and Fredrickson, D.S. (1966) Proc. Natl. Acad. Sci. USA 55, 366–369.
104a Baraton, G. and Revol, A. (1977) Clin. Chim. Acta 76, 339–343.
105 Fowler, S. (1969) Biochim. Biophys. Acta 191, 481–484.
106 Weinreb, N.J., Brady, R.O. and Tappel, A.L. (1968) Biochim. Biophys. Acta 159, 141–146.
107 Gatt, S., Herzl, A. and Barenholz, Y. (1973) FEBS Lett. 30, 281–285.
108 Gatt, S. (1976) Biochem. Biophys. Res. Commun. 76, 235–241.
109 Gatt, S., Dinur, T. and Kopolovic, J. (1978) J. Neurochem. 31, 547–550.
110 Rao, B.G. and Spence, M.W. (1976) J. Lipid Res. 17, 506–515.
111 Spence, M.W. and Burgess, J.K. (1978) J. Neurochem. 30, 917–919.
112 Spence, M.W., Burgess, J.K., Sperker, E.R., Hamed, L. and Murphy, M.G. (1981) in Lysosomes and Lysosomal Storage Diseases (Callahan, J.W. and Lowden, J.A., eds.), pp. 219–228, Raven Press, New York.
113 Hostetler, K.Y. and Yazaki, P.J. (1979) J. Lipid Res. 20, 456–462.
114 Yamaguchi, S. and Suzuki, K. (1978) J. Biol. Chem. 253, 4090–4092.
115 Gatt, S., Dinur, T. and Leibovitz-Ben Gershon, Z. (1978) Biochim. Biophys. Acta 531, 206–214.
116 Pick, L. (1933) Am. J. Med. Sci. 185, 601–616.
117 Gatt, S. and Gottediner, T. (1976) J. Neurochem. 26, 421–422.
118 Pantchev, P.G., Brady, R.O., Gal, A.E. and Hibbert, S.R. (1977) Biochim. Biophys. Acta 488, 312–321.
119 Callahan, J.W., Shankaran, P., Khalil, M. and Gerrie, J. (1978) Can. J. Biochem. 56, 885–891.
120 Yamaguchi, S. and Suzuki, K. (1977) J. Biol. Chem. 252, 3805–3813.
121 Yamanaka, T., Hanada, E. and Suzuki, K. (1981) J. Biol. Chem. in press.
122 Callahan, J.W., Khalil, M. and Gerrie, J. (1974) Biochem. Biophys. Res. Commun. 58, 384–390.
123 Callahan, J.W., Khalil, M. and Philippart, M. (1975) Pediat. Res. 9, 908–913.
124 Harzer, K., Anzil, A.P. and Schuster, I. (1977) J. Neurochem. 29, 1155–1157.
125 Yedgar, S. and Gatt, S. (1976) Biochemistry 15, 2570–2573.

126 Yedgar, S. and Gatt, S. (1980) Biochem. J. 185, 749–754.
126a Gatt, S., Dinur, T., Yedgar, S. and Leibovitz-Ben Gershon, Z. (1978) Adv. Exp. Med. Biol. 101, 487–500.
127 Yedgar, S., Cohen, R., Gatt, S. and Barenholz, Y. (1981) Biochem. J. 201, 597–603.
128 Rooij, R.E., Liem, K.O., Brouwer-Schipper, J.W., Husmann, G.G. and Hooghwinkel, G.J.M. (1975) Clin. Chim. Acta 59, 71–79.
129 Poulos, A. and Pollard, A.C. (1976) Clin. Chim. Acta 72, 327–335.
130 Poulos, A. and Pollard, A.C. (1976) Clin. Chim. Acta 73, 353–356.
131 Maziere, J.C., Maziere, C., Hosli, P. and Polonovski, J. (1979) Biomedicine 31, 104.
132 Maziere, J.C., Maziere, C. and Polonovski, J. (1979) Clin. Chim. Acta 95, 447–451.
133 Seidel, D., Klenk, J., Fischer, G. and Pilz, H. (1978) J. Clin. Chem. Clin. Biochem. 16, 407–411.
134 Fensom, A.H., Benson, P.F., Babarik, A.W., Grant, A.R. and Jacobs, L. (1977) Biochem. Biophys. Res. Commun. 74, 877–883.
135 Gal, A., Brady, R.O., Hibbert, S.R. and Penchev, P.G. (1975) New England J. Med. 293, 632–636.
136 Gal, A.E. and Fash, F.J. (1976) Chem. Phys. Lipids 76, 71–79.
137 Gatt, S., Dinur, T. and Barenholz, Y. (1978) Biochim. Biophys. Acta 530, 503–507.
138 Gatt, S., Barenholz, Y., Goldberg, R., Dinur, T., Besley, G., Leibovitz-Ben Gershon, Z., Rosenthal, J., Desnick, R.J., Devine, E.A., Shafit-Zogardo, B. and Tsuruki, F. (1981) in Methods in Enzymology (Lowenstein, J. ed.), Vol. 72, pp. 351–374, Academic Press, New York.
139 Pellkofer, R. and Sandhoff, K. (1980) J. Neurochem. 34, 988–992.
140 Hirshfeld, D. and Loyter, A. (1975) Arch. Biochem. Biophys. 167, 186–192.
141 Record, M., Loyter, A. and Gatt, S. (1980) Biochem. J. 187, 115–121.
142 Barranger, J.A., Pantchev, P.G., Furbish, F.S., Steer, C.J., Jones, E.A. and Brady, R.O. (1978) Biochem. Biophys. Res. Commun. 83, 1055–1060.
143 Bowser, P.A. and Gray, G.M. (1978) J. Invest. Dermatol. 70, 331–335.
144 Eisenberg, S., Stein, Y. and Stein, O. (1969) J. Clin. Invest. 48, 2320–2329.
145 Eisenberg, S., Stein, Y. and Stein, O. (1969) Biochim. Biophys. Acta 176, 557–569.
146 Klein, F. and Vincendon, G. (1979) C.R. Soc. Biol. 173, 137–141.
147 Lentskii, K.M. and Matsepa, R.L. (1973) Ukr. Biochim. Zh. 45, 7–12.
148 Nilsson, A. (1968) Biochim. Biophys. Acta 164, 575–584.
149 Rachmilewitz, D., Eisenberg, S., Stein, Y. and Stein, O. (1967) Biochim. Biophys. Acta 144, 624–632.
150 Wenger, D.A., Wharton, C. and Seeds, N.W. (1979) Life Sci. 24, 679–684.
151 Zitman, D., Chazan, S. and Klibansky, C. (1978) Clin. Chim. Acta 86, 37–43.
152 Zwaal, R.F., Roelofsen, B., Comfurius, P. and Van Deenen, L.L.M. (1971) Biochim. Biophys. Acta 233, 474–479.
153 Pagtan, I., Macchia, V. and Katzen, R. (1968) J. Biol. Chem. 243, 3750–3755.
154 Zwaal, R.F.A., Roelofsen, B. and Colley, C.M. (1973) Biochim. Biophys. Acta 300, 159–182.
155 Zwaal, R.F.A., Roelofsen, B., Comfurius, P. and Van Deenen, L.L.M. (1975) Biochim. Biophys. Acta 406, 83–96.
156 Cohen, R. and Barenholz, Y. (1978) Biochim. Biophys. Acta 509, 181–187.
157 Smyth, C.J., Mollby, R. and Wadstrom, T. (1975) Infect. Immun. 12, 1104–1111.
158 Mollby, R. (1976) Zentr. Bakteriol. Parasit. Infekt. Hyg. 665–677.
159 Ikezawa, H., Mori, M., Ohyabu, T. and Taguchi, R. (1978) Biochim. Biophys. Acta 528, 247–256.
160 Bazovska, S. (1978) Csk. Epidemiol. Mikrobiol. Immunol. 27, 137–143.
161 Soucek, A., Michelec, C. and Souckova, A. (1971) Biochim. Biophys. Acta 227, 116–128.
162 Linder, R. and Bernheimer, A.W. (1978) Biochim. Biophys. Acta 530, 236–246.
163 Bernheimer, A.W., Linder, R. and Avigad, L.S. (1980) Infect. Immun. 29, 123–137.
164 Niemann, A. (1914) Jahrb. Kinderheilk. 79, 1–10.
165 Pick, L. (1926) Ergebn. Med. Kinderheilk. 29, 519–627.
166 Klenk, E. (1937) Z. Physiol. Chem. 229, 151–156.
167 Fredrickson, D.S. and Sloan, H.R. (1972) in The Metabolic Basis of Inherited Disease (Stanbury, J.B., Wijngaarden, J.B. and Fredrickson, D.S., eds.), pp. 783–807, McGraw-Hill, New York.

168 Brady, R.O. (1973) Adv. Enzymol. 38, 293–315.
169 Brady, R.O. (1975) Ann. Int. Med. 82, 257–261.
170 Brady, R.O. (1978) in The Metabolic Basis of Inherited Disease (Stanbury, J.B., Wijngaarden, J.B. and Fredrickson, D.S., eds.), pp. 718–730, 4th ed., McGraw-Hill, New York.
171 Wenger, D.A. (1977) in Practical Enzymology of the Sphingolipidoses (Glew, R.H. and Peters, S.P., eds.), pp. 39–70, Liss, New York.
172 Galjaard, H. (1980) Genetic, Metabolic Disease, Elsevier, Amsterdam.
173 Gajdos, A. (1972) Nouv. Presse Med. 1, 1789–1792.
174 Suzuki, K. (1978) Methods Enzymol. 50, 456–488.
175 Dacremont, G., Kint, J.A. and Cocquit, G. (1973) New Eng. J. Med. 289, 592–593.
176 Schwarzmuller, B. (1978) Med. Welt 29, 849–852.
177 Callahan, J.W. and Khalil, M. (1976) Adv. Exp. Med. Biol. 68, 397–478.
178 Callahan, J.W., Jones, C.S., Shankaran, P. and Gerrie, J. (1981) in Lysosomes and Lysosomal Storage Diseases (Callahan, J.W. and Lowden, J.A., eds.), pp. 205–218, Raven, New York.
179 Hers, H.G. and Van Hoof, F. (1973) Lysosomes and Storage Diseases, Academic Press, New York.
180 Crocker, A.C. (1961) J. Neurochem. 7, 69–80.
181 Schneider, E.L., Pentchev, P.G., Hibbert, S.R., Sawitsky, A. and Brady, R.O. (1978) J. Med. Genet. 370–374.
182 Martin, J.J., Philippart, M., Van Hauwaert, J. and Callahan, J.W. (1972) Arch. Neurol. 27, 45–51.
183 Kampine, J.P., Brady, R.O., Kanfer, J.N., Feld, M. and Shapiro, D. (1966) Science 155, 86–88.
184 Callahan, J.W. and Khalil, M. (1975) Pediat. Res. 75, 914–918.
184a Besley, G.T.N. (1977) FEBS Lett. 80, 71–74.
185 Wenger, D.A. (1978) Adv. Exp. Med. Biol. 101, 707–717.
186 Sloan, H.R., Uhlendorf, B.W., Kanfer, J.N., Brady, R.O. and Fredrickson, D.S. (1969) Biochem. Biophys. Res. Commun. 34, 582–588.
187 Sloan, H.R. (1970) Chem. Phys. Lipids 5, 250–260.
188 Minami, R., Matsura, Y., Nakamura, F., Kudoh, T., Sogawa, H., Oyanagi, K., Sukegawa, K., Orii, T., Maruyama, K. and Nakao, T. (1979) Hum. Genet. 47, 159–167.
188a Gautier, M., Rachman, F., Carreau, J.P., Garçon, E. and Raulin, J. (1972) Arch. Fr. Pediatr. 29, 635–639.
189 Besley, G.T.N. (1978) Clin. Chim. Acta 40, 269–278.
190 Besley, G.T.N., Hoogeboom, A.J.M., Hoogeveen, A., Kleijer, W.I. and Galjaard, H. (1980) Hum. Genet. 54, 409–412.
191 Christomanou, N. (1980) Z. Physiol. Chem. 361, 1489–1502.
192 Svennerholm, L., Hakansson, G., Mansson, J.E. and Vanier, M.T. (1979) Clin. Chim. Acta 92, 53–64.
193 Chazan, S., Zitman, D. and Klibansky, C. (1978) Clin. Chim. Acta 86, 45–49.
194 Patrick, A.D., Young, E., Kleijer, W.J. and Niermeijer, M.F. (1977) Lancet ii, 144, 8029.
195 Radwan, M. and Litwin, J.A. (1976) Pediat. Pol. 51, 1351–1354.
196 Jungalwala, F.B. and Milunsky, A. (1978) Pediat. Res. 12, 655–659.
197 Adachi, M., Volk, B.W. and Schneck, L. (1976) Am. J. Pathol. 85, 229–231.
198 Sakuragawa, N., Sakuragawa, M., Kuwabara, T., Pentchev, P.G., Barranger, J.A. and Brady, R.O. (1977) Science 197, 317–319.
199 Aubert-Tulkens, G., Van Hoof, F. and Tulkens, P. (1979) Lab. Invest. 40, 481–491.
200 Daloze, P., Delvin, E.E., Glorieux, F.H., Corman, J.L., Bettez, P. and Toussi, T. (1977) Am. J. Med. Genet. 1, 229—239.
201 Small, D.M. (1970) Fed. Proc. Am. Soc. Exp. Biol. 29, 1320–1326.
202 Sundaralingam, M. (1972) Ann. N.Y. Acad. Sci. 195, 324–355.
203 Pascher, I. (1976) Biochim. Biophys. Acta 455, 433–451.
204 Chapman, D. (1973) in Form and Function of Phospholipids (Ansell, G.B., Dawson, R.M.C. and Hawthorne, J.N., eds.), pp. 117–142, Elsevier, Amsterdam.
205 Rouser, G., Kitchevsky, G. and Yamamoto, A. (1972) Adv. Lipid. Res. 10, 261–360.
206 Pauling, L. (1960) The Nature of the Chemical Bond, p. 449, Cornell Univ. Press, Ithaca, NY.

207 Huang, C. (1976) Nature 259, 242–244.
208 Abrahamson, S., Dahlen, B., Lofgren, H., Pascher, I. and Sundell, S. (1977) in Structure of Biological Membranes (Abrahamson, S. and Pascher, I., eds.), pp. 1–23, Plenum, New York.
209 Lee, A.G. (1975) Prog. Biophys. Mol. Biol. 29, 3–56.
210 Pearson, R.H. and Pascher, I. (1979) Nature 281, 499–501.
211 Lofgren, H. and Pascher, I. (1977) Chem. Phys. Lipids 20, 263–271.
212 Sozoka, F. and Papahajopoulos, D. (1980) Annu. Rev. Biophys. Bioeng. 9, 467–508.
213 Bangham, A.D. (1969) Prog. Biophys. Mol. Biol. 18, 29–95.
214 Huang, C. (1969) Biochemistry 8, 344–351.
215 Sheetz, M.P. and Chan, S.I. (1972) Biochemistry 11, 4573–4581.
216 Thompson, T.E., Lentz, B.R. and Barenholz, Y. (1977) in Biochemistry of Membrane Transport, FEBS Symp. No. 42 (Smenza, G. and Carafoli, E., eds.), pp. 47–71, Springer, New York.
217 Lichtenberg, D., Freire, E., Schmidt, C.F., Barenholz, Y. and Thompson, T.E. (1981) Biochemistry 19, 3662–3665.
218 Barenholz, Y., Suurkuusk, J.E., Mountcastle, D., Thompson, T.E. and Biltonen, R.L. (1976) Biochemistry 15, 2441–2447.
219 Calhoun, W.I. and Shipley, G.G. (1979) Biochim. Biophys. Acta 555, 436–441.
220 Hertz, R. and Barenholz, Y. (1975) Chem. Phys. Lipids 15, 138–456.
221 Cooper, V.G., Yedgar, S. and Barenholz, Y. (1974) Biochim. Biophys. Acta 363, 86–97.
222 Lee, A.G., Birdsall, N.J.M. and Metcalf, J.C. (1974) In Methods in Membrane Biology (Korn, E.D., ed.), Vol. 2, pp. 1–156, Plenum, New York.
223 Gershfeld, N.L. (1976) Annu. Rev. Phys. Chem. 27, 349–376.
224 Jalal, I. and Zografi, G. (1979) J. Coll. Int. Sci. 68, 196–198.
225 Landau, F.M., Cadenhead, D.A. and Kellner, B.M.J. (1980) J. Coll. Int. Sci. 73, 264–266.
226 Jain, M.K. and Wagner, R.C. (1980) Introduction to Biological Membranes, Wiley, New York.
227 Shah, D.O. and Schulman, J.H. (1967) Lipids 2, 21–27.
228 Raper, J.H., Gammack, D.B. and Sloane-Stanley, G.H. (1966) Biochem. J. 98, 21p.
229 Shah, D.O. and Schulman, J.H. (1967) Biochim. Biophys. Acta 135, 184–187.
230 Shah, D.O. and Schulman, J.H. (1965) J. Lipid Res. 6, 341–349.
231 Colacicco, C. (1973) Chem. Phys. Lipids 10, 66–72.
232 Tausk, R.J.M., Karmigyelt, J., Oudshoorn, C. and Overbeek, J.Th.G. (1974) Biophys. Chem. 1, 175.
233 Schmidt, C.F., Barenholz, Y., Huang, C. and Thompson, T.E. (1977) Biochemistry 16, 3948–3954.
234 Hershberg, R.D., Reed, G.H., Slotboom, A.J. and DeHaas, G.H. (1976) Biochim. Biophys. Acta 424, 73–81.
235 Reiss-Husson, F. (1967) J. Mol. Biol. 25, 363–382.
236 Shipley, G.G., Avecilla, L.S. and Small, D.M. (1974) J. Lipid Res. 15, 124–131.
237 Luzzati, V. (1968) in Biological Membranes (Chapman, D., ed.), Vol. 1, pp. 71–123, Academic Press, New York.
238 Shipley, G.G. (1973) in Biological Membranes (Chapman, D. and Wallach, D.F.H., eds.), Vol. II, pp. 1–89, Academic Press, New York.
239 Tardieu, A.V., Luzzati, V. and Remen, F.C. (1973) J. Mol. Biol. 75, 711–733.
240 Cullis, P.R. and Hope, M.J. (1980) Biochim. Biophys. Acta 597, 533–542.
241 Yeagle, P.L., Hutton, W.C. and Martin, R.B. (1978) Biochemistry 17, 5745–5749.
242 Hue, S.W., Stewart, T.P. and Yeagle, P.L. (1980) Biochim. Biophys. Acta 601, 271–281.
243 Van Dijck, P.W.M., Van Zoelen, E.J.J., Seldensdijk, R., Van Deenen, L.L.M. and De Gier, J. (1976) Chem. Phys. Lipids 336–343.
244 Calhoun, W.I. and Shipley, G.G. (1979) Biochemistry 18, 1717–1722.
245 Mabrey, S. and Sturtevant, J.M. (1978) in Methods in Membrane Biology (Korn, E.D., ed.), Vol. 9, pp. 237–274, Plenum, New York.
246 Ladbrooke, B.D., Williams, R.M. and Chapman, D. (1968) Biochim. Biophys. Acta 150, 333–340.
247 Estep, T.N., Calhoun, W.I., Barenholz, Y., Biltonen, R.L., Shipley, G.G. and Thompson, T.E. (1980) Biochemistry 19, 20–24.

248 Sheridan, J.P., Merajver, S.D., Gaber, B.P. and Barenholz, Y. (1981) Biophys. J. 33, 157a.
249 Dagan, A., Cohen, R. Gatt, S. and Barenholz, Y. (1981) in preparation.
250 Lentz, B.R., Barenholz, Y. and Thompson, T.E. (1976) Biochemistry 15, 4521–4528.
250a Thompson, T.E., Huang, C. and Litman, B.J. (1974) in Cell Surface in Development (Moscona, A.A., ed.), p. 1, Wiley, New York.
251 Schmidt, C.F., Barenholz, Y., Huang, C. and Thompson, T.E. (1978) Nature 271, 775–777.
252 Barton, P.G. and Gunstone, F.D. (1975) J. Biol. Chem. 250, 4470–4476.
253 Lentz, B.R., Barenholz, Y. and Thompson, T.E. (1976) Biochemistry 15, 4529–4537.
254 Chapman, D., Urbina, J. and Keough, K.M. (1974) J. Biol. Chem. 249, 2512–2521.
255 Horwitz, A.F. (1972) in Membrane Molecular Biology (Fox, L.F. and Keith, A.D., eds.), pp. 164–192, Sinauer, Stanford.
256 Seelig, J. (1977) Quart. Rev. Biophys. 10, 353–418.
257 Seelig, J. (1978) Biochim. Biophys. Acta 515, 105–140.
258 Gaffney, B.J. and Chen, S.C. (1977) in Methods in Membrane Biology (Korn, E.D., ed.), pp. 291–358, Plenum, New York.
259 Seelig, J. (1976) in Spin Labelling (Berliner, L.J., ed.), pp. 373–409, Academic Press, New York.
260 Smith, I.C.P. and Butler, K.W. (1976) in Spin Labelling (Berliner, L.J., ed.), pp. 411–451, Academic Press, New York.
261 Griffith, O.H. and Jost, P.C. (1976) in Spin Labelling (Berliner, L.J., ed.), pp. 453–523, Academic Press, New York.
262 McConnell, W.M. (1976) Spin Labelling (Berliner, L.J., ed.), pp. 525–558, Academic Press, New York.
263 Radda, G.K. (1975) in Methods in Membrane Biology (Korn, E.D., ed.), pp. 97–188, Plenum, New York.
264 Shinitzky, M. and Barenholz, Y. (1978) Biochim. Biophys. Acta 515, 367–394.
265 Gaber, B.P. and Peticolas, L. (1977) Biochim. Biophys. Acta 465, 260–274.
266 Wallach, D.F.H., Verma, S.P. and Fookson, J. (1979) Biochim. Biophys. Acta 559, 153–208.
267 De Kruijff, B., Cullis, P.R. and Verkleij, A.J. (1980) Trends Biochem. Sci. 5, 79–81.
268 Untracht, S.H. and Shipley, G.G. (1977) J. Biol. Chem. 252, 4449–4457.
269 Borchov, C., Shinitzky, M. and Barenholz, Y. (1979) Cell Biophys. 1, 219–228.
270 Rintoul, D.A., Chou, S. and Silbert, D.F. (1979) J. Biol. Chem. 254, 10070–10077.
271 Lentz, B.R., Hoechli, M. and Barenholz, Y. (1981) Biochemistry 20, 6803–6809.
272 Schmidt, C.F., Barenholz, Y. and Thompson, T.E. (1977) Biochemistry 16, 2649–2656.
273 Berden, J.A., Cullis, P.R., Hoult, D.I., McLaughlin, A.C., Radda, G.K. and Richards, R.E. (1974) FEBS Lett. 46, 55–58.
274 Berden, J.A., Barker, R.W. and Radda, G.K. (1975) Biochim. Biophys. Acta 375, 186–208.
275 Castellino, F.J. (1978) Arch. Biochem. Biophys. 184, 465–470.
276 Nes, W.R. (1974) Lipids 9, 596–610.
277 Green, C. (1977) in International Reviews of Biochemistry, Biochemistry of Lipids (Goodwin, T.W., ed.), Vol. 14, pp. 101–152, University Park Press, Baltimore.
278 Demel, R.A. and De Kruijff, B. (1976) Biochim. Biophys. Acta 457, 109–132.
279 Huang, C. (1977) Lipids 12, 348–356.
280 Vandenheuvel, F.A. (1963) J. Am. Chem. Oil Soc. 40, 455–471.
281 Oldfield, E. and Chapman, D. (1971) Biochim. Biophys. Res. Commun. 43, 610–616.
282 Oldfield, E. and Chapman, D. (1972) FEBS Lett. 21, 303–306.
283 Lee, A.G. and Metcalf, J.C. (1972) FEBS Lett. 23, 203–207.
284 Kunishita, T. and Taketomi, T. (1979) Japan. J. Exp. Med. 49, 151–156.
285 Lyman, E.M., Knowles, C.L. and Sribney, M. (1976) Can. J. Biochem. 54, 358–364.
286 Estep, T.N., Mountcastle, D.B., Biltonen, R.L. and Thompson, T.E. (1978) Biochemistry 17, 1984–1989.
287 Estep, T.N., Mountcastle, D.B., Barenholz, Y., Biltonen, R.L. and Thompson, T.E. (1979) Biochemistry 18, 2112–2117.

288 Demel, R.A., Jansen, J.W.O., Van Dijck, P.W.M. and Van Deenen, L.L.M. (1977) Biochim. Biophys. Acta 465, 1–10.
289 Van Dijck, P.W.M. (1979) Biochim. Biophys. Acta 555, 89–101.
290 Burns, C.H. and Rothblatt, G.H. (1969) Biochim. Biophys. Acta 176, 616–625.
291 Pal, R., Barenholz, Y. and Wagner, R.R. (1981) Biochemistry 20, 530–539.
292 Helenius, A. and Simons, K. (1975) Biochim. Biophys. Acta 415, 29–79.
293 Kagawa, Y. (1972) Biochim. Biophys. Acta 265, 297–338.
294 Tanford, C. and Reynolds, J.A. (1976) Biochim. Biophys. Acta 457, 133–170.
295 Gatt, S. and Barenholz, Y. (1973) Annu. Rev. Biochem. 42, 61–90.
296 Yedgar, S., Barenholz, Y. and Cooper, V.G. (1974) Biochim. Biophys. Acta 363, 98–111.
297 Yedgar, S., Hertz, R. and Gatt, S. (1974) Chem. Phys. Lipids 13, 404–414.
298 Lichtenberg, D., Yedgar, S., Cooper, V.G. and Gatt, S. (1979) Biochemistry 18, 2574–2582.
299 Hertz, R. and Barenholz, Y. (1977) J. Colloid Int. Sci. 60, 188–200.
300 Small, D.M. (1971) in The Bile Acids (Nair, P.P. and Kritchevsky, D., eds.), pp. 249–356, Plenum, New York.
301 Barenholz, Y., Gafni, A. and Eisenberg, S. (1978) Chem. Phys. Lipids 21, 179–185.
302 Roelofsen, B. and Zwall, R.F.A. (1975) in Methods in Membrane Biology (Korn, E.D., ed.), Vol. 7, pp. 147–177, Plenum, New York.
303 Bernheimer, A.W. (1974) Biochim. Biophys. Acta 344, 27–50.
304 Otnaess, A.B. (1980) FEBS Lett. 114, 202–204.
305 Rao, B.G. and Spence, M.W. (1977) Ann. Neurol. 1, 385–392.
306 Kramer, R., Schlatter, C. and Zahler, P. (1972) Biochim. Biophys. Acta 282, 146–156.
307 Widnell, C.C. and Unkless, J.C. (1968) Proc. Natl. Acad. Sci. USA 61, 1050–1057.
308 Widnell, C.C. (1972) Meth. Enzymol. 32, 368–374.
309 Evans, W.H. and Gurd, J.W. (1973) Biochem. J. 133, 189–199.
310 Sandermann, H. (1978) Biochim. Biophys. Acta 515, 209–237.
311 Sood, C.K., Sweet, C. and Zull, J.E. (1972) Biochim. Biophys. Acta 282, 429–434.
312 Linder, R., Bernheimer, A.W. and Kim, K. (1977) Biochim. Biophys. Acta 282, 429–434.
313 Watkins, M.W., Hitt, A.S. and Bulger, J.E. (1977) Biochem. Biophys. Res. Commun. 79, 640–647.
314 Herbert, P.N., Gotto, A.M. and Fredrickson, D.S. (1978) in The Metabolic Basis of Inherited Disease (Stanbury, J.B., Wijngaarden, J.B. and Fredrickson, D.S., eds.), pp. 544–588, 4th ed., McGraw-Hill, New York.
315 Assmann, G., Sokoloski, E.A. and Brewer, H.B. (1974) Proc. Natl. Acad. Sci. USA 71, 549–553.
316 Yeagle, P.L., Hutton, W.C., Huang, C. and Martin, R.B. (1975) Proc. Natl. Acad. Sci. USA 72, 3477–3481.
317 Assmann, G., Hight, R.J., Sokoloski, E.A. and Brewer, H.B. (1974) Proc. Natl. Acad. Sci. USA 71, 3701–3705.
318 Stoffel, W., Zierenberg, O., Tunggal, B. and Schreiber, E. (1974) Proc. Natl. Acad. Sci. USA 71, 3696–3700.
319 see [14].
320 Arnon, R. and Teitelbaum, D. (1974) Chem. Phys. Lipids 13, 352–366.
321 Morrisett, J.D., Jackson, R.L. and Gotto, A.M. (1975) Annu. Rev. Biochem. 44, 183–207.
322 Eisenberg, S. and Levy, R.I. (1975) Adv. Lipid. Res., 13, 1–89.
323 Ellory, J.C. and Lew, V.L. (1977) Membrane Transport in Red Cells, Academic Press, New York.
324 Lux, S., Marchesi, V.T. and Fox, C.F., eds. (1974) Normal and Abnormal Red Cell Membranes, Liss, New York.
325 Cooper, R.A., Durocher, J.R. and Leslie, M.H. (1977) J. Clin. Invest. 60, 115–121.
326 Barenholz, Y., Yechiel, E., Cohen, R. and Deckelbaum, R.L. (1981) Cell Biophys. 13, 115–122.
327 Mills, G.L., Taylor, C.E. and Chapman, M.J. (1976) Clin. Sci. Mol. Med. 51, 221–231.
328 Deckelbaum, R.J., Eisenberg, S., Fainaru, M., Barenholz, Y. and Olivercrone, T. (1979) J. Biol. Chem. 254, 6079–6087.
329 Thompson, T.E. (1978) in Molecular Specialization and Symmetry in Membrane Function (Solomon, A.K. and Karnovsky, M., eds.), pp. 78–98, Harvard Univ. Press, Cambridge, MA.

330 Rothman, J.E. and Lenard, J. (1977) Science 195, 743–753.
331 Etemadi, A. (1980) Biochim. Biophys. Acta 604, 423–425.
332 Verkleij, A.J., Zwall, R.F.A., Roelofsen, B., Comfurius, P., Kastelijn, D. and Van Deenen, L.L.M. (1973) Biochim. Biophys. Acta 323, 178–193.
333 Gazitt, Y., Ohad, I. and Loyter, A. (1975) Biochim. Biophys. Acta 382, 65–72.
334 Bloj, B. and Zilversmit, D.B. (1976) Biochemistry 15, 1277–1283.
335 Crain, R.C. and Zilversmit, D.B. (1980) Biochemistry 19, 1440–1447.
336 Bretcher, M.S. (1973) Science 181, 622–629.
337 Gordesky, S.E., Marinetti, G.V. and Love, R. (1975) J. Membrane Biol. 20, 111–132.
338 Whiteley, N.M. and Berg, H.C. (1974) J. Mol. Biol. 87, 541–561.
339 Renooij, W., Van Golde, L.M.G., Zwaal, R.F.A. and Van Deenen, L.L.M. (1976) Eur. J. Biochem. 61, 53–58.
340 Billington, D., Coleman, R. and Tusak, Y.A. (1977) Biochim. Biophys. Acta 466, 526–530.
341 Sandara, A. and Pagano, R.E. (1978) Biochemistry 17, 332–338.
342 Patzer, E.J., Moore, N.F., Barenholz, Y., Shaw, J.M. and Wagner, R.R. (1978) J. Biol. Chem. 253, 4544–4550.
343 Rothman, J.E., Tsai, D.K., Davidowicz, E.A. and Lenard, J. (1976) Biochemistry 15, 2361–2370.
344 Klenk, H.D. (1973) in Biological Membranes (Chapman, D. ed.), Vol. 2, pp. 145–183, Academic Press, London.
345 Patzer, E.J., Wagner, R.R. and Dubovi, E.J. (1979) CRC Crit. Rev. Biochem. 165.
346 Lenard, J. and Compans, R.W. (1974) Biochim. Biophys. Acta 344, 51–94.
347 Blough, H.A. and Tiffany, J.M. (1973) Adv. Lipid Res. 11, 267–339.
348 Rouser, G. and Solomon, R.D. (1969) Lipids, 4, 232–234.
349 Smith, E.B. and Cantab, B.A. (1960) Lancet 1, 799–803.
350 Smith, E.B. (1974) Adv. Lipid Res. 11, 267–339.
351 Böttcher, C.J.F. and Van Gent, J. (1961) J. Atheroscler. Res. 1, 36–46.
352 Steth, S.K. and Newman, H.A.I. (1975) Circ. Res. 36, 294–299.
353 Portman, O.W., Alexander, M. and Maruff, C.A. (1967) Arch. Biochem. Biophys. 122, 344–353.
354 Portman, O.W. (1969) Ann. N.Y. Acad. Sci. 162, 120–136.
355 Parker, F. and Odland, G.F. (1966) Am. J. Pathol. 48, 197–240.
356 Bondjers, S. and Bjorkerad, S. (1973) Arteriosclerosis 17, 71–83.
357 Bierman, E.L., Eisenberg, S., Stein, O. and Stein, Y. (1973) Biochim. Biophys. Acta 329, 163–169.
358 Eisenberg, S. and Rachmilewitz, D. (1955) J. Lipid Res. 16, 341–351.
359 Schneider, E.L., Pentchev, P.G. Hibbert, S.R., Sawitsky, A. and Brady, R.O. (1978) J. Med. Genet. 370–374.
360 Bierman, E.L. and Albers, J.J. (1975) Biochim. Biophys. Acta 388, 198–202.
361 Broekhuyse, R.M. (1969) Biochim. Biophys. Acta 187, 354–365.
362 Broekhuyse, R.M. (1971) Biochim. Biophys. Acta 218, 546–548.
363 Feldman, G.L., Feldman, L.S. and Rouser, G. (1966) Lipids 1, 161.
364 Bruce, A. (1974) J. Lipid Res. 15, 103–107.
365 Owens, R. and Hughes, B.D. (1970) J. Lipid Res. 11, 486–495.
366 Gottfried, E.L. (1967) J. Lipid Res. 8, 321–327.
367 Gottfried, E.L. (1971) J. Lipid Res. 12, 531–537.
368 Van Blitterswijk, W.J., Van der Sluis, P.J., Hilgers, J., Hilkmann, H.A.M., Feltkamp, C.A. and Emmelot, P. (1978) in Advances in Comparative Leukemia Research (Bentvelzen Hilgers, J. and Yohn, D.S., eds.), pp. 341–344, Elsevier, Amsterdam.
369 Koizumi, K., Kano-Tanaka, K., Shimizu, S., Nishida, K., Yamanaka, N. and Ota, K. (1980) Biochim. Biophys. Acta 619, 344–352.
370 Bergelson, L.D., Dyatlovitskaya, E.V., Sorokina, I.B. and Gorkova, N.P. (1974) Biochim. Biophys. Acta 360, 361–365.
371 Hostetler, K.Y., Zenner, B.D. and Morris, H.P. (1976) Biochim. Biophys. Acta 441, 231–238.
372 Barsukov, L.I., Kulikov, V.I., Sinakova, I.M., Tikhenova, G.V., Ostrovski, D.N. and Bergelson, L.D. (1978) Eur. J. Biochem. 90, 331–336.

373 Crain, R.C. and Zilversmit, D.B. (1980a) Biochemistry 19, 1433–1439.
374 Yamakawa, T., Veta, N. and Irie, R. (1962) Japan. J. Exp. Med. 32, 289–296.
375 Wood, R. (1973) Tumor Lipids, American Oil Chem. Soc., Rees Campaign, IL.
376 Carrol, K.K. (1975) Prog. Biochem. Pharmacol. Lipids Tumors, Karger, Basel.
377 Wallach, D.F.H. (1975) Membrane Mol. Biol. of Neoplastic Cells, Elsevier, Amsterdam.
378 Obara, Y., Cotlier, E., Kim, J.O., Lueck, K. and Tao, R. (1976) Invest. Ophthalmol. 15, 966–969.
379 Cotlier, E., Obara, Y. and Toftness, B. (1978) Biochim. Biophys. Acta 530, 267–278.
380 Colley, C.M., Zwaal, R.F.A., Roelofsen, B. and Van Deenen, L.L.M. (1973) Biochim. Biophys. Acta 307, 74–82.
381 see [163].
382 Wenger, D.A., Barth, G. and Githens, J.H. (1977) Am. J. Dis. Child 13, 955–961.
383 Coleman, R., Lowe, P.J. and Billington, D. (1980) Biochim. Biophys. Acta 599, 294–300.
384 Borchov, H., Zahler, P., Wilbrundt, W. and Shinitzky, M. (1977) Biochim. Biophys. Acta 470, 382–388.
385 Lanetsberger, F.R., Compans, R.W., Choppin, P.W. and Lenard, J. (1973) Biochemistry 12, 4498–4502.
386 Moore, N.F., Barenholz, Y., McAllister, P.E. and Wagner, R.R. (1976) J. Virol. 19, 275–278.
387 Moore, R., Barenholz, Y. and Wagner, R.R. (1976) J. Virol. 19, 126–135.
388 Fredrickson, D.S., Sloane, H.R. and Hansen, C.T. (1969) J. Lipid Res. 10, 288–293.
389 Fettiplace, R. (1978) Biochim. Biophys. Acta 513, 1–10.
390 Kirk, G. (1977) Biochim. Biophys. Acta 464, 157–164.
391 Joiner, C.H. and Lauf, P.K. (1979) Biochim. Biophys. Acta 552, 540–545.
392 Neuringer, L.J., Sears, B., Jungalwala, F.B. and Shviver, E.K. (1979) FEBS Lett. 104, 173–175.

Abbreviations:

C16 SPM, *N*-palmitoyl sphingosylphosphorylcholine; C18 SPM, *N*-stearoyl sphingosylphosphorylcholine; C24 SPM, *N*-lignoceryl sphingosylphosphorylcholine; C24 : 1 SPM, *N*-nervonyl sphingosylphosphorylcholine; C16 DHSPM, *N*-palmitoyldihydrosphingosylphosphorylcholine; CDPC, cytidine diphosphorylcholine; CMC, critical micellar concentration; DPH, 1,6-diphenylhexatriene; DMPC, dimyristoylphosphatidylcholine; DOPC, dioleyl phosphatidylcholine; DPPC, dipalmitoyl phosphatidylcholine; HDL, high density lipoprotein; LCB, long chain base (sphingosine base); LDL, low density lipoprotein; MLV, multilamellar large vesicles (liposomes); PC, phosphatidylcholine; PE, phosphatidylethanolamine; POPC, 1-palmitoyl,2-oleyl phosphatidylcholine; SPC, sphingosylphosphorylcholine; SPM, sphingomyelin; SUV, small unilamellar vesicles (liposomes); VLDL, very low density lipoprotein.

CHAPTER 5

Phosphatidate metabolism and its relation to triacylglycerol biosynthesis

DAVID N. BRINDLEY and R. GRAHAM STURTON

Department of Biochemistry, University of Nottingham Medical School, Queen's Medical Centre, Nottingham NG7 2UH, U.K.

1. Introduction

The formation of glycerolipids is one of the major metabolic fates of fatty acids and phosphatidate (1,2-diacyl-*sn*-glycero-3-phosphate) is an important intermediate in this synthesis. This lipid is the precursor of the major phospholipids which provide important structural units in biological membranes, and which are needed for the transport of fat. Alternatively, the phosphatidate can be channelled into the production of triacylglycerol which enables many organisms to store energy in a very concentrated form. The deposits of triacylglycerol in adipose tissue also serve as heat insulation and protection. In addition, triacylglycerols are incorporated into lipoproteins in higher animals and this enables fatty acids to be transported from the small intestine and the liver to other organs.

Phosphatidate is synthesized by the esterification of *sn*-glycerol-3-phosphate (glycerophosphate) or dihydroxyacetone phosphate. The former route was first described in the 1950s and it appears to have a ubiquitous distribution in the animal and plant kingdoms. In the late 1960s it was discovered that dihydroxyacetone phosphate could also serve as an acyl-acceptor for fatty acids in phosphatidate synthesis and that acyldihydroxyacetone phosphate is an obligatory intermediate in the synthesis of alkyl- and alkenyl-glycerolipids. This chapter will attempt to review the characteristics and control of the enzymes that are involved in the synthesis of phosphatidate and its subsequent metabolism. Particular attention will be placed upon the conversion of phosphatidate to triacylglycerol.

A number of reviews that deal with glycerolipid metabolism in general or in relation to a specific organ have already appeared and these will provide the reader with further background information [1–5].

2. Biosynthesis of phosphatidate

(a) From glycerophosphate

Kornberg and Pricer [6] first showed that palmitate and glycerophosphate could be incorporated into phosphatidate, which was later shown to be the precursor of

Hawthorne/Ansell (eds.) Phospholipids

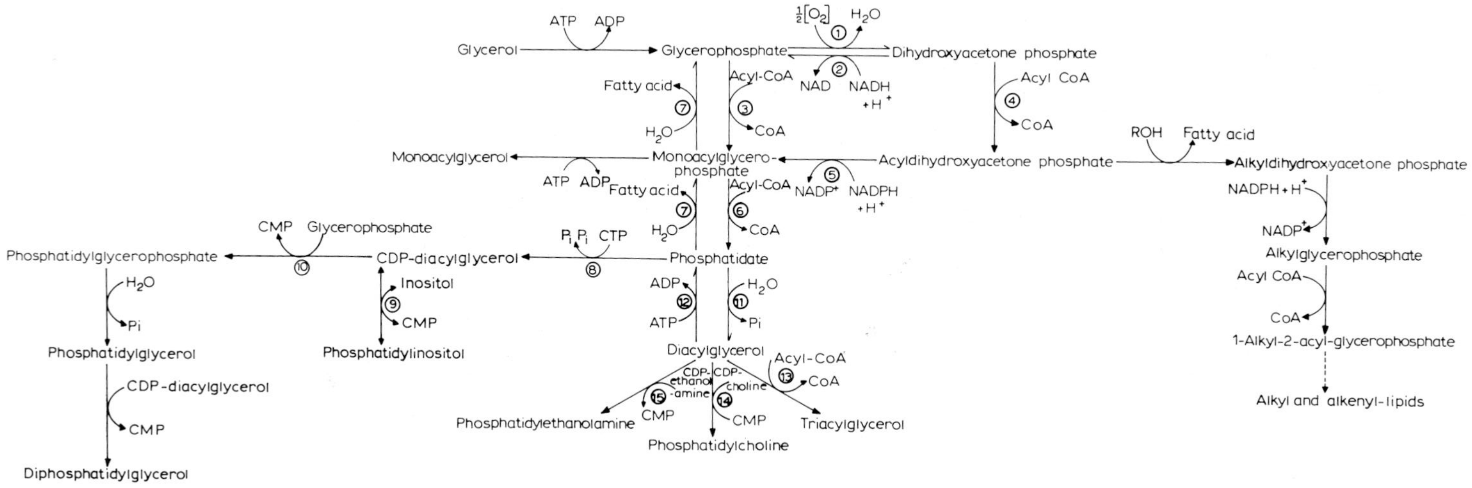

Fig. 1. Pathways of glycerolipid synthesis and phosphatidate metabolism. Enzyme activities and their abbreviations, where used, are indicated as follows: (1) Glycero-3-phosphate dehydrogenase, EC 1.1.99.5; (2) Glycero-3-phosphate dehydrogenase (NAD^+) EC 1.1.1.8; (3) GPAT, glycerophosphate acyltransferase EC 2.3.1.15; (4) DHAPAT, dihydroxyacetone phosphate acyltransferase EC 2.3.1.42; (5) Acyldihydroxyacetone phosphate reductase; (6) Monoacyl-GPAT, monoacyl-glycerophosphate acyltransferase; (7) Phosphatidate deacylase system: phospholipase A type activities; (8) PACT, phosphatidate cytidylyltransferase EC 2.7.7.41; (9) CDP diacylglycerol-inositol 3-phosphatidyltransferase EC 2.7.8.11; (10) Glycerophosphate phosphatidyltransferase EC 2.7.8.5; (11) PAP, phosphatidate phosphohydrolase, EC 3.1.3.4; (12) Diacylglycerol kinase EC 2.7.1.–; (13) DGAT, diacylglycerol acyltransferase EC 2.3.1.20; (14) Choline phosphotransferase EC 2.7.8.2; (15) Ethanolamine phosphotransferase EC 2.7.8.1.

triacylglycerol and various phospholipids [7,8]. This route of biosynthesis (Fig. 1) has been demonstrated in a wide variety of species [1–5]. The first reaction is catalysed by GPAT * which produces 1-monoacyl-*sn*-glycero-3-phosphate [9–11]. This activity can be stimulated by Ca^{2+}, Mg^{2+} and Mn^{2+} [9,10]. Partially purified preparations of this enzyme from mitochondrial and microsomal fractions of rat liver are stimulated by phospholipids [9,10]; a mixture of phosphatidylserine, phosphatidylinositol and phosphatidylethanolamine is particularly effective [12]. Phosphatidylglycerol is a good activator in preparations from *E. coli* [13]. The subsequent esterification of monoacylglycerophosphate to phosphatidate is catalysed by a different enzyme from that which acylates glycerophosphate. This is concluded since the two activities can be physically separated from each other during purification [9,10,14], and from the observation that a mutant of *E. coli* contained a heat-labile GPAT and a normal monoacyl-GPAT [15]. The acyl-donors for these reactions in mammalian systems are acyl-CoA esters, whereas acyl-ACP and acyl-CoA esters can be used by what appears to be an identical enzyme in *E. coli* [16]. Acyl-ACP esters seem to be the preferred precursors for glycerolipid synthesis in *Clostridium butyricum* [17] and *Rhodopseudomonas speroides* [18].

GPAT in rat liver is located on the outer mitochondrial membrane [19–21] on the inner surface [22], and it is also found in the endoplasmic reticulum. A predominant localization in rough endoplasmic reticulum has been reported for GPAT [23], whereas in another report the specific activities in the rough and smooth endoplasmic reticulum fractions were similar [24]. By contrast, it has also been claimed that GPAT is primarily situated in the smooth membranes of the endoplasmic reticulum, whereas the specific activity of monoacyl-GPAT was similar in the two membrane fractions [25]. Treatment of rats with phenobarbital gave a pronounced increase in the activity of GPAT in the smooth endoplasmic reticulum [23]. The acyltransferases are found on the cytoplasmic side of the endoplasmic reticulum [26].

The main product obtained after the esterification of glycerophosphate by microsomal fractions is phosphatidate [24,27–29], whereas mitochondria produce mainly 1-acyl-glycero-3-phosphate (lysophosphatidate) [10,21,24,28,30,31]. This difference is probably caused by the relatively low activity of monoacyl-GPAT in the outer mitochondrial membrane [32,33]. The relatively low rate of conversion of phosphatidate to diacylglycerol by particulate fractions is partly explained by the removal of a portion of the phosphatidate phosphohydrolase into the soluble fraction after conventional centrifugation (Section 6). Lysophosphatidate is not completely recovered in the lipid phase of some extraction procedures and this may account for why some authors have claimed that the mitochondrial activity could be explained by microsomal contamination [28,30,32].

The fact that the mitochondrial GPAT has different properties from the microsomal activity also confirms the separate identity of the mitochondrial system. The

* The abbreviations given to enzymes and their position in glycerolipid metabolism are shown in Figs. 1 and 2.

GPAT activity in mitochondria is more resistant to inhibition by sulphydryl-reagents [31,34–37], proteolytic enzymes [38], and heat [36] than the microsomal activity. It is stimulated by acetone whereas the microsomal activity is inhibited [31]. The pH profile of microsomal monoacyl-GPAT was reported to have a sharp optimum at pH 7, whereas the mitochondrial activity remained relatively constant between pH 6.6 and 8.5 [33]. The mitochondrial GPAT has a lower apparent K_m for glycerophosphate [36] and for acyl-CoA esters [34,37] than the microsomal enzyme, and the specificities for acyl-CoA esters are different. The mitochondrial GPAT shows a distinct preference for saturated long-chain fatty acyl-CoA esters, e.g. palmitoyl-CoA [10,21,31,35,39,40], whereas the microsomal enzyme is able to use a variety of saturated and unsaturated fatty acids [27,34,40]. The rate of esterification of palmitate relative to that of other fatty acids in mitochondria is increased when their concentration is relatively high, whereas the relative rate in microsomal fractions is decreased and specificity is lost [34,40]. Acylation at position-2 of glycerophosphate is fairly specific for unsaturated fatty acids in both mitochondrial and microsomal fractions [34,40,41]. The overall fatty acid specificity of the acyltransferase activities in phosphatidate synthesis appears to be important in controlling the predominant distribution of saturated fatty acids in the 1-position of glycerolipids and unsaturated fatty acids in the 2-position. However, their acyl specificity is by no means absolute and the fatty acid composition of newly synthesised phosphatidate will to some extent reflect the fatty acids that are available in the cell. The fatty acid composition of glycerolipids that are derived from phosphatidate can then be modified by a deacylation-reacylation cycle [41,42]. For example, 1-stearoyl-2-oleoylglycero-3-phosphocholine can be converted to 1-stearoylglycero-3-phosphocholine by the action of phospholipase A_2 (Chapter 9). The enzyme responsible for the reacylation of this lyso-derivative specifically uses arachidonoyl-CoA.

A further difference between the mitochondrial and microsomal GPAT is seen in their specificities for the acyl-acceptor. Dihydroxyacetone phosphate can substitute for glycerophosphate in the microsomal system, and these two acceptors are mutually competitive. By contrast, the mitochondrial GPAT cannot use dihydroxyacetone phosphate and it is not inhibited by it (Section 2b).

About half of the total GPAT activity of rat, guinea pig, rabbit and bovine liver is associated with the mitochondrial fraction [31,32,36], and about 30% of the activity in embryonic liver and 1-day-old hepatocytes from chickens is mitochondrial [35]. However, about 90% of the mitochondrial activity is lost from hepatocytes of 5–8-day-old chickens [35]. It will also be seen in Section 4 that the mitochondrial and microsomal GPAT of rat liver appear to respond differently to changes in metabolic state. In organs such as heart, kidney, adrenal glands [31] and adipose tissue [43] about 10% of the total activity is mitochondrial, whereas little if any mitochondrial GPAT was found in Ehrlich cells [31]. The mitochondrial activity was also not detected in secondary cultures of chicken embryo fibroblasts, and in baby hamster kidney cells it contributed 5–8% of the acylating capacity [35].

The function of the high GPAT activity in the mitochondria, particularly of liver, is by no means clear. Mitochondria are known to synthesize diphosphatidylglycerol (cardiolipin) from phosphatidate (Fig. 1) and this lipid is characteristic of the inner mitochondrial membrane [1,5]. The conversion of phosphatidate to triacylglycerol, phosphatidylcholine and phosphatidylethanolamine is normally considered not to take place in mitochondria, or to occur at very low rates [1,5]. Consequently, the subsequent fate of the majority of the lysophosphatidate is not certain. It could be cycled back to glycerophosphate (Section 7), or it (or phosphatidate) could be transferred to the endoplasmic reticulum for further metabolism. This assumes that the high mitochondrial capacity is expressed. From theoretical considerations this should be so since the K_m's for glycerophosphate and for acyl-CoA esters in mitochondria are lower than for the microsomal system [34,36,37]. The changes in fatty acid composition of glycerolipids in cultured cells and changes in mitochondrial GPAT also indicate that the latter activity contributes significantly to glycerolipid synthesis [35]. The possible function of mitochondrial acylation in controlling the balance between triacylglycerol synthesis and ketogenesis will be discussed in Section 4.

(b) From dihydroxyacetone phosphate

The previous Section described the synthesis of phosphatidate from glycerophosphate, but in addition to this dihydroxyacetone phosphate can also act as an acyl-acceptor. The acyldihydroxyacetone phosphate that is formed is reduced to 1-acylglycerophosphate and a second esterification reaction produces phosphatidate (Fig. 1). The biosynthesis of acyldihydroxyacetone phosphate was first observed with preparations from guinea pig liver that were then referred to as mitochondrial fractions [44–46]. Activity was also detected in the microsomal fraction [45] and from later work with other tissues it appears that this particular activity is catalysed by GPAT. This conclusion relies on the observation that glycerophosphate and dihydroxyacetone phosphate are mutually competitive for their esterifications [45,47,48], and that the two acylations have similar pH optima, chain length specificities for acyl-CoA esters and similar profiles of inhibition by heat, *N*-ethylmaleimide, trypsin and detergents [47–49]. Since this GPAT has a ubiquitous distribution in cells, this also implies that most cells have the potential for esterifying dihydroxyacetone phosphate.

There is also a second enzyme that esterifies dihydroxyacetone phosphate which is distinguished from the GPAT in that glycerophosphate is neither a substrate nor an inhibitor [45,50]. This specific DHAPAT activity is also resistant to proteolysis except in the presence of detergents [51], and is either not affected by *N*-ethylmaleimide [51], or is stimulated [37,43]. This DHAPAT activity cannot be accounted for by the activity of the mitochondrial GPAT since dihydroxyacetone phosphate does not inhibit the latter activity [45,52].

Clofenapate (sodium 4-(4′-chlorophenyl)phenoxyisobutyrate) inhibits the specific DHAPAT to a greater extent than GPAT [50,52,53]. The specific DHAPAT activity in brain and lung has a lower pH optimum than that of GPAT [54,55] and it is stimulated by sodium cholate whereas GPAT is inhibited [54]. Much greater levels of stimulation of up to 20-fold in the activity of the specific DHAPAT were recorded with cholate and deoxycholate in preparations from Harderian gland [51].

The specific DHAPAT in liver was initially thought to be localized in mitochondria [44–46,52] in the outer membrane [56]. However, more detailed studies of subcellular fractions have shown that its activity can be separated from the mitochondrial GPAT and that the distribution of activity parallels that of uricase [37,57–59]. This indicates a peroxisomal localization. The DHAPAT has a higher reaction rate with saturated fatty acyl-CoA esters (C_{14}–C_{18}) than with oleoyl and linoleoyl-CoA [45]. The specific activities of DHAPAT and GPAT in parenchymal cells of rat liver were respectively 7 and 41 times higher than those in non-parenchymal cells [37]. It is possible that the activity of the specific DHAPAT that has been measured in the microsomal fractions of brain [54] and Harderian gland [51] might also have a peroxisomal origin.

Acyldihydroxyacetone phosphate reductase, which catalyses the next reaction in the synthesis of glycerolipids, is enriched in peroxisomal fractions of liver although some activity was also identified in the endoplasmic reticulum [58]. This enzyme has also been reported to occur in a number of other mammalian tissues in fractions that were described as mitochondrial and microsomal [60]. It seems likely that the same reductase is responsible for the reduction of both acyl- and alkyldihydroxyacetone phosphate [61]. NADPH rather than NADH is a specific cofactor [60–62] and the hydrogen atom is transferred from position 4 of the B-side of the nicotinamide ring [60].

(c) From monoacylglycerols and diacylglycerols

Sections 2a and b were concerned with the synthesis de novo of phosphatidate from glycerophosphate and dihydroxyacetone phosphate. Phosphatidate can also be derived by the action of diacylglycerol kinase from diacylglycerol that is formed by the degradation of other glycerolipids. In particular the kinase functions in the cycle of events that involves the stimulated breakdown and resynthesis of phosphatidylinositol (Chapter 7). This series of reactions is widely distributed amongst mammalian cell types. The diacylglycerol that is derived from phosphatidylinositol has a relatively high content of arachidonate and the activity of the kinases probably accounts for the greater concentration of this acid in phosphatidate than would be expected from the specificities of the enzymes involved in esterifying glycerophosphate and dihydroxyacetone phosphate [63]. However, the specificity of the kinase towards fatty acids is not so strict as to account for the predominantly 1-stearoyl-2-arachidonoyl species of phosphatidylinositol [64].

Diacylglycerol kinase occurs in both particulate and soluble fractions of liver [63] and brain [64,65]. It requires Mg^{2+} and its activity is stimulated by deoxycholate [63–66]. The activity in brain exceeds that of GPAT and other enzymes involved in the synthesis of phosphatidylinositol, and it is unlikely that the kinase is rate-limiting in this synthesis [65]. In erythrocyte ghosts the activity of the kinase was 2500 times greater than that of GPAT [67]. These results indicate that in some tissues the kinase could provide an important route for phosphatidate synthesis.

Diacylglycerol kinase of rat liver is stereospecific for *sn*-1,2-diacylglycerols and the rate with *sn*-1,3-diacylglycerols is very low [63]. Fractions of small intestinal mucosa also failed to phosphorylate *sn*-1,3-dioleoylglycerol [68]. Monoacylglycerols of the *sn*-1- or 2-configurations can act as substrates, but the rates are less than with *sn*-1,2-diacylglycerols [63,64,67,69]. The lysophosphatidate that is formed from monoacylglycerols provides an additional substrate for phosphatidate synthesis into which it is rapidly converted [69]. This pathway could provide an alternative route for the synthesis of triacylglycerols from monoacylglycerols during fat absorption by the small intestine [68], but the physiological importance of this is still uncertain [2].

Diacylglycerol kinase from rat liver did not phosphorylate ceramide (*N*-acylsphingosine) or some other long-chain alcohols [63], whereas that from *E. coli* was able to do so [66]. It was suggested that the broad specificity of the latter enzyme could mean that it serves to phosphorylate a variety of acceptors in addition to diacylglycerol [66].

3. The relative contribution of the glycerophosphate and dihydroxyacetone phosphate pathways to the synthesis of glycerolipids

The quantitative role of acyldihydroxyacetone phosphate in glycerolipid metabolism is very much in dispute. It is generally agreed that this substrate is an obligatory intermediate in the synthesis of alkyl- and alkenyl-lipids (Fig. 1), and so the physiological need for the acylation of dihydroxyacetone phosphate is established. There is also no question that the enzymic capacity to convert acyldihydroxyacetone phosphate into the major glycerolipids exists in most mammalian cells. What is in contention is the extent to which this series of reactions operates in vivo. In yeast the synthesis of glycerolipids from acyldihydroxyacetone phosphate may not be quantitatively important since its reduction to lysophosphatidate could not be demonstrated [49].

It was estimated from the kinetic properties of the microsomal acylation of glycerophosphate and dihydroxyacetone phosphate in rat liver that the synthesis of phosphatidate from glycerophosphate should be more than 84 times greater than from dihydroxyacetone phosphate in vivo [48]. Similarly, it was concluded that glycerophosphate is the predominant precursor for glycerolipid synthesis in adipose tissue [47,70]. However, these conclusions were partly based on the assumption that the K_m, K_i and V_{max} values from studies in vitro can be extrapolated to conditions within the cell, and the activities of the specific GPAT in mitochondria and the specific DHAPAT of peroxisomes were not taken into account. These latter two activities have lower apparent K_m values for acyl-CoA esters than that of the endoplasmic reticulum [37], and so they may be primarily responsible for the esterification of fatty acids when their supply is limiting. This again assumes that these differences of affinity operate in vivo. Pollock et al. [71] measured the direct incorporation of dihydroxyacetone phosphate into phosphatidate by the homogenates of a variety of tissues and compared this rate with that via glycerophosphate. The rate of glycero-3-phosphate dehydrogenase (NAD^+) and GPAT always exceeded that of DHAPAT. However, this potential was not expressed and there was a predominant synthesis of glycerolipids from dihydroxyacetone phosphate [71]. When equimolar concentrations of glycerol phosphate and dihydroxyacetone phosphate were incubated with a "mitochondrial" fraction of rat liver the rate of synthesis of complex lipids was approximately equal from these two precursors [52]. With fractions of rabbit lung incubated with equimolar concentrations of these compounds, 41% of the esterification directly involved dihydroxyacetone phosphate [55]. Agranoff and Hajra [72] determined the relative contributions of the glycerophosphate and dihydroxyacetone phosphate pathways in tissue homogenates by measuring respectively the incorporation of NAD[^{3}H] and NADP[^{3}H] into the C_2 position of glycerolipids. They concluded that the dihydroxyacetone phosphate pathway plays a significant part in esterification by homogenates of mouse liver and a dominant role in homogenates of Ehrlich ascites tumour cells.

Attempts have been made to estimate the relative contributions of the acylation of dihydroxyacetone phosphate and glycerophosphate in whole cells by a variety of methods. All of these have disadvantages which include problems of isotope effects, the possible existence of different pools of substrate, the choice of substrate and the specificities for reduced pyridine nucleotides. Furthermore, these techniques are complicated and difficult to apply in vivo. Consequently, the conclusions from this type of work permit us to make estimates for the potential to synthesize glycerolipids by both routes of metabolism, but as yet a detailed knowledge of how the importance of these routes alters under different physiological states is still not available.

The first method that was employed involved incubating cells with a mixture of [^{14}C]glycerol and [2-^{3}H]glycerol [53,73–79]. Conversion of glycerophosphate to dihydroxyacetone phosphate (Fig. 1) liberates the ^{3}H and therefore lipids synthesized from this precursor contain only ^{14}C. Both ^{3}H and ^{14}C will be incorporated equally by the glycerophosphate pathway. Therefore a decrease in the $^3H/^{14}C$ ratio indicates the extent of the esterification of dihydroxyacetone phosphate. Surprisingly, increases in this ratio were found and some authors concluded that the dihydroxyacetone phosphate pathway is not important in rat liver and in *Clostridium butyricum* [73,75]. With *E. coli* no change in the isotopic ratio was found and because unlabelled dihydroxyacetone phosphate also failed to modify the labelling pattern of lipids by homogenates, it was again concluded that the major synthesis was directly from glycerophosphate [76]. Plackett and Rodwell [74] using *Mycoplasma* strain Y recognised that the increase in the $^3H/^{14}C$ ratio was caused by the isotope effect of glycero-3-phosphate dehydrogenase (EC 1.1.99.5). Subsequent work with rat liver slices demonstrated that this effect could be large. It is therefore essential to calculate the contributions of the two pathways by comparing the $^3H/^{14}C$ ratio in lipid with that in glycerophosphate and not that in glycerol [52,53,77].

When this was done, 50–75% of the glycerolipid synthesized by rat liver slices was calculated to have been derived by the acylation of dihydroxyacetone phosphate [53,77]. In contrast, it has been claimed that synthesis of triacylglycerols by this route in isolated hepatocytes is of minor importance [78]. However, by the authors' admission, the conditions used in their Experiment 1 were unsuitable for this determination since glycerolipid was not synthesized during the time period chosen for the study. In Experiment 2, the rate of glycerolipid measured with ^{14}C was approximately constant for 40 min and in this case the $^{3}H/^{14}C$ ratio was lower than that in glycerophosphate, but the authors failed to calculate the contributions of the two pathways [78]. The use of 10 mM glycerol in this experiment should also theoretically have increased the ratio of NADH/NAD in the cytoplasm and thus favoured the glycerophosphate pathway by increasing the ratio of glycerophosphate: dihydroxyacetone phosphate.

Mason [79] adopted a slightly different approach in his work with type II alveolar cells of lung. He argued that the $^{3}H/^{14}C$ ratio in the head group glycerol of phosphatidylglycerol would have been identical to that in glycerophosphate throughout the incubations. By comparing the former ratio with that in the acylglycerol of phosphatidylglycerol he concluded that about 56% of the latter was derived by acylation of (dihydroxyacetone) phosphate. The equivalent value for phosphatidylcholine was 64%.

An alternative approach to the measurement of the two pathways in whole cells that synthesize ether lipids is to use a mixture of D-[U-^{14}C] and D-[3-^{3}H]glucose as precursors [80,81]. This procedure generates [4-^{3}H]NADPH which is incorporated into the 2-position of glycerolipids when acyl-dihydroxyacetone phosphate is reduced (Fig. 1). It is then possible to calculate the relative activities of the two pathways by comparing the $^{3}H/^{14}C$ ratio at position C_2 of saponifiable lipids with that in alkyl and alk-1′-enyl lipids which are assumed to be synthesized entirely from dihydroxyacetone phosphate. A correction factor is also required to compensate for some labelling of glycerophosphate at position 2. It was estimated that between 49 and 61% of the glycerolipid synthesized by BHK-21 and BHK-Ts-a/1b-2 cells was formed by direct esterification of dihydroxyacetone phosphate [80,81].

A further argument in favour of a significant involvement of the dihydroxyacetone phosphate pathway in glycerolipid synthesis depends upon the theoretical considerations of cofactor requirements. Most anabolic pathways use NADPH as a reductive cofactor and the synthesis of glycerolipids from glucose by the direct acylation of dihydroxyacetone phosphate (Fig. 1) fits this requirement [71,82]. The conversion of dihydroxyacetone phosphate to glycerophosphate requires NADH which is normally involved in reductive degradation. In liver, glycerophosphate dehydrogenase, which catalyses this reaction, is also involved in permitting gluconeogenesis from glycerol to take place and in maintaining the redox state. The existence of the dihydroxyacetone phosphate pathway and the ability to synthesize glycerolipids directly from glycerophosphate derived from glycerol means that the dehydrogenase is not an obligatory step in glycerolipid biosynthesis [82].

4. Control of phosphatidate synthesis

One of the obvious sites for the regulation of phosphatidate synthesis is at the level of GPAT, since this is the first committed reaction in the synthesis of glycerolipids

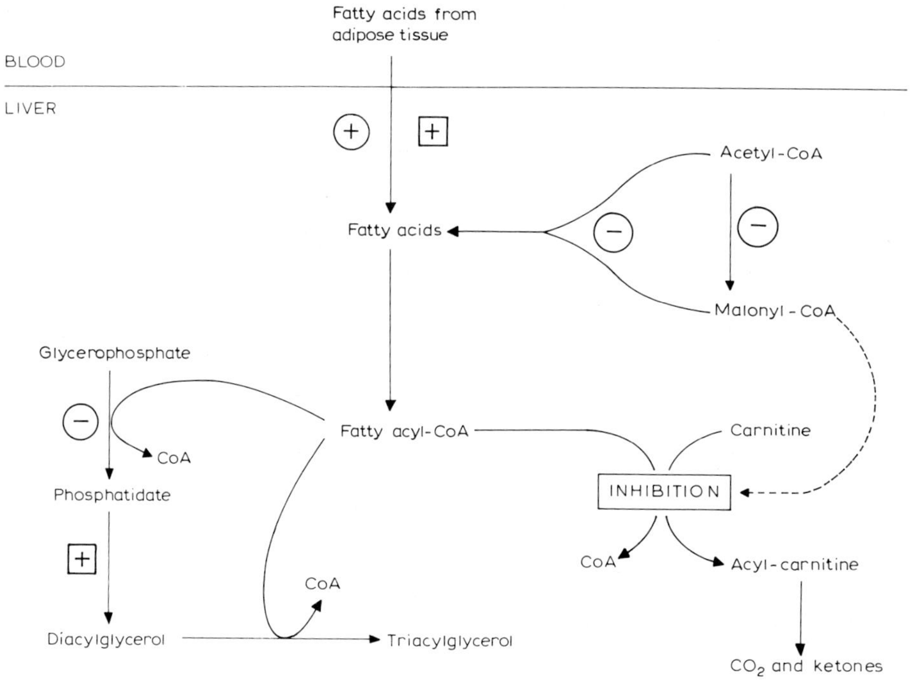

Fig. 2. Effects of insulin, glucagon and glucocorticoids on the metabolism of fatty acids in the liver. The effects of a low insulin : glucagon ratio is shown by the circles and those of glucocorticoids by squares. Increased rates are indicated by, +, and decreases by, −. A low insulin : glucagon ratio decreases the rate of fatty acid synthesis and the concentration of malonyl-CoA, which relieves the inhibition of CAT (carnitine acyltransferase, EC 2.3.1.21). This promotes β-oxidation, and the competition for acyl-CoA esters by mitochondrial GPAT decreases. However, in these conditions, the supply of fatty acids from adipose tissue may exceed the requirement for β-oxidation and the excess acids are esterified by the microsomal GPAT. PAP activity is high because of increased glucocorticoid concentrations and this facilitates triacylglycerol synthesis. Details are given in sections 4 and 9.

(Fig. 2). GPAT activity is rate-limiting in phosphatidate synthesis in mitochondrial [30] and microsomal fractions [29] of rat liver. The provision of activated fatty acids for this synthesis does not appear to be limited by the activity of acyl-CoA synthetase, although in other organs e.g. small intestine, the excess of activating capacity does not appear to be so great [83]. At present there is no clear evidence that special pools of acyl-CoA esters are preferentially channelled into esterification, oxidation, elongation and desaturation, and the indications are that there is competition for acyl-CoA esters by the initial enzymes of the respective pathways. The control of the partition of fatty acids between ketogenesis and triacylglycerol synthesis in the liver has received particular attention.

When the concentration ratio of insulin to glucagon in the blood falls the partitioning of fatty acids into β-oxidation increases whereas that into triacylglycerols decreases in relative terms [84,85]. Part of this change is brought about by the acute regulation of CAT by malonyl-CoA (Fig. 2). The activity of acetyl-CoA carboxylase decreases in response to the lowered insulin : glucagon ratio and thus the concentration of malonyl-CoA in the liver falls. The action of malonyl-CoA in inhibiting CAT is thereby decreased and the flux of fatty acids to β-oxidation is increased [86]. There are also long term increases in hepatic CAT activity in starvation [87,88], and decreases in GPAT activity have been reported for mitochondrial [37,88,89] and microsomal fractions [88–90]. Starvation also decreases the microsomal DHAPAT activity [91], (which is probably identical to the GPAT activity; Section 2b), and the total DHAPAT activity [37]. However, in other experiments there was no significant change in microsomal GPAT activity after starvation [37,92,93], or in diabetic rats [94,95]. The mitochondrial GPAT activity in the liver seemed to be far more responsive to starvation [37,88], and it was significantly decreased in diabetes [95]. It seems likely that the mitochondrial GPAT activity is acutely regulated by insulin [95].

The mechanism whereby these changes are brought about is not completely established. It has been proposed that the accumulation of Ca^{2+} in the endoplasmic reticulum may be involved since this correlates with the decrease in the rate of phosphatidate synthesis [96,97]. The latter was lowered in microsomal fractions obtained from rat livers that had been perfused with dibutyryl-cyclic AMP [97]. The uptake of Ca^{2+} by microsomal fractions was also higher in male than female rats, whereas the synthesis of triacylglycerol and the secretion of very low density lipoproteins was higher in the female [98]. An alternative mechanism that has been suggested could involve the phosphorylation of GPAT by a cyclic AMP-dependent protein kinase. Evidence for this has been obtained from experiments with rat adipose tissue and the activity was restored by incubating the GPAT with alkaline phosphatase [99]. It has also been shown that GPAT activity in adipocytes is increased after incubation with insulin and decreased after incubation with adrenalin [100]. This action could help to prevent the re-esterification of fatty acids during lipolysis in adipose tissue, but the physiological importance of this apparent control of GPAT needs to be established. By contrast, the synthesis of phosphatidate in the heart is increased in diabetes [101].

The fatty acids that are mobilized from adipose tissue in catabolic conditions are taken up by the liver and preferentially oxidized. This appears to be facilitated by the increased activity of CAT and the decreased activity of GPAT, especially in the mitochondrial fraction. This is particularly important when the supply of fatty acids is low. However, in many of these conditions (e.g. starvation, diabetes, stress) the total synthesis of triacylglycerols can increase in response to the increased supply of fatty acids from adipose tissue. Glycerophosphate concentrations can also increase in vivo in these conditions [86,89]. When the requirement for β-oxidation is satisfied, the excess fatty acids and acyl-CoA esters are converted to triacylglycerols [102]. The high capacity for phosphatidate synthesis may be provided by the microsomal GPAT which as discussed in Section 2a, has a high K_m for glycerophosphate and fatty acyl-CoA esters. The maintenance of the high capacity to synthesize phosphatidate in starvation and in diabetes is also demonstrated by the effects of (+)-carnitine. This blocks the oxidation of fatty acids which are then immediately converted to triacylglycerols and phospholipids [86,103]. It has been shown that the decreased capacity of hepatocytes from starved rats to synthesize triacylglycerols probably results from their decreased content of glycerophosphate. This may be partly related to the depletion of glycogen. When glycerophosphate concentrations are increased by adding precursors to the medium, the differences in the rates of triacylglycerol synthesis between the hepatocytes of fed and starved rats disappear, demonstrating that the esterification system is not defective [103].

There is some evidence that glucocorticoids may be involved in controlling the mitochondrial and microsomal GPAT since adrenalectomy leads to a decrease in these activities [37]. In the latter case the decrease was seen in fasted but not fed rats. Administering cortisol to adrenalectomized rats increased the GPAT activities [104]. Similarly, the injection of corticotropin has been shown to increase GPAT and possibly DHAPAT activities after 6 h [105]. However, the increases were small by comparison to those seen for PAP. The control of glycerolipid synthesis by glucocorticoids will be discussed further in Section 9.

The effect of modifying the composition of the diet on the activity of the microsomal GPAT of the liver has been measured using glycerophosphate or dihydroxyacetone phosphate as acyl-acceptors (Section 2b). Feeding diets rich in glucose or fructose was expected to increase triacylglycerol synthesis and these treatments resulted in about 2-fold increases in activity as measured with both acyl-acceptors [92,106,107]. However, in other experiments the inclusion of sucrose in the diet of rats did not significantly alter these activities [105,108]. The effect of dietary fat is very confusing since this has been reported to increase [92,109], decrease [87,91,110] or not to change [93,105,108] the microsomal rate of esterification. The degree of unsaturation of the fat cannot explain these discrepancies.

The ingestion of ethanol produces marked increases in the rate of hepatic triacylglycerol synthesis, but it takes about 6 weeks of ethanol feeding to increase the microsomal GPAT activity in rats [109]. No increase is observed 6 h after a single dose of ethanol [111], or after 10 days of chronic ethanol feeding [109].

We know practically nothing about whether the specific DHAPAT in peroxisomes responds to physiological stimuli. It may be significant that this organelle also has its own complement of enzymes capable of β-oxidation [112], and the competition for acyl-CoA esters with DHAPAT may be a factor in regulating their direction of metabolism. The peroxisomal system may be particularly important in hepatic metabolism when animals are fed on high fat diets or with clofibrate, when peroxisome number and capacity for β-oxidation can increase [112–115]. Feeding clofibrate did increase the activity of the *N*-ethylmaleimide-insensitive DHAPAT by about 2-fold which was similar to the increases in activity of the mitochondrial and microsomal GPAT [116]. The increases for CAT [117] and acyl-CoA oxidase [116] were even greater and this ought to promote β-oxidation relative to fatty acid esterification. The only experiments so far which have determined the effects of high fat diets on the peroxisomal DHAPAT failed to demonstrate an increase [105].

However, there were also no significant changes in the activity of acyl-CoA oxidase which indicates that peroxisomal proliferation is probably not an obligatory consequence of feeding a high fat diet.

In rat adipose tissue only about 17% of the total DHAPAT was *N*-ethylmaleimide-insensitive and presumably this may also represent peroxisomal activity [43]. This activity was not significantly changed when rats were fed on diets enriched with sucrose, corn oil, or beef tallow rather than with starch [118].

The subsequent conversion of phosphatidate to triacylglycerol, phosphatidylcholine and phosphatidylethanolamine is normally considered to take place in the endoplasmic reticulum [1,4,5]. Therefore, at present it is difficult to understand how the synthesis of phosphatidate in mitochondria and peroxisomes could contribute to these processes. It is feasible that phosphatidate could be transported to the endoplasmic reticulum by phospholipid exchange proteins, but a significant synthesis of glycerolipids via this process has yet to be demonstrated. Another possibility is that acyl-dihydroxyacetone phosphate could itself act as a carrier of acyl-groups among subcellular compartments. This compound could then act as the acyl-donor in the synthesis of a number of glycerolipids [119]. It may be significant that hepatic peroxisomes are seen in association with lipid droplets and that they are situated close to the smooth endoplasmic reticulum with which they have connections [120,121]. The suggestion that peroxisomes are involved in lipid synthesis and turnover [121] is now supported by more recent studies of their enzymic composition.

The effects of drugs in modifying hepatic glycerolipid synthesis have also been investigated. For example, phenobarbital injections can increase the rate of triacylglycerol synthesis and secretion [122], and the total microsomal GPAT activity is increased 12 h after this treatment [23]. However, this activity subsequently declines to control values after 2 days of treatment [23,109]. Hypolipidaemic drugs related to clofibrate (ethyl *p*-chlorophenoxyisobutyrate) were expected to decrease the rate of hepatic triacylglycerol synthesis. They do have the ability to inhibit directly GPAT activity [123,124] but *p*-chlorophenoxyisobutyrate itself was not very potent compared to the more hydrophobic derivatives such as halofenate* and clofenapate [125]. Clofenapate also seemed to be more effective in inhibiting the esterification of dihydroxyacetone phosphate than that of glycerophosphate [52,53].

The type of regulation considered so far involves either changes in the concentration, or state of activation of the acyltransferases concerned in phosphatidate synthesis. A further level of control could be exerted by modifying the availability of substrates, or their physical form. Acyl-CoA esters are potential detergents and it is thought that they are attached to fatty acid-binding proteins within cells. Such complexes can increase the rate of acyltransfer [126], and it has been postulated that the concentration of binding proteins is under physiological control [127,128]. For

---

* The derivative 2-(*p*-chlorophenyl)-2-(*m*-trifluoromethylphenoxy) acetate was used.

instance, this concentration is higher in the livers of female than of male rats and this correlates with an increased rate of triacylglycerol synthesis [127]. Polyamines also interact with acyl-CoA esters and in so doing they stimulate the activities of GPAT and DHAPAT. At the same time they decrease the rate of hydrolysis of these thioesters [129,130].

A further mechanism that has been proposed to control phosphatidate synthesis is a feed-back inhibition by monoacylglycerols (or in experimental work their ether analogues). The rationale for this is that the concentration of monoacylglycerols in cells will increase during periods of active lipolysis, and the inhibition could prevent the recycling of fatty acids back to triacylglycerol [131]. Although such inhibitions can be demonstrated in vitro, the kinetic interpretation is very difficult because of the lipid nature of these compounds [50], and the physiological importance of this mechanism is still in doubt.

The work that has been described in this Section demonstrates that the rate of phosphatidate synthesis can be controlled by regulating acyltransferase activities. This could be particularly important when the supply of fatty acids to the liver is low. However, the magnitudes of many of the changes in acyltransferase activity are relatively small compared to those in PAP which catalyse the subsequent flux to diacylglycerol. Furthermore, a number of situations exist in which large changes in hepatic triacylglycerol synthesis occur, but in which little if any alteration is observed in GPAT activity. It is therefore likely that a second point of control occurs at the level of PAP and that this may be particularly important in regulating the flux to triacylglycerol when the rate of fatty acid supply to the liver is high. This will be discussed in Section 9.

5. Conversion of phosphatidate to CDP-diacylglycerol

Sections 5–7 are concerned with the possible metabolic fates of phosphatidate. First it can be converted to CDP-diacylglycerol by a cytidylyltransferase action (PACT) [132] involving CTP (Fig. 1). It has also been demonstrated that CDP-diacylglycerol can be formed by the reaction of CMP with phosphatidylinositol since the CDP-diacylglycerol-inositol 3-phosphatidyltransferase reaction is reversible [133]. CDP-diacylglycerol can be regarded as an activated form of phosphatidate which can serve as the precursor of phosphatidylinositol [134], phosphatidylglycerol [135] and diphosphatidylglycerol [136] in mammalian systems. In *E. coli* it has been shown to be an intermediate in the synthesis of the three major phospholipids: phosphatidylserine, phosphatidylglycerol and phosphatidylethanolamine [137,138] (see also Chapter 11).

When PACT was assayed using aqueous dispersions of phosphatidate the activity in the microsomal fraction of guinea pig liver showed a requirement for Mg^{2+} [132]. Mn^{2+} could substitute for Mg^{2+}, but the maximum rate was only about 50% that with Mg^{2+}. With a preparation from embryonic chick brain the highest rate was obtained with 18 mM $MnCl_2$ [139]. PACT has also been assayed with phosphatidate that had been incorporated into the microsomal membranes themselves and again a requirement for bivalent cations e.g. Mg^{2+} was observed [140–142].

The effect of detergents on the activity of PACT is important since a number of workers have used these compounds in order to disperse the phosphatidate used in their assays [132,139,143]. The cationic and anionic detergents appear to alter the charge on the phosphatidate emulsion and thereby change the activity of PACT. These effects are dependent upon the concentrations of bivalent cations that are present, and they will be dealt with in Section 8.

PACT activity in particulate fractions from *Micrococcus cerificans* [144], or when purified from *Saccharomyces cerevisiae* [145] had an absolute requirement for non-ionic detergent for activity. By contrast, Triton-X 100 inhibited the PACT activity in the microsomal fraction of rat liver [141]. This enzyme in the microsomal fraction of cerebral cortex was inhibited by 93% at 0.5% (w/v) Triton-X 100, but the mitochondrial activity was relatively unaffected [146]. This effect may be responsible for some of the disagreement concerning the subcellular distrubition of PACT. Several groups have claimed a predominantly mitochondrial localization [139,143,147], whereas others showed that the majority of the activity was in the endoplasmic reticulum [132,148,149]. Undoubtedly, tissue and species differences will exist and PACT activity appears to be associated exclusively with the particulate fractions of the cell [132,146]. The consensus of opinion seems to be that PACT can be a true constituent of mitochondria [139,143,147,148], the endoplasmic reticulum [132,148,149], and of nuclei [132,150].

PACT from both prokaryotes and eukaryotes can catalyse the reaction of both r-CTP and d-CTP with phosphatidate. In *E. coli* the cytosine-liponucleotide pool contained an equal mixture of r-CDP- and d-CDP-diacylglycerols and both of these appeared to serve as precursors of phosphatidylglycerol and phosphatidylserine [151]. It was suggested that a single enzyme was responsible for the formation of CDP-diacylglycerol from the two forms of CTP since these substrates were mutually competitive, and during the purification of PACT there was no change in their relative effectiveness as precursors [152]. d-CTP can also be used by PACT from mammalian sources [150,153–155]. The resulting d-CDP-diacylglycerol can be used by rat liver mitochondria to form phosphatidylglycerol [153], and by neuronal nuclei in the synthesis of phosphatidylinositol [155]. Studies on the cation optima for the incorporation of r-CTP and d-CTP into CDP-diacylglycerol and the mutual competitiveness of these substrates again indicate that a single enzyme is involved in this reaction [155]. Despite these findings d-CDP-diacylglycerol was not detected in the liponucleotide fractions isolated from bovine liver [156], bovine brain [157], or rat pineal gland [158]. Possibly this reflects the low concentrations of d-CTP in vivo. The importance of the formation of d-CDP-diacylglycerol in the synthesis of acidic lipids in eukaryotes is not yet known.

In rat liver phosphatidylinositol has a markedly different fatty acid composition to that of phosphatidate. It is rich in tetraenoic acids e.g. arachidonate, and it contains only low concentrations of mono- and dienoic acids [159], whereas the converse is true for phosphatidate [160,161]. CDP-diacylglycerol isolated from bovine liver or brain [156,157] had a similar fatty acid composition to the corresponding phosphatidylinositol. It was therefore suggested that PACT might be relatively specific for the tetraenoic forms of phosphatidate [140]. However, PACT showed little fatty acid specificity towards phosphatidate when this was presented as an aqueous dispersion [162], or bound to membranes [140]. An alternative mechanism is that CDP-diacylglycerol undergoes a deacylation-reacylation cycle in which the acyltransferase shows a marked preference for arachidonoyl-CoA [163].

CDP-diacylglycerol was found to be present in beef liver at a concentration of 5–17 μmol/kg, whereas the concentration of phosphatidate was about 780 μmol/kg [156]. In addition, CDP-diacylglycerol was barely detectable in rat pineal gland unless propranolol was added [158]. These observations suggest that the formation of CDP-diacylglycerol may be rate-limiting in acidic lipid synthesis [156–158,164], and that PACT may be a regulatory enzyme. However, the low concentration of CDP-diacylglycerol could result from its rapid deacylation. There is indirect evidence that this occurs in mammalian tissues [140], and a hydrolase activity has been partially purified from *E. coli* [165].

Little is known about the physiological regulation of PACT. Studies in vitro have shown that its activity is sensitive to the concentration of ions and this will be discussed further in Section 8. GTP can stimulate PACT activity, whereas ATP and F^- inhibit [166], but the importance of this in control is uncertain.

Regulatory enzymes are frequently identified by their ability to change activity in response to physiological stimuli. PACT activity in the microsomal fraction of the livers of diabetic rats was unchanged, whereas the inositol phosphatidyltransferase which is responsible for converting CDP-diacylglycerol to phosphatidylinositol (Fig. 1, reaction 9) was significantly decreased [94]. PACT activity was also unchanged when rats were fed acutely with ethanol [111], or chronically on diets enriched with sucrose, lard or corn oil [108]. By contrast Fallon et al. [167] reported a 25% decrease in PACT activity in rats fed a diet rich in fructose, and this was accompanied by an increase in PAP activity. It has been reported that PACT activity is 3.6 times higher in the mitochondria from 7777 hepatomas than in those from normal rat livers, and this was associated with a 62% decrease in the microsomal activity of the tumours compared to the normal livers [164]. These changes could be partly responsible for the increases of 96% and 46% respectively in the concentration of phosphatidylinositol and diphosphatidylglycerol of the mitochondria in the tumours [168].

6. *Conversion of phosphatidate to diacylglycerol*

The major route of phosphatidate metabolism in the liver is its conversion to diacylglycerol by PAP (Fig. 1). This lipid then serves as the precursor for the major zwitterionic phospholipids, phosphatidylethanolamine and phosphatidylcholine, as well as for the synthesis of triacylglycerol. PAP activity was first demonstrated in plant tissue [169], and it has subsequently been detected in a large variety of mammalian tissues (for review, see 170).

There is general agreement that PAP activity is found in the particulate fractions of mammalian cells. In liver the highest specific activity was recorded in the lysosomal fraction, but plasma membranes, mitochondria and the endoplasmic reticulum also contained activity that could not be attributed to contamination [170]. Presumably, the lysosomal activity is concerned with the degradation of phospholipids rather than with their synthesis. It is difficult to draw absolute conclusions about the distribution of PAP since many of the assays involved the determination of phosphate release from phosphatidate. As will be discussed in Section 7, this can also occur via the action of phospholipase A activities on phosphatidate followed by the dephosphorylation of the glycerophosphate. It is therefore important to know the extent to which these reactions could contribute to the total release of phosphate in any given subcellular fraction.

PAP in the microsomal fraction of rat liver has been solubilized and fractionated into two distinct activities [171]. One fraction (F_A) was non-specific in that it hydrolysed a number of phosphate esters and had a high K_m for phosphatidate. It was also not inhibited by diacylglycerol. The second fraction (F_B) was specific for phosphatidate or lysophosphatidate. It had a low K_m for these substrates, and it was inhibited non-competitively by diacylglycerol. The authors suggested that F_A contained a non-specific phosphomonoesterase and the activity in F_B was probably involved in glycerolipid synthesis [171].

It was also shown that PAP activity in rat liver mitochondria could be separated into activity which was readily extracted by repeated freezing and thawing, and another activity which was insoluble. This could not be separated from the particulate material by a variety of techniques [172]. Subsequent partial purification of the solubilized activity gave fractions that could hydrolyse hexadecylphosphate, glycero-2-phosphate and ATP in addition to phosphatidate [173]. However, the authors presented evidence that hexadecylphosphate and phosphatidate were dephosphorylated by different enzymes. Phosphatidylcholine, phosphoinositide, and *rac*-glycero-3-phosphate were not hydrolysed by the partially purified PAP [173].

The occurrence of PAP in mitochondrial and microsomal fractions ought to mean that they should be able to efficiently convert phosphatidate into diacylglycerol. Furthermore, the latter compound should be rapidly metabolized to triacylglycerol in the microsomal fraction by the action of DGAT. It was therefore difficult in the late 1950s and the first half of the 1960s to understand why this did not occur. Normally phosphatidate accumulated as the major product of esterification of glycerophosphate and the addition of the soluble fraction was necessary before this was converted to triacylglycerol. Thus a number of papers appeared which referred to soluble stimulating factors (for reviews, see 170 and 174). These factors consisted of unsaturated fatty acids [175], and a variety of proteins [174,176], including a heat-stable protein with a relative molecular mass in the range 8000–16000 [177,178]. However, the major effect was produced by a heat-labile protein which was identified as a soluble PAP [179,180]. The activity was determined by measuring the rate of diacylglycerol formation from membrane-bound phosphatidate that had been synthesized in the membranes from glycerophosphate. This PAP activity had not previously been recognised, since its rates with emulsions of phosphatidate were very low. It was therefore thought that the soluble PAP specifically acted upon phosphatidate that was a part of a biological membrane, and that artificial emulsions could not serve as substrates.

Subsequent work has shown that this strict specificity does not appear to exist, and different groups of workers have recently reported high rates of activity for the soluble PAP when using phosphatidate emulsions [181–184]. The lack of activity in earlier work with phosphatidate emulsions was probably caused by the failure to remove the Ca^{2+} which had bound to the phosphatidate during its preparation. This cation inhibits the soluble and microsomal PAP activity [171,184–186]. It is advisable to add EGTA to the assays and then to adjust the concentration of Mg^{2+} to give maximum rates. The use of mixed emulsions with phosphatidylcholine can stimulate the soluble PAP activity up to 7-fold [181,183]. This probably results from a better interaction of PAP with its substrate, and it is probably caused by the decrease in the negative charge density of the phosphatidate in the artificial membrane. The accumulated evidence shows that phosphatidate emulsions (especially when mixed with phosphatidylcholine) provide an alternative form of the substrate to membrane-bound phosphatidate. The characteristics and physiological response of PAP observed with these two types of substrate preparation are compatible [182,184, Section 9]. Phosphatidate emulsions also have the advantage of providing a more clearly defined assay system, and a better control of kinetic parameters. However, care should be exercised when interpreting the activities of PAP measured by different workers with different substrates. Recent experiments with rat lung have shown that the activities obtained with aqueous dispersions of phosphatidate and membrane-bound phosphatidate could be partially resolved by gel filtration [187].

The relative activities of the soluble and microsomal PAP vary from species to species. Most experimental work has been performed with the livers of male rats in which these specific activities were similar [183,184,188]. However, the microsomal activities in the livers of male rabbits [122] and guinea pigs [189] were respectively 20- and 50-fold greater than the soluble activity. It appears likely that the soluble PAP in the cell is closely associated with the endoplasmic reticulum membranes in which the phosphatidate is synthesized. These species differences could indicate that this association is stronger in rabbit and guinea-pig livers than in rat liver. Alternatively, there may be a much higher activity of distinct soluble enzyme in the rat.

It was also claimed that the specific activity of the soluble PAP in normal female rats was lower by an order of magnitude than in males [188]. However, another group reported that the soluble activity was higher in the female than in males [190]. Both groups agreed that the microsomal PAP was higher in females.

This soluble activity of rat liver and adipose tissue is characterized by the large stimulation in activity that is produced with Mg^{2+} [181,184,186,191]. This is not an absolute requirement for Mg^{2+} since activity is still retained in the presence of large quantities of EDTA, and amphiphilic cations can substitute for Mg^{2+} [186] as will be discussed further in Section 8. The effect of Mg^{2+} in stimulating the microsomal PAP activity in rat liver [184], and the mitochondrial activity of rat adipose tissue

[191] is far less than that with the equivalent soluble activity. Fluoride inhibited both the microsomal [171,173] and soluble [180] activities. The pH optimum of the microsomal PAP was reported to be about 5.8 [173], whereas that of the soluble enzyme was in the region of 6.8–7.6 [173,181]. Differences have also been reported with respect to the fatty acid composition of the phosphatidate used as substrate. When membrane-bound phosphatidate was synthesized from myristate or palmitate, its rate of hydrolysis by the soluble enzyme was greater than when laurate and stearate were used [192]. Furthermore, the phosphatidate synthesized from a mixture of palmitate and oleate was a better substrate than that obtained when these acids were used separately. By contrast, the rate of hydrolysis by the microsomal PAP was greatest when the phosphatidate was prepared from stearate, oleate, or a mixture of palmitate and oleate [192]. However, it is doubtful whether PAP exhibits much selectivity for fatty acids during the synthesis of glycerolipid in the liver in vivo [160,193].

The differences that have been reported in the properties of PAP from various sites in the cell suggest that a number of different proteins exhibit this activity. However, it is possible that different properties could result from different physical states of the enzyme e.g. whether or not it is membrane-bound. A final answer to this question must depend on the immunochemical characterisation of the different activities. There is also some controversy related to the physiological importance of the different PAP activities in triacylglycerol synthesis. Both the microsomal and soluble enzymes appear to be involved and this will be discussed further in Section 9.

7. Deacylation of phosphatidate

Evidence is presented in Section 9 that PAP is involved in the regulation of triacylglycerol synthesis. This enzyme has also been claimed to have the lowest activity of the microsomal enzymes that are responsible for this synthesis, and it was therefore suggested that it is rate-limiting in the pathway [185]. By contrast, it has been claimed that the microsomal PAP activity as determined by P_i release has a large reserve capacity [142]. Furthermore, in experiments with isolated hepatocytes [^{14}C]glycerol was incorporated rapidly into triacylglycerol and phosphatidylcholine with little activity accumulating in phosphatidate [194]. In addition, the concentration of phosphatidate in microsomal fractions of rat liver is only about a quarter that of diacylglycerol [167]. These findings appear to contradict the proposition that PAP is rate-limiting in triacylglycerol synthesis. These apparent discrepancies can be reconciled by the observation that phosphatidate can also be degraded by phospholipase A type activities (Fig. 1). These activities have been detected in mitochondrial [180,192], microsomal [183,184,195,196] and soluble fractions [183,184,196] of rat liver.

Lysophosphatidate was not detected as an intermediate in the hydrolysis of phosphatidate emulsions, or membrane-bound phosphatidate by microsomal and soluble fractions of rat liver [183,184]. Presumably this is because the lysophospholipase activity is relatively high. The specific activity of phosphatidate deacylase was about four times greater in the microsomal fraction from rat liver than in the soluble fraction when these were measured with a phosphatidate emulsion [184]. The product formed by the soluble system was entirely glycerophosphate, whereas the major product with microsomal fractions was glycerol and P_i [183]. Therefore, the determination of the capacity of PAP in the microsomal fraction is not reliably measured by P_i release from either aqueously dispersed or membrane-bound phosphatidate [183,184].

It is not yet known whether the deacylase activities described in liver are specific for phosphatidate. Such a specific phospholipase A_2 has been demonstrated in platelets, but this activity required Ca^{2+} [197]. By contrast the hepatic activities are stimulated by Mg^{2+} rather than by Ca^{2+} [184]. The physiological function of this activity is also not certain. It has been suggested that the deacylation of phosphatidate to lysophosphatidate is an obligatory step in the synthesis of triacylglycerols [195]. It could also operate to prevent the excessive accumulation of phosphatidate in membranes [184,196], which if allowed to occur might profoundly alter their properties. The activities of the deacylase systems in microsomal and soluble fractions are of a similar magnitude to those of PAP. If these activities are mutually competitive for substrate then this could be important in regulating the subsequent route of phosphatidate metabolism. This idea was tested by feeding rats with a single dose of fructose, sorbitol, glycerol or ethanol in order to increase PAP activity [196]. The only significant change in deacylase activity in the microsomal and soluble fractions was a small increase produced by ethanol in the former fraction. However, in these experiments there were highly significant correlations between (a) the microsomal deacylase and microsomal PAP activity and (b) the soluble deacylase and PAP activities. All four treatments increase the ratio of PAP: deacylase activity, and this is consistent with the ability of these nutrients to stimulate hepatic triacylglycerol synthesis.

8. Effects of ions on the direction of phosphatidate metabolism

A large variety of drugs including some phenothiazine neuroleptics, imipramine antidepressants, local anaesthetics, anorectics, hypolipidaemic agents, propranolol and some other β-adrenoreceptor blockers, morphine and levorphanol share the ability to redirect the route of glycerolipid metabolism. Despite the diversity of pharmacological function, these compounds have two structural features in common: they possess a hydrophobic region and a primary or substituted amine that can bear a positive charge. They promote the accumulation of the acidic phospholipids, phosphatidate, CDP-diacylglycerol, phosphatidylinositol, phosphatidylglycerol, diphosphatidylglycerol and lysobisphosphatidate. The particular pattern of acidic lipids formed depends upon the tissue and the conditions that prevail. At the same time the synthesis of phosphatidylcholine, phosphatidylethanolamine and triacylglycerol is decreased (for reviews, see refs. 198–200). When animals are treated chronically with the more hydrophobic cationic drugs that have long biological half-lives, a phospholipidosis occurs. This is characterised by the general accumulation of phospholipids in lysosomes, but in relative terms there is a specific accumulation of acidic phospholipids. Part of the reason for this seems to be the increased production of these lipids which bind the cationic drug (see also Chapter 6, Section 9c). These complexes subsequently accumulate in the lysosomes and they are not readily degraded by phospholipase action [199–203].

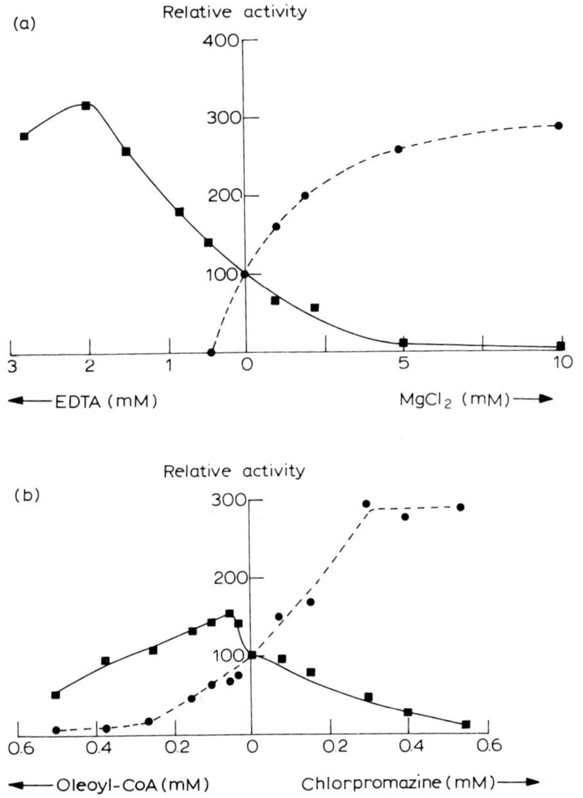

Fig. 3. The effect of ions and EDTA on phosphatidate metabolism. The figure shows the effect of adding $MgCl_2$, EDTA, chlorpromazine or oleoyl-CoA on the conversion of membrane-bound phosphatidate to diacylglycerol (■) or CDP-diacylglycerol (●). The amount of CDP-diacylglycerol formed in the absence of any addition was 0.48 nmol. The membrane-bound activity of PAP was supplemented by the addition of enzyme that had been partially purified from the soluble fraction from rat liver. The amount of diacylglycerol formed in the absence of any addition ranged from 3.5 nmol to 8.05 nmol depending on the amount of soluble PAP added [141]. The membrane-bound phosphatidate used in these preparations already contained some Mg^{2+} which was derived during its preparation. Oleoyl-CoA was shown to interact with phosphatidate in other experiments using phosphatidate emulsions. Its partition coefficient was 24000±2300 (mean±S.D. from three independent experiments), and the method used to determine this is described in [186].

One of the major effects of amphiphilic cations on glycerolipid synthesis is the redirection of phosphatidate metabolism. Their action depends on the concentration of Mg^{2+} that is present (Fig. 3). In these experiments membrane-bound phosphatidate was used and this was prepared in the presence of Mg^{2+}. The concentration of this cation was slightly greater than that required to give optimum rates for PAP and this can be seen by the effects of adding EDTA [141] (Fig. 3). The Mg^{2+} concentration was about optimum for the activity of the phosphatidate deacylase system [184], whereas it was sub-optimum for PACT [141] (Fig. 3). Under these conditions, the addition of chlorpromazine and norfenfluramine decreased the rate of synthesis of diacylglycerol and stimulated that of CDP-diacylglycerol [141,204] (Fig. 3). This result is compatible with the effects of amphiphilic cations in promoting the synthesis and accumulation of acidic phospholipids, and in decreasing the synthesis of those lipids derived from diacylglycerol. Phosphatidate deacylation was less susceptible to inhibition by amphiphilic cations than was PAP [184]. This could provide the cell with the ability to cycle some of the phosphatidate back to glycerophosphate provided that the concentration of drug does not rise too high.

The mechanisms of these effects have been investigated using phosphatidate emulsions. The inhibition of PAP by the cationic drugs is of a competitive type probably resulting from their interaction with phosphatidate rather than with PAP itself. This explains why the potency of the drugs is related to their abilities to partition into the artificial phosphatidate membranes [186]. In addition to the hydrophobic interaction the positive charge on the amine is attracted to the negative charge on the phosphate of the phosphatidate. When Mg^{2+} is absent the addition of chlorpromazine stimulates both the dephosphorylation and deacylation of phosphatidate and it replaces the Mg^{2+} requirement [184,186]. Amphiphilic cations can also substitute for part of the Mg^{2+} requirement of PACT, but this enzyme appears to have a specific need for bivalent cations [141]. This is probably required for the formation of a complex with CTP.

The interaction of Mg^{2+} and amphiphilic cations with membranes containing phosphatidate alters their packing arrangement and electrical potential [205–207]. The enzymes that metabolize phosphatidate have different sensitivities to these changes and this explains why the direction of glycerolipid metabolism is modified by these cations. Similarly, amphiphilic anions such as clofenapate (results not shown) and oleoyl-CoA (Fig. 3) also partition into phosphatidate emulsions and they have opposite effects to the amphiphilic cations.

The effects of Ca^{2+} on phosphatidate metabolism are not identical to those of Mg^{2+}. They inhibited the action of PAP on membrane-bound phosphatidate that contained Mg^{2+}, and had little effect on the deacylase activity [184]. However, Ca^{2+} did not stimulate PACT activity [208]. Ca^{2+} was also less effective in stimulating the activity of PAP on emulsions of potassium phosphatidate than was Mg^{2+} [186]. It is known that Ca^{2+} ions interact with membranes containing acidic lipids and alter their physical properties including their transition temperature, aggregation and permeability. However, the effects are different from those observed with Mg^{2+} in membranes containing phosphatidate [206], and particularly phosphatidylserine

[209]. Such observations probably account for some of the different effects that these bivalent cations exhibit on phosphatidate metabolism.

It is not yet clear whether the availability of inorganic cations is a factor in regulating the direction of phosphatidate metabolism in vivo. However, the events described in this Section do provide a reasonable explanation for the redirection of glycerolipid synthesis observed with amphiphilic cationic drugs.

9. Physiological control of PAP activity and of triacylglycerol synthesis

There is considerable evidence that an increase in the capacity of the liver to synthesize triacylglycerol is normally accompanied by an increase in PAP activity. Regulation at this point is reasonable since phosphatidate lies at an important branch-point in metabolism (Fig. 1) [102,210,211]. The diacylglycerol that is formed by PAP is preferentially incorporated into phosphatidylcholine and phosphatidylethanolamine when the supply of fatty acids is low. One rate-limiting factor in the formation of these lipids is the availability of CDP-choline and CDP-ethanolamine [212]. When the supply of fatty acids to the liver is high, the formation of diacylglycerol is increased and the capacity to synthesize phosphatidylcholine and phosphatidylethanolamine is exceeded. The remaining diacylglycerol can then be converted to triacylglycerol and this is facilitated by the ready availability of acyl-CoA esters. DGAT has a relatively high specific activity in the liver and although some changes in its activity have been reported in different physiological conditions, these are less than those observed with PAP [211].

The hypothesis that PAP is important in controlling hepatic triacylglycerol synthesis is supported by the observations that the accumulation of phosphatidate in liver is inversely related to the rate of triacylglycerol synthesis [108,167,182,213,214]. However, the accumulation of phosphatidate is probably limited by the action of the deacylase system (Section 7). The factors that are responsible for controlling PAP activity are not yet clearly established, but the availability of glucocorticoids seems to be of major importance [102,214–216]. These hormones stimulate the synthesis, accumulation and secretion of triacylglycerol by the liver [214,217,218] and they also increase the activity of PAP in isolated perfused livers [215] or in isolated hepatocytes [219]. The increases in PAP activity were seen after about 4 h and they were blocked by actinomycin D and cycloheximide [215,219]. The glucocorticoids are therefore probably promoting the synthesis of PAP, and the magnitude of the increase was similar to that observed for tyrosine aminotransferase [219]. These effects of glucocorticoids on PAP activity appear to contribute to many of the observed changes in hepatic triacylglycerol synthesis, but other factors are likely to be involved. For instance, the effect of corticosterone in stimulating the PAP activity in isolated hepatocytes can be suppressed by insulin [271].

The control of triacylglycerol synthesis in the liver differs from that of fatty acid biosynthesis in which the circulating concentrations of insulin and glucagon are of prime importance. Glucocorticoids have a permissive effect on the stimulation of fatty acid biosynthesis by insulin [220,221], and the rates of fatty acid and triacylglycerol synthesis are both high when animals are fed on diets rich in carbohydrate. The soluble PAP activity in the livers of rats fed on this type of diet reached a peak 2 h after dark, and this was probably caused by the peak in the concentration of corticosterone which preceded it by 4 h [222]. Thus the capacity to synthesize triacylglycerols is co-ordinated with the period of most active feeding and fatty acid biosynthesis.

The inclusion of fructose, sorbitol, glycerol and ethanol in the diet stimulates hepatic triacylglycerol synthesis. Rats fed by stomach tube, or injected with these nutrients show a marked increase in the microsomal and soluble PAP activities of the liver when compared with controls treated with saline or glucose [111,182,196,223]. The activities of the other enzymes of triacylglycerol synthesis, including DGAT, were not significantly increased by this acute treatment with ethanol [111]. Part of the increase in PAP activity is mediated by the increased concentration of circulating corticosterone which is produced by these nutrients in the absence of an insulin response [224]. Adrenalectomised rats that had been maintained by providing saline in their drinking water only showed a 1.7-fold increase in soluble PAP activity 7 h after feeding with ethanol. The equivalent increase for the control rats was 6.9-fold [224]. The adrenalectomy itself produced a 25% decrease in PAP activity. The glucocorticoid involvement in increasing hepatic triacylglycerol synthesis is also compatible with the observation that an intact pituitary–adrenal axis is necessary for ethanol to produce a fatty liver [225–228]. Further evidence for the involvement of glucocorticoids in the production of a fatty liver by ethanol comes from work with the hypotriglyceridaemic drug, benfluorex. Chronic treatment of rats with this compound decreases the duration of the ethanol-induced increase in corticosterone [229,230], the extent of the increase in the activity of PAP [111] and the rates of synthesis and accumulation of hepatic triacylglycerols [213,229]. It does not alter the rate of absorption or oxidation of ethanol [213].

A slightly different technical approach to studying the involvement of glucocorticoids is to maintain adrenalectomised rats by injecting cortisol or corticosterone. The liver's metabolism can therefore be influenced by glucocorticoids, whereas ethanol feeding should produce no further increase in this effect. In these adrenalectomised rats, ethanol produced a 1.7-fold increase in soluble PAP activity after 12 h, compared with a 1.6-fold increase in the control rats [231]. However, this former increase is smaller than has been observed in other studies [111,182,196]. Although adrenalectomy followed by glucocorticoid injections did not appear to modify the ethanol-induced increase in the soluble PAP activity, it did abolish the increase in the microsomal fraction [231]. This may indicate that the activities in these two fractions are under a different control, even though their activities do increase in parallel after feeding rats with ethanol, sorbitol, fructose or glycerol [196].

The portion of the ethanol-induced increase in PAP activity that does not appear to be caused by glucocorticoids [224,231] may result from changes in the metabolic environment of the liver. The concentration of glycerophosphate and the changed redox state, i.e. the increased NADH/NAD^+ ratio, have been proposed as important factors [223,231]. Some support for this suggestion is derived from the use of pyrazole which inhibits ethanol oxidation and prevents part of the ethanol-induced increase in PAP activity [231,232]. The mechanism whereby the changed redox state affects the increase is not known, but it may also involve increased enzyme synthesis [232].

The redox change only appears to be responsible for about 15% of the total increase in PAP activity that was observed when ethanol was fed to rats that had been adrenalectomised and maintained on saline [224]. It is also relevant to note that feeding fructose, which should not significantly alter the hepatic redox state, increased the soluble PAP activity to the same extent as did sorbitol, which could do so [196]. The general conclusions from this type of work are that glucocorticoids are important in controlling the activity of PAP, but this effect may be modified by other hormones, or by the changes in substrate supply in the liver.

The effects of nutrients on hepatic PAP that have been discussed so far refer to short-term studies in which single doses of the nutrients were administered. However, it is also known that the high activity of PAP is maintained when hamsters are fed chronically on ethanol [233]. Rats that were fed on artificial diets enriched with fructose or glucose had high PAP activities [106], and high fructose diets also increased the activity of DGAT when this was measured with membrane-bound diacylglycerol [167]. It is not known whether this chronic effect of fructose is also caused by an increased availability of glucocorticoids, although there is evidence that diets rich in sucrose might cause this to happen [234,235]. Since glucocorticoids are so important in regulating the activity of PAP, it is important to assess whether the stress of dietary modification rather than the replacement of a specific dietary component is responsible for the observed changes. In experiments where rats were fed for three weeks on pelleted diets containing 40% by weight of sucrose rather than starch the activity of the soluble PAP was not significantly different [105].

The type and quantity of fat in the diet can also alter the rate of hepatic triacylglycerol synthesis. Rats that were fed a powdered diet enriched with lard synthesized relatively more triacylglycerol than did those on the starch diet, and they also had a slightly higher PAP activity in their livers [108]. However, in subsequent work we employed pelleted diets enriched with beef tallow or corn oil rather than with starch. The energy densities relative to protein were approximately equal. In these rats the basal activities of the soluble PAP in the livers were not significantly different [105]. It has also been shown that the addition of mustard-seed oil or corn oil to the diet of rats did not alter PAP activity [236]. However, this activity is increased in essential fatty acid deficiency [237].

Animals that have been fed on high fat diets can show abnormal stress responses to cold [238], nembutal narcosis [239] and to corticotropin injection [239], although the latter may not be a universal phenomenon [105]. They also have a prolonged corticosterone response after being force-fed with fructose [102,240]. This also provokes an increased PAP activity in the liver compared to the rats fed a high carbohydrate diet [240]. It is also known that high fat diets exaggerate the effects of fructose [241,242] and ethanol [243–245] in stimulating hepatic triacylglycerol synthesis, and this could be partly explained by the changed glucocorticoid status [102,240]. The maintenance or increase in the capacity to synthesize triacylglycerols in the liver when animals are fed on a high fat diet contrasts with the decreased ability to synthesize fatty acids.

A similar disparity between these two pathways of lipogenesis also occurs in stress conditions in which the major supply of fatty acids to the liver is by mobilization from adipose tissue. These are conditions in which the influence of glucocorticoids in controlling metabolism is increased and in which hepatic PAP activity is raised. The latter effect occurs in starvation [90,246,247], mildly ketotic [94] and severely ketotic diabetes [248], hypoxia [247], after surgical stress including subtotal hepatectomy [90], and during the accumulation of triacylglycerol in the liver after injecting hydrazine [249] and morphine [250]. Phenobarbital injections also increased PAP activity in rabbits [122] and guinea-pigs [189], but not in rats [188]. In one report the activity of PAP in the livers of starved rats decreased rather than increased [251], but this discrepancy may relate to how the rats were handled. These rats [251] were not put on grid-bottomed cages, and they were allowed to eat saw-dust and faeces which could have induced less stress than the complete absence of food. It is also important to taken account of the period of starvation since there is a natural circadian rhythm for PAP [222]. This can complicate the interpretation of results unless appropriate controls are used.

The high PAP activity in the livers of diabetic rats was normalised when these rats were injected with insulin [248]. Evidence has also been presented that the PAP activity in diabetic livers is less susceptible to feed-back inhibition by very-low-density lipoproteins, and that this could be a factor in the increased triacylglycerol synthesis that can be observed in ketotic diabetes [252].

The effects of ethanol in producing a stress response and a fatty liver have already been discussed. Part of the action of hydrazine [253] and morphine [250] in producing the increase in PAP activity could result from the increase in circulating glucocorticoids. However, both of these compounds also appeared to have a direct effect in increasing PAP activity in isolated hepatocytes [249,250]. Similar direct effects have been reported with carbon tetrachloride [254], and bromobenzene [255], and these stimulations are dependent on the presence of Ca^{2+}.

It may seem paradoxical that the capacity for triacylglycerol synthesis in the liver should increase in catabolic conditions in which the concentration of glucagon, catecholamines and glucocorticoids relative to insulin is increased. The partitioning of fatty acids into β-oxidation rather than triacylglycerol synthesis is favoured in these conditions (Section 4), and the diacylglycerol that is produced is preferentially incorporated into phosphatidylcholine and phosphatidylethanolamine [85,256,257]. This may occur because the affinity of DGAT for diacylglycerol is less than that of the choline and ethanolamine phosphotransferases (Fig. 1). It has also been reported that glucagon decreases the activity of DGAT in hepatocytes without altering the activity of choline phosphotransferase [258]. However, cyclic-AMP analogues do inhibit phosphatidylcholine synthesis in hepatocytes, and this probably occurs by the decrease in activity of CTP phosphocholine cytidylyltransferase [259]. One of the major factors in controlling the synthesis of triacylglycerols is the availability of fatty acids and precursors for the glyceride-glycerol backbone (see Section 4). In many catabolic conditions large quantities of fatty acids are mobilised from adipose tissue and this supply to the liver exceeds the requirements of β-oxidation and phospholipid synthesis.

The provision of a high capacity for triacylglycerol synthesis enables the liver to protect itself against the accumulation of toxic concentrations of fatty acids and acyl-CoA esters. This also ensures that CoA is regenerated. The increased activity of PAP appears to facilitate this increased triacylglycerol synthesis possibly because it is at a rate-controlling step in the pathway. Alternatively, the increase may be designed to ensure that PAP does not become rate-limiting when the supply of fatty acids is increased and the demand for triacylglycerol synthesis rises. This situation is seen in ketotic diabetes [94,248]. It is important to note that DGAT activity is also raised in ketotic diabetes [248,260], and that it is normal in mildly ketotic diabetes [94]. This implies that any effect of glucagon in decreasing DGAT activity can be over-ridden in these situations.

The triacylglycerol that is formed can be temporarily stored in the liver or secreted in very-low-density lipoprotein. The heart is able to take up the fatty acids from VLDL since lipoprotein lipase in this organ is controlled by glucocorticoids rather than by insulin [261]. The ability of the heart to synthesize triacylglycerols is high in ketotic diabetes. There were indications that this might result from an increased activity of PAP rather than GPAT [262], but the opposite conclusion was reached in a later paper [101]. The high capacity to synthesize triacylglycerol enables the heart to cope with the increased supply of fatty acids that are ultimately destined to provide a source of energy.

The evidence that PAP regulates the rate of triacylglycerol synthesis in adipose tissue is less complete than for liver and the results are less clear-cut. PAP activity has been reported to increase in both adipose tissue and livers of genetically obese (ob/ob) mice [210,263], and the adipose tissue of obese human beings [264]. There were indirect indications that the inclusion of sucrose in the diets of rats might increase the PAP activity [265]. However, rats fed on a slightly different diet that was enriched with sucrose failed to show significantly increased PAP activity when this was assayed directly [118]. It was also found that feeding glucose, fructose, sorbitol, glycerol or ethanol did not alter the PAP activity in adipose tissue 6 h later. This was the time at which the activity in liver was markedly increased by the latter four nutrients [196]. It may also be significant in terms of the control of phosphatidate metabolism that no phosphatidate deacylase activity could be detected in adipose tissue under conditions where it could be readily demonstrated in liver fractions [118].

Differences between these two tissues are expected in catabolic conditions in which the synthesis of triacylglycerols in adipose tissue should decrease, whereas the capacity in liver may be maintained or increased. A fall of 62%, 52% and 36% in the soluble PAP activity in rat adipose tissue has been reported after 24, 48 and 72 h of starvation, respectively [266]. However, these activities were expressed relative to DNA and the decreases were in the range of 15–22% when expressed relative to protein. By contrast, other authors failed to show a significant change [118,267]. Evidence in favour of a regulation of PAP by catabolic hormones has been provided by work with isolated adipocytes. The presence of noradrenalin produced a rapid decrease in the total Mg^{2+}-stimulated PAP activity which was blocked by propranolol and reversed by insulin [268,269]. In other work, lipolytic agents including adrenalin, cyclic-AMP analogues, and theophylline were reported to decrease the soluble PAP activity, but to increase that in the microsomal fraction [270]. By contrast corticotropin, which was also lipolytic, increased both the microsomal and soluble PAP activities [270]. The opposite effect was seen after injecting corticotropin in vivo, since the activity of the soluble PAP was decreased by this treatment [118]. However, in these experiments other hormones could have been released in response to the injection which might also have influenced metabolism in adipose tissue.

10. Conclusion

This chapter has attempted to describe the enzymes that are responsible for the synthesis and metabolism of phosphatidate. Particular emphasis has been placed on the relationship of this to the synthesis of triacylglycerol in the liver. Many of the characteristics of the enzymes are known and the pathways for the metabolism of various glycerolipids have been established. What is missing is a knowledge of how these enzymes interact and how they are controlled in different tissues, since this is likely to vary. We do not even know the relative importance of glycerophosphate and dihydroxyacetone phosphate as precursors for the back-bone of glycerolipids in any mammalian tissue. The contribution of the mitochondrial and peroxisomal esterification systems in controlling triacylglycerol synthesis in addition to the microsomal system is also uncertain. Some information is available about the acute and chronic hormonal control of glycerolipid synthesis, but this is far from complete. The present description has dealt with the effects of insulin, glucagon and glucocorticoids. The action of insulin and glucagon in regulating the relative flux of fatty acids through the GPAT and CAT reactions provides a rational explanation for the co-ordinated control of fatty acid synthesis, β-oxidation and esterification (Fig. 2). The effects of glucocorticoids in increasing PAP activity enable the liver to increase its synthesis of triacylglycerols in catabolic conditions when the fatty acid supply is high. The liver can then export this potential energy to other organs in the form of VLDL. In this sense triacylglycerol synthesis may fulfil a similar function to gluconeogenesis and ketogenesis in these catabolic conditions [216,219]. It is likely that there is a similar and co-ordinated control of these pathways.

Although the activities of the enzymes of triacylglycerol synthesis do change in response to physiological stimuli, the mechanisms which produce these changes are largely unknown. These investigations will require the purification of the enzymes and the raising of antibodies to them. Relatively little progress has been made in this respect, largely because of the intrinsic difficulties of working with membrane-bound enzymes that act on lipid substrates. One of the big challenges of glycerolipid metabolism will be to overcome these problems and to understand the details of how the control of this metabolism is co-ordinated with the rest of intermediary metabolism.

References

1 Van den Bosch, H. (1974) Ann. Rev. Biochem. 43, 243–277.
2 Brindley, D.N. (1974) in D.H. Smyth (Ed.), Biomembranes, Vol. 4B, Plenum, New York, pp. 621–671.
3 Snyder, F. (1977) Lipid Metabolism in Mammals, Vols. 1 and 2, Plenum, New York.
4 O'Doherty, P.J.A. (1978) in A. Kuksis (Ed.), Handbook of Lipid Research, Vol. 1, Plenum, New York, pp. 289–339.
5 Bell, R.M. and Coleman, R.A. (1980) Annu. Rev. Biochem. 49, 459–487.
6 Kornberg, A. and Pricer, W.E. (1953) J. Biol. Chem. 204, 345–358.
7 Smith, S.W., Weiss, S.B. and Kennedy, E.P. (1957) J. Biol. Chem. 228, 915–922.
8 Stein, Y. and Shapiro, B. (1957) Biochim. Biophys. Acta 24, 197–198.
9 Yamashita, S. and Numa, S. (1972) Eur. J. Biochem. 31, 565–573.
10 Monroy, G., Kelker, H.C. and Pullman, M.E. (1973) J. Biol. Chem. 248, 2845–2852.
11 Tamai, Y. and Lands, W.E.M. (1974) J. Biochem. 76, 847–860.
12 Kelker, H.C. and Pullman, M.E. (1979) J. Biol. Chem. 254, 5364–5371.
13 Kito, M., Ishinaga, M. and Nishihara, M. (1978) Biochim. Biophys. Acta 529, 237–249.
14 Yamashita, S., Hosaka, K. and Numa, S. (1972) Proc. Natl. Acad. Sci. USA 69, 3490–3492.
15 Ray, T.K., Cronan, J.E., Mavis, R.D. and Vagelos, P.R. (1970) J. Biol. Chem. 245, 6442–6448.
16 Ray, T.K. and Cronan, J.E. (1975) J. Biol. Chem. 250, 8422–8427.
17 Goldfine, H., Ailhaud, G.P. and Vagelos, P.R. (1967) J. Biol. Chem. 242, 4466–4475.
18 Lineking, D.R. and Goldfine, H. (1975) J. Biol. Chem. 250, 8530–8535.
19 Shephard, E.H. and Hübscher, G. (1969) Biochem. J. 113, 429–440.
20 Zborowski, J. and Wojtczak, L. (1969) Biochim. Biophys. Acta 187, 73–84.
21 Daae, L.N.W. (1972) Biochim. Biophys. Acta 270, 23–31.
22 Nimmo, H.G. (1979) FEBS Lett. 101, 262–264.
23 Higgins, J.A. and Barnett, R.J. (1972) J. Cell. Biol. 55, 282–298.
24 Nachbaur, J., Colbeau, A. and Vignais, P.M. (1971) C.R. Acad. Sci. Paris, D 272, 1015–1018.
25 Yamashita, S., Hasaka, K., Taketo, M. and Numa, S. (1973) FEBS Lett. 29, 235–238.
26 Coleman, R. and Bell, R.M. (1978) J. Cell. Biol. 76, 245–253.
27 Abou-Issa, H.M. and Cleland, W.W. (1969) Biochim. Biophys. Acta 176, 692–698.
28 Daae, L.N.W. (1972) FEBS Lett. 27, 46–48.
29 Lloyd-Davies, K.A. and Brindley, D.N. (1975) Biochem. J. 152, 39–49.
30 Sánchez, M., Nicholls, D.G. and Brindley, D.N. (1973) Biochem. J. 132, 697–706.
31 Halder, D., Tso, W.-W. and Pullman, M.E. (1979) J. Biol. Chem. 254, 4502–4509.
32 Bremer, J., Bjerve, K.S., Borrebaek, B. and Christiansen, R. (1976) Mol. Cell. Biochem. 12, 113–125.
33 Halder, D. (1978) Fed. Proc. 37, 1494.
34 Yamada, K. and Okuyama, H. (1978) Arch. Biochem. Biophys. 190, 409–420.
35 Stern, W. and Pullman, M.E. (1978) J. Biol. Chem. 253, 8047–8055.
36 Nimmo, H.G. (1979) Biochem. J. 177, 283–288.
37 Bates, E.J. and Saggerson, E.D. (1979) Biochem. J. 182, 751–762.

38 Haldar, G., Carroll, M., Morris, P., Grosjean, C. and Anzalone, T. (1980) Fed. Proc. 39, 1992.
39 Monroy, G., Rola, F.H. and Pullman, M.E. (1972) J. Biol. Chem. 247, 6884–6894.
40 Bjerve, K.S., Daae, L.N.W. and Bremer, J. (1976) Biochem. J. 158, 249–254.
41 Yamashita, S., Hosaka, K. and Numa, S. (1973) Eur. J. Biochem. 38, 25–31.
42 Hill, E.E. and Lands, W.E.M. (1968) Biochim. Biophys. Acta 152, 645–648.
43 Saggerson, E.D., Carpenter, C.A., Cheng, C.H.K. and Sooranna, S.R. (1980) Biochem. J. 190, 183–189.
44 Hajra, A.K. and Agranoff, B.W. (1968) J. Biol. Chem. 243, 1617–1622.
45 Hajra, A.K. (1968) J. Biol. Chem. 243, 3458–3465.
46 Hajra, A.K. and Agranoff, B.W. (1968) J. Biol. Chem. 243, 3542–3543.
47 Schlossman, D.M. and Bell, R.M. (1976) J. Biol. Chem. 251, 5738–5744.
48 Schlossman, D.M. and Bell, R.M. (1977) Arch. Biochem. Biophys. 182, 732–742.
49 Schlossman, D.M. and Bell, R.M. (1978) J. Bact. 133, 1368–1376.
50 Dodds, P.F., Gurr, M.I. and Brindley, D.N. (1976) Biochem. J. 160, 693–700.
51 Rock, C.O., Fitzgerald, V. and Snyder, F. (1977) J. Biol. Chem. 252, 6363–6366.
52 Bowley, M., Manning, R. and Brindley, D.N. (1973) Biochem. J. 136, 421–427.
53 Bowley, M. and Brindley, D.N. (1976) Int. J. Biochem. 7, 141–147.
54 Hajra, A.K. and Burke, C. (1978) J. Neurochem. 31, 125–134.
55 Fisher, A.B., Huber, G.A., Furia, L., Bassett, D. and Rabinowitz, J.L. (1976) J. Lab. Clin. Med. 87, 1033–1040.
56 Jones, C.L. and Hajra, A.K. (1976) Fed. Proc. 35, 1724.
57 Jones, C.L. and Hajra, A.K. (1977) Biochem. Biophys. Res. Commun. 76, 1138–1143.
58 Hajra, A.K., Burke, C.L. and Jones, C.L. (1979) J. Biol. Chem. 254, 10896–10900.
59 Jones, C.L. and Hajra, A.K. (1980) J. Biol. Chem. 255, 8289–8295.
60 LaBelle, E.F. and Hajra, A.K. (1972) J. Biol. Chem. 247, 5825–5834.
61 LaBelle, E.F. and Hajra, A.K. (1972) J. Biol. Chem. 249, 6936–6944.
62 Rao, G.A. and Abraham, S. (1978) Lipids 13, 95–98.
63 Kanoh, H. and Åkesson, B. (1978) Eur. J. Biochem. 85, 225–232.
64 Bishop, H.H. and Strickland, K.P. (1980) Lipids 15, 285–291.
65 Lapetina, E.G. and Hawthorne, J.N. (1971) Biochem. J. 122, 171–179.
66 Schneider, E.G. and Kennedy, E.P. (1976) Biochim. Biophys. Acta 441, 201–212.
67 Hokin, L.E. and Hokin, M.R. (1963) Biochim. Biophys. Acta 67, 470–484.
68 Paris, R. and Clément, G. (1969) Proc. Soc. Exp. Biol. Med. 131, 363–365.
69 Pieringer, R.A. and Hokin, L.E. (1962) J. Biol. Chem. 237, 653–658.
70 Molaparast, F., Shrago, E. and Elson, C.E. (1979) J. Nutr. 109, 1560–1569.
71 Pollock, R.J., Hajra, A.K. and Agranoff, B.W. (1975) Biochim. Biophys. Acta 380, 421–436.
72 Agranoff, B.W. and Hajra, A.K. (1971) Proc. Natl. Acad. Sci. USA 68, 411–415.
73 Hill, E.E. and Lands, W.E.M. (1970) Biochim. Biophys. Acta 202, 209–211.
74 Plackett, P. and Rodwell, A.W. (1970) Biochim. Biophys. Acta 210, 230–240.
75 Okuyama, H. and Lands, W.E.M. (1970) Biochim. Biophys. Acta 218, 376–377.
76 Benns, G. and Proulx, P. (1972) Canad. J. Biochem. 50, 16–19.
77 Manning, R. and Brindley, D.N. (1972) Biochem. J. 130, 1003–1012.
78 Rognstad, R., Clark, R.G. and Katz, J. (1974) Biochem. J. 140, 249–257.
79 Mason, R.J. (1978) J. Biol. Chem. 253, 3367–3370.
80 Pollock, R.J., Hajra, A.K., Folk, W.R. and Agranoff, B.W. (1975) Biochem. Biophys. Res. Commun. 65, 658–664.
81 Pollock, R.J., Hajra, A.K. and Agranoff, B.W. (1976) J. Biol. Chem. 251, 5149–5154.
82 Harding, J.W., Pyeritz, E.A., Copeland, E.S. and White, H.B. (1975) Biochem. J. 146, 223–229.
83 Brindley, D.N. (1973) Biochem. J. 132, 707–715.
84 Ontko, J.A. (1972) J. Biol. Chem. 247, 1788–1800.
85 Groener, J.E.M. and Van Golde, L.M.G. (1977) Biochim. Biophys. Acta 487, 105–114.
86 McGarry, J.D. and Foster, D.W. (1980) Ann. Rev. Biochem. 49, 395–420.

87 Aas, M. and Daae, L.N.W. (1971) Biochim. Biophys. Acta 239, 208–216.
88 Van Tol, A. (1974) Biochim. Biophys. Acta 357, 14–23.
89 Zammit, V.A. (1981) Biochem. J. 198, 75–83.
90 Mangiapane, E.H., Lloyd-Davies, K.A. and Brindley, D.N. (1973) Biochem. J. 134, 103–112.
91 Rao, G.A., Sorrels, M.F. and Reiser, R. (1971) Lipids 6, 88–92.
92 Fallon, H.J. and Kemp, E.L. (1968) J. Clin. Invest. 47, 712–719.
93 Wiegand, R.D., Rao, G.A. and Reiser, R. (1973) J. Nutr. 103, 1414–1424.
94 Whiting, P.H., Bowley, M., Sturton, R.G., Pritchard, P.H., Brindley, D.N. and Hawthorne, J.N. (1977) Biochem. J. 168, 147–153.
95 Bates, E.J. and Saggerson, E.D. (1977) FEBS Lett. 84, 229–232.
96 Soler-Argilaga, C., Russell, R.L. and Heimberg, M. (1977) Biochem. Biophys. Res. Commun. 78, 1053–1059.
97 Soler-Argilaga, C., Russell, R.L., Werner, H.V. and Heimberg, M. (1978) Biochem. Biophys. Res. Commun. 85, 249–256.
98 Soler-Argilaga, C., Russell, R.L. and Heimberg, M. (1978) Biochem. Biophys. Res. Commun. 83, 869–873.
99 Nimmo, H.G. and Houston, D. (1978) Biochem. J. 176, 607–610.
100 Soorana, S.R. and Saggerson, E.D. (1976) FEBS Lett. 64, 36–39.
101 Murthy, V.K. and Shipp, J.C. (1980) J. Mol. Cell. Cardiol. 12, 299–309.
102 Brindley, D.N., Cooling, J. and Burditt, S.L. (1979) in G. Ailhaud (Ed.), Obesity — Cellular and Molecular Aspects, Vol. 87, Inserm, Paris, pp. 251–262.
103 Debeer, L.J., Declercq, P.E. and Mannaets, G.P. (1981) FEBS Lett. 124, 31–34.
104 Bates, E.J. and Saggerson, E.D. (1981) FEBS Lett. 128, 230–232.
105 Lawson, N., Jennings, R.J., Pollard, A.D., Sturton, R.G., Ralph, S.J., Marsden, C.A., Fears, R. and Brindley, D.N. (1981) Biochem. J. 200, 265–273.
106 Lamb, R.G. and Fallon, H.J. (1974) Biochim. Biophys. Acta 348, 179–188.
107 Lamb, R.G. and Fallon, H.J. (1976) J. Lipid Res. 17, 406–411.
108 Glenny, H.P., Bowley, M., Burditt, S.L., Cooling, J. Pritchard, P.H., Sturton, R.G. and Brindley, D.N. (1978) Biochem. J. 174, 535–541.
109 Joly, J.-G., Fieman, H., Ishii, H. and Lieber, C.S. (1973) J. Lipid Res. 14, 337–343.
110 Iritani, N. and Fukuda, E. (1980) J. Nutr. 110, 1138–1143.
111 Pritchard, P.H., Bowley, M., Burditt, S.L., Cooling, J., Glenny, H.P., Lawson, N., Sturton, R.G. and Brindley, D.N. (1977) Biochem. J. 166, 639–642.
112 Lazarow, P.B. and de Duve, C. (1976) Proc. Natl. Acad. Sci USA 73, 2043–2046.
113 Ishii, H., Fukomori, N., Horie, S. and Suga, T. (1980) Biochim. Biophys. Acta 617, 1–11.
114 Neat, C.E., Thomassen, M.S. and Osmundsen, H. (1980) Biochem. J. 186, 369–371.
115 Neat, C.E., Thomassen, M.S. and Osmundsen, H. (1981) Biochem. J. 196, 149–159.
116 Pollard, A.D. and Brindley, D.N. (1982) Biochem. Pharmacol., 31, 1650–1652.
117 Daae, L.N.W. and Aas, M. (1973) Atherosclerosis 17, 389–400.
118 Lawson, N., Pollard, A.D., Jennings, R.J., Gurr, M.I. and Brindley, D.N. (1981) Biochem. J. 200, 285–294.
119 Hajra, A.K. (1974) Biochem. Biophys. Res. Commun. 57, 668–674.
120 Novikoff, P.M., Novikoff, A.B., Quintana, N. and Davies, C. (1973) J. Histochem. Cytochem. 21, 540–558.
121 Novikoff, A.B. and Novikoff, P.M. (1973) J. Histochem. Cytochem. 21, 963–966.
122 Goldberg, D.M., Roomi, M.W., Yu, A. and Roncari, D.A.K. (1980) Biochem. J. 192, 165–175.
123 Fallon, H.J., Adams, L.L. and Lamb, R.G. (1972) Lipids 7, 106–109.
124 Lamb, R.G. and Fallon, H.J. (1972) J. Biol. Chem. 247, 1281–1287.
125 Brindley, D.N. and Bowley, M. (1975) Biochem. J. 148, 461–469.
126 Miskin, S. and Turcotte, R. (1974) Biochem. Biophys. Res. Commun. 60, 376–381.
127 Ockner, R.K., Burnett, D.A., Lysenko, N. and Manning, J.A. (1979) J. Clin. Invest. 64, 172–181.
128 Iritani, N., Fukuda, E. and Indguchi, K. (1980) J. Nutr. Sci. Vitamin. 26, 271–277.

129 Jamdar, S.C. (1977) Arch. Biochem. Biophys. 182, 723–731.
130 Jamdar, S.C. (1979) Arch. Biochem. Biophys. 195, 81–94.
131 Schultz, F.M., Wylie, M.B. and Johnston, J.M. (1971) Biochem. Biophys. Res. Commun. 45, 246–250.
132 Carter, J.R. and Kennedy, E.P. (1966) J. Lipid Res. 7, 678–683.
133 Hokin, M., Sadeghian, K., Harris, D.W. and Merrin, J.S. (1977) Biochem. Biophys. Res. Commun. 78, 364–371.
134 Paulus, H. and Kennedy, E.P. (1960) J. Biol. Chem. 235, 1303–1311.
135 Kiyasu, J.Y., Pieringer, R.A., Paulus, H. and Kennedy, E.P. (1963) J. Biol. Chem. 238, 2293–2298.
136 Hostetler, K.Y., Van den Bosch, H. and Van Deenen, L.L.M. (1972) Biochim. Biophys. Acta 260, 507–513.
137 Kanfer, J. and Kennedy, E.P. (1964) J. Biol. Chem. 239, 1720–1726.
138 Chang, Y.-Y. and Kennedy, E.P. (1967) J. Lipid Res. 8, 447–455.
139 Petzold, G.L. and Agranoff, B.W. (1967) J. Biol. Chem. 242, 1187–1191.
140 Holub, B.J. and Piekarski, J. (1976) Lipids 11, 251–257.
141 Sturton, R.G. and Brindley, D.N. (1977) Biochem. J. 162, 25–32.
142 Van Heusden, G.P.H. and Van den Bosch, H. (1978) Eur. J. Biochem. 84, 405–412.
143 Cotman, C.W., McCaman, R.E. and Dewhurst, S.A. (1971) Biochim. Biophys. Acta 249, 395–405.
144 McCaman, R.E. and Finnerty, W.R. (1968) J. Biol. Chem. 243, 5074–5080.
145 Belendiuk, G., Mangnall, D., Tung, B., Westley, J. and Getz, G.S. (1978) J. Biol. Chem. 253, 4555–4565.
146 Thompson, R.J. (1977) Biochem. Soc. Trans. 5, 49–51.
147 Vorbeck, M.L. and Martin, A.P. (1970) Biochem. Biophys. Res. Commun. 40, 901–908.
148 Davidson, J.B. and Stanacev, N.Z. (1974) Can. J. Biochem. 52, 936–939.
149 Van Golde, L.M.G., Raben, J., Batenburg, J.J., Fleischer, B., Zambrano, F. and Fleischer, S. (1974) Biochim. Biophys. Acta 360, 179–192.
150 Thompson, R.J. (1977) J. Neurochem. 29, 387–391.
151 Raetz, C.R.H. and Kennedy, E.P. (1973) J. Biol. Chem. 248, 1098–1105.
152 Langley, K.E. and Kennedy, E.P. (1978) J. Bacteriol. 136, 85–95.
153 Ter Schegget, J., Van den Bosch, H., Van Baak, M.A., Hostetler, K.Y. and Borst, P. (1971) Biochim. Biophys. Acta 239, 234–242.
154 Yamada, M., Aucker, J. and Weissbach, A. (1976) Arch. Biochem. Biophys. 177, 461–467.
155 Thompson, R.J. (1977) J. Neurochem. 29, 383–395.
156 Thompson, W. and Macdonald, G. (1975) J. Biol. Chem. 250, 6779–6785.
157 Thompson, W. and Macdonald, G. (1976) Eur. J. Biochem. 65, 107–111.
158 Hauser, G. and Eichberg, J. (1975) J. Biol. Chem. 250, 105–112.
159 Holub, B.J. and Kuksis, J. (1971) Can. J. Biochem. 49, 1347–1356.
160 Åkesson, B., Elovson, J. and Arvidson, G. (1970) Biochim. Biophys. Acta 210, 15–27.
161 Possmayer, F., Scherphof, G.L., Dubbelman, T.M.A.R., Van Golde, L.M.G. and Van Deenen, L.L.M. (1969) Biochim. Biophys. Acta 176, 95–110.
162 Bishop, H.H. and Strickland, K.P. (1970) Can. J. Biochem. 54, 249–260.
163 Thompson, W. and Macdonald, G. (1978) J. Biol. Chem. 253, 2712–2715.
164 Hostetler, K.Y., Zenner, B.D. and Morris, H.P. (1976) Biochem. Biophys. Res. Commun. 72, 418–425.
165 Raetz, C.R.H., Dowhan, W. and Kennedy, E.P. (1976) J. Bacteriol. 125, 855–863.
166 Sribney, M., Dove, J.L. and Lyman, E.M. (1977) Biochem. Biophys. Res. Commun. 79, 749–755.
167 Fallon, H.J., Barwick, J., Lamb, R.G. and Van den Bosch, H. (1975) J. Lipid Res. 16, 107–115.
168 Hostetler, K.Y., Zenner, B.D. and Morris, H.P. (1976) Biochim. Biophys. Acta 441, 231–238.
169 Kates, M. (1956) Can. J. Biochem. Physiol. 34, 967–980.
170 Hübscher, G. (1970) in S.J. Wakil (Ed.), Lipid Metabolism, Academic Press, New York, pp. 279–370.
171 Caras, I. and Shapiro, B. (1975) Biochim. Biophys. Acta 409, 201–211.
172 Sedgwick, B. and Hübscher, G. (1965) Biochim. Biophys. Acta 106, 63–77.
173 Sedgwick, B. and Hübscher, G. (1967) Biochim. Biophys. Acta 144, 397–408.

174 Hübscher, G., Brindley, D.N., Smith, M.E. and Sedgwick, B. (1967) Nature (Lond.) 216, 449–453.
175 Brindley, D.N., Smith, M.E., Sedgwick, B. and Hübscher, G. (1967) Biochim. Biophys. Acta 144, 285–295.
176 Tzur, R. and Shapiro, B. (1964) J. Lipid Res. 5, 542–547.
177 Roncari, D.A.K. and Mack, E.Y.W. (1975) Biochem. Biophys. Res. Commun. 67, 790–796.
178 Roncari, D.A.K. and Mack, E.Y.W. (1980) Clin. Res. 28, 693A.
179 Johnston, J.M., Rao, G.A., Lowe, P.A. and Schwarz, B.E. (1967) Lipids 2, 14–20.
180 Smith, M.E., Sedgwick, B., Brindley, D.N. and Hübscher, G. (1967) Eur. J. Biochem. 3, 70–77.
181 Hosaka, K., Yamashita, S. and Numa, S. (1975) J. Biochem. 77, 501–509.
182 Savolainen, M.J. (1977) Biochem. Biophys. Res. Commun. 75, 511–518.
183 Sturton, R.G. and Brindley, D.N. (1978) Biochem. J. 171, 263–266.
184 Sturton, R.G. and Brindley, D.N. (1980) Biochim. Biophys. Acta 619, 494–505.
185 Lamb, R.G. and Fallon, H.J. (1974) Biochim. Biophys. Acta 348, 166–178.
186 Bowley, M., Cooling, J., Burditt, S.L. and Brindley, D.N. (1977) Biochem. J. 165, 447–454.
187 Casola, P.G. and Possmayer, F. (1981) Biochim. Biophys. Acta 664, 298–315.
188 Goldberg, B.J., Roomi, M.W., Yu, A. and Roncari, D.A.K. (1981) Biochem. J. 196, 337–346.
189 Goldberg, C.J.B., Yu, A., Roomi, N. and Roncari, D.A.K. (1980) Can. J. Biochem. 59, 48–53.
190 Savolainen, M.J., Lehtonen, M.A., Ruokenen, A. and Hassinen, I.E. (1981) Metabolism 30, 706–711.
191 Jamdar, S.C. and Fallon, H.J. (1973) J. Lipid Res. 14, 517–524.
192 Mitchell, M.P., Brindley, D.N. and Hübscher, G. (1971) Eur. J. Biochem. 18, 214–220.
193 Åkesson, B., Elovson, J. and Arvidson, G. (1970) Biochim. Biophys. Acta 218, 44–56.
194 Lamb, R.G., Wood, C.K., Landa, B.M., Guzelian, P.S. and Fallon, H.J. (1977) Biochim. Biophys. Acta 489, 318–329.
195 Tzur, R. and Shapiro, B. (1976) Eur. J. Biochem. 64, 301–305.
196 Sturton, R.G., Pritchard, P.H., Han, L.-Y. and Brindley, D.N. (1978) Biochem. J. 174, 667–670.
197 Billah, M.M., Lapetina, E.G. and Cuatrecasas, P. (1981) J. Biol. Chem. 256, 5399–5403.
198 Brindley, D.N., Allan, D. and Michell, R.H. (1975) J. Pharm. Pharmacol. 27, 462–464.
199 Brindley, D.N., Bowley, M., Sturton, R.G., Pritchard, P.H., Burditt, S.L. and Cooling, J. (1978) in S. Garattini and R. Samanin (Eds.), Central Mechanisms of Anorectic Drugs, Raven Press, New York, pp. 301–317.
200 Brindley, D.N. (1983) in P.B. Curtis-Prior (Ed.), Biochemical Pharmacology of Metabolic Disease States, Vol. 1, Obesity, Elsevier/North-Holland Biomedical Press, Amsterdam, in press.
201 Lüllmann, H., Lüllmann-Rauch, R. and Wasserman, O. (1975) Crit. Rev. Toxicol. 4, 185–218.
202 Michell, R.H., Allan, D., Bowley, M. and Brindley, D.N. (1976) J. Pharm. Pharmacol. 28, 331–332.
203 Lüllmann, H., Lüllmann-Rauch, R. and Wasserman, O. (1978) Biochem. Pharmacol. 27, 1103–1108.
204 Brindley, D.N., Bowley, M., Sturton, R.G., Pritchard, P.H., Burditt, S.L. and Cooling, J. (1977) Biochem. Soc. Trans. 5, 40–43.
205 Bangham, A.D., Standish, M.M. and Miller, N. (1965) Nature 208, 1295–1297.
206 Ito, T. and Ohnishi, S. (1974) Biochim. Biophys. Acta 352, 29–37.
207 Papahadjopoulos, D., Jacobson, K., Poste, G. and Shepherd, G. (1975) Biochim. Biophys. Acta 394, 504–519.
208 Brindley, D.N., Bowley, M., Sturton, R.G., Pritchard, P.H., Cooling, J. and Burditt, S.L. (1978) Adv. Exp. Biol. Med. 101, 227–234.
209 Newton, C., Pangborn, W., Nir, S. and Papahadjopoulos, D. (1978) Biochim. Biophys. Acta 506, 281–287.
210 Fallon, H.J., Lamb, R.G. and Jamdar, S.C. (1977) Biochem. Soc. Trans. 5, 37–40.
211 Brindley, D.N. (1977) in R. Dils and J. Knudsen (Eds.), Regulation of Fatty Acid and Glycerolipid Metabolism, Pergamon, Oxford, pp. 31–40.
212 Åkesson, B. and Sundler, R. (1977) Biochem. Soc. Trans. 5, 43–48.
213 Pritchard, P.H. and Brindley, D.N. (1977) J. Pharm. Pharmacol. 29, 343–349.
214 Glenny, H.P. and Brindley, D.N. (1978) Biochem. J. 176, 777–784.
215 Lehtonen, M.A., Savolainen, M.J. and Hassinen, I.E. (1979) FEBS Lett. 99, 162–165.

216 Brindley, D.N. (1981) Clin. Sci. 61, 129–133.
217 Klausner, H. and Heimberg, M. (1967) Am. J. Physiol. 212, 1236–1246.
218 Reaven, E.P., Kolterman, O.G. and Reaven, G.M. (1974) J. Lipid Res. 15, 74–83.
219 Jennings, R.J., Lawson, N., Fears, R. and Brindley, D.N. (1981) FEBS Lett. 133, 119–122.
220 Diamant, S. and Shafrir, E. (1975) Eur. J. Biochem. 53, 541–546.
221 Kirk, C.J., Verrinder, T.R. and Hems, D.A. (1976) Biochem. J. 156, 593–602.
222 Knox, A.M., Sturton, R.G., Cooling, J. and Brindley, D.N. (1979) Biochem. J. 180, 441–443.
223 Savolainen, M.J. and Hassinen, I.E. (1978) Biochem. J. 176, 885–892.
224 Brindley, D.N., Cooling, J., Burditt, S.L., Pritchard, P.H., Pawson, S. and Sturton, R.G. (1979) Biochem. J. 180, 195–199.
225 Mallov, S. and Bloch, J.L. (1956) Am. J. Physiol. 184, 29–34.
226 Brodie, B.B. and Maiekel, R.P. (1963) Ann. N.Y. Acad. Sci. 104, 1049–1058.
227 Maiekel, R.P. and Brodie, B.B. (1963) Ann. N.Y. Acad. Sci. 104, 1059–1064.
228 Maling, H.M., Wakabayashi, M. and Horning, M.G. (1963) Adv. Enzyme Regul. 1, 247–257.
229 Brindley, D.N., Sturton, R.G., Pritchard, P.H., Cooling, J. and Burditt, S.L. (1979) Curr. Med. Res. Opin. 6 (Suppl. 1) 91–100.
230 Pritchard, P.H., Cooling, J., Burditt, S.L. and Brindley, D.N. (1978) J. Pharm. Pharmacol. 31, 406–407.
231 Savolainen, M.J. and Hassinen, I.E. (1980) Arch. Biochem. Biophys. 201, 640–645.
232 Wood, C.K. and Lamb, R.G. (1979) Biochim. Biophys. Acta 572, 121–131.
233 Lamb, R.G., Wood, C.K. and Fallon, H.J. (1979) J. Clin. Invest. 63, 14–20.
234 Yudkin, J. and Szanto, S. (1971) Br. Med. J. 1, 349.
235 Bruckdorfer, K.R., Kang, S.S. and Yudkin, J. (1973) Proc. Nutr. Soc. 32, 12A.
236 Kako, K.J. and Peckett, S.D. (1981) Lipids 16, 23–29.
237 Stewart, J.H. and Briggs, G.M. (1981) Biochem. J. 198, 413–416.
238 Carroll, K.K. and Noble, R.L. (1952) Endocrinology 51, 476–486.
239 Hülsmann, W.C. (1978) Mol. Cell. Endocrinol. 12, 1–8.
240 Brindley, D.N., Cooling, J., Glenny, H.P., Burditt, S.L. and McKechnie, I.S. (1981) Biochem. J. 200, 275–283.
241 MacDonald, I. (1971) Proc. Nutr. Soc. 30, 72A–73A.
242 Bruckdorfer, K.R., Kari-Kari, B.P.B., Khan, I.H. and Yudkin, J. (1972) Nutr. Metab. 14, 228–237.
243 Jones, D.P. and Greene, E.A. (1966) Am. J. Clin. Nutr. 18, 350–357.
244 Carrol, C. and Williams, L. (1971) J. Nutr. 101, 997–1012.
245 Chen, N.S.C., Chen, N.C., Johnson, R.J., McGinnis, J. and Dyer, I.A. (1977) J. Nutr. 107, 1114–1119.
246 Vavrečka, M., Mitchell, M.P. and Hübscher, G. (1969) Biochem. J. 115, 139–145.
247 Kinnula, V.L., Savolainen, M.J. and Hassinen, I.E. (1978) Acta Physiol. Scand. 104, 148–155.
248 Murthy, V.K. and Shipp, J.C. (1979) Diabetes 28, 472–478.
249 Lamb, R.G. and Banks, W.L. (1979) Biochim. Biophys. Acta 574, 440–447.
250 Lamb, R.G. and Dewey, W.L. (1981) J. Pharmacol. Exp. Ther. 216, 496–499.
251 Sturton, R.G., Butterwith, S.C., Burditt, S.L. and Brindley, D.N. (1981) FEBS Lett. 126, 297–300.
252 Murthy, V.K. and Shipp, J.C. (1981) J. Clin. Invest. 67, 923–930.
253 Cooling, J., Burditt, S.L. and Brindley, D.N. (1979) Biochem. Soc. Trans. 7, 1051–1053.
254 Schwertz, D.W. and Lamb, R.G. (1979) Fed. Proc. 38, 539.
255 Lamb, R.G. and Cabral, F.M. (1980) Fed. Proc. 39, 768.
256 Iritani, N., Yamashita, S. and Numa, S. (1976) J. Biochem. 80, 217–222.
257 Geelen, M.J.H., Groener, J.E.M., de Haas, C.G.M., Wisserhof, T.A. and Van Golde, L.M.G. (1978) FEBS Lett. 90, 57–60.
258 Haagsman, H.P., de Haas, C.G.M., Geelen, M.J.H. and Van Golde, L.M.G. (1981) Biochim Biophys. Acta 664, 74–81.
259 Pelech, S.L., Pritchard, P.H. and Vance, D.E. (1981) J. Biol. Chem. 256, 8283–8286.
260 Young, D.L. and Lynen, F. (1969) J. Biol. Chem. 244, 377–383.

261 Friedman, G., Stein, O. and Stein, Y. (1978) Biochim. Biophys. Acta 531, 222–232.
262 Murthy, V.K. and Shipp, J.C. (1977) Diabetes 26, 222–229.
263 Jamdar, S.C., Shapiro, D. and Fallon, H.J. (1976) Biochem. J. 158, 327–334.
264 Belfiore, F., Rabinazzo, A.M., Borzi, V. and Iannello, S. (1978) Diabetologia 15, 218.
265 Dodds, P.F., Brindley, D.N. and Gurr, M.I. (1976) Biochem. J. 160, 701–706.
266 Moller, F., Green, P. and Harkness, E.J. (1977) Biochim. Biophys. Acta 486, 359–368.
267 Daniel, A.M. and Rubinstein, H.S. (1968) Can. J. Biochem. 46, 1039–1045.
268 Cheng, C.H.K. and Saggerson, E.D. (1978) FEBS Lett. 87, 65–68.
269 Cheng, C.H.K. and Saggerson, E.D. (1978) FEBS Lett. 93, 120–124.
270 Moller, F., Wong, K.H. and Green, P. (1981) Can. J. Biochem. 59, 9–15.
271 Lawson, N., Jennings, R.J., Fears, R. and Brindley, D.N. (1982) FEBS Lett. 143, 9–12.

CHAPTER 6

Polyglycerophospholipids: phosphatidylglycerol, diphosphatidylglycerol and *bis*(monoacylglycero)phosphate

KARL Y. HOSTETLER

Department of Medicine (Metabolic Diseases), University of California, San Diego and the VA Medical Center, La Jolla, CA 92093, U.S.A.

1. Introduction

This chapter will deal with the biochemistry of the polyglycerophospholipids which include phosphatidylglycerol, diphosphatidylglycerol (cardiolipin) and bis(monoacylglycero)phosphate (lysobisphosphatidic acid). While the discussion will consider primarily the biochemistry of these lipids in animal tissues, some information regarding the occurrence and metabolism of these glycerophospholipids in plants and microorganisms will also be presented. In this chapter generic names will be used, e.g., diphosphatidylglycerol and bis(monoacylglycero)phosphate, rather than the respective trivial names, cardiolipin and lysobisphosphatidic acid. IUPAC-IUB nomenclature will be employed and in discussions of stereochemistry, the stereospecific numbering system will be used [1,2].

In this class of phospholipids, either two or three molecules of glycerol are present, joined by phosphodiester linkage. Two, three or four long-chain fatty acid groups may be present in ester linkage. Structural formulae of the major polyglycerophospholipids are shown in Fig. 1. These structures are intended only for general orientation and do not indicate the stereochemical configuration of these compounds.

The discovery, structural and stereochemical data, distribution in nature, pathways of synthesis and degradation and subcellular localisation of the polyglycerophospholipids will be emphasized. In addition, several special topics involving the role of two of these compounds in pulmonary surfactant and lipid storage diseases will be considered. However, several important areas of investigation have not been covered due to considerations of space. These are studies on the physical properties and protein–lipid interactions of phosphatidylglycerol and diphosphatidylglycerol, the probable existence of separate physical and metabolic pools of these lipids in bacteria and the role of phosphatidylglycerol as a precursor of complex cell wall components in microorganisms.

Hawthorne/Ansell (eds.) Phospholipids

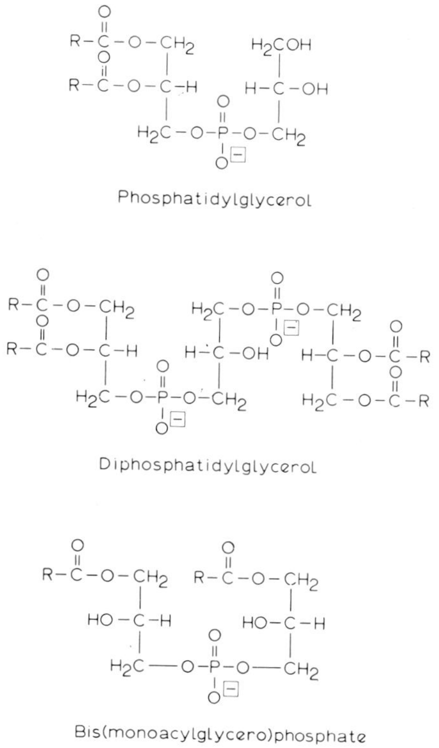

Fig. 1. Structural formulae of the polyglycerophospholipids.

2. Discovery of the polyglycerophospholipids

(a) Diphosphatidylglycerol

Diphosphatidylglycerol (cardiolipin) was the first of the polyglycerophospholipids to be discovered. In 1942, Pangborn isolated diphosphatidylglycerol from lipid extracts of beef heart by solvent fractionation of the cadmium complexes of the phospholipids. A non-nitrogen containing phospholipid was isolated which proved to be reactive in the Wasserman serological test for syphilis [3]. After improving the method of isolation and purification [4,5], she demonstrated that alkaline hydrolysis of cardiolipin led to the production of fatty acids, primarily linoleic and oleic in a ratio of 5 : 1, and a polyester of glycerophosphoric acid and glycerol [6]. Based on these findings, several structures were proposed for cardiolipin which was suggested to be a "complex phosphatidic acid". Although the structures proposed by Pangborn did not include the correct one for diphosphatidylglycerol, her studies represent a significant achievement, having been accomplished without the aid of silicic acid column chromatography, thin-layer chromatography with silica gel or paper chromatography of the water-soluble products of mild alkaline hydrolysis, these being basic analytical techniques of present-day phospholipid biochemistry.

(b) Phosphatidylglycerol

Phosphatidylglycerol was discovered in 1958 by Benson and Mauro in the alga, *Scendesmus.* These authors isolated ^{32}P-labelled glycerophosphoglycerol from the lipids of *Scendesmus,* and on acid hydrolysis they found glycerol and glycerophosphate. Periodate oxidation showed the presence of vicinal hydroxyl groups in both glycerols indicating an α,α'-diglycerophosphate configuration [7]. Since substantial amounts of glycerophosphoglycerol were found in *Scendesmus* cells, a further examination of the phospholipids was undertaken. Phospholipids from *Scendesmus* cells labelled with ^{32}P were separated by paper chromatography and a component representing 41% of the lipid phosphorus was identified as phosphatidylglycerol; paper chromatography of the water-soluble products of alkaline hydrolysis gave material identical in chromatographic behaviour to synthetic glycerophosphoglycerol. The presence of vicinal hydroxyl groups in intact phosphatidylglycerol was established by the formation and characterisation of the derivatives formed with benzoylchloride and with acetone [8]. Phosphatidylglycerol was also identified in clover, barley and tobacco where it represented 24.9, 22.6 and 22.0% respectively of the total phospholipid [8].

(c) Bis(monoacylglycero)phosphate

Bis(monoacylglycero)phosphate (lysobisphosphatidic acid) was isolated from pig lung by Body and Gray in 1967 [9]. They prepared a total lipid extract of pig lung and fractionated the lipids by silicic acid column chromatography. After elution of neutral lipids with chloroform/methanol (98:2, v/v), they found diphosphatidylglycerol, phosphatidylglycerol and an unknown lipid in successive fractions of chloroform/methanol (9:1, v/v). The unknown compound had an R_F on silicic acid-impregnated paper and on thin layers of silica gel which was greater than that of phosphatidylglycerol. The compound contained phosphorus, glycerol and fatty acid esters in a ratio of 1:1.8:2.3 and yielded glycerophosphoglycerol as the water-soluble product of alkaline hydrolysis; treatment of the intact lipid with periodate did not result in production of formaldehyde, indicating the absence of adjacent free hydroxyl groups. Acetolysis of the intact phospholipid gave only one glyceroacetate derivative identified by thin-layer chromatography as diacetylmonoacylglycerol. Taken together, the above evidence established the compound as bis(monoacylglycero)phosphate [9]. Later in 1967, Body and Gray also isolated and characterised acylphosphatidylglycerol (semilysobisphosphatidic acid) from rabbit lung essentially as described above, except that acetolysis of the intact phospholipid gave equal amounts of diacetylmonoacylglycerol and monoacetyldiacylglycerol, establishing the unknown as a triacylated derivative of glycerophospho-

glycerophosphoglycerol [10]. In 1974 the tetracylated derivative of glycerophosphoglycerol, bis(diacylglycero)phosphate (bis-phosphatidic acid), was isolated from cultured baby hamster kidney cells and its structure was demonstrated by Brotherus and Renkonen [11].

3. *Structural and stereochemical investigations*

(a) Diphosphatidylglycerol

Although cardiolipin was discovered by Pangborn in 1942, many years were required for the elucidation of its structure. Based on her analytical data, Pangborn proposed a structure for cardiolipin consisting of four glycerols connected by phosphodiester linkage [6]. However, McKibbin and Taylor found a ratio of glycerol to phosphate of 3:2 in cardiolipin isolated from canine liver [12]; Faure and Morelec-Coulon isolated cardiolipin from heart muscle and found a ratio of glycerol : phosphorus : fatty acid esters of 3:2:4 [13]. MacFarlane and Gray confirmed this finding and proposed a diphosphatidylglycerol structure [14,15]. Glycerol diphosphate was subsequently shown to be a degradation product of cardiolipin [15–17]. It was demonstrated that the water-soluble polyglycerophosphate backbone of cardiolipin obtained by mild alkaline hydrolysis gave one mole of formaldehyde per mole of phosphate when it was reacted with sodium metaperiodate [18]. Cardiolipin gave two moles of diacylglycerol and one mole of glycerodiphosphate upon hydrolysis in acetic acid, favouring the diphosphatidylglycerol structure [19]. In this type of hydrolysis, it was demonstrated that a free hydroxyl adjacent to the phosphate was required to give diacylglycerol, a finding which strongly supported the diphosphatidylglycerol structure for cardiolipin [20].

A final structural proof was provided when de Haas, Bonsen and van Deenen synthesized diphosphatidylglycerol by a condensation between the silver salt of α-stearoyl-β-oleoyl-L-α-glycerobenzyl phosphate and α,γ-diiodoglycero-β-*tert.*butylether. After removal of the benzyl and *tert.* butyl protecting groups, diphosphatidylglycerol was obtained [21,22]. Pure ox heart cardiolipin was compared with the synthetic diphosphatidylglycerol and was found to be identical with regard to chromatographic behaviour, melting point, optical rotation, and infrared absorption. Incubation with phospholipase A gave rise to two lysocompounds containing three and two acyl chains per molecule. This can be taken to represent an early proof that the phosphatidyl groups of cardiolipin have the *sn*-glycero-3-phosphate configuration in view of the stereospecificity of this enzyme. Finally, an important proof of the structure resulted from the incubation of ox heart cardiolipin and synthetic diphosphatidylglycerol with phospholipase C. In short incubations, diacylglycerol and phosphatidylglycerophosphate were produced, and upon more prolonged incubation, diacylglycerol and glycerodiphosphate were found in both cases (due to the further action of phospholipase C on phosphatidylglycerol phosphate). Finally, the synthetic diphosphatidylglycerol was shown to substitute for cardiolipin in the serologic test for syphilis (VDRL) [22].

The studies of Rose [23] generally supported the diphosphatidylglycerol structure but questioned the presence of a free hydroxyl group on the interior glycerol moiety. Alternative structures for cardiolipins were proposed by Courtade et al. [24] in which a fatty acid or vitamin A is esterified to a phosphate group. However, it was subsequently shown by Nielsen that the findings leading to these suggested structures could be accounted for by cation effects and the autoxidation of linoleic acid residues, providing further support for the diphosphatidylglycerol structure [25].

(b) Phosphatidylglycerol

After the discovery of phosphatidylglycerol in 1958, the elucidation of its structure and stereochemistry followed rapidly. In 1961 Benson and Miyano isolated ^{14}C-labelled phosphatidylglycerol from *Chlorella* which had been allowed to grow in the presence of $^{14}CO_2$. They purified phosphatidylglycerol and deacylated it by mild alkaline hydrolysis. ^{14}C-labelled glycerophosphoglycerol was oxidised to [^{14}C]glyceric acid and recrystallised from either D,L-glyceric acid (100% retention of ^{14}C), D-glyceric acid (50% retention) or L-glyceric acid (50% retention) establishing that the two glycerol moieties of phosphatidylglycerol have an opposite stereochemical configuration [26]. However, the results of this study did not allow assignment of stereoconfiguration to the acylated glycerol versus the free glycerol.

In 1962 Haverkate, Houtsmuller and van Deenen isolated ^{32}P-labelled phosphatidylglycerol from *Bacillus cereus* and tested its susceptibility to phospholipases. Phospholipase A converted phosphatidylglycerol to lysophosphatidylglycerol; phospholipase C gave diacylglycerol and glycerophosphate and phospholipase D gave phosphatidic acid and glycerol [27], essentially confirming the structure proposed by Benson and Mauro [8]. Since phospholipase A_2 (*Crotalus adamanteus*) is inactive toward *sn*-glycero-1-phosphate derivatives, the authors suggested that the acylated glycerol must have the *sn*-glycero-3-phosphate configuration.

Subsequently, Haverkate and van Deenen isolated and purified phosphatidylglycerol from spinach leaves by silicic acid column chromatography. In an elegant study, phosphatidylglycerol was subjected to hydrolysis with phospholipase C and phospholipase D. The glycerophosphate produced by phospholipase C action was isolated and was essentially unreactive when tested with the stereospecific enzyme, *sn*-glycero-3-phosphate dehydrogenase (EC 1.1.99.5). A similar examination of the glycerophosphate obtained by mild alkaline hydrolysis of the phosphatidic acid produced by phospholipase D, showed that this glycerophosphate moiety had the *sn*-glycero-3-phosphate configuration. The structure of phosphatidylglycerol was thus shown to be *sn*-1,2-diacylglycero-3-phospho-*sn*-1′-glycerol [28,29]. Subsequently, Op den Kamp and coworkers isolated glucosaminylphosphatidylglycerol from *Bacillus megaterium*; after conversion to phosphatidylglycerol an approach similar to that above showed that this bacterial phosphatidylglycerol also has the *sn*-1,2-diacylglycero-3-phospho-*sn*-1′-glycerol configuration [30]. The stereochemistry of phosphatidylglycerol isolated from *Pseudomonas BAL-31* and from its bacteriophage PM2 was also studied using phospholipases C and D and *sn*-glycero-

3-phosphate dehydrogenase. In these studies, the phosphatidylglycerols obtained from the bacterium and the phage were found to have the *sn*-1,2-diacylglycero-3-phospho-*sn*-1′-glycerol configuration [31].

Phosphatidylglycerol may be produced during the phospholipase D hydrolysis of phosphatidylcholine if glycerol is present [32,33]. Based on the liberation of racemic glycerophosphate from this phosphatidylglycerol by treatment with acetic acid, it was concluded that the free glycerol moiety is racemic [32]. Subsequent studies of natural and synthetic phosphatidylglycerols using circular dichroism spectra concluded that phosphatidylglycerol formed from egg lecithin by the action of phospholipase D in the presence of glycerol has the *sn*-3-phosphatidyl-*sn*-1′-glycerol configuration [34]. However, Joutti and Renkonen later prepared phosphatidylglycerol from egg lecithin and glycerol by transphosphatidylation (see Chapter 9) with phospholipase D and examined the glycerophosphate released by phospholipase C treatment using *sn*-glycero-3-phosphate dehydrogenase. The released glycerophosphate was found to be a racemic mixture indicating that the phosphatidylglycerol formed by transphosphatidylation with phospholipase D is an equimolar mixture of *sn*-3-phosphatidyl-*sn*-1′-glycerol and *sn*-3-phosphatidyl-*sn*-3′-glycerol [35]. Thus, the weight of the evidence supports the presence of a racemic free glycerol moiety in phosphatidylglycerol formed by phospholipase D-catalysed transphosphatidylation.

(c) Bis(monoacylglycero)phosphate and related compounds

The chemistry and primary structure of bis(monoacylglycero)phosphate were essentially fully described in the original work of Body and Gray [9,10], but the position of the fatty acyl esters was not defined. In 1973 Wherrett and Huterer isolated bis(monoacylglycero)phosphate from the liver of a patient who had died of an undefined type of lipid storage disease. Using NMR spectroscopy, they provided evidence that the fatty acyl esters are positioned on the primary hydroxyl groups of the glycerol moieties [36].

Phosphatidylglycerol, which has the same glycerophosphoglycerol backbone as bis(monoacylglycero)phosphate, has the *sn*-3-glycerophospho-*sn*-1′-glycerol structure as noted above. Renkonen, Fisher and co-workers made the important discovery that the stereochemistry of the glycerophosphoglycerol backbone of bis(monoacylglycero)phosphate differs from that of phosphatidylglycerol [37]. These workers isolated bis(monoacylglycero)phosphate from cultured hamster fibroblasts; after purification by aluminium oxide column chromatography, they subjected the compound to strong alkaline hydrolysis cleaving the glycerol-containing phosphate diesters to glycerol and α-glycerophosphate or β-glycerophosphate via a cyclic phosphate intermediate. The resulting α-glycerophosphate was analysed using the stereospecific enzyme, *sn*-glycero-3-phosphate dehydrogenase, to determine the configuration of the glycerols in the parent lipid. Using this method they showed that bis(monoacylglycero)phosphate from BHK-21 cells has the *sn*-1-glycerophospho-*sn*-1′-glycerol configuration while phosphatidylglycerol from a bacterium, *Streptococcus lactis,* gave the expected *sn*-3-glycerophospho-*sn*-1′-glycerol configuration; egg phosphatidylcholine was shown to have a glycerophosphate residue having the anticipated *sn*-glycero-3-phosphate configuration [37]. Subsequently, these findings were extended to bis(monoacylglycero)phosphate from rat liver and rabbit and pig lung [38]. The stereochemistry of this interesting compound will be considered further in Section 5 on the biosynthesis of the polyglycerophospholipids.

Acylphosphatidylglycerol was isolated from *Salmonella typhimurium* by Olsen and Ballou [39]; mild alkaline hydrolysis gave glycerophosphoglycerol as the water-soluble product and it had a fatty acyl ester:glycerol:phosphorus ratio of 3:2:1. After more vigorous alkaline hydrolysis the glycerophosphate fragments from acylphosphatidylglycerol were examined using *sn*-glycero-3-phosphate dehydrogenase; evidence was provided showing that the glycerols have the opposite stereoconfiguration. Based on data obtained from NMR and comparative rates of triphenylmethylation, the acyl ester was assigned to the primary hydroxyl of the "free glycerol" moiety of phosphatidylglycerol [39].

Bis(diacylglycero)phosphate (bisphosphatidic acid) was isolated from a marine bacterium (MB 45) by McAllister and De Siervo in 1975 and extensive structural studies were carried out [40]. It also had glycerophosphoglycerol as its water-soluble backbone after mild alkaline hydrolysis; on acetolysis only diacylmonoacetyl glycerol was identified. The ratio of glycerol:phosphorus:fatty acyl esters was 2.0:1.1:3.6 and its molecular weight was 1280. Stereochemical studies were not reported but this compound presumably has the *sn*-glycero-3-phospho-*sn*-1′-glycerol configuration like bacterial phosphatidylglycerol and acylphosphatidylglycerol [40].

4. Distribution and properties of polyglycerophospholipids in animals, plants and microorganisms

(a) Distribution in nature

Polyglycerophospholipids are widely distributed in nature and have been found in animals, plants and microorganisms. Table 1 shows the polyglycerophospholipid content reported for a number of animal tissues and cells [41–57]. Diphosphatidylglycerol (cardiolipin) is most plentiful in cardiac muscle, the source from which it was originally isolated [3]; in human heart muscle it comprises 9.0% of total lipid phosphorus [41], versus 12.6 and 14.7% in bovine heart and rat heart, respectively [41,42]. Other muscle tissues which are not so highly specialised for oxidative metabolism have lesser amounts of diphosphatidylglycerol ranging from 1.4–2.1% in the rat [41,43] to 6.6 and 8.9% in human and bovine skeletal muscle, respectively [41]. The diphosphatidylglycerol content of the brain is very low, representing 0.2% of total lipid phosphorus [44]. The diphosphatidylglycerol content of lung is also low as shown in Table 1, and this lipid is absent from fetal and adult lung washings (pulmonary surfactant) [45,46].

In animal tissues, phosphatidylglycerol is present only in trace amounts and

TABLE 1

Polyglycerophospholipid content of animal cells and tissues

Source	Ref.	% Total lipid phosphorus			
		DPG	PG	BMP	Other
Alveolar type II cell, rat	49		10.3	1.8	
Alveolar type II cell, rabbit	50		4.2	0.8	
Alveolar type II cell, rat	51	1.3	10.4	0.8	
Alveolar macrophage, rabbit	52	0.9		16.9	
Alveolar macrophage, rabbit	50		1.6	14.0	
Alveolar macrophage, rabbit	53	1.4	1.7	17.9	2.6 [a]
Baby hamster kidney cell	55	3.2		1.7	
Baby hamster kidney cell, degenerating	54	3.4		2.9	0.3 [a], 0.2 [b]
Brain, human	44	0.2		0.1	
Kidney, rat	56	6.5	0.3	0.1	
Kidney, rat	42	7.0	1.5	1.2	
Kidney, human	56	4.2	0.6	0.1	
Kidney, bovine	56	6.5	0.4	0.2	
Liver, rat	56	4.5	0.3	0.2	
Liver, rat	42	5.7	0.6	0.4	
Liver, human	56	3.7	0	1.0	
Lung, rat	47	1.1	2.2	0.3	
Lung, rat	45		5.0		
Lung, rat	48	1.4	3.3	1.0	
Lung, rat	42	0.8	4.1	0.8	
Lung, human	47	1.0	2.5	1.5	
Pulmonary surfactant	57		10.0		
Pulmonary surfactant	45		11.0		
Pulmonary surfactant, healthy newborn	46		6.0	0.8	
Pulmonary surfactant, respiratory distress syndrome	46		<0.1	0	
Muscle, diaphragm, rat	43	5.6	0.7		
Muscle, skeletal, rat	41	1.4	0.9	0.3	
Muscle, skeletal, rat	43	2.1	0.7		
Muscle, skeletal, human	41	6.6	1.0	tr	
Muscle, skeletal, bovine	41	8.9	0.3	0.1	
Muscle, heart, rat	41	11.2	1.0	0	
Muscle, heart, rat	42	14.7	1.5	0	
Muscle, heart, human	41	9.0	0.6	0.2	
Muscle, heart, bovine	41	12.6	0.2	0	
Polymorphonuclear leucocyte, guinea pig	52	1.2		0	
Spleen, rat	56	2.1	0.8	0.6	
Spleen, rat	42	2.9	1.1	3.0	
Spleen, human	56	1.0	0.3	0.3	

Abbreviations: DPG, diphosphatidylglycerol; PG, phosphatidylglycerol; BMP, bis(monoacylglycero)-phosphate.

[a] acylphosphatidylglycerol.

[b] bis(diacylglycero)phosphate.

TABLE 2

Polyglycerophospholipid composition of plants *

Source	Reference	% Total lipid phosphorus	
		DPG	PG
Leaves			
Barley	8		23
Clover, sweet	8		25
Clover, white	58	10	4
Lettuce	58	16	12
Maidenhair tree	58	7	23
Maize	58	16	31
Moss	58	2	18
Pumpkin	59		28
Rye-grass	58	7	25
Sycamore	60	2	5
Tobacco	8		22
Fruit, root or tuber			
Apple	61	2	7
Parsnip	58	22	8
Potato	62	2	3
Sugar beet	63	2	6
Green algae			
Chlorella vulgaris	64	1	27
Euglena gracilis	65	14	10
Hydrodictyon africanum	66	4	25
Scendesmus obliquus	8	tr	42
Blue-green algae			
Anabena variabilis	67		100
Anacystis nidulans	6		100

* Abbreviations as in Table 1.

usually represents less than 1% of tissue total lipid phosphorus (Table 1). However, in the lung, phosphatidylglycerol is present in substantially greater quantities where it represents from 2.2 to 5.0% of total lipid phosphorus [42,45,47,48]. Phosphatidylglycerol is an important component of pulmonary surfactant, comprising 6–11% of the total lipid phosphorus. In the alveolar type II cell, which is thought to be the site of synthesis, storage and secretion of pulmonary surfactant, phosphatidylglycerol represents 4.2–10.4% of the total phospholipid [49–51].

In normal animal tissues, bis(monoacylglycero)phosphate is found in substantial quantities only in the alveolar macrophage where it represents 14–18% of total lipid phosphorus [50,52,53]. In other tissues this compound is present in trace quantities seldom representing more than 1.0% of total phospholipids (Table 1). Large amounts of bis(monoacylglycero)phosphate may be present in diseased tissues in certain inherited or drug-induced lipidoses which are discussed in Section 9 below. Small

TABLE 3

Polyglycerophospholipid content in microorganisms

Source	Ref.	% Total lipid phosphorus		
		DPG	PG	Other
Gram-negative bacteria				
Acholeplasma laidlawii	69		100	
Agrobacter tumefaciens	87		29	
Azobacter agilis	87	2	27	
Brucella abortus	89	6	16	
Caulobacter crescentus	72	9	69	10 [a]
Enterobacter aerogenes	87	3	21	
Escherichia coli	87	1	20	
Escherichia coli (wild)	90	3	22	
Escherichia coli (mutant 11-2)	90	3	1	
Hemophilus parainfluenzae	91	3	18	
Marine bacteria – MB 45	84	0	28	2 [b]
Mycoplasma gallisepticum	92		30	
Mycoplasma hominis	70		87	
Nesseria gonorrhea	93	1	22	
Nesseria gonorrhea	94	2	19	
Proteus vulgaris	87	3	17	
Pseudomonas aeruginosa	87	9	15	
Pseudomonas fluorescens	88	8	1	
Pseudomonas fluorescens (Mg^{2+}-limited)	88	50	8	
Rhodopseudomonas capsulatum			41	
Rhodospirilum rubrum	96	5	10	
Salmonella typhimurium	97	3	18	
Salmonella typhimurium	39	4	11	2 [c]
Serratia marcescens	87	17	14	
Thiobacillus thiooxidans	86	7	37	
Gram-positive bacteria				
Bacillus amyloliquefaciens	78	17	48	
Bacillus cereus	98	5–25	25–32	1– 5 [d]
Bacillus megaterium (pH 7)	85		34–45	8–14 [c]
Bacillus megaterium (pH 5)	85		5–10	30–35 [e]
Bacillus subtilis	81	38	13	10 [d]
Methylosinus trichosporum	73	1	58	
Micrococcus lysodeikticus	71	4	72	
Pneumococcus, I-192R	83	70	25	
Staphylococcus aureus	74	11–14	63	16–18 [d]
Staphylococcus aureus	75	0–20	10–60	18–80 [d]
Staphylococcus aureus	76	5	76	14 [d], 0.7 [f]
Staphylococcus aureus (Tazaki)	77	18	43	35 [d]
Staphylococcus aureus, L-form	77	79	12	6 [d]
Streptococcus, group B	82	59	12	7 [d]; 19 [g]
Rickettsiae				
Rickettsia prowazeki	99	3	20	

TABLE 3 (continued)

Source	Ref.	% Total lipid phosphorus		
		DPG	PG	Other
Spirochetes				
Treponema pallidum	80	13		
Treponema pallidum	79	4	7	

Abbreviations as in Table 1.
[a] PGX: Positive ninhydrin reaction; polyphosphoglycerol backbone.
[b] bis(diacylglycero)phosphate.
[c] acylphosphatidylglycerol.
[d] lysylphosphatidylglycerol.
[e] glucosaminylphosphatidylglycerol.
[f] phosphatidylglucose.
[g] diphosphatidyl(glucosyl)glycerol.

amounts of acylphosphatidylglycerol have been reported in rabbit alveolar macrophages [53,54], and in the degenerating baby hamster kidney cell where traces of bis(diacylglycero)phosphate are also present [54].

Table 2 shows the polyglycerophospholipid composition of some plants [8,58–68]. In contrast to animal tissues, where phosphatidylglycerol is generally present only in trace quantities, this phosphoglyceride is often a major constituent of the leaves of plants, representing 20–30% of total lipid phosphorus in such diverse plants as barley, sweet clover, maidenhair tree, maize, pumpkin, rye-grass and tobacco [8,58–60]. The phospholipids of white clover and sycamore leaves contain only 4–5% phosphatidylglycerol while lettuce and moss are intermediate with 12 and 18%, respectively [58,60]. Phosphatidylglycerol is also present in fruits, roots and tubers, where it represents 2–8% of total lipid phosphorus [58,61–63]. As shown in Table 2, algae also contain substantial amounts of phosphatidylglycerol; in fact, in several blue-green algae, phosphatidylglycerol is the only glycerophospholipid present [8,64–68].

Diphosphatidylglycerol is often found in plants. The leaves of maize (16%), lettuce (16%) and white clover (10%), the parsnip root (22%) [58] and the green alga, *Euglena gracilis* (14%) [65] contain the highest percentages of diphosphatidylglycerol. Bis(monoacylglycero)phosphate, acylphosphatidylglycerol and bis(diacylglycero)phosphate have not been found in plants.

Microorganisms, like plants, contain substantial amounts of polyglycerophospholipids as shown in Table 3 [39,69–99]. Phosphatidylglycerol is a major phospholipid component of bacteria often accounting for 10–35% of the total lipid phosphorus in both Gram-positive and Gram-negative organisms. It is worthy of note that several organisms have phosphatidylglycerol as the only phosphoglyceride, *Acholeplasma laidlawii* [69], or as the major glycerophospholipid, *Mycoplasma hominis*, 87% [70], *Micrococcus lysodeikticus*, 72% [71], *Caulobacter crescentus*,

69% [72], *Methylosinus trichosporum,* 58% [73], *Staphylococcus aureus,* as high as 60–68% [74–77] and *Bacillus amyloliquefaciens,* 48% of total lipid phosphorus [78].

Although some microorganisms lack diphosphatidylglycerol, it comprises 2–25% of total lipid phosphorus in the great majority of instances in both Gram-negative and Gram-positive bacteria as shown in Table 3. In *Treponema pallidum,* the causative agent of syphilis, diphosphatidylglycerol represents 4–13% of total lipid phosphorus [79,80]; this is of interest since diphosphatidylglycerol has been shown to be antigenic in the serological test for syphilis (VDRL) [3,22]. Several organisms contain substantially greater proportions of diphosphatidylglycerol; e.g., *Bacillus subtilis,* 38% [81], *Streptococcus,* group B, 59% [82]. *Pneumococcus,* 70% [83] and the L-form of *Staphylococcus aureus,* 79% of total lipid phosphorus [77].

The related compound, acylphosphatidylglycerol, has been identified and quantitated in *Salmonella typhimurium* where it represents 2% of total lipid phosphorus [39]. Bis(diacylglycero)phosphate (bisphosphatidic acid) has been identified in a Gram-negative marine bacterium, MB-45, by DiSiervo and Reynolds [84] where it represents 2% of total lipid phosphorus.

Gram-positive bacteria often contain substantial amounts of lysylphosphatidylglycerol, e.g., *Bacillus cereus, Bacillus megaterium, Bacillus subtilis,* and *Staphylococcus aureus* (Table 3). Glucosaminylphosphatidylglycerol has been demonstrated in *Bacillus megaterium* at low medium pH [85]. Recently, diphosphatidyl(glucosyl)glycerol has been demonstrated by Fischer in a strain of group B *Streptococci,* where it represents 19% of total lipid phosphorus [82].

The metabolic state of the organism may strongly influence the relative proportions of polyglycerophospholipids in bacteria. It has been shown in a number of instances that log growth phase bacteria contain more phosphatidylglycerol and less diphosphatidylglycerol than do stationary phase organisms. Decreases of about 50% in the phosphatidylglycerol content have been demonstrated in *Thiobacillus thiooxidans* [86] and in *Azobacter agilis, Agrobacter tumifaciens, Escherichia coli* and *Proteus vulgaris* [87]. Diphosphatidylglycerol was not detected in log phase *Agrobacter tumifaciens* but represented 19% of total lipid phosphorus in stationary cultures [87]. Increases of 7–11-fold were seen in the percentage of diphosphatidylglycerol in stationary cultures of *Escherichia coli* and *Azobacter agilis* [87] while smaller increases of 2–5-fold were found in *Thiobacillus thiooxidans* [86] and *Proteus vulgaris* [87]. Lowered pH was shown to result in reduced percentages of phosphatidylglycerol and lysylphosphatidylglycerol in *Bacillus megaterium* whereas the amount of glucosaminylphosphatidyl was greatly increased [85]. Limitation of magnesium ion in the culture medium resulted in marked increases (6–8-fold) in the amount of diphosphatidylglycerol and phosphatidylglycerol in *Pseudomonas fluorescens* [88].

(b) Fatty acid composition of polyglycerophospholipids from some mammalian sources

The fatty acid composition of phosphatidylglycerol was first determined in animal tissues by Gray [100]. Phosphatidylglycerol was extracted from rat liver mitochondria where it represented 0.4% of the total lipid phosphorus; its major fatty acids are oleic acid 21%, linoleic acid 20%, stearic acid 14% and palmitic acid 12% [100]. Several workers isolated phosphatidylglycerol from lung surfactant [45,57]. The fatty acid composition of lung surfactant phosphatidylglycerol is of note for its very high content of saturated fatty acids; palmitic acid represents 45–58% of total fatty acids in humans [45] or dogs [57], respectively. Phosphatidylglycerol from the alveolar type II cell, which is thought to be the source of pulmonary surfactant, is also very greatly enriched in palmitic acid which represents 56% of the total fatty acids [50].

Diphosphatidylglycerol (cardiolipin) from ox heart, rat liver and rat kidney is unusual for its marked enrichment in esters of linoleic acid. MacFarlane, who first reported the fatty acid composition of cardiolipin in 1957, found it to contain 72% linoleic acid [101]. Rose and Gray independently reported 77 and 84% linoleic acid in purified rat liver diphosphatidylglycerol [23,100]. Rat kidney diphosphatidylglycerol is also highly enriched in linoleic acid which represents 61% of the total fatty acids [24]. However, the diphosphatidylglycerol purified from rat brain, lung and testis do not exhibit such high degrees of unsaturation in their fatty acids. In these tissues diphosphatidylglycerol contains much less linoleic acid, ranging from 8 to 15%, and has a much greater content of the fully saturated palmitic and stearic acids, which together represent 60–71% of total fatty acids [24].

Bis(monoacylglycero)phosphate isolated from the pulmonary alveolar macrophage, where it represents 14–18% of total lipid phosphorus (Table 1) contains predominantly oleic (49–58%) and linoleic acid (22–26%) [50,52]. Rat liver tritosomes [36] and rat liver (after treatment with the phospholipidosis-inducing agent, chlorphentermine) [102] contain bis(monoacylglycero)phosphate highly enriched in esters of docosahexaenoic acid (C 22:6) which represents 64–69% of total fatty acids; much lower amounts of this fatty acid are found in rat spleen [102] and in the liver of a patient with an uncharacterised lipid storage disease [36]. In these instances, the predominant fatty acids in bis(monoacylglycero)phosphate are oleic (37–57%) and linoleic (10–19%) [36,102].

Phosphatidylglycerol is the immediate metabolic precursor of diphosphatidylglycerol, as will be discussed in detail below. In addition, both phosphatidylglycerol and diphosphatidylglycerol can be converted to bis(monoacylglycero)phosphate. Nevertheless, it is readily apparent (Table 4) that the fatty acid composition of these related compounds is strikingly different. For example, the linoleic acid content of rat liver mitochondrial phosphatidylglycerol is 20% of total fatty acids versus 77–94% in diphosphatidylglycerol and only 5–8% in bis(monoacylglycero)phosphate from rat liver [23,36,100,102]. Furthermore, docosahexaenoic acid, which represents 64–69% of the total fatty acids in bis(monoacylglycero)phosphate from rat liver [36,102] is absent in diphosphatidylglycerol and phosphatidylglycerol. Finally, palmitic acid represents 12% of phosphatidylglycerol fatty acids but comprises only 1–3% of the fatty acids of diphosphatidylglycerol and bis(monoacylglycero)phosphate [23,36,100,102]. The metabolic reasons for this finding are still unclear and will be discussed below in the section on the biosynthesis of polyglycerophospholipids. In contrast, the fatty acids of diphosphatidylglycerol in bacteria, although not identical, usually bear a resemblance to those of the metabolic precursor, phosphatidylglycerol [73,93,103,104].

5. Biosynthesis of the polyglycerophospholipids

(a) Phosphatidylglycerol synthesis

Kennedy et al. elucidated the biosynthesis of phosphatidylglycerol using membrane preparations from chicken liver [105]. It had been shown previously that CDP-diacylglycerol and inositol could be converted to phosphatidylinositol in a reaction catalysed by chicken liver microsomes [106]. It seemed logical that glycerol might react with CDP-diacylglycerol in an analogous manner to give phosphatidylglycerol. Surprisingly, this was not found to be the case; rather *sn*[1,3′-^{14}C]-glycero-3-phosphate reacted with CDP-diacylglycerol to produce a radioactive lipid [105]. No radioactive lipid was formed if labelled glycerol was substituted for *sn*-glycero-3-phosphate. A careful analysis of the reaction products by silicic acid column chromatography revealed the presence of two peaks; the major peak representing more than 90% of the total counts was eluted first in lower concentrations of methanol in chloroform while the minor peak required higher methanol concentrations for elution. The early peak was identified as phosphatidylglycerol while the other peak was found to be phosphatidylglycerophosphate. The formation of phosphatidylglycerol was inhibited by Hg^{2+} ions which caused an accumulation of phosphatidylglycerophosphate, suggesting that the latter compound is an intermediate in phosphatidylglycerol biosynthesis. The pH optimum with chicken liver

TABLE 4

Fatty acid composition of polyglycerophospholipids from animal sources

Fatty acid	Phosphatidylglycerol				Diphosphatidylglycerol			
	Rabbit alveolar Type II cell	Rat liver mito.	Dog lung surfactant	Human lung surfactant	Rat brain	Ox heart	Rat kidney	Rat liver
16:0	56	12	58	45	14	1	8	1
16:1	5	2	5	7		5		1
18:0	3	14	11	5		1	14	7
18:1	27	21	20	37	46	11	16	8
18:2	5	20	3	5	28	72	61	77
18:3	1				8	8		1
20:4		7				2		4
22:6								
Reference	100	57	45	50	101	100	23	24

mitochondria was 8.0; divalent cations were not required for activity. Of importance was the finding that the reaction exhibited a very high degree of specificity for *sn*-glycero-3-phosphate over *sn*-glycero-1-phosphate. The biosynthetic pathway is shown below [105]:

CDP-diacylglycerol + *sn*-glycero-3-P → phosphatidylglycerophosphate + CMP (1)

phosphatidylglycerophosphate → phosphatidylglycerol + Pi (2)

The stereoconfiguration of the phosphatidylglycerol formed by this series of reactions would be *sn*-1,2-diacylglycero-3-phospho-*sn*-1′-glycerol which is the same as that reported for natural phosphatidylglycerols (see Section 3, above).

Using a similar approach, this pathway was also shown to be active in *E. coli*. The major differences noted were that the reaction requires a divalent cation such as Mn^{2+} and that a non-ionic detergent, Triton X-100, was necessary for optimal activity [107]. The enzyme which catalyzes the synthesis of phosphatidylglycerophosphate was subsequently isolated from *E. coli* and partially purified 30-fold over the starting material. The preparation was free of phosphatidylglycerophosphate phosphatase activity; the pH optimum was 8.5, and the enzyme required Mg^{2+} or Mn^{2+} for activity. It was not inhibited by sulphhydryl reagents and required Triton X-100 for optimal activity [108]. Phosphatidylglycerol synthesis from *sn*-glycero-3-phosphate and CDP-diacylglycerol was also found in *Bacillus megaterium* [109].

Evidence for the biosynthesis of phosphatidylglycerol by the steps shown in

			Bis(monoacylglycero)phosphate					
Rat liver mito.	Rat lung	Rat testis	Alveolar macro-phage	Alveolar macro-phage	Rat liver lyso-somes	Human liver	Rat liver	Rat spleen
1	23	55	3	4	3	6	3	9
3			1	1	1	2	1	3
tr	44	16	7	4	1	5	3	2
10	15	15	58	49	5	57	8	37
84	15	13	22	26	6	20	7	19
				1				
tr			7	9				
					69	9	64	6
24	24	24	52	36	36	102	102	50

reactions (1) and (2) above, was also demonstrated in rat brain [110], sheep brain [111,112] and in rat heart mitochondria [113]. Evidence suggesting lipid dependence of the enzymes of phosphatidylglycerol biosynthesis in beef heart mitochondria and microsomes was suggested by the following. Mitochondria and microsomes lost substantial activity after delipidation with organic solvent mixtures, 70 and 29%, respectively. However, the activity could not be restored by incubation with sonicated dispersions of phospholipid, raising the possibility that the loss of activity might have been due to denaturation of the enzymes [114]. In lung microsomes and mitochondria, evidence for phosphatidylglycerol synthesis by the mechanism shown in reactions (1) and (2) above was provided by several groups of workers [48,115,116] (see also Section 8).

In plants, phosphatidylglycerol synthesis was first demonstrated in cauliflower inflorescence mitochondria by Dounce and Dupont [117]. Marshall and Kates demonstrated the presence of the same pathway in a microsomal fraction from spinach leaves [118]. However, unlike the mammalian enzyme, the plant system required a divalent cation (Mg^{2+} or Mn^{2+}) and the conversion of phosphatidylglycerophosphate to phosphatidylglycerol was not substantially inhibited by Hg^{2+}.

Some evidence has been provided indicating the possibility of hormonal regulation of mammalian phosphatidylglycerol synthesis. In the rat prostate, conversion of [^{3}H]glycero-3-phosphate to phosphatidylglycerol in the presence of CDP-diacylglycerol was reduced by 48% in the homogenate of castrated rats. Phosphatidylglycerol synthesis could be restored to normal by testosterone treatment [119]. A smaller decrease (29%) was noted in the activity of the mitochondrial fraction of prostate, suggesting the possibility that the predominance of the hormone effect is on extramitochondrial sources of phosphatidylglycerol synthesis.

Hormonal effects have also been shown in lung and related tissues. In foetal lung, Rooney et al. showed that cortisol treatment of foetal rabbits increased the rate of phosphatidylglycerol synthesis in lung homogenates by 53% while several other enzymes of phospholipid synthesis were unaffected [120]. Cortisol also stimulated the incorporation of several radioactive precursors into phosphatidylglycerol in isolated alveolar type II cells while thyroxine had no effect [121]. Some evidence has also been presented indicating that cyclic AMP increases the incorporation of radioactive precursors into phosphatidylglycerol in foetal rabbit lung slices [122]. Finally, in cultured alveolar carcinoma cells, prolactin was found to stimulate the incorporation of radioactive glycerol into phosphatidylglycerol and other phospholipids [123]. Thus, a role for cortisol and possibly prolactin in the regulation of phosphatidylglycerol synthesis in lung appears to be likely. However, the doses of these agents required to produce the effects have often been in the pharmacological range and the physiological relevance of these observations is as yet unclear.

The specificity of the nucleotide diphosphate diacylglycerol for phosphatidylglycerol synthesis in mammalian systems has been investigated by several groups. In 1971 it was noted that radioactive deoxyCTP was incorporated into acid-insoluble material in liver mitochondria at a rate much greater than other deoxynucleotide triphosphates. This was found to be due to the formation of deoxyCDP-diacylglycerol which was shown to support the synthesis of phosphatidylglycerol in mitochondria at a rate 16% of the rate of CDP-diacylglycerol [124]. Poorthuis and Hostetler found that ADP-diacylglycerol and UDP-diacylglycerol were also effective substrates in rat liver mitochondrial phosphatidylglycerol synthesis with maximal velocities of 4.2 and 5.4 nmol · $mg^{-1} \cdot h^{-1}$ versus 7.0 for CDP-diacylglycerol. However, it seemed unlikely that ADP- and UDP-diacylglycerol contribute significantly to phosphatidylglycerol synthesis in vivo since the microsomal and mitochondrial enzymes of liponucleotide synthesis were highly specific for CTP [125].

Purification of phosphatidylglycerophosphate synthetase from *Bacillus licheniformis* and *Escherichia coli* was reported by Dowhan and co-workers [126–128]. In these elegant studies, phosphatidylglycerophosphate synthetase (PGPS) was solubilised by treatment of the membranous bacterial pellets with Triton X-100. The enzyme was isolated from the solubilised material using affinity chromatography with CDP-diacylglycerol linked covalently with adipic acid to Sepharose. PGPS binds to the CDP-diacylglycerol-Sepharose and can be eluted with buffers containing hydroxylamine. A 6000-fold purification has been achieved for the *E. coli* enzyme [127,128]. The purified enzyme interacts with large amounts of detergent and its M_r-value is estimated to be 200000. On SDS disc gel electrophoresis, one major band is present which accounts for 85% of the activity. The enzyme requires Mg^{2+} and Triton X-100 is necessary for optimal activity [128].

McMurray and Jarvis have solubilised PGPS from rat or pig liver mitochondrial membranes with Triton X-100 or Nonidet P-40 and have achieved partial purification by gel filtration. The resulting enzyme preparation had a specific activity 6-fold greater than that of intact mitochondria [129]. In contrast to the PGPS in intact mitochondria, the partially purified enzyme was found to require divalent cations. PGPS could be stimulated by about 2.5-fold by dispersions of phospholipid (Asolectin). When purified phospholipids were studied, only phosphatidylethanolamine seemed to stimulate the enzyme. No information was provided as to the degree of purity of the enzyme preparation [129].

(b) Phosphatidylglycerophosphatase (phosphatidylglycerophosphate phosphohydrolase, EC 3.1.3.27)

The biosynthesis of phosphatidylglycerol proceeds through phosphatidylglycerophosphate as noted above (reaction 2). Most of the studies of phosphatidylglycerol biosynthesis have not specifically examined the phosphatidylglycerophosphatase step as it is generally not rate limiting. Several authors have carried out experiments designed to measure this reaction alone.

Chang and Kennedy [130] were the first to study phosphatidylglycerophosphatase. They determined its intracellular localisation in sonically-disrupted *E. coli* cells and found that it was associated with the particulate membranous fraction. It was strongly activated by Triton X-100 and could be extracted from the particulate fraction with this non-ionic detergent. The enzyme was partially purified by ion-exchange chromatography in Triton X-100-containing buffers to a specific activity 10.4

times that of the whole sonicate. The preparation exhibited a pH optimum of 7.5 and had a requirement for Mg^{2+}. The apparent K_m for phosphatidylglycerophosphate was 83 μM. The enzyme was inhibited by Hg^{2+}, *N*-ethylmaleimide and fluoride ions. The enzyme preparation did not hydrolyse glycero-3-phosphate; minor activity against phosphatidic acid was noted which was felt to be due to contamination [130].

Phosphatidylglycerophosphate usually does not accumulate during phosphatidylglycerol synthesis as noted above. However, in mitochondria isolated from BHK-21 cells, Lipton and McMurray [131,132] found phosphatidylglycerophosphate to be the predominant product over phosphatidylglycerol (92/5 versus 7/91 in rat liver mitochondria). Further investigation showed this anomalous situation to be due to the fact that phosphatidylglycerophosphatase is a soluble enzyme in these cells. Addition of cell supernatant restored phosphatidylglycerophosphate conversion to phosphatidylglycerol and allowed for diphosphatidylglycerol synthesis [131,132].

Phosphatidylglycerol is an important component of pulmonary surfactant as noted above. Several studies of phosphatidylglycerophosphatase have been carried out in the lung. In lamellar bodies isolated from pig lung, Johnson and co-workers [133] demonstrated the presence of a phosphatase capable of hydrolysing both phosphatidic acid and phosphatidylglycerophosphate. The pattern of inhibition of the enzyme activity by heat and mercuric ions was virtually identical for both substrates. The presence of phosphatidic acid inhibited the hydrolysis of phosphatidylglycerophosphate and vice versa; the apparent K_m for phosphatidic acid was 65 μM and for phosphatidylglycerophosphate, 20 μM [133]. Benson isolated a phosphatidic acid phosphatase from pulmonary surfactant which had an identical apparent K_m for phosphatidic acid, 67 μM [134]. Although kinetic studies were not done, the enzyme isolated from surfactant was also able to hydrolyse phosphatidylglycerophosphate. Hallman and Gluck have shown evidence that phosphatidylglycerophosphatase in lung microsomes and lamellar bodies increases greatly just prior to gestation in rabbit foetuses [135].

Solubilisation and partial purification of phosphatidylglycerophosphatase have been reported by MacDonald and McMurray [136]. The enzyme was released from whole rat liver mitochondria by hypotonic swelling followed by sonication in hypertonic glycerol. After gel filtration the activity was found in a peak near the void volume which had a specific activity 10.8 times greater than that of the starting material. Phosphatidylglycerophosphatase was inhibited by sulphydryl reagents, fluoride ion, by many divalent cations and by detergents. The apparent K_m for phosphatidylglycerophosphate was 2 μM and substrate concentrations above 0.1 mM were inhibitory. No information on the degree of purity of the enzyme was provided [136].

(c) Diphosphatidylglycerol biosynthesis

The biosynthesis of diphosphatidylglycerol was first demonstrated in 1967 using a particulate fraction prepared from *E. coli* [137]. Phosphatidyl[2-^{3}H]glycerol was the substrate for the synthesis of diphosphatidylglycerol. Since the presence of CDP-diacylglycerol stimulated the reaction 2.4-fold, it was at first assumed that the reaction proceeded as follows in bacteria:

$$\text{phosphatidylglycerol} + \text{CDP-diacylglycerol} \rightarrow \text{diphosphatidylglycerol} + \text{CMP} \tag{3}$$

Early experiments in mammalian mitochondria failed to demonstrate the biosynthesis of diphosphatidylglycerol under conditions which might have been expected to lead to the reactions shown in reaction 3, above [105,110–113]. In 1971 Davidson and Stanacev made the observation that mitochondria from guinea pig liver, although seemingly unable to convert phosphatidylglycerol to diphosphatidylglycerol in the presence of added CDP-diacylglycerol, could incorporate *sn*-[2-^{3}H]glycero-3-phosphate into [^{3}H]diphosphatidylglycerol in the presence of added CTP, ATP, CoA and fatty acid. However, since many radioactive lipids were formed, including phosphatidylglycerophosphate, phosphatidylglycerol, diacylglycerol, phosphatidic acid and phosphatidylcholine, the steps involved in the formation of diphosphatidylglycerol remained unclear [138,139].

Hostetler et al. [140] found that diphosphatidylglycerol could be formed by rat liver mitochondria incubated with exogenous CDP-diacylglycerol and *sn*-[2-^{3}H]glycero-3-phosphate in the presence of 10 mM Mg^{2+} and without detergents; phosphatidylglycerol was the main product. Direct conversion of exogenous phosphatidyl[2-^{3}H]glycerol to [^{3}H]diphosphatidylglycerol was demonstrated to occur in the presence of CDP-diacylglycerol; since the rate of diphosphatidylglycerol formation in the absence of CDP-diacylglycerol was only 8% of that found with the liponucleotide present, it seemed likely that CDP-diacylglycerol was also a substrate in the reaction. Evidence was presented showing that the results cannot be explained by bacterial contamination [140]. Subsequently, it was shown that non-ionic detergents such as Tween 20 and Triton X-100 strongly inhibit the biosynthesis of diphosphatidylglycerol [141] probably accounting for the absence of diphosphatidylglycerol formation in many of the earlier studies.

The mechanism of mammalian mitochondrial diphosphatidylglycerol synthesis was elucidated in 1972 by Hostetler, van den Bosch and van Deenen [142]. Radioactive phosphatidylglycerol was prepared which was labelled with tritium in the acylated glycerol and with ^{14}C in the polar head group glycerol. When this doubly-labelled substrate was converted to diphosphatidylglycerol, the product had the same ^{3}H : ^{14}C ratio as that of the substrate. Furthermore, no [^{14}C]glycerol was released, eliminating the possibility of the condensation of two molecules of phosphatidylglycerol as shown below:

$$2\ \text{phosphatidylglycerol} \rightarrow \text{diphosphatidylglycerol} + \text{glycerol} \tag{4}$$

In 1973, further support for this mechanism (reaction 3) was provided by Stanacev and co-workers [143] who directly demonstrated the incorporation of ^{14}C from CDP-diacyl[^{14}C]glycerol into diphosphatidylglycerol. The CDP-diacylglycerol pathway of diphosphatidylglycerol synthesis has also been shown to operate in pig heart mitochondria [144], in guinea pig liver mitochondria [145], in mitochondria from several rat hepatomas [146,147] and in mitochondria from BHK-21 cells [132].

Although it was initially thought that diphosphatidylglycerol synthesis in bacteria also proceeded by the CDP-diacylglycerol pathway, several observations pointed towards synthesis of diphosphatidylglycerol by the condensation of two molecules of phosphatidylglycerol. For example, it was found in *E. coli* and later in *S. aureus* that diphosphatidylglycerol formation from phosphatidylglycerol proceeded even under conditions where energy metabolism is greatly limited [148,149]. Substantial degrees of conversion of exogenous [^{32}P]phosphatidylglycerol to [^{32}P]diphosphatidylglycerol (up to 90%) were found in membrane or soluble preparations from *M. lysodeikticus* in the absence of CDP-diacylglycerol [150].

The mechanism of bacterial diphosphatidylglycerol biosynthesis was established independently by several groups of investigators. Using a membrane fraction from *S. aureus,* Short and White [151] found that radioactive phosphatidylglycerol could be converted to diphosphatidylglycerol in the absence of CDP-diacylglycerol. ^{14}C from CDP-diacyl[^{14}C]glycerol was not incorporated into the product, in contrast to the findings in mammalian mitochondria [143]. Incubation of a doubly-labelled phosphatidylglycerol containing [^{14}C]fatty acids and ^{32}P with *S. aureus* membranes gave diphosphatidylglycerol having the same ratio of $^{14}C/^{32}P$. These authors found that the stoichiometry of the release of [^{14}C]glycerol from [^{14}C]phosphatidylglycerol was that predicted by reaction 4. Independent proof of the presence of the "phosphatidylglycerol condensation pathway" was also provided in *E. coli* by Hirschberg and Kennedy [152] who demonstrated the one to one stoichiometry of glycerol release and diphosphatidylglycerol formation. They employed phosphatidylglycerol doubly-labelled with ^{32}P and with ^{3}H in the free glycerol. During conversion of this substrate to diphosphatidylglycerol, the ratio of $^{32}P/^{3}H$ doubled in accordance with reaction 4. Similar findings were obtained independently by Hostetler et al. [142]. Tunaitis and Cronan [153] confirmed that endogenous CDP-diacylglycerol is not a substrate for bacterial diphosphatidylglycerol synthesis. Subsequently, the phosphatidylglycerol condensation pathway for diphosphatidylglycerol synthesis (reaction 4) was also demonstrated in *L. plantarum* and *M. smegmatis* [154,155]. Thus, bacterial synthesis of diphosphatidylglycerol takes place by a different mechanism than that found in mammalian cells.

Bacterial diphosphatidylglycerol synthesis by condensation of two molecules of phosphatidylglycerol does not require divalent cations [150,151,154] and is inhibited by detergents at high concentrations [150,154]. The pH optimum is acidic in *S. aureus* (4.4) [151] and *L. plantarum* (5.1) [154] or neutral (7.0) in *M. lysodeikticus* [150]. Interestingly, bacterial diphosphatidylglycerol synthetase is strongly inhibited by the reaction products, glycerol and diphosphatidylglycerol, and by phosphatidic acid [156].

In contrast, mammalian mitochondrial biosynthesis of diphosphatidylglycerol by the CDP-diacylglycerol pathway requires a divalent cation such as Co^{2+}, Mn^{2+} or

Mg^{2+} [157]. The reaction is inhibited by Ca^{2+} and by non-ionic detergents [141,157]. The reaction exhibits higher activity at an alkaline pH, e.g. 8.0–8.4 in Tris buffer (K.Y. Hostetler, unpublished). Although ADP-diacylglycerol, UDP-diacylglycerol and deoxyCDP-diacylglycerol have been shown to substitute for CDP-diacylglycerol in the reaction, it seems likely that the latter liponucleotide is the only one of physiological importance [125,141].

As noted above, diphosphatidylglycerol from heart, liver and kidney is unusual for its enrichment in esters of linoleic acid (Table 4). Several authors have investigated possible metabolic explanations for this finding. Eichberg [158] prepared di(lysophosphatidyl)glycerol using phospholipase A_2 and studied its reacylation by mitochondrial and microsomal acyltransferases. In both microsomes and mitochondria the order of activity of acyl CoA esters at optimal conditions was stearoylCoA > oleoylCoA ≫ linoleoylCoA. Thus, no preference for linoleoylCoA was apparent in the reacylation of lysodiphosphatidylglycerol [158]. Similarly, CDP-dilinoleoylglycerol, although a satisfactory substrate, was not especially preferred in the de novo synthesis of diphosphatidylglycerol [157]. Thus, the problem remains unresolved. However, the finding that the turnover of [^{14}C]linoleoyl esters of diphosphatidylglycerol is more rapid than that of other fatty acids suggests that deacylation of diphosphatidylglycerol followed by reacylation with linoleic acid could be an important factor [159]. In addition, the nature of the CDP-diacylglycerols produced near the site of diphosphatidylglycerol synthesis might also affect the ultimate fatty acid composition.

McMurray and Jarvis [160] have recently solubilised diphosphatidylglycerol synthetase from rat or pig liver mitochondria. The enzyme was not extracted by procedures which remove peripheral membrane proteins. The enzyme was released by treatment with 1% Miranol H2M and partially purified by gel filtration; the specific activity of the purified preparation was 4.5-fold higher than that of intact mitochondria. The solubilised enzyme required Co^{2+}, Mn^{2+} or Mg^{2+} and in the presence of optimal amounts of Co^{2+}, a number of other divalent cations including Ca^{2+}, Ba^{2+}, Hg^{2+}, Cu^{2+} and Ni^{2+}, were inhibitory. Like the bacterial enzyme, mammalian diphosphatidylglycerol synthetase was strongly inhibited by its product, diphosphatidylglycerol [160].

Finally, diphosphatidylglycerol may also be formed during the action of phospholipase D on phosphatidylglycerol [161,162]. The major product, however, is phosphatidic acid (96%), while less than 2% appears to be converted to diphosphatidylglycerol. This reaction does not appear to be taking place in bacteria where 2 moles of phosphatidylglycerol condense to form glycerol and diphosphatidylglycerol since no phosphatidic acid formation is apparent [150].

(d) Biosynthesis of bis(monoacylglycero)phosphate and acylphosphatidylglycerol

The synthesis of bis(monoacylglycero)phosphate was first demonstrated in vitro in 1975 by Poorthuis and Hostetler [163] using a crude preparation of mitochondria from rat liver. During the biosynthesis of phosphatidylglycerol from CDP-diacylglycerol and *sn*-[1,3-^{14}C]glycero-3-phosphate, small amounts of [^{14}C]bis(monoacylglycero)phosphate and [^{14}C]acylphosphatidylglycerol were formed. In addition, phosphatidyl[1′,3′-^{14}C]glycerol could be converted directly to these two products in the presence of protein; the reaction was abolished by heating, indicating the enzymatic nature of the reaction [163]. In a subsequent study, the formation of bis(monoacylglycero)phosphate in liver was found to be optimal at pH 4.4. Both of the lysophosphatidylglycerols were converted to bis(monoacylglycero)phosphate, ruling out the possibility that acylphosphatidylglycerol is an obligatory intermediate. The acylation of the free glycerol moiety of phosphatidylglycerol was shown to be independent of acylCoA-dependent acyltransferases, which is not surprising since this glycerol moiety has the *sn*-1 configuration and acyltransferases are stereospecific for *sn*-glycero-3-phosphate residues. The phospholipase A activity of lysosomes, as reflected by generation of [^{14}C]lysophosphatidylglycerol from [^{14}C]phosphatidylglycerol, is distinct from bis(monoacylglycero)phosphate synthetase in that the former enzyme had a greater heat stability and was unaffected by sulphydryl reagents and Triton X-100 which inhibited bis(monoacylglycero)phosphate synthetase [164]. Furthermore, it was demonstrated that chloroquine at 25 mM inhibited lysosomal phospholipase A by 50% while the synthesis of bis(monoacylglycero)phosphate was stimulated by 40% [165].

In addition to phosphatidylglycerol, diphosphatidyl[^{14}C]glycerol can be converted to [^{14}C]bis(monoacylglycero)phosphate when incubated with lysosomes. This appears to take place via lysophosphatidylglycerol, an important product of lysosomal diphosphatidylglycerol hydrolysis [166].

Since acyltransferases are not involved in the introduction of an acyl group on the *sn*-glycero-1-phosphate moiety of phosphatidylglycerol, it appeared that a transacylation reaction might be involved in bis(monoacylglycero)phosphate synthesis. It was found that bis(monoacylglycero)phosphate synthetase could be solubilised from lysosomes by repeated freezing and thawing (> 90%) with little or no loss of activity [167]. However, after removal of endogenous lipid from the soluble preparation with *n*-butanol, bis(monoacylglycero)phosphate synthesis was found to be nearly absent. The loss of activity could be fully restored by adding a sonicated dispersion of lysosomal phospholipids [168]. Of the phospholipids, only phosphatidylinositol and bis(monoacylglycero)phosphate itself, could restore the activity [167,169]. Evidence was presented establishing the transfer of an acyl group from [G-^{3}H]phosphatidylinositol to the free glycerol of [^{14}C]phosphatidylglycerol [167]. In the foregoing experiments, the stereoconfiguration of the substrate was *sn*-1,2-(diacyl)glycero-3-phospho-*sn*-1′-glycerol but the stereoconfiguration of the products, bis(monoacylglycero)phosphate and acylphosphatidylglycerol, was not determined.

Bis(monoacylglycero)phosphate synthesis has also been demonstrated in homogenates of rabbit and rat alveolar and peritoneal macrophages and in human white blood cells in vitro by Huterer and Wherrett [53]. In these studies *sn*-[U-^{14}C]glycero-3-phosphate was converted to the major product, phosphatidyl[^{14}C]glycerol, in the presence of CDP-diacylglycerol at pH 7.4, confirming the role of phosphatidylglycerol as the precursor of bis(monoacylglycero)phosphate [53]. Interestingly, in intact pulmonary macrophages the turnover of the fatty acyl moieties was found to be many times greater than that of the components of the glycerophosphoglycerol backbone, and polyunsaturated fatty acid incorporation into bis(monoacylglycero)phosphate was much greater than that of saturated fatty acids [53].

From the important work of Renkonen, Fischer and co-workers, it has been known since 1974 that naturally-occurring bis(monoacylglycero)phosphate has a different stereoconfiguration than that of its precursor, phosphatidylglycerol. These workers showed that the natural stereoconfiguration is almost exclusively *sn*-(monoacyl)glycero-1-phospho-*sn*-1′-(monoacyl)glycerol in material isolated from BHK cells, rat liver, rabbit lung and pig lung [37,38]. This group subsequently developed a micromethod which allowed the determination of the stereoconfiguration of very small quantities of radioactive glycerophospholipids [170]. Using this method, it was shown in rat liver lysosomes that phosphatidyl-*sn*-rac-[U-^{14}C]glycerol as well as [^{32}P]diphosphatidylglycerol are converted in vitro to radioactive bis(monoacylglycero)phosphate having the natural *sn*-glycero-1-phospho-*sn*-1′-glycerol configuration after a prolonged incubation of 12 h [170]. However, in BHK 21 cells incubated with ^{32}Pi, it was found that bis(monoacylglycero)[^{32}P]phosphate formed early (i.e., at 5–6 h) has a substantial proportion of *sn*-glycero-3-phosphate residues, but after 60 h most of the residues have the *sn*-glycero-1-phosphate configuration [171]. Somerharju and Renkonen subsequently demonstrated directly that both *sn*-(monoacyl)glycero-3-[^{32}P]phospho-*sn*-rac-glycerol and *sn*-(monoacyl)glycero-3-[^{32}P]phospho-*sn*-1′-glycerol were converted in BHK cells to bis(monoacylglycero)[^{32}P]phosphate having substantial amounts of *sn*-glycero-3-phosphate in the early phase [172]. However, upon prolonged incubation for 20 h the glycerol residues assumed primarily the *sn*-glycero-1-phosphate configuration. These results are consistent with the hypothesis that natural phosphatidylglycerol, i.e., having the *sn*-glycero-3-phospho-*sn*-1′-glycerol stereoconfiguration, is incorporated into bis(monoacylglycero)phosphate by acyl transfer as noted above, followed by an unknown reaction which results in the *sn*-glycero-1-phospho-*sn*-1′-glycerol configuration previously shown by Renkonen, Fischer and co-workers [37,38,170–172]. An intramolecular rearrangement involving the *sn*-glycero-3-phosphate moiety seems most likely at present since radioactivity from both the *sn*-glycero-1-phosphate and the *sn*-glycero-3-phosphate residues of phosphatidylglycerol appears to be retained in the product [163–169,172].

To date, phosphatidylglycerol, diphosphatidylglycerol and lysophosphatidylglycerol have been shown to act as precursors of bis(monoacylglycero)phosphate both in vitro and in vivo as noted above. These compounds appear to be the only phospholipids which give rise to bis(monoacylglycero)phosphate, since Somerharju and Renkonen [172] injected dispersions of [^{32}P]phosphatidylcholine, [^{32}P]phosphatidylethanolamine, [^{32}P]sphingomyelin and [^{32}P]phosphatidic acid into rats in vivo and did not find incorporation of ^{32}P into bis(monoacylglycero)phosphate. However, [^{32}P]phosphatidylglycerol and [^{32}P]diphosphatidylglycerol were excellent precursors of bis(monoacylglycero)phosphate in these circumstances [172].

In bacteria, the synthesis of acylphosphatidylglycerol was first observed by Proulx and co-workers. They observed that [^{14}C]phosphatidylglycerol was converted to a less polar phospholipid by a particulate preparation from *E. coli.* The unknown compound had the glycerophosphoglycerol backbone, was not attacked by phospholipase D, and co-chromatographed with synthetic bis(diacylglycero)phosphate [173]. Subsequently, in more detailed analytical studies, the product was shown to be acylphosphatidylglycerol [174]. The reaction has a pH optimum of 7.0, requires Ca^{2+}, and is inhibited by several other divalent cations and Triton X-100. It was suggested that the reaction does not require acylCoA and that phosphatidylglycerol and phosphatidylethanolamine serve as acyl donors [175].

Nishijima et al. [176] subsequently provided evidence that acylphosphatidylglycerol formation in *E. coli* requires a detergent-resistant phospholipase A_1 which appears to generate 2-acyl-*sn*-glycero-3-phosphoglycerol or 2-acyl-*sn*-glycero-3-phosphoethanolamine. These compounds then donate an acyl group to the free glycerol of phosphatidylglycerol in a reaction not requiring Ca^{2+} which is catalysed by a heat-labile factor present in the *E. coli* particulate fraction [176]. Bis(monoacylglycero)phosphate synthesis has not been observed in bacteria or plants.

6. *Degradation of polyglycerophospholipids*

(a) Phosphatidylglycerol

Phosphatidylglycerol is susceptible to the actions of the phospholipases A, phospholipase C and phospholipase D. The latter two enzymes were used in the studies of Haverkate and van Deenen to establish the structure and stereochemistry of this glycerophospholipid [27–29]. The general subject of phospholipase action on glycerophospholipids is covered in more detail in Chapter 9.

(b) Diphosphatidylglycerol

In contrast to phosphatidylglycerol, diphosphatidylglycerol is hydrolysed slowly or not at all by most phospholipases C; under some conditions, it may be hydrolysed by certain of the phospholipases D. Since both of its phosphatidyl moieties have the *sn*-glycero-3-phosphate configuration, diphosphatidylglycerol is also subject to degradation by various phospholipases A.

In mammalian liver the turnover of diphosphatidylglycerol has been found to be much less rapid than that of other phospholipid classes, based on data obtained with ^{32}Pi [159,177,178] or with radioactive glycerol [159]. Beyond this, information on regulation of diphosphatidylglycerol degradation in mammalian tissue is rather limited.

Waite and Sisson [179] isolated and partially purified phospholipase A_2 from rat liver mitochondria. In the presence of Ca^{2+}, this enzyme hydrolysed exogenous diphosphatidylglycerol readily although it was not attacked as rapidly as phosphatidylethanolamine, phosphatidylserine or phosphatidylcholine. Hostetler et al.

[147] isolated mitochondria prelabelled in vivo with ^{32}Pi from normal rat liver and the 7777 rat hepatoma. Upon incubation with 5 mM Ca^{2+} under conditions suitable for endogenous mitochondrial phospholipase A_2 activity, rapid disappearance of [^{32}P]diphosphatidylglycerol was noted. After 3 h, roughly 50% of the [^{32}P]diphosphatidylglycerol had been hydrolysed in both normal and tumour mitochondria demonstrating that the endogenous mitochondrial phospholipase A is capable of degrading substantial amounts of diphosphatidylglycerol in situ [147].

Diphosphatidylglycerol degradation has also been studied using lysosomes from rat liver. These experiments are of importance since it is likely that the ultimate degradation of mitochondria involves lysosomal hydrolases. Hambrey and Mellors [180] showed that [^{32}P]diphosphatidylglycerol was degraded by sequential deacylation to the monoacyl derivative in lysosomes at pH 5 in the presence of Triton X-100. The monoacyl derivative was cleaved to lysophosphatidylglycerol and glycerol-phosphate (90%) while a small proportion (10%) was completely deacylated. These findings were generally confirmed by Poorthuis and Hostetler [166] who observed in addition that if the incubation was carried out at pH 4.4 with a low concentration of Triton X-100 (0.05 mg/ml) diphosphatidyl[^{14}C]glycerol was also converted to [^{14}C]bis(monoacylglycero)phosphate.

In bacteria, a diphosphatidylglycerol-specific phospholipase D was demonstrated in homogenates of *Haemophilus parainfluenza* by Ono and White [181]. This enzyme converted [^{32}P]diphosphatidylglycerol to phosphatidic acid and phosphatidylglycerol; the enzyme required Mg^{2+} and had a pH optimum of 7.5–8.0. Phosphatidylcholine, phosphatidylethanolamine and phosphatidylglycerol were not hydrolysed [181]. Astrachan [182] analysed the phosphatidic acid and phosphatidylglycerol products of [^{32}P]diphosphatidylglycerol hydrolysis with phospholipase C and alkaline hydrolysis to glycerophosphates which were tested for reaction with the stereospecific enzyme glycero-3-phosphate dehydrogenase. The glycerophosphate residue released from phosphatidylglycerol by phospholipase C contained only *sn*-glycero-1-phosphate, demonstrating that the diphosphatidylglycerol-specific phospholipase D attacks only the *sn*-glycero-3′-phospho-*sn*-3′-glycerol bond of the substrate [182]. Diphosphatidylglycerol-specific phospholipase D was subsequently also reported in *Escherichia coli* [183,184] and in *Proteus vulgaris, Salmonella typhimurium* and *Pseudomonas aeruginosa* [185]. However, it was absent from the Gram-positive bacteria, *Bacillus subtilis* and *Staphylococcus aureus,* as well as *Saccharomyces cerevisiae* and rat liver mitochondria [185].

A phospholipase A_1 has been described in *Acinetobacter* H01-N which actively hydrolyses diphosphatidylglycerol in the absence of metal ions [186]. Interestingly, this bacterial species has been shown to contain substantial amounts of the triacyl derivative of diphosphatidylglycerol which represents 5–7% of total lipid phosphorus in log growth phase and 12% in stationary cultures [187]. The substrate specificity of this enzyme is unknown.

(c) Bis(monoacylglycero)phosphate

This glycerophospholipid has been reported only in mammalian cells, in contrast to the other polyglycerophospholipids which are ubiquitously distributed in nature. As will be discussed in detail below, bis(monoacylglycero)phosphate is thought to be confined to mammalian lysosomes [188]. The only studies of its degradation to date have been carried out in lysosomes. Weglicki and co-workers [189] studied the degradation of endogenous lysosomal phospholipids by incubating lysosomes in isotonic sucrose buffered at pH 5.0 at 37°C for varying periods of time. After a 1 h incubation, 40–43% of the endogenous phosphatidylcholine and phosphatidylethanolamine was degraded but only 9% of the bis(monoacylglycero)phosphate had been hydrolysed. These studies clearly established the relative resistance of bis(monoacylglycero)phosphate to lysosomal degradation [189]. However, the mechanism of its hydrolysis was not clear.

Matsuzawa and Hostetler [169] examined this question by preparing [G-^{3}H]-bis(monoacylglycero)phosphate from natural bis(monoacylglycero)phosphate by catalytic exchange labelling. Using this substrate, it was found that the initial rate of

TABLE 5

Subcellular localisation of polyglycerophospholipids in animal tissues *

Subcellular fraction	Phosphatidylglycerol			Diphosphatidylglycerol		
	Guinea pig liver	Rat liver	Rat lung	Guinea pig brain	Pig heart	Rat kidney
Homogenate		1.0	3.3			
Nuclear fraction		1.3	0.4			
Mitochondrial fraction	2.3	1.1	1.7	11.1	18.1	20.2
Outer membrane	2.5				0.4	
Inner membrane	2.2				25.4	
Lysosomes		0				1.8
Lamellar bodies (lung)			11.2			
Phagocytic vesicles						
Microsomal fraction	1.1	1.1	1.7	0.4	11.7	2.4 [a]
Golgi						3.6
Plasma membrane			2.4			
Myelin fragments				0		
Synaptic vesicles				0.3		
Cell supernatant			5.8			
Lung wash			11.0			
Reference	190	164	48	218	222	212

* Results expressed as percentages of total lipid phosphorus.

[a] Rough endoplasmic reticulum.

hydrolysis of bis(monoacylglycero)phosphate is only 10% of that of phosphatidylcholine, in general agreement with the findings of Weglicki et al. [189]. Some degradation occurred by deacylation to lysophosphatidylglycerol, but surprisingly substantial amounts of monoacylglycerol were formed as well, indicating that a lysosomal phosphodiesterase plays an important role in the catabolism of this compound. The pH optimum was 4.4 for both deacylation and phosphodiesterase cleavage of bis(monoacylglycero)phosphate [169]. The resistance of this compound to degradation is thought to be due in large part to its *sn*-glycero-1-phospho-*sn*-1′-glycerol stereoconfiguration.

7. The subcellular localisation of polyglycerophospholipids and their biosynthetic pathways

(a) Phosphatidylglycerol

In many studies of the lipid composition of subcellular fractions from animal tissues,

					Bis(monoacylglycero)phosphate				
Guinea pig liver	Rat liver	Rat liver	Rat liver	Rat lung	Alveolar macro-phage	BHK cell	Rat liver	Rat liver	Rat lung
			6.3	1.4	16.9	1.7	0.4	1.0	1.0
				0		1.0			0
22.5	14	17.8	19.3	7.3		2.5 [b]	0.2	0.7	0.5
3.2	3								
21.5	21								
			0			19.0 [c]	7.0	23.4	
				0.3					1.5
					26.7				
0.5	0	1.1 [a]	0.7	0.2		1.1		0.1	0.7
		1.0							
		1.0		0					0
									1.0
				0.1					1.7
190	209	212	191	48	52	55	188	191	48

[b] Crude mitochondrial fraction.
[c] "Floating fraction"; greatly enriched in lysosomal marker enzymes.

phosphatidylglycerol is not reported as a component of the lipid extracts. This is undoubtedly due to the fact that it is not well separated from other phospholipids in many thin-layer chromatographic systems. Specific methods may be required to reproducibly measure this usually trace component of the phospholipids of mammalian tissues; the failure to report phosphatidylglycerol should not be taken as an assurance of its absence [42]. Thus, the following discussion will concentrate on the studies which have specifically reported the presence of this compound.

Phosphatidylglycerol was first reported in mitochondria by Gray [100] who found that it comprised 0.4% of rat liver mitochondrial phospholipids. Parsons et al. [190] and McMurray and Dawson [178] reported 2.3 and 4.0% phosphatidylglycerol in mitochondria, but subsequent reports using a method developed specifically for the measurement of phosphatidylglycerol indicated the presence of lower amounts of this lipid in mitochondria, ranging from 0.7 to 1.1% of total lipid phosphorus [164,191]. As summarised in Table 5, phosphatidylglycerol was detected in both the inner and outer mitochondrial membranes [178,190]. In the liver, phosphatidylglycerol was also present in microsomes, where it represented 0.5–1.1% of lipid phosphorus, and nuclei, 1.3% [164,190,191], but it was absent from purified liver lysosomes isolated from rats treated with Triton WR-1339 [191]. However, in lysosomes isolated from rats treated with chloroquine or diethylaminoethoxyhexestrol to induce hepatic phospholipidosis, phosphatidylglycerol represented 0.5% of total lipid phosphorus [191]. One group reported the presence of 4.8% phosphatidylglycerol in liver plasma membranes [192]; in our own studies (K.Y. Hostetler and L.B. Hall, unpublished), 0.5% phosphatidylglycerol was found in rat liver plasma membranes using a method developed specifically for phosphatidylglycerol analysis [42].

In rat lung, phosphatidylglycerol was most enriched in the lamellar body fraction where it represented 8.1–11.2% of lipid phosphorus [48,116]. It was also present in nuclei, mitochondria, microsomes, plasma membrane and in the cell supernatant as shown in Table 5 [48]. In the lung wash (pulmonary surfactant), phosphatidylglycerol represented 6.2–11.0% of total phospholipid [48,116]. No information is available on the subcellular distribution of phosphatidylglycerol in other mammalian tissues.

In *Saccharomyces cerevisiae, Neurospora crassa,* and in the cauliflower, phosphatidylglycerol has been reported in both the mitochondrial and microsomal fractions [193–195]. In mitochondria isolated from sycamore leaves, phosphatidylglycerol represented 2.5 and 4.5% of phospholipids in the inner and outer membranes, respectively [60], but in the potato, phosphatidylglycerol was confined to the inner mitochondrial membrane where it represented 5% of lipid phosphorus [196].

The subcellular localisation of phosphatidylglycerol biosynthesis has been studied most extensively in liver. Kiyasu et al. [105] first examined the subcellular localisation of this reaction in chicken liver and found that the specific activity of the enzyme was greatest in the mitochondrial fraction, 2.8 nmol $\cdot$ mg^{-1} $\cdot$ h^{-1}, while the specific activity of the microsomal and nuclear fractions was 0.6 and 0.4 nmol $\cdot$ mg^{-1} protein $\cdot$ h^{-1}, respectively; there was no activity in the cell supernatant. Van Golde

TABLE 6

Subcellular localisation of polyglycerophospholipid biosynthesis in mammalian tissues

Subcellular fraction	Phosphatidylglycerol synthesis [a]						Diphosphatidylglycerol synthesis			Bis(monoacylglycero)phosphate synthesis [b]	
	Rat brain	Rat liver	Rat liver	Rat liver	Rat lung	Rat lung	Rat liver	Morris 7777 hepatoma [b]	*S. cerevisiae* [c]	Rat liver	Rat liver
Homogenate	2.3			2.0	1.7	1.1				0.2	0.3
Nuclear fraction	2.5			0.4	<0.1	1.8					
Mitochondrial fraction									0.9	0.4	1.4
Purified mitochondria	18.2	10.5	2.4	8.6	8.8	8.0	3.0	5.5		<0.1	0.1
Outer membrane			1.1	4.5			0.2				
Inner membrane			3.8	8.6			2.8				
Interstitial soluble protein			<0.1				0				
Purified lysosomes										51.3	61.0
Lamellar body fraction					2.4	5.2					
Microsomal fraction	1.5		0.9	0.7	4.9	4.4	<0.1	0.1	<0.1	0.1	0.3
Rough endoplasmic reticulum		1.4			7.4						
Smooth endoplasmic reticulum		1.3			8.0						
Golgi		1.1		0.4							
Plasma membrane				0.9	1.9						
Supernatant	0.5			<0.1	<0.1				<0.1		
Reference	112	197	141	199	48	116	141	147	193	164	191

[a] $nmol \cdot mg^{-1} \cdot h^{-1}$.
[b] $pmol \cdot mg^{-1} \cdot h^{-1}$.
[c] nmol glycerol-3-phosphate incorporated $\cdot mg^{-1}$ protein $\cdot h^{-1}$.

et al. [197] found similar results in rat liver and determined that both rough and smooth endoplasmic reticulum and Golgi synthesized phosphatidylglycerol with specific activities ranging from 1.1 to 1.4 nmol · mg^{-1} prot. · h^{-1} (Table 6). In mitochondria, Hostetler and van den Bosch [141] showed that both the inner and outer membranes could synthesize this phospholipid, the former being more active, 3.8 versus 1.1 nmol · mg^{-1} prot. · h^{-1}. The liver plasma membrane is also capable of phosphatidylglycerol synthesis as first demonstrated by Victoria et al. [198]. Most of these findings were also confirmed by Jelsema and Morre [199] as shown in Table 6.

Several groups have studied the subcellular localisation of phosphatidylglycerol synthesis in lung where it is present in more than trace amounts by virtue of its role as a component of pulmonary surfactant. Both Hallman and Gluck [48] and Rooney et al. [116] found a substantial capacity for phosphatidylglycerol synthesis in lung mitochondria (8.0–8.8 nmol · mg^{-1} prot. · h^{-1}) and microsomes (4.4–4.9 nmol · mg^{-1} prot. · h^{-1}), in contrast to liver where the microsomes are much less active than mitochondria. Lamellar bodies isolated from lung also appear to have the capacity to synthesize phosphatidylglycerol since this fraction was said to be free of mitochondrial contamination; the activity observed (5.2 nmol · mg^{-1} prot. · h^{-1}) could not be accounted for on the basis of the degree of microsomal contamination present in this fraction [116].

In heart [114] and brain [110–112], phosphatidylglycerol biosynthesis was primarily mitochondrial although measurable activity was also present in the microsomal fraction. In contrast to mammalian systems, Marshall and Kates [118] found that in spinach leaves, the microsomal fraction was responsible for most of the cellular phosphatidylglycerol synthesis.

(b) Diphosphatidylglycerol

As noted previously, phosphatidylglycerol is present in many intracellular sites in the tissues where the problem has been carefully studied. In contrast, diphosphatidylglycerol is discretely localised to mitochondria in most mammalian tissues. A number of early studies in liver, kidney and heart showed that this phospholipid was associated primarily with the mitochondrial fraction [200–204]. However, in many of these studies the contamination of the respective subcellular fractions with mitochondria was not assessed. As the techniques for obtaining purified subcellular fractions became more sophisticated and estimation of purity by marker enzyme measurements and electron microscopy came into common use, it was possible to pinpoint the subcellular localisation of enzymes and phospholipids more exactly. Results of some selected studies of the subcellular localisation of diphosphatidylglycerol in mammalian tissues are shown in Table 5.

In liver, which has been studied most extensively, it was first shown by Parsons et al. [190] that diphosphatidylglycerol is localised exclusively to mitochondria, especially to the inner mitochondrial membrane where it represents 21% of total lipid phosphorus, while the liver microsomes were essentially devoid of this lipid. These findings were confirmed by several other groups [178,205–210]. Diphosphatidylglycerol was also found to be essentially absent from the Golgi apparatus, the plasma membranes of liver [211–212] and several rat hepatomas [213], and from purified nuclear membranes from rat liver [214]. Diphosphatidylglycerol was reported to be present in the plasma membranes and microsomal fraction of the Zadjela hepatoma [215]. However, in other studies, diphosphatidylglycerol was found to be localised to mitochondria and was not present in significant quantities in the microsomes of seven other hepatoma lines nor in the microsomes of regenerating and fetal rat liver [216,217].

In guinea pig brain, diphosphatidylglycerol was also localised to the mitochondrial fraction, as shown by Eichberg and Dawson [218]. As can be seen in Table 5, this lipid represented 11% of brain mitochondrial lipid phosphorus but was essentially absent from microsomes, synaptic vesicles and myelin fragments. In the kidney, diphosphatidylglycerol was also mitochondrial, constituting 9–20% of total lipid phosphorus; its levels are very low in the microsomes, representing 0–2.4% of total lipid phosphorus [202,205,208,212]. The Golgi apparatus and plasma membrane of rat kidney have very little diphosphatidylglycerol as shown in Table 5 [212]. Diphosphatidylglycerol also has a mitochondrial localisation in rat lung [48]. It has an inner mitochondrial membrane localisation in *Neurospora crassa* [194], potato mitochondria [196], cauliflower mitochondria [195], and in mitochondria from sycamore leaves [60].

Although the earliest study in pig heart muscle suggested that little diphosphatidylglycerol was present in microsomes [200], subsequent reports suggested that microsomes from human, ox, bovine and rat heart contained substantial amounts of diphosphatidylglycerol ranging from 9 to 14% of total lipid phosphorus [208,219–221]. In several of these studies, the purity of the microsomal fraction was examined by electron microscopy and it was concluded that the findings were unlikely to be due to mitochondrial contamination. As shown in Table 5, Comte et al. [222] found 11.7% diphosphatidylglycerol in pig heart microsomes which were free of mitochondrial contamination as demonstrated by the absence of cytochrome oxidase. Diphosphatidylglycerol accounted for 18.1% of the total lipid phosphorus of mitochondria and was predominately located in the inner mitochondrial membrane where it represented 25.4% of the total versus only 0.4% in the outer membrane [222]. Thus, heart appears to be the only tissue with extramitochondrial diphosphatidylglycerol.

Hostetler and van den Bosch [141] first examined the subcellular localisation of diphosphatidylglycerol biosynthesis by measuring the conversion of radioactive phosphatidylglycerol to diphosphatidylglycerol in the presence of CDP-diacylglycerol and Mg^{2+}. As shown in Table 6, diphosphatidylglycerol biosynthesis in liver was limited to the mitochondrial inner membrane while the outer mitochondrial membrane, interstitial soluble protein and the microsomal fraction were essentially inactive. Intact guinea pig heart mitochondria were also shown to convert phosphatidylglycerol to diphosphatidylglycerol [144]. Diphosphatidylglycerol synthesis also had a mitochondrial localisation in the Jensen sarcoma, the rat hepatoma 27 [146] and in the Morris 7777 hepatoma [147]. In subcellular fractions from the yeast,

Saccharomyces cerevisiae, Cobon et al. [193] measured the incorporation of radioactive glycero-3-phosphate into diphosphatidylglycerol in the presence of a complex incubation mixture which contained additions to support the acylation of glycerophosphate to phosphatidic acid and for its subsequent conversion to CDP-diacylglycerol. They found that phosphatidylglycerol was a prominent product only in the mitochondrial fraction; some of this material was converted to diphosphatidylglycerol (Table 6). The latter reaction was not observed in the other subcellular fractions, indicating that diphosphatidylglycerol synthesis is probably also a mitochondrial process in yeast [193].

(c) Bis(monoacylglycero)phosphate

Wherrett and Huterer [188] first demonstrated the lysosomal localisation of bis(monoacylglycero)phosphate in the liver of rats treated with Triton WR-1339 which allows the isolation of a highly purified population of secondary lysosomes. This glycerophospholipid, which represented only 0.4% of homogenate phospholipids, accounted for 7.0% of the phospholipids of the lysosomal fraction as shown in Table 5 [188]. In the alveolar macrophage, a phagocytic vesicle fraction was found to be enriched in bis(monoacylglycero)phosphate over that of the cellular homogenate [52]. In the baby hamster kidney cell (BHK cell), Brotherus and Renkonen [55] isolated a lysosomal fraction which contained 19% bis(monoacylglycero)phosphate compared with 1.0% in the nuclei, 1.1% in the microsomes, and 1.7% in the homogenate, respectively. Debuch and co-workers [223,224] showed that in liver a number of different types of secondary lysosomes are rich in bis(monoacylglycero)phosphate. However, these authors did not examine other subcellular fractions for the presence of bis(monoacylglycero)phosphate. Matsuzawa and Hostetler [191] studied the subcellular localisation of bis(monoacylglycero)phosphate in the liver of rats treated with Triton WR-1339 and several drugs which cause phospholipidosis. As shown in Table 5, purified lysosomes contained 23.4% bis(monoacylglycero)phosphate compared with only 1% in the homogenate, 0.7% in purified mitochondria, and 0.1% in microsomes. The small amounts of bis(monoacylglycero)phosphate in other parts of the cell were consistent with lysosomal contamination, confirming that this phospholipid is an exclusive component of lysosomes [191].

Poorthuis and Hostetler [164] first demonstrated the lysosomal localisation of bis(monoacylglycero)phosphate biosynthesis as shown in Table 6. Rat liver lysosomes converted radioactive phosphatidylglycerol to bis(monoacylglycero)phosphate at a rate of 51 $pmol \cdot mg^{-1}$ $prot. \cdot h^{-1}$ versus 0.2 in the homogenate, <0.1 in the purified mitochondria and 0.1 in microsomes, respectively [164]. This finding was extended when similar results were obtained for lysosomes isolated from the liver of rats treated with two drugs which greatly increase the cellular content of bis(monoacylglycero)phosphate [191]. The subcellular localisation of bis(monoacylglycero)-phosphate synthesis has not yet been studied in tissues other than liver.

To summarize, phosphatidylglycerol and the enzymes of its biosynthesis are present in many different intracellular sites at least in tissues such as lung and liver where the problem has been carefully studied. In contrast, diphosphatidylglycerol and bis(monoacylglycero)phosphate are products of phosphatidylglycerol metabolism which are specifically localised to mitochondria and lysosomes, respectively. The enzymes of the biosynthesis of these two compounds are located at the corresponding intracellular sites, at least in liver, the tissue which has been studied most extensively. A possible exception appears to occur in heart where extramitochondrial diphosphatidylglycerol may be present.

8. Phosphatidylglycerol in pulmonary surfactant and amniotic fluid

Polyglycerophospholipids are not transported in the plasma lipoproteins and are not important components of red blood cells; they are generally essentially absent from urine and urine sediments. However, as noted previously, phosphatidylglycerol is a quantitatively important component of pulmonary surfactant. The following discussion will review briefly the biological role of phosphatidylglycerol in pulmonary surfactant and amniotic fluid. The role of phosphatidylcholine is considered in Chapter 1.

The lung secretes a surface-active material consisting largely of phospholipid which lines the pulmonary alveoli, reducing the surface tension at the air/water interface and preventing the collapse of the alveoli at the end of expiration. Phosphatidylcholine, especially the dipalmitoyl molecular species, is a major component of pulmonary surfactant which contributes to the lowering of the surface tension [225] (see Ch. 1). Phosphatidylglycerol is also an important component of pulmonary surfactant, representing 7–11% of the total lipid phosphorus [57,116,226]. Phosphatidylglycerol isolated from pulmonary surfactant, like surfactant phosphatidylcholine, has fatty acyl chains which are highly saturated [45,226–228]. Respiratory distress syndrome in prematurely-born infants has been linked to a deficiency of lung surfactant [229–231].

Pulmonary surfactant is synthesized in alveolar type II cells, stored in intracellular lamellar bodies and then secreted by exocytosis into the alveolar space where the stored phospholipid becomes part of the lung surfactant [49]. Alveolar macrophages may play a role in the degradation of surfactant; it is interesting to note that the phospholipids of this cell type contain 16% bis(monoacylglycero)phosphate [49]. This phospholipid is synthesized from phosphatidylglycerol [164] which comprises about 10% of surfactant phospholipids.

The biosynthesis of phosphatidylglycerol in lung subcellular fractions has been discussed in detail above. It is also worth noting that evidence has been provided demonstrating the importance of the acyldihydroxyacetone pathway which accounts for 56 and 64% of the formation of phosphatidylglycerol and phosphatidylcholine in isolated alveolar type II cells [232]. In addition, two intracellular transfer proteins have been isolated from rat lung which are capable of transferring phosphatidylglycerol; one is specific for phosphatidylglycerol while the other can also transfer phosphatidylcholine and phosphatidylethanolamine from phospholipid vesicles to lung mitochondria [233]. These proteins may conceivably be involved in transfer of phosphatidylglycerol from its intracellular site of synthesis to the lamellar body. Finally, a number of studies using radioactive precursors of phosphatidylglycerol have shown this phospholipid to be metabolically active in lung [226,234–236] and in alveolar type II cells [49,50,237,238].

Since the foetal lung contributes surfactant lipids to the amniotic fluid in utero, it became apparent that the phospholipid composition of the amniotic fluid reflected the metabolic and developmental status of foetal lung. The pioneering studies of Louis Gluck and co-workers established that the maturity of foetal lung was related to the ratio of surface-active phosphatidylcholine/sphingomyelin, with lung immaturity possible at ratios less than 2.0. Using this approach it became apparent that respiratory distress syndrome in the neonate could be excluded with a high degree of accuracy prior to birth [239,240].

Although a mature ratio of surface-active phosphatidylcholine to sphingomyelin (>2.0) was highly accurate in predicting the absence of respiratory distress syndrome (>98%), ratios less than 2.0 did not correlate very well with the presence of respiratory distress syndrome [241]. Additional factors were suggested to be of importance. As shown in Fig. 2 taken from the work of Gluck and co-workers, acidic phospholipids appear in amniotic fluid late in the course of pregnancy [242]. Phosphatidylinositol reaches a peak at 35 weeks of gestation and declines thereafter. Phosphatidylglycerol does not appear before 35 weeks and rises thereafter to levels of about 7–10% at birth [241,242]. An important discovery was made when Hallman et al. [46] noted that phosphatidylglycerol was nearly absent in the lung effluent from infants with respiratory distress syndrome. These results are shown in Fig. 3,

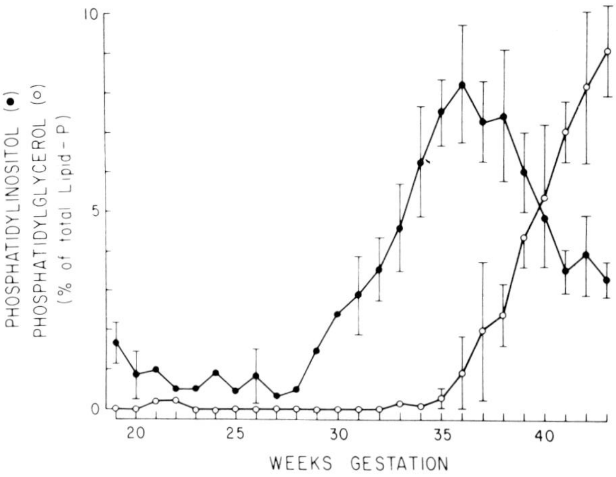

Fig. 2. The content of phosphatidylinositol (●) and phosphatidylglycerol (○) in amniotic fluid during normal gestation. (Reproduced from Hallman, M., Kulovich, M., Kirkpatrick, E., Sugarman, R.G. and Gluck, L. (1976) Am. J. Ob. Gyn. 125, 613–617, with permission.)

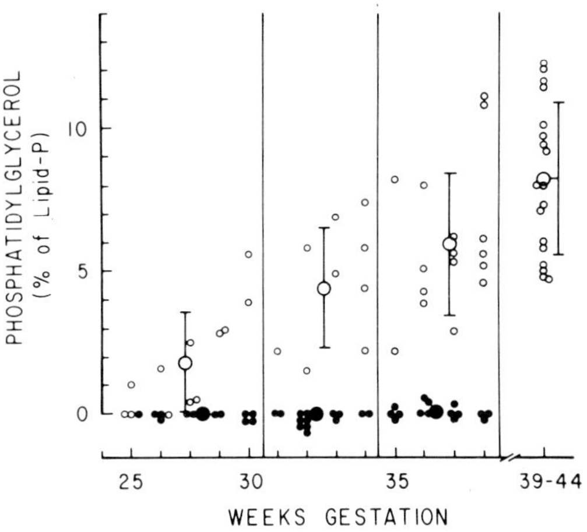

Fig. 3. The content of phosphatidylglycerol in lung effluent of newborns from 1 to 48 h after birth. ○, controls; ●, cases of respiratory distress syndrome. (Reproduced from Hallman, M., Feldman, B.H., Kirkpatrick, E. and Gluck, L. (1977) Pediatr. Res. 11, 714–720, with permission.)

which is taken from the paper by Hallman et al. [46]. After 28 weeks of gestation, phosphatidylglycerol was always present in control newborns but was absent in newborns with respiratory distress syndrome. This finding has led to the determination of phosphatidylglycerol in amniotic fluid as an important adjunctive test for lung maturity. Thus, the presence in amniotic fluid of phosphatidylglycerol representing 3.0% or more of total phospholipids taken together with a mature ratio of surface-active phosphatidylcholine to sphingomyelin (>2.0) form the basis for predicting the maturity of foetal lung, a subject of great practical importance in clinical obstetrics and neonatology [241,243,244].

The exact role of phosphatidylglycerol in lung surfactant is not known. As noted previously its acyl chains are highly saturated like those of surfactant phosphatidylcholine; phosphatidylglycerol isolated from surfactant has been shown to have surface-active properties [227]. It has also been suggested that phosphatidylglycerol may stabilise pulmonary surfactant [245]. Interestingly, in purified surfactant isolated from infants with respiratory distress syndrome, minimum surface tensions were higher (17.2 ± 1.9 dynes/cm^2) than those in surfactant isolated from control subjects (12.1 ± 2.0) [46]. Further evidence that phosphatidylglycerol enhances the surface-active properties of dipalmitoylphosphatidylcholine was suggested by studies in surfactant-depleted lung [246] and in studies of the behaviour of dipalmitoylphosphatidylcholine and phosphatidylglycerol mixtures at the air/water interface [247].

9. Lipid storage diseases and bis(monoacylglycero)phosphate metabolism

(a) Congenital conditions

Rouser et al. [248] were the first to point out that bis(monoacylglycero)phosphate levels were greatly increased in the liver tissue of patients with Niemann-Pick disease (Type A) representing 8–14% of lipid phosphorus compared with only 0.8% in normal controls. Since this report, a number of groups have reported storage of this lysosomal phospholipid in Niemann-Pick disease and its variant forms, shown in Table 7. In Niemann-Pick disease and its variants, the degree of bis(monoacylglycero)phosphate accumulation varies widely in various tissues from as little as 2% to as much as 14% of total lipid phosphorus [248–256]. In addition to Niemann-Pick disease and its variant forms, bis(monoacylglycero)phosphate accumulation has also been reported in the sea-blue histiocyte syndrome [257,258], in adult neurovisceral lipidosis [259] and in juvenile dystonic lipidosis [260].

In Type A and B Niemann-Pick disease, tissue levels of sphingomyelinase are low or absent (see Chapter 4) but in the other disorders the principal metabolic error is uncertain. The metabolic basis for the accumulation of bis(monoacylglycero)phosphate is unknown, although it has been suggested that this is a non-specific finding due to increased numbers of tissue lysosomes [188,260]. However, this appears highly unlikely in view of the fact that Rouser et al. [248] did not find increased

TABLE 7

Inherited diseases associated with storage of bis(monoacylglycero)phosphate in body tissues

Lipid storage disease	Reference
Niemann-Pick disease, Type A [a]	Rouser et al., 1968 [248]
	Callahan and Phillipart, 1971 [249]
	Elleder et al., 1980 [250]
Niemann-Pick disease, Type B	Elleder et al., 1980 [250]
Niemann-Pick disease, Type C	Callahan and Phillipart, 1971 [249]
	Harzer et al., 1978 [251]
	Elleder et al., 1978 [252]
Niemann-Pick disease, Type D	Rao and Spence, 1977 [253]
Niemann-Pick disease, "Variant"	Seng et al., 1971 [254]
	Elleder et al., 1975 [255]
	Debuch and Wiedemann, 1978 [256]
Sea blue histiocyte syndrome	Silverstein et al., 1970 [257]
	Gauthier et al., 1977 [258]
Adult neurovisceral lipidosis	Wherrett and Rewcastle, 1969 [259]
Juvenile dystonic lipidosis [b]	Karpati et al., 1977 [260]
Murine partial deficiency of sphingomyelinase and glucocerebrosidase	Pentchev et al., 1980 [261]

[a] Niemann-Pick classification according to Crocker [280].

[b] Possibly a Niemann-Pick disease "Variant" case.

levels of bis(monoacylglycero)phosphate in the liver of patients with Tay-Sachs disease, Gaucher's disease, metachromatic leukodystrophy, and juvenile amaurotic familial idiocy, conditions which are characterised by increased numbers of lysosomes and a high degree of lysosomal storage. In addition, no substantial increase in bis(monoacylglycero)phosphate in liver or brain was found in the lysosomal storage conditions, infantile neuronal ceroid lipofucinosis, Spielmeyer-Sjögren type of neuronal ceroid lipofucinosis and aspartylglycosaminuria [44]. Several possible mechanisms which might lead to bis(monoacylglycero)phosphate storage will be discussed below.

Finally, Pentchev et al. [261] have described an autosomal recessive inborn error of metabolism in mice characterised by lipid storage in various body tissues. The activity in liver of two lysosomal enzymes, sphingomyelinase and glucocerebrosidase, was noted to be 20–30% of normal. Bis(monoacylglycero)phosphate was noted to accumulate in the liver of affected mice (qualitative data only), but the most prominent finding was storage of sphingomyelin, cholesterol, glucocerebroside and lactosylceramide [261].

In some cases of Niemann-Pick disease and sea-blue histiocyte syndrome, tissue levels of bis(monoacylglycero)phosphate have not been reported to be increased. This may be due to heterogeneity in the patient population, or more likely, to inadequate methodology for quantifying bis(monoacylglycero)phosphate in mixtures of complex lipids.

(b) Acquired lipidoses

In addition to the inherited lipid storage diseases (Table 7), marked tissue accumulation of bis(monoacylglycero)phosphate also occurs in the acquired lipidoses caused by certain cationic amphiphilic drugs. Since the discovery in Japan by Yamamoto et al. [262–264] of a number of cases of acquired foam cell lipidosis in man in 1970, interest in this disorder, which can be reproduced by administration of cationic amphiphilic drugs to experimental animals, has been heightened [265–267]. This human lipid storage disease was caused by the coronary vasodilator drug, 4,4′bis(diethylaminoethoxy)α,β-diethyldiphenylethane (diethylaminoethoxyhexestrol), and was characterised clinically by hepatosplenomegaly, jaundice and fever and the presence in body tissues, especially liver and spleen, of large numbers of phospholipid-rich multilamellar inclusions [262–267]. All phospholipids and bis(monoacylglycero)phosphate were greatly increased; in human liver bis(monoacylglycero)phosphate accounted for as much as 29.4% of total lipid phosphorus [264]. In rats, the disorder was less pronounced with bis(monoacylglycero)phosphate representing 5.7% in spleen and 7.0% of total lipid phosphorus in liver [265]. Total tissue cholesterol was also increased in man and animals [264–267]. Further, it was noted that phosphatidylinositol, another acidic phospholipid, was greatly increased and that the tissue accumulation of this drug was closely correlated with the level of these two phospholipids [268]. In animals the disease caused by 4,4′-bis(diethylaminoethoxy)α,β-diethyldiphenylethane was less severe due to the presence of a hydroxylation pathway for elimination of the drug which is apparently absent in man [268].

Many other cationic amphiphilic drugs are capable of causing cellular phospholipid and bis(monoacylglycero)phosphate storage. However, a detailed treatment of this subject is beyond the scope of this chapter. Excellent reviews have been published by Lüllmann et al. [269] and Lüllmann-Rauch [270].

The phospholipid accumulation in drug-induced lipidosis was shown to be localised to lysosomes in the liver of rats treated with 4,4′-bis(diethylaminoethoxy)α,β-diethyldiphenylethane, and the drug itself is also highly concentrated in lysosomes [191]. Matsuzawa and Hostetler have shown that this agent is a highly effective inhibitor of lysosomal phospholipases A and the recently discovered phospholipase C [165,271]. The mechanism of the drug inhibition is unknown at the present time.

(c) Possible mechanisms of bis(monoacylglycero)phosphate storage

Nothing is known about the rate of tissue bis(monoacylglycero)phosphate synthesis and breakdown in the genetic diseases associated with the storage of this lipid (Table 7). However, based on knowledge obtained from studies in the liver of normal rats and in rats with drug-induced lipidosis several factors may be of importance.

Bis(monoacylglycero)phosphate synthesis in liver occurs in the lysosomes [164,191] only from phosphatidylglycerol, diphosphatidylglycerol or lysophosphatidylglycerol [164,166,167,191]. Indeed, it has been shown that other phospholipids do not serve as the substrate in this reaction [172]. In unmodified lysosomes, bis(monoacylglycero)phosphate levels are low (4.0% of lipid phosphorus) and phosphatidylglycerol and diphosphatidylglycerol are not present [272]. In order for synthesis of bis(monoacylglycero)phosphate to occur, the substrates must be supplied to lysosomes. This presumably occurs by autophagy involving intracellular membranes which contain phosphatidylglycerol or diphosphatidylglycerol. Lipids are also delivered to lysosomes after adsorptive endocytosis of lipoproteins [273], but since lipoproteins do not contain significant amounts of phosphatidylglycerol and diphosphatidylglycerol, it seems unlikely that the latter route would be of significance in explaining the accumulation of bis(monoacylglycero)phosphate. After conversion of lysophosphatidylglycerol to bis(monoacylglycero)phosphate by acyl transfer [167], the compound acquires the *sn*-glycero-1-phospho-*sn*-1′-glycerol stereoconfiguration by an unknown reaction or series of reactions [37,38]. This compound is resistant to degradation by lysosomal phospholipases, as its initial rate of hydrolysis is only 10% of that of phosphatidylcholine [169,189]. Although cationic amphiphilic drugs strongly inhibit lysosomal phospholipases A and C, they have little effect on the formation of bis(monoacylglycero)phosphate from phosphatidylglycerol [165]. Thus, it seems apparent that anything which would lead to an increased delivery of substrate to lysosomes might result in increased synthesis and accumulation of bis(monoacylglycero)phosphate.

Increased synthesis of phosphatidylglycerol might also be a factor of importance (see also Chapter 5, Section 8). Cationic amphiphilic drugs may redirect phospholipid synthesis toward the phosphatidylglycerol and other acidic phospholipids by inhibition of phosphatidate phosphohydrolase, resulting in increased synthesis of CDP-diacylglycerol and its products, phosphatidylglycerol and phosphatidylinositol [274–278]. Since both of these compounds are precursors of bis(monoacylglycero)-phosphate [167], accumulation of this lysosomal lipid might be noted. In agreement with this schema, increased incorporation of [^{3}H]glycerol into phosphatidylglycerol was demonstrated in vivo in liver mitochondria and microsomes from rats treated with 4,4′-bis(diethylaminoethoxy)α,β-diethyldiphenylethane [279]. These studies also suggested that there may be increased transfer of newly-synthesized phospholipids to lysosomes consistent with accelerated autophagy. Also, phosphatidylglycerol levels in liver homogenates of drug-treated rats were significantly higher than in controls [191]. Thus, both increased synthesis of phosphatidylglycerol and its increased delivery to lysosomes by autophagy might lead to the accumulation of bis(monoacylglycero)phosphate, given its demonstrated slow rate of degradation [169,189].

Finally, lysosomal storage of bis(monoacylglycerophosphate could, in principle, result from inhibition or genetic deletion of the phospholipases A and C which are responsible for its catabolism in lysosomes [169,271]. At present, no information is available on these potential causes in patients (Table 7). Nevertheless, it seems likely that the storage of bis(monoacylglycero)phosphate in these inherited conditions will ultimately be shown to be due to specific enzymic alterations and effects rather than being due to non-specific factors as was previously suggested. Inhibition of lysosomal phospholipases is thought to be an important mechanism in the production of drug-induced lipidosis [165].

10. Concluding remarks

In ending this chapter, a few brief comments about possible functions of polyglycerophospholipids in mammalian cells are in order.

Phosphatidylglycerol is generally present only as a minor component of tissue phospholipids, representing less than 1% of total lipid phosphorus. The locations of phosphatidylglycerol and the enzymes which catalyse its biosynthesis are ubiquitous. In liver and lung which have been most extensively studied, most subcellular membranes contain phosphatidylglycerol and have the capacity to synthesize this lipid. Although it is unusually difficult to state the role of individual phospholipids, it is readily apparent that this lipid is the precursor of both diphosphatidylglycerol and bis(monoacylglycero)phosphate. A specific function of phosphatidylglycerol has been established in the lung where this lipid is an important component of pulmonary surfactant representing about 10% of surfactant total lipid phosphorus. Phosphatidylglycerol is widely distributed in nature and is a major component of the membrane phospholipids of many bacteria and plants.

Diphosphatidylglycerol is generally confined to mammalian mitochondria in contrast to its precursor, phosphatidylglycerol, which is found in many intracellular locations. The biosynthesis of diphosphatidylglycerol takes place in the inner mitochondrial membrane and this phospholipid is therefore not subject to exchange catalysed by phospholipid exchange proteins. The specific function of diphosphatidylglycerol is not known with certainty. However, it has been shown that cytochrome c oxidase, the terminal electron transport complex of the inner mitochondrial membrane, contains bound phospholipid. A portion of this phospholipid is very tightly bound and consists primarily of diphosphatidylglycerol [281,282]. These tightly bound molecules of diphosphatidylglycerol are essential for maximal activity of cytochrome oxidase and cannot be replaced by other phospholipids [283,284]. Thus, there appears to be a specific role for diphosphatidylglycerol in the activity of cytochrome c oxidase.

As noted above, diphosphatidylglycerol is a major component of the phospholipids of the mitochondrial inner membrane comprising 20–25% of the total lipid phosphorus. Diphosphatidylglycerol is the only phospholipid component of the inner membrane which adopts the hexagonal H_{II} phase in the presence of Ca^{2+} [285,286]. De Kruijff et al. have proposed a possible role for diphosphatidylglycerol in mitochondrial Ca^{2+} transport based on the observation that divalent cation transport is facilitated when hexagonal H_{II} phase lipid is present [287]. In support of this idea, they found that ruthenium red, a potent inhibitor of mitochondrial Ca^{2+} transport, also inhibits the Ca^{2+}-induced formation of the diphosphatidylglycerol hexagonal H_{II} phase [287]. Finally, diphosphatidylglycerol appears to interact specifically with cytochrome c to produce the hexagonal H_{II} phase which may be important for the function of this protein and the cytochrome c oxidase system [287].

Bis(monoacylglycero)phosphate is also localised to a specific subcellular compartment, the lysosome. Like diphosphatidylglycerol, it is not subject to exchange between membranes catalysed by phospholipid exchange proteins. The biosynthesis of this lipid takes place in lysosomes and it appears to represent the only example of a compound which is synthesized in lysosomes. The function of bis(monoacylglycero)phosphate is unknown. It has been shown by several groups that this lipid is quite resistant to degradation by lysosomal phospholipases, possibly due in part to its unique glycero-1-phosphate stereochemical configuration. Thus, a potential role may be to stabilize the lysosomal phospholipid bilayer against degradation by potentially lytic endogenous phospholipases.

Acknowledgements

This work was supported in part by N.I.H. Grant GM 24979 and by the Research Service of the San Diego Veterans Administration Medical Center. During the preparation of this chapter the author was a Fellow of the John Simon Guggenheim Foundation. Dr. H. van den Bosch kindly reviewed the manuscript and Drs. L. Gluck, M. Hallman, W. Dowhan and W.C. McMurray provided preprints of articles in press.

References

1 IUPAC-IUB Committee on Biochemical Nomenclature (1977) Proc. Natl. Acad. Sci. 74, 2222–2230.
2 IUPAC-IUB Commission on Biochemical Nomenclature (1977) Lipids 12, 455–468.
3 Pangborn, M.C. (1942) J. Biol. Chem. 143, 247–256.
4 Pangborn, M.C. (1945) J. Biol. Chem. 161, 71–82.
5 Pangborn, M.C. (1944) J. Biol. Chem. 153, 343–348.
6 Pangborn, M.C. (1947) J. Biol. Chem. 168, 351–361.
7 Maruo, B. and Benson, A.A. (1957) J. Am. Chem. Soc. 79, 4564–4565.
8 Benson, A.A. and Maruo, B. (1958) Biochim. Biophys. Acta 27, 189–195.
9 Body, D.R. and Gray, G.M. (1967) Chem. Phys. Lipids 1, 254–263.
10 Body, D.R. and Gray, G.M. (1967) Chem. Phys. Lipids 1, 424–448.
11 Brotherus, J. and Renkonen, O. (1974) Chem. Phys. Lipids 13, 11–20.
12 McKibbin, J.M. and Taylor, W.E. (1952) J. Biol. Chem. 196, 427–436.
13 Faure, M. and Morelec-Coulon, M.-J. (1956) Ann. Inst. Pasteur Paris 104, 246–263.
14 MacFarlane, M.G. and Gray, G.M. (1957) Biochem. J. 67, 25–26p.
15 Gray, G.M. and MacFarlane, M.G. (1958) Biochem. J. 70, 409–425.
16 Faure, M. and Morelec-Coulon, M.-J. (1958) Compt. Rend. Acad. Sci. 245, 2181–2183.
17 LeCoq, J. and Ballou, C.E. (1964) Biochem. 3, 976–980.
18 MacFarlane, M.G. (1958) Nature (Lond.) 182, 946.
19 MacFarlane, M.G. and Wheeldon, L.W. (1959) Nature 183, 1808.
20 Morelec-Coulon, M.-J., Faure, M. and Marechal, J. (1960) Bull. Soc. Chim. Biol. 42, 867–876.
21 De Haas, G.H. and Van Deenen, L.L.M. (1965) Nature 206, 935.
22 De Haas, G.H., Bonsen, P.P.M. and Van Deenen, L.L.M. (1966) Biochim. Biophys. Acta 116, 114–124.
23 Rose, H. (1964) Biochim. Biophys. Acta 84, 104–127.
24 Courtade, S., Marinetti, G.V. and Stotz, E. (1967) Biochim. Biophys. Acta 137, 121–134.
25 Nielsen, H. (1971) Biochim. Biophys. Acta 231, 370–384.
26 Benson, A.A. and Miyano, M. (1961) Biochem. J. 81, 31p.
27 Haverkate, F., Houtsmuller, U.M.T. and Van Deenen, L.L.M. (1962) Biochim. Biophys. Acta 63, 547–549.
28 Haverkate, F. and Van Deenen, L.L.M. (1964) Biochim. Biophys. Acta 84, 106–108.
29 Haverkate, F. and Van Deenen, L.L.M. (1965) Biochim. Biophys. Acta 106, 78–92.
30 Op den Kamp, J.A.F., Bonsen, P.P.M. and Van Deenen, L.L.M. (1969) Biochim. Biophys. Acta 176, 298–305.
31 Ruettinger, R.T. and Brewer, G.J. (1978) Biochim. Biophys. Acta 529, 181–185.
32 Yang, S.F., Freer, S. and Benson, A.A. (1967) J. Biol. Chem. 242, 477–484.
33 Dawson, R.M.C. (1967) Biochem. J. 102, 205–210.
34 Batrakov, S.G., Panosyan, A.G., Kogan, G.A. and Bergelson, L.D. (1975) Biochem. Biophys. Res. Commun. 66, 755–762.
35 Joutti, A. and Renkonen, O. (1976) Chem. Phys. Lipids 17, 264–266.
36 Wherrett, J.R. and Huterer, S. (1973) Lipids 8, 531–533.
37 Brotherus, J., Renkonen, O., Herrmann, J. and Fischer, W. (1974) Chem. Phys. Lipids 13, 178–182.
38 Joutti, A., Brotherus, J., Renkonen, O., Laine, R. and Fischer, W. (1976) Biochim. Biophys. Acta 450, 206–209.
39 Olsen, R.W. and Ballou, C.E. (1971) J. Biol. Chem. 246, 3305–3313.
40 McAllister, D.J. and DiSiervo, A.J. (1975) J. Bacteriol. 123, 302–307.
41 Simon, G. and Rouser, G. (1969) Lipids 4, 607–614.
42 Poorthuis, B.J., Yazaki, P.J. and Hostetler, K.Y. (1976) J. Lipid Res. 17, 433–437.
43 Okano, G., Matsuzaka, H. and Shimojo, T. (1980) Biochim. Biophys. Acta 619, 167–175.
44 Kahma, K., Brotherus, J., Haltia, M. and Renkonen, O. (1976) Lipids 11, 539–544.
45 Rooney, S.A., Canavan, P.M. and Motoyama, E.K. (1974) Biochim. Biophys. Acta 360, 56–67.

46 Hallman, M., Feldman, B.H., Kirkpatrick, E. and Gluck, L. (1977) Pediatr. Res. 11, 714–720.
47 Baxter, C.F., Rouser, G. and Simon, G. (1969) Lipids 4, 243–244.
48 Hallman, M. and Gluck, L. (1975) Biochim. Biophys. Acta 409, 172–191.
49 Mason, R.J., Dobbs, L.G., Greenleaf, R.D. and Williams, M.C. (1977) Fed. Proc. 36, 2697–2702.
50 Smith, F.B. and Kikkawa, Y. (1979) Lab. Invest. 40, 172–177.
51 Mason, R.J. and Williams, M.C. (1980) Biochim. Biophys. Acta 617, 36–50.
52 Mason, R.J., Stossel, T.P. and Vaughan, M. (1972) J. Clin. Invest. 51, 2399–2407.
53 Huterer, S. and Wherrett, J. (1979) J. Lipid Res. 20, 966–973.
54 Somerharju, P. (1979) Biochim. Biophys. Acta 574, 461.
55 Brotherus, J. and Renkonen, O. (1977) J. Lipid Res. 18, 191–202.
56 Rouser, G., Simon, G. and Kritchevsky, G. (1969) Lipids 4, 599–606.
57 Pfleger, R.C., Henderson, R.F. and Waide, J. (1972) Chem. Phys. Lipids 9, 51–68.
58 Roughan, P.G. and Batt, R.D. (1969) Phytochemistry 8, 363–369.
59 Roughan, P.G. (1970) Biochem. J. 117, 1–8.
60 Bligny, R. and Douce, R. (1980) Biochim. Biophys. Acta 617, 254–263.
61 Galliard, T. (1968) Phytochemistry 7, 1715–1922.
62 Ben Abdelkader, A. (1968) Physiol. Vegetale 6, 417–442.
63 Beiss, U. (1969) Landwirtsch. Forsch. 23, 198–206.
64 Sastry, P.S. and Kates, M. (1963) Can. J. Biochem. 43, 1445–1453.
65 Calvayrac, R. and Douce, R. (1970) FEBS Letters 7, 259–262.
66 Bailey, D.S. and Northcote, D.H. (1976) Biochem. J. 156, 295–300.
67 Sato, Naoki, Murata, N., Miura, Y. and Ueta, N. (1979) Biochim. Biophys. Acta 572, 19–28.
68 Hirayama, O. (1967) J. Biochem. (Tokyo) 61, 179–185.
69 Bevers, E.M., Singal, S.A., Op den Kamp, J.A.F. and Van Deenen, L.L.M. (1977) Biochemistry 16, 1290–1295.
70 Schiefer, H.G., Gerhardt, U. and Brunner, H. (1975) Z. Physiol. Chem. 356, 559–565.
71 MacFarlane, M.G. (1961) Biochem. J. 80, 45p.
72 Contreras, I., Shapiro, L. and Henry, S. (1978) J. Bacteriol. 135, 1130–1136.
73 Weaver, T.L., Patrick, M.A. and Dugan, P.R. (1975) J. Bacteriol. 124, 602–605.
74 Houtsmuller, U.M.T. and Van Deenen, L.L.M. (1965) Biochim. Biophys. Acta 106, 564–576.
75 Gould, R.M. and Lennarz, W.J. (1970) J. Bacteriol. 104, 1135–1144.
76 Short, S.A. and White, D.C. (1971) J. Bacteriol. 108, 219–226.
77 Hayami, M., Okabe, A., Kariyama, R., Abe, M. and Kanemasa, Y. (1979) Microbiol. Immunol. 23, 435–442.
78 Paton, J.C., May, B.K. and Elliott, W.H. (1978) J. Bacteriol. 135, 393–401.
79 Meyer, H. and Meyer, F. (1971) Biochim. Biophys. Acta 231, 93–106.
80 Matthews, H.M., Yang, T.K. and Jenkin, H.M. (1979) Infect. Immunol. 24, 713–719.
81 Bishop, D.G., Rutberg, L. and Samuelsson, B. (1967) Eur. J. Biochem. 2, 448–453.
82 Fischer, W. (1977) Biochim. Biophys. Acta 487, 74–88.
83 Brundish, D.E., Shaw, N. and Baddiley, J. (1967) Biochem. J. 104, 205–211.
84 DiSiervo, A.J. and Reynolds, J.W. (1975) J. Bacteriol. 123, 294–301.
85 Op den Kamp, J.A.F., Houtsmuller, U.M.T. and Van Deenen, L.L.M. (1965) Biochim. Biophys. Acta 106, 438–441.
86 Shively, J.M. and Benson, A.A. (1967) J. Bacteriol. 94, 1679–1683.
87 Randle, C.L., Albro, P.W. and Dittmer, J.C. (1969) Biochim. Biophys. Acta 187, 214–220.
88 Minnikin, D.E. and Abdolrahimzadeh, H. (1974) FEBS Letters 43, 257–260.
89 Thiele, O.W., Busse, D. and Hoffmann, K. (1968) Eur. J. Biochem. 5, 513–519.
90 Nishijima, M. and Raetz, C.R. (1979) J. Biol. Chem. 254, 7837–7844.
91 White, D.C. (1968) J. Bacteriol. 96, 1159–1170.
92 Rottem, S. and Markowitz, O. (1979) Biochemistry 18, 2930–2935.
93 Sud, I.J. and Feingold, D.S. (1975) J. Bacteriol. 124, 713–717.
94 Senff, L.M., Wegener, W.S., Brooks, G.F., Finnerty, W.R. and Makula, R.A. (1976) J. Bacteriol. 127, 874–880.

95 Steiner, S., Sojka, G.A., Conti, S.F., Gest, H. and Lester, R.L. (1970) Biochim. Biophys. Acta 203, 571–574.
96 Hirayama, O. (1968) Agr. Biol. Chem. 32, 34–41.
97 Ames, G.F. (1968) J. Bacteriol. 95, 833–843.
98 Lang, D.R. and Lundgren, D.G. (1970) J. Bacteriol. 101, 483.
99 Winkler, H.H. and Miller, E.T. (1978) J. Bacteriol. 136, 175–178.
100 Gray, G.M. (1964) Biochim. Biophys. Acta 84, 35–40.
101 MacFarlane, M.G. and Gray, G.M. (1957) Biochem. J. 67, 25–26p.
102 Khan, M.U. and Williams, J.P. (1977) J. Chromatogr. 140, 179–185.
103 Plackett, P., Smith, P.F. and Mayberry, W.R. (1970) J. Bacteriol. 104, 798–807.
104 Fischer, W. (1977) Biochim. Biophys. Acta 487, 89–104.
105 Kiyasu, J.Y., Pieringer, R.A., Paulus, H. and Kennedy, E.P. (1963) J. Biol. Chem. 238, 2293–2298.
106 Paulus, H. and Kennedy, E.P. (1960) J. Biol. Chem. 235, 1303–1311.
107 Kanfer, J.N. and Kennedy, E.P. (1964) J. Biol. Chem. 239, 1720–1724.
108 Chang, Ying-Ying and Kennedy, E.P. (1967) J. Lipid Res. 8, 447–455.
109 Patterson, P.H. and Lennarz, W.J. (1971) J. Biol. Chem. 246, 1062–1072.
110 Possmayer, F., Balakrishnan, G. and Strickland, K.P. (1968) Biochim. Biophys. Acta 164, 79–87.
111 Stanacev, N.Z., Isaac, D.C. and Brookes, K.B. (1968) Biochim. Biophys. Acta 152, 806–808.
112 Davidson, J.B. and Stanacev, N.Z. (1970) Can. J. Biochem. 48, 633–642.
113 Stanacev, N.Z., Stuhne-Sekalec, L., Brookes, K.B. and Davidson, J.B. (1969) Biochim. Biophys. Acta 176, 650–653.
114 Stuhne-Sekalec, L. and Stanacev, N.Z. (1970) Can. J. Biochem. 48, 1214–1221.
115 Hallman, M. and Gluck, L. (1974) Biochem. Biophys. Res. Commun. 60, 1–7.
116 Rooney, S.A., Page-Roberts, B.A. and Motoyama, E.K. (1975) J. Lipid Res. 16, 418–425.
117 Douce, R.L. and Dupont, J. (1969) C.R. Acad. Sci. Paris Ser. D. 268, 1657–1660.
118 Marshall, M.O. and Kates, M. (1972) Biochim. Biophys. Acta 260, 558–570.
119 Stanacev, N.Z., Stuhne-Sekalec, L. and Anderson, K.M. (1970) Endocrinology 86, 1205–1211.
120 Rooney, S.A., Gross, I., Gassenheimer, L.N. and Motoyama, E.K. (1975) Biochim. Biophys. Acta 398, 433–441.
121 Post, M., Batenburg, J.J. and Van Golde, L.M.G. (1980) Biochim. Biophys. Acta 618, 308–317.
122 Hallman, M. (1977) Biochem. Biophys. Res. Commun. 77, 1094–1102.
123 Porreco, R.P., Merritt, T.A. and Gluck, L. (1980) Am. J. Ob. Gyn. 136, 1071–1074.
124 Ter Schegget, J., Van den Bosch, H., Van Baak, M.A., Hostetler, K.Y. and Borst, P. (1971) Biochim. Biophys. Acta 239, 234–242.
125 Poorthuis, B.J.H.M. and Hostetler, K.Y. (1976) Biochim. Biophys. Acta 431, 408–415.
126 Larson, T.J., Hirabayshi, T. and Dowhan, W. (1976) Biochemistry 15, 974–979.
127 Hirabayshi, T., Larson, T.J. and Dowhan, W. (1976) Biochemistry 15, 5205–5211.
128 Dowhan, W. and Hirabayshi, T. (1981) Methods Enzymol. 71, 555–560.
129 McMurray, W.C. and Jarvis, E.C. (1978) Can. J. Biochem. 56, 414–419.
130 Chang, Y.Y. and Kennedy, E.P. (1967) J. Lipid Res. 8, 456–462.
131 Lipton, J.H. and McMurray, W.C. (1976) Biochem. Biophys. Res. Commun. 73, 300–305.
132 Lipton, J.H. and McMurray, W.C. (1977) Biochim. Biophys. Acta 486, 228–242.
133 Johnston, J.M., Reynolds, G., Wylie, M.B. and MacDonald, P.C. (1978) Biochim. Biophys. Acta 531, 65–71.
134 Benson, B.J. (1980) Proc. Natl. Acad. Sci. U.S.A. 77, 808–811.
135 Hallman, M. and Gluck, L. (1980) Ped. Res. 14, 1250–1259.
136 MacDonald, P.M. and McMurray, W.C. (1980) Biochim. Biophys. Acta 620, 80–89.
137 Stanacev, N.Z., Chang, Y. and Kennedy, E.P. (1967) J. Biol. Chem. 242, 3018–3019.
138 Davidson, J.B. and Stanacev, N.Z. (1971) Biochem. Biophys. Res. Commun. 42, 1191–1199.
139 Davidson, J.B. and Stanacev, N.Z. (1971) Can. J. Biochem. 49, 1117–1124.
140 Hostetler, K.Y., Van den Bosch, H. and Van Deenen, L.L.M. (1971) Biochim. Biophys. Acta 239, 113–119.

141 Hostetler, K.Y. and Van den Bosch, H. (1972) Biochim. Biophys. Acta 260, 380–386.
142 Hostetler, K.Y., Van den Bosch, H. and Van Deenen, L.L.M. (1972) Biochim. Biophys. Acta 260, 507–513.
143 Stanacev, N.Z., Davidson, J.B., Stuhne-Sekalec, L. and Domazet, Z. (1973) Can. J. Biochem. 51, 286–304.
144 Domazet, Z., Stuhne-Sekalec, L., Davidson, J.B. and Stanacev, N.Z. (1973) Can. J. Biochem. 51, 274–285.
145 Davidson, J.B. and Stanacev, N.Z. (1974) Can. J. Biochem. 52, 936–939.
146 Dyatlovitskaya, E.V., Hostetler, K.Y., Einisman, L.I. and Gorkova, N.P. (1976) Biokhimiya 76, 1421–1425.
147 Hostetler, K.Y., Zenner, B.D. and Morris, H.P. (1978) J. Lipid Res. 19, 553–560.
148 Rampini, C., Barbu, E. and Polonovski, J. (1970) C.R. Acad. Sci. Paris Ser. D 270, 882–885.
149 Polonovski, J., Wald, R., Paysant, M., Rampini, C. and Barbu, E. (1971) Ann. Inst. Pasteur Paris 120, 589–598.
150 DiSiervo, A.J. and Salton, M.R.J. (1971) Biochim. Biophys. Acta 239, 280–292.
151 Short, S.A. and White, D.C. (1972) J. Bacteriol. 109, 820–826.
152 Hirshberg, C.B. and Kennedy, E.P. (1972) Proc. Natl. Acad. Sci. U.S.A. 69, 648–651.
153 Tunaitis, E. and Cronan, J.E.J. (1973) Arch. Biochem. Biophys. 155, 420–427.
154 Burritt, M.F. and Henderson, T.O. (1975) J. Bacteriol. 123, 972–977.
155 Mathur, A.K., Murthy, P.S., Saharia, G.S. and Venkitasubramanian, T.A. (1976) Can. J. Microbiol. 22, 354–358.
156 DiSiervo, A.J. (1975) Can. J. Biochem. 53, 1031–1034.
157 Hostetler, K.Y., Galesloot, J.M., Boer, P. and Van den Bosch, H. (1975) Biochim. Biophys. Acta 380, 382–389.
158 Eichberg, J. (1974) J. Biol. Chem. 249, 3423–3429.
159 Landriscina, C., Megli, F.M. and Quagliariello, E. (1976) Lipids 11, 61–66.
160 McMurray, W.C. and Jarvis, E.C. (1980) Can. J. Biochem. 58, 771–776.
161 Stanacev, N.S. and Stuhne-Sekalec, L. (1970) Biochim. Biophys. Acta 210, 350–352.
162 Stanacev, N.Z., Stuhne-Sekalec, L. and Domazet, Z. (1973) Can. J. Biochem. 51, 747–753.
163 Poorthuis, B.J.H.M. and Hostetler, K.Y. (1975) J. Biol. Chem. 250, 3297–3302.
164 Poorthuis, B.J.H.M. and Hostetler, K.Y. (1976) J. Biol. Chem. 251, 4596–4502.
165 Matsuzawa, Y. and Hostetler, K.Y. (1980) J. Biol. Chem. 255, 5190–5194.
166 Poorthuis, B.J.H.M. and Hostetler, K.Y. (1978) J. Lipid Res. 19, 309–315.
167 Matsuzawa, Y., Poorthuis, B.J.H.M. and Hostetler, K.Y. (1978) J. Biol. Chem. 253, 6650–6653.
168 Hostetler, K.Y. and Poorthuis, B.J.H.M. (1978) in Cyclitols and Phosphoinositides (Wells, W.W. and Eisenberg Jr., F., eds.), Academic Press, New York, pp. 585–597.
169 Matsuzawa, Y. and Hostetler, K.Y. (1979) J. Biol. Chem. 254, 5997-6001.
170 Joutti, A. and Renkonen, O. (1979) J. Lipid Res. 20, 840–847.
171 Joutti, A. (1979) Biochim. Biophys. Acta 575, 10–15.
172 Somerharju, P. and Renkonen, O. (1980) Biochim. Biophys. Acta 618, 407–419.
173 Benns, G. and Proulx, P. (1971) Biochem. Biophys. Res. Commun. 44, 382–389.
174 Cho, K.S., Benns, G. and Proulx, P. (1973) Biochim. Biophys. Acta 326, 355–360.
175 Cho, K.S., Hong, S.D., Cho, J.M., Chang, C.S. and Lee, K.S. (1977) Biochim. Biophys. Acta 486, 47–54.
176 Nishijima, M., Sa-eki, T., Tamori, Y., Doi, O. and Nojima, S. (1978) Biochim. Biophys. Acta 528, 107–118.
177 Taylor, C.B., Bailey, E. and Bartley, W. (1967) Biochem. J. 105, 605–609.
178 McMurray, W.C. and Dawson, R.M.C. (1969) Biochem. J. 112, 91–108.
179 Waite, M. and Sisson, P. (1971) Biochem. 10, 2377–2383.
180 Hambrey, P.N. and Mellors, A. (1975) Biochem. Biophys. Res. Commun. 62, 939–945.
181 Ono, Y. and White, D.C. (1970) J. Bacteriol. 103, 111–115.
182 Astrachan, L. (1973) Biochim. Biophys. Acta 296, 79–88.

183 Cole, R., Benns, G. and Proulx, P. (1974) Biochim. Biophys. Acta 337, 325–332.
184 Rampini, C. (1975) C.R. Acad. Sci. Paris Ser. D 281, 1431–1433.
185 Cole, R. and Proulx, P. (1975) J. Bacteriol. 124, 1148–1152.
186 Torregrossa, R.E., Makula, R.A. and Finnerty, W.R. (1977) J. Bacteriol. 131, 493–498.
187 Torregrossa, R.E., Makula, R.A. and Finnerty, W.R. (1977) J. Bacteriol. 131, 486–492.
188 Wherrett, J.R. and Huterer, S. (1972) J. Biol. Chem. 247, 4114–4120.
189 Weglicki, W.B., Ruth, R.C. and Owens, K. (1973) Biochem. Biophys. Res. Commun. 51, 1077–1082.
190 Parsons, D.F., Williams, G.R., Thompson, W., Wilson, D.F. and Chance, B. (1967) in Mitochondrial Structure and Compartmentation, (Quagliariello, E., Papa, S., Slater, E.C. and Tager, J.M., eds.) Adriatica Editrice, Bari, pp. 29–70.
191 Matsuzawa, Y. and Hostetler, K.Y. (1980) J. Lipid Res. 21, 202–214.
192 Ray, T.K., Skipski, V.P., Barclay, M., Essner, E. and Archibald, F.M. (1969) J. Biol. Chem. 244, 5528–5536.
193 Cobon, G.S., Crowfoot, P.D. and Linnane, A.W. (1974) Biochem. J. 144, 265–275.
194 Hallermeyer, G. and Neupert, W. (1974) Hoppe-Seyler's Z. Physiol. Chem. 355, 279–288.
195 Moreau, F., Dupont, J. and Lance, C. (1974) Biochim. Biophys. Acta 345, 294–304.
196 McCarty, R.E., Dounce, R. and Benson, A.A. (1973) Biochim. Biophys. Acta 316, 266–270.
197 Van Golde, L.M.G., Fleischer, B. and Fleischer, S. (1971) Biochim. Biophys. Acta 249, 318–330.
198 Victoria, E.J., Van Golde, L.M.G., Hostetler, K.Y., Scherphof, G.L. and Van Deenen, L.L.M. (1971) Biochim. Biophys. Acta 239, 443–457.
199 Jelsema, C.L. and Morré, D.J. (1978) J. Biol. Chem. 253, 7960–7971.
200 Marinetti, G.V., Erbland, J. and Stotz, E. (1958) J. Biol. Chem. 233, 562–565.
201 MacFarlane, M.G., Gray, G.M. and Wheeldon, L.W. (1960) Biochem. J. 77, 626–631.
202 Strickland, E.H. and Benson, A.A. (1960) Arch. Biochem. Biophys. 88, 344–348.
203 Collins, F.D. and Shotlander, V.L. (1961) Biochem. J. 79, 321–324.
204 Getz, G.S., Bartley, W., Stirpe, F., Notton, B.M. and Renshaw, A. (1962) Biochem. J. 83, 181–191.
205 Fleischer, S., Rouser, G., Fleischer, B., Cash, A. and Kritchevsky, G. (1967) J. Lipid Res. 8, 170–180.
206 Getz, G.S., Bartley, W., Lurie, D. and Notton, B. (1968) Biochim. Biophys. Acta 152, 325–339.
207 Levy, M. and Sauner, M.-T. (1968) Chem. Phys. Lipids 2, 291–295.
208 Rouser, G., Nelson, G.J., Fleischer, S. and Simon, G. (1968) in Biological Membranes (Chapman, D., ed.), Academic Press, New York, pp. 5–69.
209 Stoffel, W. and Schiefer, H.-G. (1968) Z. Physiol. Chem. 349, 1017–1026.
210 Colbeau, A., Nachbaur, J. and Vignais, P.M. (1971) Biochim. Biophys. Acta 249, 462–492.
211 Keenan, T.W. and Morré, D.J. (1970) Biochemistry 9, 19–25.
212 Zambrano, F., Fleischer, S. and Fleischer, B. (1975) Biochim. Biophys. Acta 380, 357–369.
213 Van Hoeven, R.P. and Emmelot, P. (1972) J. Memb. Biol. 9, 105–126.
214 Kleinig, H. (1970) J. Cell Biol. 46, 396–402.
215 Bergelson, L.D., Dyatlovitskaya, E.V., Torkhovskaya, T.I., Sorokina, I.B. and Gorkova, N.P. (1970) Biochim. Biophys. Acta 210, 287–298.
216 Hostetler, K.Y., Zenner, B.D. and Morris, H.P. (1976) Biochim. Biophys. Acta 441, 231–238.
217 Hostetler, K.Y., Zenner, B.D. and Morris, H.P. (1979) Cancer Res. 39, 2978–2983.
218 Eichberg, J., Jr., Whittaker, V.P. and Dawson, R.M.C. (1964) Biochem. J. 92, 91–100.
219 Wheeldon, L.W., Schumert, Z. and Turner, D.A. (1965) J. Lipid Res. 6, 481–489.
220 Gloster, J. and Harris, P. (1969) Cardiovasc. Res. 3, 45–51.
221 Gloster, J. and Harris, P. (1970) Cardiovasc. Res. 4, 1–5.
222 Comte, J., Maisterrena, B. and Gautheron, D.C. (1976) Biochim. Biophys. Acta 419, 271–284.
223 Stremmel, W. and Debuch, H. (1976) Z. Physiol. Chem. 357, 803–810.
224 Tjiong, H.B., Lephtin, J. and Debuch, H. (1978) Z. Physiol. Chem. 359, 63–67.
225 Goerke, J. (1974) Biochim. Biophys. Acta 344, 241–261.
226 Sanders, R.L. and Longmore, W.J. (1975) Biochemistry 14, 835–840.
227 Henderson, R.F. and Pfleger, R.C. (1972) Lipids 7, 492–494.
228 Okano, G. and Akino, T. (1979) Lipids 14, 541–546.

229 Avery, M.E. and Mead, J. (1959) Am. J. Dis. Children 97, 517–523.
230 Chu, J., Clements, J.A., Cotton, E.K., Klaus, M.H., Sweet, A.Y. and Tooley, W.H. (1967) Pediatrics 40, 709–782.
231 Gluck, L., Kulovich, M.V., Eidelman, A.I., Cordero, L. and Khazin, A.F. (1972) Pediat. Res. 6, 81–99.
232 Mason, R.J. (1978) J. Biol. Chem. 253, 3367–3370.
233 Van Golde, L.M.G., Oldenborg, V., Post, M., Batenburg, J.J., Poorthuis, B.J.H.M. and Wirtz, K.W.A. (1980) J. Biol. Chem. 255, 6011–6013.
234 Longmore, W.J. and Mourning, J.T. (1977) J. Lipid Res. 18, 309–313.
235 Jobe, A., Kirkpatrick, E. and Gluck, L. (1978) J. Biol. Chem. 253, 3810–3816.
236 Jobe, A., Ikegami, M. and Sarton-Miller, E. (1980) Biochim. Biophys. Acta 617, 65–75.
237 Eagle, M.J., Sanders, R.L. and Douglas, W.H.J. (1980) Biochim. Biophys. Acta 617, 225–236.
238 Mason, R.J. and Dobbs, L.G. (1980) J. Biol. Chem. 255, 5101–5107.
239 Gluck, L., Kulovich, M.V., Borer, R.C., Jr., Brenner, P.H., Anderson, G.G. and Spellacy, W.N. (1971) Am. J. Ob. Gyn. 109, 440–445.
240 Borer, R.C., Jr., Gluck, L., Freeman, R.K. and Kulovich, M.V. (1971) Pediat. Res. 5, 655–661.
241 Kulovich, M.V., Hallman, M.B. and Gluck, L. (1979) Am. J. Ob. Gyn. 135, 57–63.
242 Hallman, M., Kulovich, M., Kirkpatrick, E., Sugarman, R.G. and Gluck, L. (1976) Am. J. Ob. Gyn. 125, 613–617.
243 Gluck, L. (1978) Clin. Ob. Gyn. 21, 547–559.
244 Kulovich, M.V. and Gluck, L. (1979) Am. J. Ob. Gyn. 135, 64–70.
245 Hallman, M. and Gluck, L. (1976) J. Lipid Res. 17, 257–262.
246 Ikegani, M., Silverman, J. and Adams, F.H. (1979) Pediat. Res. 13, 777–780.
247 Bangham, A.D., Morley, C.J. and Phillips, M.C. (1979) Biochim. Biophys. Acta 573, 552–556.
248 Rouser, G., Kritchevsky, G., Yamamoto, A., Knudson, A.G. and Simon, G. (1968) Lipids 3, 287–290.
249 Callahan, J.W. and Phillipart, M. (1971) Neurology 21, 442.
250 Elleder, M., Smid, F., Harzer, K. and Cihula, J. (1980) Virchows Arch. A 385, 215–231.
251 Harzer, K., Scholte, W., Pfeiffer, J., Benz, H.U. and Anzil, A.P. (1978) Acta Neuropathol. 43, 97–104.
252 Elleder, M., Jirasek, A., Smid, F., Harzer, K. and Schelegerova, D. (1978) Virchows Arch. A 377, 329–338.
253 Rao, B.G. and Spence, M.W. (1977) Ann. Neurol. 1, 385–392.
254 Seng, P.N., Debuch, H., Witter, B. and Wiedemann, H.-R. (1971) Z. Physiol. Chem. 352, 280–288.
255 Elleder, M., Smid, F. and Kohn, R. (1975) Virchows Arch. A 365, 239–255.
256 Debuch, H. and Wiedemann, H.R. (1978) Eur. J. Pediatr. 129, 99–101.
257 Silverstein, M.N., Ellefson, R.D. and Ahern, E.J. (1970) New Engl. J. Med. 282, 1–4.
258 Gautier, M., Raulin, J., Lapous, D., Loriette, C., Carreau, J.P. and Scotto, J. (1977) Biomedicine 26, 52–60.
259 Wherrett, J.R. and Rewcastle, N.B. (1969) Clin. Res. 17, 665.
260 Karpati, G., Carpenter, S., Wolfe, L.S. and Andermann, F. (1977) Neurology 27, 32–42.
261 Pentchev, P.G., Gal, A.E., Booth, A.D., Omedeo-Sale, F., Fouks, J., Neumeyer, B.A., Quirk, J.M., Dawson, G. and Brady, R.O. (1980) Biochim. Biophys. Acta 619, 669–679.
262 Yamamoto, A., Adachi, S., Ishibe, T., Shinji, Y., Kaki-Uchi, Y., Seki, K. and Kitani, T. (1970) Lipids 5, 566–571.
263 Yamamoto, A., Adachi, S., Kitani, T., Shinji, Y., Seki, K., Nasu, T. and Nishikawa, M. (1971) J. Biochem. 69, 613–615.
264 Yamamoto, A., Adachi, S., Ishikawa, K., Yokomura, T., Kitani, T., Nasu, T., Imoto, T. and Nishikawa, M. (1971) J. Biochem. 70, 775–784.
265 Adachi, S., Matsuzawa, Y., Yokomura, T., Ishikawa, K. and Uhara, S. (1972) Lipids 7, 1–7.
266 Kasama, K., Yoshida, K., Takeda, S., Akeda, S. and Kawai, K. (1974) Lipids 9, 235–243.
267 Yamamoto, A., Adachi, S., Matsuzawa, Y., Kitani, T., Hiraoka, A. and Seki, K.-I. (1976) Lipids 11, 616–622.

268 Matsuzawa, Y., Yamamoto, A., Adachi, S. and Nishikawa, M. (1977) J. Biochem. 82, 1369–1377.
269 Lüllmann, H., Lüllmann-Rauch, R. and Wassermann, O. (1975) CRC Crit. Rev. Toxicol. 4, 185–218.
270 Lüllmann-Rauch, R. (1979) in Lysosomes in Applied Biology and Medicine, Vol. 6, (Dingle, J.T., Jacques, P.J. and Shaw, I.H., eds.), North-Holland, Amsterdam, pp. 49–130.
271 Matsuzawa, Y. and Hostetler, K.Y. (1980) J. Biol. Chem. 255, 646–652.
272 Bleistein, J., Heidrich, H.G. and Debuch, H. (1980) Hoppe-Seyler's Z. Physiol. Chem. 361, 595–597.
273 Goldstein, J.L. and Brown, M.S. (1977) Ann. Rev. Biochem. 46, 897–930.
274 Brindley, D.N. and Bowley, M. (1975) Biochem. J. 148, 461–469.
275 Michell, R.H., Allan, D., Bowley, M. and Brindley, D.N. (1976) J. Pharm. Pharmac. 28, 331–332.
276 Sturton, R.G. and Brindley, D.N. (1977) Biochem. J. 162, 25–32.
277 Brindley, D.N., Bowley, M., Sturton, R.G., Pritchard, P.H., Burditt, S.L. and Cooling, J. (1977) Biochem. Soc. Trans. 5, 40–43.
278 Eichberg, J., Gates, J. and Hauser, G. (1979) Biochim. Biophys. Acta 573, 90–106.
279 Matsuzawa, Y. and Hostetler, K.Y. (1980) Biochim. Biophys. Acta 620, 592–602.
280 Crocker, A.C. (1961) J. Neurochem. 7, 69–80.
281 Awasthi, Y.C., Chuang, T.F., Keenan, T.W. and Crane, F.L. (1971) Biochim. Biophys. Acta 226, 42–52.
282 Robinson, N.C. and Capaldi, R.A. (1977) Biochemistry 16, 375–381.
283 Robinson, N.C., Strey, F. and Talbert, L. (1980) Biochemistry 19, 3656–3661.
284 Fry, M. and Green, D.E. (1980) Biochem. Biophys. Res. Commun. 93, 1238–1246.
285 Cullis, P.R. and De Kruijff, B. (1979) Biochim. Biophys. Acta 559, 399–420.
286 De Kruijff, B., Cullis, P.R. and Verkleij, A.J. (1980) Trends Biochem. Sci. 5, 79–81.
287 De Kruijff, B., Verkleij, A.J., Van Echteld, C.J.A., Gerritsen, W.J., Noordam, P.C., Mombers, C., Rietveld, A., De Gier, J., Cullis, P.R., Hope, M.J. and Nayar, R. (1981) in International Cell Biology 1980–1981 (Schweiger, H.G., ed.), Springer-Verlag, Berlin, pp. 559–571.

CHAPTER 7

Inositol phospholipids

J.N. HAWTHORNE

Department of Biochemistry, University Hospital and Medical School, Queen's Medical Centre, Nottingham NG7 2UH, U.K.

1. Discovery

Inositol was first reported as a lipid constituent by Anderson and Roberts in 1930 [1], who obtained it from a phospholipid of avian tubercle bacillus. Klenk and Sakai [2] discovered the first inositol lipid from a plant source, soybean oil. In 1942 Folch and Woolley [3] described a brain phospholipid containing inositol and the subsequent work of Folch, who introduced the term phosphoinositide, stimulated wider interest in these compounds. Of the possible stereoisomers, only *myo*-inositol has been found in the naturally occurring phosphoinositides.

2. Chemistry

(a) Phosphatidylinositol and its phosphates

The most widely distributed phosphoinositide is phosphatidylinositol (structure I), in which a phosphatidic acid residue is attached to the 1-hydroxyl of *myo*-inositol. Although *myo*-inositol is optically inactive, the substitution in the 1-position pro-

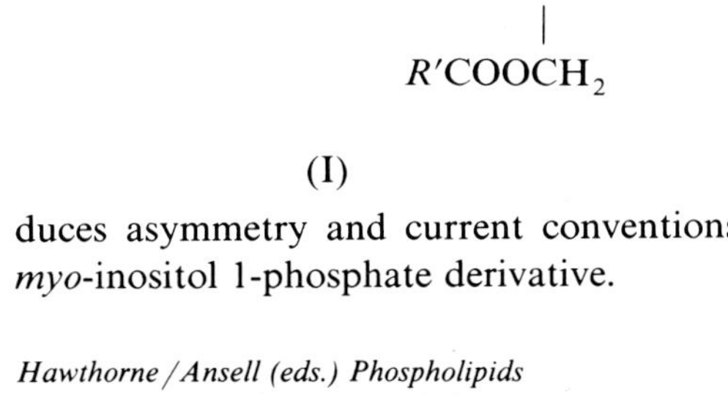

(I)

duces asymmetry and current conventions describe phosphatidylinositol as a 1D-*myo*-inositol 1-phosphate derivative.

Hawthorne/Ansell (eds.) Phospholipids

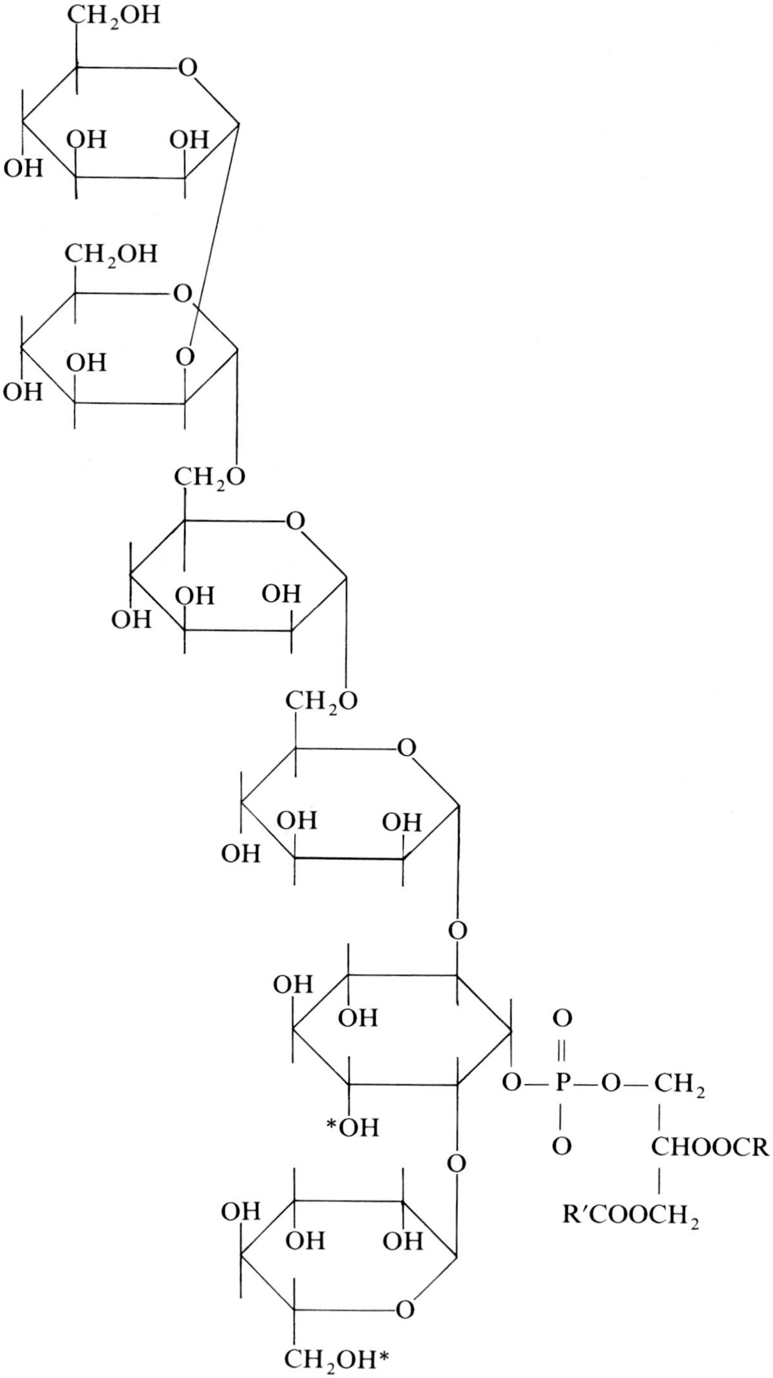
CH₂OH
O
OH
OH
OH
CH₂OH
O
OH
O
OH
CH₂O
O
OH
OH
OH
CH₂O
O
OH
OH
OH
O
OH
OH
O
O—P—O—CH₂
O
*OH
CHOOCR
O
R′COOCH₂
OH
OH
OH
O
CH₂OH*

(II)

Appreciable quantities of diphosphoinositide and triphosphoinositide occur in nervous tissue and smaller concentrations in other tissues and certain micro-organisms [4]. The structures are clear from the preferred nomenclature [5]: phosphatidylinositol 4-phosphate (diphosphoinositide) and phosphatidylinositol 4,5-bis(phosphate). Some versions of the IUPAC-IUB recommendations have numbering errors for these compounds which are corrected on page 19 of the reference quoted [5].

(b) Phosphatidylinositol mannosides

The phospholipid fraction from *Mycobacterium tuberculosis* studied by Anderson et al. [1] proved to contain both *myo*-inositol and D-mannose. The work of Vilkas and Lederer [6], Nojima [7] and Ballou and Lee [8] showed that a family of phosphatidylinositol derivatives containing from one to five mannose residues occurred in this organism. The structure of the parent pentamannoside is shown on p. 264 (II). A mannotetraose is linked to the 6-position of the inositol and a further mannose to the 2-position, all glycosidic bonds being in the α-conformation. Lipid from *M. phlei* contains a hexamannoside in which one more mannose is linked α-1,2 to the mannotetraose. It seems likely that in their natural form the phosphatidylinositol mannosides may have one or two additional fatty acids [9]. There is evidence to suggest that these fatty acids may be attached to the hydroxyls indicated by asterisks in structure II (see review of Ambron and Pieringer [10]). The phosphatidylinositol

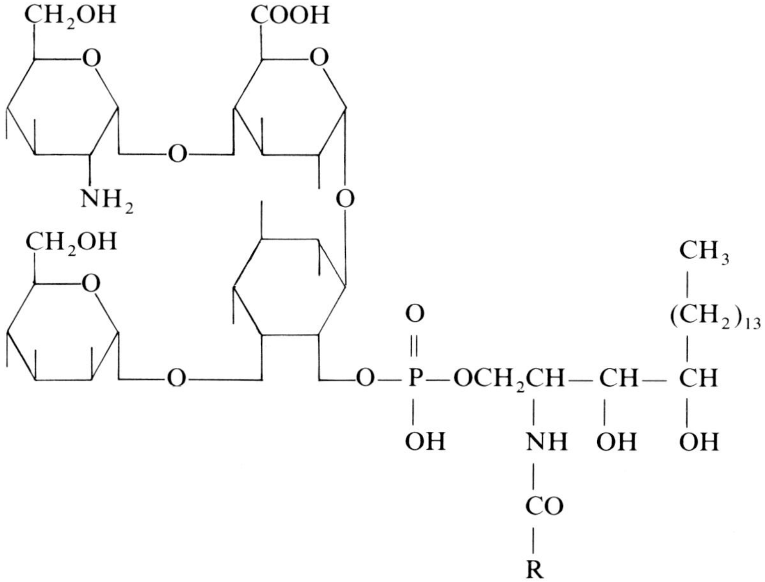

III

mannosides occur not only in the mycobacteria but also in corynebacteria, propionibacteria and actinomycetes. In *M. tuberculosis*, palmitic and tuberculostearic (10 methyl octadecanoic) acids accounted for 90% of the fatty acids of the mannosides.

(c) Sphingolipids containing inositol

These compounds have been reviewed by Lester et al. [11]. The pioneer work came from H.E. Carter and his group, who used the name phytoglycolipid for substances containing phytosphingosine, inositol, phosphate and sugars. Phytoglycolipid was obtained from bean leaves and various seed oils. Structure III was suggested [12] for a compound from corn and flax oils. The α-glycosidic linkage of mannose to the 2-hydroxyl of the inositol is identical with that in the bacterial phosphatidylinositol mannosides. There is a further similarity in the attachment of the other sugars, D-glucuronic acid and D-glucosamine, to the 6-hydroxyl of the inositol. The fatty acid in the ceramide portion of the molecule can be a C24 or C26 α-hydroxy compound.

Lester et al. [13,14] have isolated six novel phosphoinositol-containing sphingolipids of a similar type from tobacco leaves. A major component had structure IV and the equivalent compound without the *N*-acetyl residue was also present.

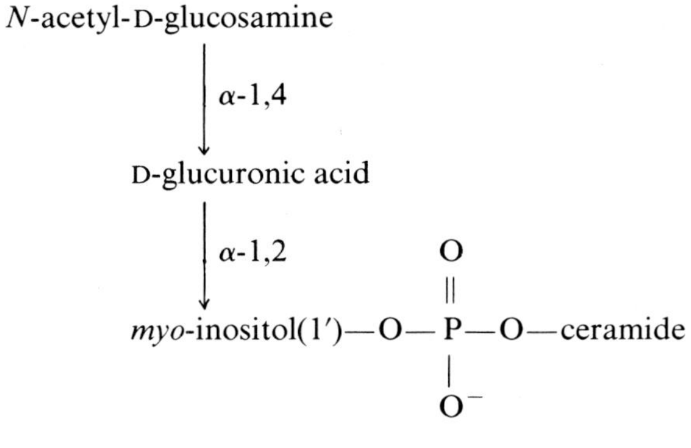

(IV)

Again the similarity to structure III is apparent, but in this case the glucuronic acid is linked to inositol at the 2-position, not the 6–OH. As with the seed oil compounds, hydroxy-acids were found in the ceramide. The more complex derivatives in tobacco leaves contained up to four arabinose molecules, one or two galactoses and in one case a mannose as well. These various sugars are attached to structure IV in a way as yet unknown.

About 40% of the lipid inositol in yeast (*S. cerevisiae*) occurs in sphingolipids. Two compounds have been partially characterised. One of them (V) has a single inositol phosphate residue linked to ceramide [15] and another is a mannoside of V, the mannose being linked to the inositol. This is probably related in structure to compound III. Another [16] has two inositol phosphate residues and the available information suggests the structure mannose-(inositol phosphate)$_2$-ceramide.

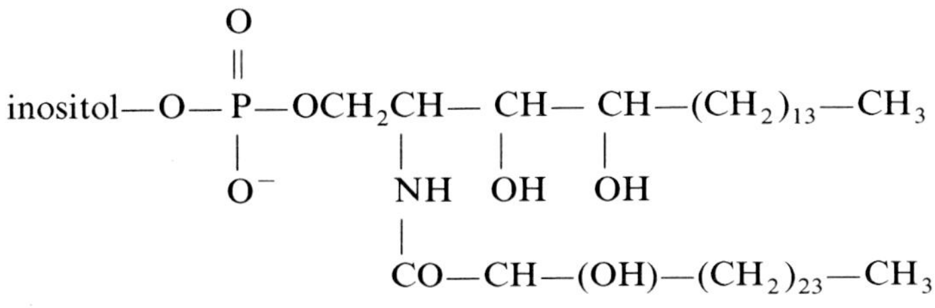

(V)

3. *Distribution in tissues and fatty acid composition*

(a) Distribution

Although there are exceptions, in most mammalian tissues the inositol phospholipids do not represent more than about 8% of the lipid phosphorus. Comprehensive tables of analytical data have been prepared by White [17]. The major inositol lipid in mammalian tissues is phosphatidylinositol, though appreciable quantities of diphosphoinositide and triphosphoinositide are also found in nervous tissue, adrenal medulla and kidney. Because of rapid post-mortem hydrolysis of triphosphoinositide to diphosphoinositide and phosphatidylinositol, accurate analyses are not easy to obtain. Figures most closely resembling the situation in vivo are obtained when rats are rapidly killed by microwave irradiation. By this method, recoveries of triphosphoinositide were 550 nmol/g brain with 45-day-old rats [18]. The corresponding figure for diphosphoinositide was 135 nmol/g. Earlier figures for brain are given by Hawthorne and Kai [19] and include the following as percentages of total lipid phosphorus for guinea-pig brain: phosphatidylinositol 3.0, diphosphoinositide 0.58 and triphosphoinositide 2.58. Corresponding percentages for phosphatidylinositol in rat tissues are as follows [4]: heart 3.7, liver 7.2, kidney 5.9, skeletal muscle 8.9, intestinal mucosa 4.1, lung 3.9. This lipid appears to be distributed uniformly among the subcellular membranes, though diphosphoinositide and triphosphoinositide may be localised in plasma membrane and some storage vesicles such as adrenal chromaffin granules [20] and the secretory granules of rat parotid [21].

Some plants and micro-organisms contain the more complex inositol lipids described in the previous section. Many plant tissue phospholipid fractions contain high proportions of phosphatidylinositol. The highest recorded [22] is 48% of total lipid phosphorus in leaves of a cold-sensitive variety of alfalfa grown at 30°C, though several other leaves and fruits give figures around 20% [4].

Phosphatidylinositol is rarely found in bacteria [23] though the mycobacteria are exceptions. As outlined in Section 2, several species have been studied and shown to contain both phosphatidylinositol and various glycosylated derivatives. Protozoa generally contain phosphatidylinositol in quantities representing about 5% of the total lipid phosphorus. In one species, *Crithidia fasciculata* [24], there was much more phosphatidylinositol (16.3% lipid P); diphosphoinositide (1.8% lipid P) and triphosphoinositide (0.7% lipid P) were also present. Figures are for cells in the logarithmic phase of growth. Amoebae also contain inositol lipids. *Entamoeba invadeus* contains both phosphatidylinositol and ceramide phosphoinositol (V) [26]. The soil organism *Acanthamoeba castellani* can synthesize diphosphoinositide [27] and contains an inositol lipid with neither glycerol nor a long-chain base [28].

Yeasts (*Saccharomyces cerevisiae* and *S. pombe*) contain considerable amounts of phosphatidylinositol [4] (around 20% lipid phosphorus) as well as sphingolipids containing inositol and mannose [15,16]. The same compounds seem to occur in *Neurospora crassa* [25].

(b) Fatty acid composition

In brain tissue as much as 80% of the phosphatidylinositol can be the 1-stearoyl, 2-arachidonoyl species (18:0, 20:4) but small quantities of many other molecular species are present [4]. Since diphosphoinositide and triphosphoinositide are formed by phosphorylation of phosphatidylinositol they have a similar fatty acid composition. The same 18:0, 20:4 phosphatidylinositol is the dominant species in human platelets [29] but is less abundant in liver [4].

Galliard [30] gives the fatty acid composition of phosphatidylinositol in various photosynthetic plants. The major saturated acid is usually palmitic and the major unsaturated species are either 18:2 or 18:3. Phosphatidylinositol from the fission yeast *S. pombe* contained 61 mol % oleic acid and 37.6 mol % palmitic acid [31], while a sample of the same lipid from baker's yeast [32] gave the following figures (mol %): palmitic acid 25.6, palmitoleic acid 30.9, stearic acid 8.6, oleic acid 31.5. Less information is available on the fatty acid composition of bacterial phosphoinositides. The phosphatidylinositol of *M. tuberculosis* contains 56 mol % palmitic acid and 44 mol % tuberculostearic (D(−)-10-methyl stearic) acid while the phosphatidylinositol mannosides of this organism have somewhat less tuberculostearic acid and 4–8% stearic acid [9]. These phosphoinositides contain only saturated fatty acids, thus differing from all the others described.

4. Biosynthesis

(a) Phosphatidylinositol

The work of Agranoff et al. [33] and Paulus and Kennedy [34] established the following biosynthetic route for phosphatidylinositol:

phosphatidic acid + CTP → CDP-diacylglycerol + pyrophosphate (1)

CDP-diacylglycerol + inositol → phosphatidylinositol + CMP (2)

The enzyme (CDP-diacylglycerol 3-phosphatidyltransferase, EC 2.7.8.11) responsible for reaction (2) has been solubilised and purified from rat brain [35] and liver [36] microsomal fractions. It is activated by either Mg^{2+} or Mn^{2+}. Radioactive inositol is also incorporated into phosphatidylinositol by an exchange reaction which does not require CDP-diacylglycerol [34,37]. The exchange is activated by Mn^{2+} but not Mg^{2+}. The purified enzyme of reaction (2) has no exchange activity, indicating that exchange is catalysed by another enzyme. At present the reaction is ill-defined, possibly being the reversal of a phospholipase D hydrolysis.

Reversal of reaction (2) has been demonstrated in mouse pancreas [38] and rabbit lung [39]. Bleasdale et al. [39] have suggested that when biosynthesis of lung surfactant phosphatidylcholine is active, the CMP produced may stimulate the reverse reaction, thus making CDP-diacylglycerol available for surfactant phosphatidylglycerol synthesis. The expected decrease in phosphatidylinositol concentration has been observed.

In tissues such as brain and liver, phosphatidylinositol has considerably more stearic and arachidonic acids than the phosphatidic acid from which it is synthesized (reactions 1 and 2). The enrichment is produced by deacylation and reacylation cycles. Baker and Thompson [40] showed that [^{3}H]arachidonic acid was incorporated in vivo into brain phosphatidylinositol by such a cycle. The same authors [41] described the acylation of 1-acyl-glycero-3-phosphoinositol by a brain microsomal fraction, arachidonoyl CoA being the most effective acylating agent. Holub et al. [42,43] have shown that the microsomal fraction of rat liver also has the necessary acyltransferases. With 1-acyl-glycero-3-phosphoinositol as acceptor, arachidonoyl CoA is the preferred substrate. With the 2-acyl compound, stearoyl CoA is a better donor than palmitoyl CoA.

Phosphatidylinositol is likely to be synthesised by the CDP-diacylglycerol route in plants and micro-organisms. This has been shown for cauliflower inflorescence [44] and yeasts [4].

(b) Phosphatidylinositol phosphates

Sequential phosphorylation of the inositol ring is responsible for the formation of diphosphoinositide and triphosphoinositide [19] (reactions 3 and 4). Both kinases (EC 2.7.1.67 and EC 2.7.1.68 respectively) require Mg^{2+} and have been described in a variety of tissues including brain, kidney, erythrocyte, adrenal medulla and yeast.

phosphatidylinositol + ATP → phosphatidylinositol 4-phosphate + ADP (3)

phosphatidylinositol 4-phosphate + ATP
→ phosphatidylinositol 4,5-bisphosphate + ADP (4)

(c) Phosphatidylinositol mannosides

Ballou and his colleagues have studied the biosynthesis of these complex phosphoinositides and their work has been reviewed by Ambron and Pieringer [10]. Mannose residues react as their GDP derivatives (e.g. reaction 5) and the additional acyl groups of structure (II) are formed from the expected CoA compounds.

$$\text{GDP mannose} + \text{phosphatidylinositol} \rightarrow \text{GDP} + \text{phosphatidylinositol mannoside} \quad (5)$$

An alternative route has been suggested, in which an inositol mannoside is first formed, then reacting with CDP-diacylglycerol to give the phosphatidylinositol mannoside.

(d) Sphingolipids containing inositol

Little is known about the biosynthesis of these compounds but Lester et al. [11] have discussed some aspects of their metabolism in yeast.

5. Catabolic pathways

(a) Hydrolysis of phosphatidylinositol

The major pathway of phosphatidylinositol hydrolysis in animal tissues follows the phospholipase C route (reaction 6).

$$\text{phosphatidylinositol} + H_2O \rightarrow \text{diacylglycerol} + \text{inositol phosphate} \quad (6)$$

This distinguishes phosphatidylinositol from the nitrogen-containing phospholipids where catabolism begins with removal of fatty acids by phospholipase A. Phosphatidylinositol hydrolysis requires calcium ions and the enzyme (phosphatidylinositol phosphodiesterase, EC 3.1.4.10) is cytosolic. The reaction may be important in relation to the increased labelling of phosphatidylinositol seen when various plasma membrane receptors are activated [45,46]. Dawson et al. [47] have shown that the initial water-soluble reaction product is D-inositol 1,2-cyclic phosphate. Lysosomes of rat liver and brain also contain a phospholipase C specific for phosphatidylinositol [48]. The lysosomal enzyme differs from the soluble one in being inhibited by calcium ions and in producing inositol 1-phosphate, not the cyclic ester.

Guinea-pig pancreas contains a phospholipase A_1 hydrolysing phosphatidylinositol and phosphatidylcholine [49] and studies of arachidonate incorporation imply that brain and other tissues have a phospholipase A_2 which attacks phosphatidylinositol [40,42]. More direct evidence for this phospholipase A_2 in brain has been obtained using phosphatidylinositol with labelled oleic acid in the 2-position [50]. The enzyme has been purified 1600 times from a brain microsomal fraction and also hydrolyses phosphatidylcholine, phosphatidylserine, phosphatidic acid and phosphatidylethanolamine, though it is most active with phosphatidylinositol [51]. Rat gastric mucosa also contains a non-specific phospholipase A_2 hydrolysing phosphatidylinositol [52].

A phospholipase C specific for phosphatidylinositol has been purified from *Bacillus cereus* [53] and *Staphylococcus aureus* [54].

(b) Hydrolysis of polyphosphoinositides

The polyphosphoinositides can be degraded both by the phospholipase C route (reactions 7 and 8) and by the phosphomonoesterase attack (reactions 9 and 10).

$$\text{triphosphoinositide} + H_2O \rightarrow \text{diacylglycerol} + \text{inositol triphosphate} \quad (7)$$

$$\text{diphosphoinositide} + H_2O \rightarrow \text{diacylglycerol} + \text{inositol bisphosphate} \quad (8)$$

$$\text{triphosphoinositide} + H_2O \rightarrow \text{diphosphoinositide} + P_i \quad (9)$$

$$\text{diphosphoinositide} + H_2O \rightarrow \text{phosphatidylinositol} + P_i \quad (10)$$

These enzyme activities have been studied in brain (for review, see ref. 19) and kidney [55,56]. The phosphomonoesterase has been purified 430-fold from rat brain [57]. The purified enzyme hydrolysed both diphosphoinositide and triphosphoinositide and the partially purified phospholipase C of guinea-pig intestinal mucosa [58] hydrolysed phosphatidylinositol and the polyphosphoinositides. It is not clear whether this also applies to the enzyme hydrolysing phosphatidylinositol in other mammalian tissues.

(c) Hydrolysis of other inositol lipids

Almost nothing is known of the catabolism of the more complex inositol lipids in micro-organisms. The mannosides of phosphatidylinositol in mycobacteria appear to be relatively inert, metabolically [10]. The cell surface of *S. cerevisiae* contains phospholipases which deacylate phosphatidylinositol to glycerophosphorylinositol [59]. Similar activity occurs in extracts of *Penicillium notatum* [60].

6. *Subcellular localisation of metabolic pathways*

Biosynthesis of phosphatidylinositol in mammalian tissues is probably located in the endoplasmic reticulum, along with phospholipid biosynthesis generally. In rat liver, phosphatidylinositol synthesis was most active in rough and smooth endoplasmic reticulum [61]. There was also significant activity in a Golgi membrane fraction, which could not be put down to contamination.

The kinase producing diphosphoinositide is associated with plasma membrane in brain and liver [19] and is also found in the membranes of secretory granules [20,21]. In rat kidney cortex [62] the same enzyme had a distribution resembling that for marker enzymes of brush-border, endoplasmic reticulum and Golgi membranes.

The biosynthesis of triphosphoinositide occurs on the inner face of the erythrocyte membrane [63] and in the Golgi membranes of kidney [62]. In brain, the kinase occurs in soluble form [19] and is also associated with myelin [64].

The phospholipase C hydrolysing phosphatidylinositol is a soluble enzyme in rat brain [65] and probably in other tissues, though phosphatidylinositol can also be degraded by lysosomal hydrolases. Both soluble and membrane-bound forms of the phospholipase C hydrolysing polyphosphoinositides are found in brain [4,19] and an association with plasma membrane has been suggested. This enzyme is also found in the erythrocyte membrane [66]. The phosphatase attacking these lipids has been attributed to plasma membrane fractions of brain [67], though other work suggests that it is soluble [19]. In kidney the phosphatase is partly soluble and partly bound to Golgi membranes [56]. In iris muscle both phosphatase and phospholipase C hydrolysing triphosphoinositide were most active in a microsomal fraction but there was also soluble activity [68].

7. *Phosphoinositide metabolism and receptor activation*

(a) Phosphatidylinositol

The most interesting feature of phosphatidylinositol metabolism is the special response in mammalian cells to a wide variety of stimuli, an effect first shown by Hokin and Hokin [69]. Receptors on the cell surface are usually involved and an increased turnover of the inositol phosphate head-group is seen. Examples include activation of muscarinic or α-adrenergic receptors, release of insulin from islets of Langerhans, activation of lymphocytes and platelets and secretion of enzymes from pancreas or parotid gland. Several reviews of the extensive literature are available [4,45,46,70] and so the present account will be selective rather than comprehensive.

Many, though not all, workers consider that the phosphatidylinositol effect begins with hydrolysis of the lipid [46] and that the following reaction cycle accounts for the increased labelling which is usually measured:

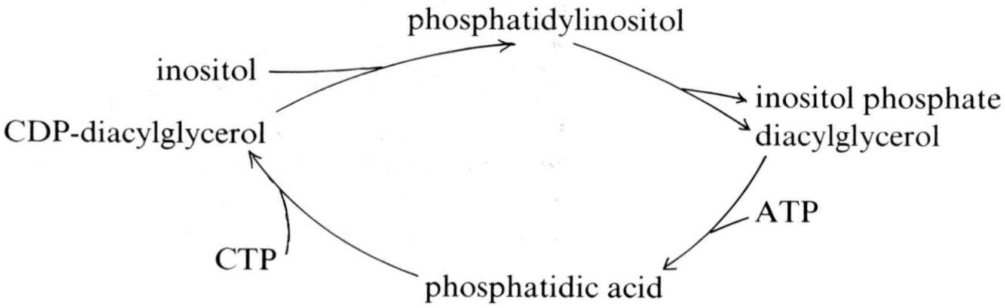

In many papers ^{32}P-labelling is used to show the effect in terms of increased specific radioactivity of phosphatidylinositol and phosphatidic acid. If the cycle above is operative any phosphatidylinositol hydrolysed as a result of receptor activation will soon be replaced. Nevertheless suitably vigorous stimulation produces a net loss of phosphatidylinositol [71,72]. Platelet activation by thrombin led to both loss of labelled phosphatidylinositol and accumulation of diacylglycerol [73]. The results were consistent with hydrolysis by the phospholipase C route. Fain and Berridge [74,75] showed that calcium transport stimulated by 5-hydroxytryptamine in blowfly salivary gland was accompanied by phosphatidylinositol breakdown. Their results supported Michell's theory that the phospholipid effect is associated with calcium gating.

(b) The calcium-gating hypothesis

Many of the hormones causing increased turnover of phosphatidylinositol also increase the entry of external calcium into the tissue. For this and other reasons, Michell [45] has suggested that the lipid changes control the permeability of the plasma membrane to calcium ions. The suggestion is that by a mechanism as yet unknown conversion of phosphatidylinositol to diacylglycerol in this membrane allows "calcium gates" to open.

One difficulty about the theory is that the phospholipase C which seems to be linked to receptor activation is a soluble enzyme, not a plasma membrane constituent as would be expected. Nor is phosphatidylinositol itself localised in this membrane. There is little evidence as yet about the precise location of the phosphatidylinositol which responds to activation. In brain, after labelling in vivo, the labelled phosphatidylinositol sensitive to electrical stimulation of synaptosomes was located in the transmitter vesicle membranes rather than the plasma membrane [76]. Stimulation of isolated rat hepatocytes by vasopressin caused loss of labelled phosphatidylinositol from all the membrane fractions which could be obtained [77].

If the phosphatidylinositol effect controls calcium gating it should be independent of external calcium ion concentrations. This is true for some tissues and was one of Michell's arguments in support of the gating theory [45]. In other tissues, however, and the number is increasing, the lipid effect is dependent on external calcium. The phosphatidylinositol response to muscarinic agonists in brain synaptosomes was abolished by removing calcium from the medium, for instance [78]. Stimulation of rabbit neutrophils by *N*-formyl-methionyl-leucyl-phenylalanine promotes secretion of β-glucuronidase through mobilisation of internal calcium ions. The process is accompanied by increased labelling of phosphatidylinositol but this response is abolished if calcium is omitted from the medium. It seems then that the receptor mobilises calcium ions from internal sources and initiates secretion without any involvement of phosphatidylinositol [79]. The early work suggested that the phosphatidylinositol effect in pancreas was independent of external calcium, but a recent study [80] showed that amylase secretion and loss of phosphatidylinositol caused by carbachol were both dependent upon calcium. Ionophore effects indicated that the phospholipid changes followed, rather than preceded, tissue changes in calcium ion concentration. Phosphatidylinositol labelling in response to cholinergic stimulation of adrenal medulla does not require external calcium. In the bovine gland the labelling follows activation of muscarinic, not nicotinic receptors [81] but only the latter enhance catecholamine secretion [82]. Secretion requires calcium influx, but this follows from nicotinic receptor activation. Activation of muscarinic receptors causes a phosphatidylinositol effect but no calcium entry [82]. It seems that pre-synaptic muscarinic receptors modulate the catecholamine secretion due to nicotinic activity.

Thus calcium gating does not provide a general explanation of the phosphatidylinositol effect. It remains to be seen whether the theory holds good in specific tissues, but there are reasons for seeking other explanations. One suggestion is that conversion of phosphatidylinositol to diacylglycerol increases membrane fluidity. Such a change might facilitate fusion between vesicle and plasma membranes in transmitter release [76]. Tubulin may be involved in many of the processes which have been discussed and *myo*-inositol can reverse the anti-mitotic effect of colchicine, which binds to tubulin. It is possible therefore that phosphatidylinositol plays a part in microtubule-plasma membrane linkage [83].

Phosphatidylinositol of mammalian tissues is rich in arachidonic acid and several authors have suggested that receptor activation might make this available as a source of prostaglandins. Hydrolysis by phospholipase A_2 would be the most convenient mechanism but other phospholipids such as phosphatidylethanolamine also contain arachidonic acid and are better substrates for the enzyme than phosphatidylinositol. Nevertheless, Marshall et al. [84] provide evidence that the inositol lipid is the source of arachidonic acid for PGE_2 biosynthesis in mouse pancreas and that 3–10 nM concentrations of prostaglandins provoke amylase secretion. Arachidonic acid is most rapidly released from phosphatidylinositol when platelets are activated by thrombin. There is evidence for the intermediate formation of diacylglycerol [73]. Thrombin causes serotonin secretion from the platelets but this and diacylglycerol production are not prevented by acetylsalicylic acid which inhibits prostaglandin synthesis. The platelet differs from the pancreas, therefore, in that secretion is not mediated by prostaglandins, though these are important in the subsequent platelet aggregation. The events are too complex for the simple conclusion that the phosphatidylinositol effect in platelets reflects prostaglandin biosynthesis.

Phosphatidylinositol hydrolysis could also affect protein phosphorylation. Kishimoto et al. [85] have shown that the diacylglycerol released by phospholipase C action activates a protein kinase by increasing its sensitivity to calcium ions. The kinase, which they call protein kinase C, occurs in several tissues, but is particularly active in brain.

(c) The role of polyphosphoinositides

A theory connecting calcium binding, polyphosphoinositide hydrolysis and membrane permeability was put forward some years ago [19] and in many ways these lipids are more attractive than phosphatidylinositol as candidates for such a role. Interconversion of triphosphoinositide, diphosphoinositide and phosphatidylinositol requires a simple phosphatase reaction and resynthesis of the polyphosphoinositides is much less complex than that of phosphatidylinositol from diacylglycerol. There is also evidence that the polyphosphoinositides occur in plasma membranes and have considerable affinity for calcium ions.

Effects of hormones on the polyphosphoinositides may have been missed because the lipids are so rapidly hydrolysed by either the phosphatase or phospholipase C route in mammalian tissues. In an attempt to avoid *post-mortem* breakdown Soukup et al. [18] killed rats by microwave irradiation. After intracisternal injection of $^{32}P_i$ or [^{3}H]inositol, carbamylcholine increased the labelling of both diphosphoinositide and triphosphoinositide in vivo over a 5-min period [86]. The effect was blocked by atropine, suggesting that muscarinic receptors were involved. Abdel-Latif et al. [87], on the other hand, showed triphosphoinositide breakdown in response to muscarinic stimulation of iris muscle. The breakdown was considered to be due to increased calcium ion influx since the enzymes of iris muscle hydrolysing triphosphoinositide are activated by this ion [88]. Entry of calcium ions into synaptosomes caused a similar loss of triphosphoinositide [89].

(d) Adrenocorticotrophic hormone (ACTH) and triphosphoinositide

Several recent papers suggest a relationship between ACTH and phosphoinositide metabolism. Injection of $ACTH_{1-24}$ into rats produced a several fold increase in polyphosphoinositide concentration of adrenal glands [90]. Polyphosphoinositides, but not phosphatidylinositol, increased pregnenolone synthesis by adrenal mitochondria. It was suggested that these compounds may mediate the ACTH-induced increase in steroid hormone synthesis, for which the cholesterol-pregnenolone conversion is rate-limiting.

ACTH inhibits the phosphorylation in vitro of a protein (B-50) from the synaptic plasma membrane [91]. Using a crude mitochondrial fraction from rat brain, ACTH was also shown [92] to decrease the labelling of diphosphoinositide and triphosphoinositide by inorganic ^{32}P. It seems possible that the kinases responsible for polyphosphoinositide synthesis might be regulated by a protein-phosphorylation system sensitive to ACTH. A protein kinase/B-50 complex from synaptosomal plasma membranes has diphosphoinositide kinase activity [93]. This kinase activity decreased with increasing phosphorylation of B-50 protein. ACTH inhibited protein phosphorylation and increased diphosphoinositide kinase activity, implying that the B-50 protein regulates the kinase or is itself the kinase. The discrepancy between these results in which ACTH increased triphosphoinositide formation and the earlier results showing the opposite effect [92] may be due to differences in calcium ion concentration.

8. Inositol lipids and diabetic neuropathy

Nerve damage is a common complication of long-standing diabetes mellitus and decreased conduction velocity in motor and sensory nerves can be detected in newly diagnosed cases. The concentration of free *myo*-inositol is roughly 30 times higher in peripheral nerve than in plasma [94]. The nerve inositol concentration is appreciably reduced in experimental diabetes [95,96] and there is also evidence that the transferase synthesizing phosphatidylinositol from CDP-diacylglycerol is less active in rats made diabetic with streptozotocin [97,98]. The decreased transferase activity is probably not due to impaired axonal transport of the enzyme [98]. Diphosphoinositide kinase was also less active in sciatic nerve of the diabetic animals. These observations and the relation between phosphoinositide metabolism and nerve impulse transmission [19] suggest that disordered inositol lipid metabolism contributes to diabetic neuropathy, at least in experimental diabetes. At present there is little information about changes in inositol concentration or phosphoinositide metabolism in relation to diabetic neuropathy in man. The subject has been reviewed by Clements [94].

9. Conclusions

Inositol has long been classed as a vitamin but its function is still not understood in biochemical terms. There is no doubt that mammalian cells which cannot synthesize it from glucose require inositol for growth and division. On the other hand, many bacterial cells lack inositol altogether.

The function of *myo*-inositol may well reside in its phospholipid derivatives. Michell has pointed out that mammalian cell surface receptors which use calcium ions as second messenger also promote the hydrolysis of phosphoinositides. A better understanding of these processes could lead to the elucidation of inositol's function as a vitamin. Whether the importance of inositol in cell division is also related to calcium fluxes remains to be seen.

The more complex lipid derivatives of inositol in mycobacteria and higher plants are even more enigmatic. We are not likely to find out much about their biological role until more is known of their metabolism.

References

1 Anderson, R.J. and Roberts, E.G. (1930) J. Biol. Chem. 89, 599–610.
2 Klenk, E. and Sakai, R. (1939) Hoppe-Seyler's Z. Physiol. Chem. 258, 33–38.
3 Folch, J. and Woolley, D.W. (1942) J. Biol. Chem. 142, 963–964.
4 Hawthorne, J.N. and White, D.A. (1975) Vitam. Horm. (New York) 33, 529–573.
5 IUPAC-IUB (1978) Biochem. J. 171, 1–19.
6 Vilkas, E. and Lederer, E. (1960) Bull. Soc. Chim. Biol. 42, 1013–1022.
7 Nojima, S. (1959) J. Biochem. (Tokyo) 46, 499–506.

8 Ballou, C.E. and Lee, Y.C. (1964) Biochemistry 3, 682–685.
9 Pangborn, M.C. and McKinney, J.A. (1966) J. Lipid Res. 7, 627–633.
10 Ambron, R.T. and Pieringer, R.A. (1973) in G.B. Ansell, R.M.C. Dawson and J.N. Hawthorne (Eds.), Form and Function of Phospholipids, Elsevier, Amsterdam, pp. 289–331.
11 Lester, R.L., Becker, G.W. and Kaul, K. (1978) in W.W. Wells and F. Eisenberg Jr. (Eds.), Cyclitols and Phosphoinositides, Academic Press, New York, pp. 83–102.
12 Carter, H.E., Strobach, D.R. and Hawthorne, J.N. (1969) Biochemistry 8, 383–388.
13 Kaul, K. and Lester, R.L. (1978) Biochemistry 17, 3569–3575.
14 Hsieh, Y.C.-Y., Kaul, K., Laine, R.A. and Lester, R.L. (1978) Biochemistry 17, 3575–3581.
15 Smith, S.W. and Lester, R.L. (1974) J. Biol. Chem. 249, 3395–3405.
16 Steiner, S., Smith, S., Waechter, C.J. and Lester, R.L. (1969) Proc. Natl. Acad. Sci. U.S.A. 64, 1042–1048.
17 White, D.A. (1973) in G.B. Ansell, R.M.C. Dawson and J.N. Hawthorne (Eds.), Form and Function of Phospholipids, Elsevier, Amsterdam, pp. 441–482.
18 Soukup, J.F., Friedel, R.O. and Shanberg, S.M. (1978) J. Neurochem. 30, 635–637.
19 Hawthorne, J.N. and Kai, M. (1970) in A. Lajtha (Ed.), Handbook of Neurochemistry, Vol. 3, Plenum Press, New York, pp. 491–508.
20 Buckley, J.T., Lefebvre, Y.A. and Hawthorne, J.N. (1971) Biochim. Biophys. Acta 239, 517–519.
21 Oron, Y., Sharoni, Y., Lefkovitz, H. and Selinger, Z. (1978) in W.W. Wells and F. Eisenberg Jr. (Eds.), Cyclitols and Phosphoinositides, Academic Press, New York, pp. 383–397.
22 Kuiper, P.J.C. (1970) Plant Physiol. 45, 684–686.
23 Kates, M. (1966) Ann. Rev. Microbiol. 20, 13–44.
24 Palmer, F.B.St.C. (1973) Biochim. Biophys. Acta 316, 296–304.
25 Lester, R.L., Smith, S.W., Wells, G.B., Rees, D.C. and Angus, W.A. (1974) J. Biol. Chem. 249, 3383–3387.
26 Van Vliet, H.H.D.M., Op den Kamp, J.A.F. and Van Deenen, L.L.M. (1975) Arch. Biochem. Biophys. 171, 55–64.
27 Buckley, J.T. (1976) Can. J. Biochem. 54, 772–777.
28 Ulsamer, A.G., Smith, F.R. and Korn, E.D. (1969) J. Cell Biol. 43, 105–114.
29 Marcus, A.J., Ullman, H.L. and Safier, L.B. (1969) J. Lipid Res. 10, 108–114.
30 Galliard, T. (1973) in G.B. Ansell, R.M.C. Dawson and J.N. Hawthorne (Eds.), Form and Function of Phospholipids, Elsevier, Amsterdam, pp. 253–288.
31 White, G.L. and Hawthorne, J.N. (1970) Biochem. J. 117, 203–213.
32 Trevelyan, W.E. (1966) J. Lipid Res. 7, 445–447.
33 Agranoff, B.W., Bradley, R.M. and Brady, R.V. (1958) J. Biol. Chem. 233, 1077–1083.
34 Paulus, H. and Kennedy, E.P. (1960) J. Biol. Chem. 235, 1303–1311.
35 Rao, R.H. and Strickland, K.P. (1974) Biochim. Biophys. Acta 348, 306–314.
36 Takenawa, T. and Egawa, K. (1977) J. Biol. Chem. 252, 5419–5423.
37 Takenawa, T., Saito, M., Nagai, Y. and Egawa, K. (1977) Arch. Biochem. Biophys. 182, 244–250.
38 Hokin-Neaverson, M., Sadeghian, K., Harris, D.W. and Merrin, J.S. (1977) Biochem. Biophys. Res. Commun. 78, 364–371.
39 Bleasdale, J.E., Wallis, P., MacDonald, P.C. and Johnston, J.M. (1979) Biochim. Biophys. Acta 575, 135–147.
40 Baker, R.R. and Thompson, W. (1972) Biochim. Biophys. Acta 270, 489–503.
41 Baker, R.R. and Thompson, W. (1973) J. Biol. Chem. 248, 7060–7065.
42 Holub, B.J. (1976) Lipids 11, 1–5.
43 Holub, B.J. and Piekarski, J. (1979) Lipids 14, 529–532.
44 Sumida, S. and Mudd, J.B. (1970) Plant Physiol. 45, 712–718.
45 Michell, R.H. (1975) Biochim. Biophys. Acta 415, 81–147.
46 Hawthorne, J.N. and Pickard, M.R. (1979) J. Neurochem. 32, 5–14.
47 Dawson, R.M.C., Freinkel, N., Jungalwala, F.B. and Clarke, N. (1971) Biochem. J. 122, 605–607.
48 Irvine, R.F., Hemington, N. and Dawson, R.M.C. (1978) Biochem. J. 176, 475–484.
49 White, D.A., Pounder, D.J. and Hawthorne, J.N. (1971) Biochim. Biophys. Acta 242, 99–107.

50 Shum, T.Y.P., Gray, N.C.C. and Strickland, K.P. (1979) Can. J. Biochem. 57, 1359–1367.
51 Gray, N.C.C. and Strickland, K.P. (1982), in press.
52 Wassef, M.K. and Horowitz, M.I. (1981) Biochim. Biophys. Acta 665, 234–243.
53 Ikezawa, H., Yamanegi, M., Taguchi, R., Miyashita, T. and Ohyabu, T. (1976) Biochim. Biophys. Acta 450, 154–164.
54 Low, M.G. and Finean, J.B. (1977) Biochem. J. 162, 235–240.
55 Lee, T.-C. and Huggins, C.G. (1968) Arch. Biochem. Biophys. 126, 206–213.
56 Cooper, P.H. and Hawthorne, J.N. (1975) Biochem. J. 150, 537–551.
57 Nijjar, M.S. and Hawthorne, J.N. (1977) Biochim. Biophys. Acta 480, 390–402.
58 Atherton, R.S. and Hawthorne, J.N. (1968) Eur. J. Biochem. 4, 68–75.
59 Angus, W.W. and Lester, R.L. (1975) J. Biol. Chem. 250, 22–30.
60 Dawson, R.M.C. (1959) Biochim. Biophys. Acta 33, 68–77.
61 Williamson, F.A. and Morre, D.J. (1976) Biochem. Biophys. Res. Commun. 68, 1201–1205.
62 Cooper, P.H. and Hawthorne, J.N. (1976) Biochem. J. 160, 97–105.
63 Garrett, R.J.B. and Redman, C.M. (1975) Biochim. Biophys. Acta 382, 58–64.
64 Deshmukh, D.S., Bear, W.D. and Brockerhoff, H. (1978) J. Neurochem. 30, 1191–1193.
65 Irvine, R.F. and Dawson, R.M.C. (1978) J. Neurochem. 31, 1427–1434.
66 Allan, D. and Michell, R.H. (1978) Biochim. Biophys. Acta 508, 277–286.
67 Sheltawy, A., Brammer, M. and Borrill, D. (1972) Biochem. J. 128, 579–586.
68 Akhtar, R.A. and Abdel-Latif, A.A. (1978) Biochim. Biophys. Acta 527, 159–170.
69 Hokin, M.R. and Hokin, L.E. (1953) J. Biol. Chem. 203, 967–977.
70 Michell, R.H. (1979) in A.T. Bull, J.R. Lagnado, J.O. Thomas and K.F. Tipton (Eds.), Companion to Biochemistry, Vol. 2, Longmans, London, pp. 205–228.
71 Hokin-Neaverson, M. (1974) Biochem. Biophys. Res. Commun. 58, 763–768.
72 Jones, L.M. and Michell, R.H. (1974) Biochem. J. 142, 583–590.
73 Rittenhouse-Simmons, S. (1979) J. Clin. Invest. 63, 580–587.
74 Fain, J.N. and Berridge, M.J. (1979) Biochem. J. 178, 45–58.
75 Fain, J.N. and Berridge, M.J. (1979) Biochem. J. 180, 655–661.
76 Pickard, M.R. and Hawthorne, J.N. (1978) J. Neurochem. 30, 145–155.
77 Kirk, C.J., Michell, R.H. and Hems, D.A. (1981) Biochem. J. 194, 155–165.
78 Griffin, H.D., Hawthorne, J.N., Sykes, M. and Orlacchio, A. (1979) Biochem. Pharmacol. 28, 1143–1147.
79 Cockroft, S., Bennett, J.P. and Gomperts, B.D. (1980) FEBS Lett. 110, 115–118.
80 Farese, R.V., Larson, R.E. and Sabir, M.A. (1980) Biochim. Biophys. Acta 633, 479–484.
81 Mohd. Adnan, N.A. and Hawthorne, J.N. (1981) J. Neurochem. 36, 1858–1860.
82 Fisher, S.K., Holz, R.W. and Agranoff, B.W. (1981) J. Neurochem. 37, 491–497.
83 Lymberopoulos, G. and Hawthorne, J.N. (1980) Exp. Cell Res. 129, 409–414.
84 Marshall, P.J., Dixon, J.F. and Hokin, L.E. (1980) Proc. Natl. Acad. Sci. U.S.A. 77, 3292–3296.
85 Kishimoto, A., Takai, Y., Mori, T., Kikkawa, U. and Nishizuka, Y. (1980) J. Biol. Chem. 255, 2273–2276.
86 Soukup, J.F., Friedel, R.O. and Schanberg, S.M. (1978) Biochem. Pharmacol. 27, 1239–1243.
87 Abdel-Latif, A.A., Akhtar, R.A. and Hawthorne, J.N. (1977) Biochem. J. 162, 61–73.
88 Akhtar, R.A. and Abdel-Latif, A.A. (1978) Biochim. Biophys. Acta 527, 159–170.
89 Griffin, H.D. and Hawthorne, J.N. (1978) Biochem. J. 176, 541–552.
90 Farese, R.V., Sabir, A.M. and Vandar, S.L. (1979) J. Biol. Chem. 254, 6842–6844.
91 Zwiers, H., Schotman, P. and Gispen, W.H. (1980) J. Neurochem. 34, 1689–1699.
92 Jolles, J., Wirtz, K.W.A., Schotman, P. and Gispen, W.H. (1979) FEBS Lett. 105, 110–114.
93 Jolles, J., Zwiers, H., Van Dongen, C.J., Schotman, P., Wirtz, K.W.A. and Gispen, W.H. (1980) Nature (Lond.) 286, 623–625.
94 Clements Jr., R.S. (1979) Diabetes 28, 604–611.
95 Green, D.G., De Jesus, P.V. and Winegrad, A.I. (1975) J. Clin. Invest. 55, 1326–1336.
96 Palmano, K.P., Whiting, P.H. and Hawthorne, J.N. (1977) Biochem. J. 167, 229–235.
97 Whiting, P.H., Palmano, K.P. and Hawthorne, J.N. (1979) Biochem. J. 179, 549–553.
98 Clements Jr., R.S. and Stockard, C.R. (1980) Diabetes 29, 227–235.

CHAPTER 8

Phosphlipid transfer proteins

JEAN-CLAUDE KADER, DOMINIQUE DOUADY
and PAUL MAZLIAK

*Laboratoire de Physiologie Cellulaire (ERA 323),
Université Pierre et Marie Curie, 4 place Jussieu 75005 Paris, France*

Membrane phospholipids undergo renewal, catabolism, biosynthesis and base exchange as described in other chapters of this book. An additional process is intermembrane exchange, catalysed by a particular category of proteins, named phospholipid transfer proteins. This chapter deals with these proteins previously described in several reviews under the name of phospholipid exchange proteins [1–4].

1. Discovery

In 1968, Wirtz and Zilversmit [5] discovered that an in vitro exchange of phospholipids occurred between microsomes and mitochondria of rat liver. These experiments were based on incubation of unlabelled mitochondria with microsomes containing [^{32}P]phospholipid, followed by re-separation of the organelles and determination of the specific radioactivity of the phospholipids. They found that the specific radioactivity of microsomal phospholipids decreased with time whereas that of mitochondrial phospholipids increased. These data were consistent with a bidirectional exchange of phospholipids between microsomes and mitochondria. In the same period, other authors obtained similar results with rat liver organelles [6–8]. The general occurrence of this exchange process to include plant cell organelles was demonstrated in 1970 [9].

In their pioneering experiments, Wirtz and Zilversmit [5] also observed that the addition of a post-microsomal supernatant enhanced the exchange of phospholipids between rat liver organelles. A similar finding was made by other workers with rat liver [7,8] and plant cells [9]. Since this active factor was heat- and protease-sensitive and was retained on dialysis, it was concluded to be a protein [10]. An additional step in the preparation consisting of adjusting the pH of a post-mitochondrial supernatant to pH 5.1 was introduced in order to eliminate the residual membranes [10] by moderate centrifugation. Whereas 95% of the phospholipids and 40% of the proteins were removed, almost all the exchange activity remained in the supernatant.

It was tempting to isolate the active protein. Wirtz and Zilversmit [11] succeeded in purifying the first phospholipid transfer protein starting from beef-heart cytosol.

Hawthorne/Ansell (eds.) Phospholipids

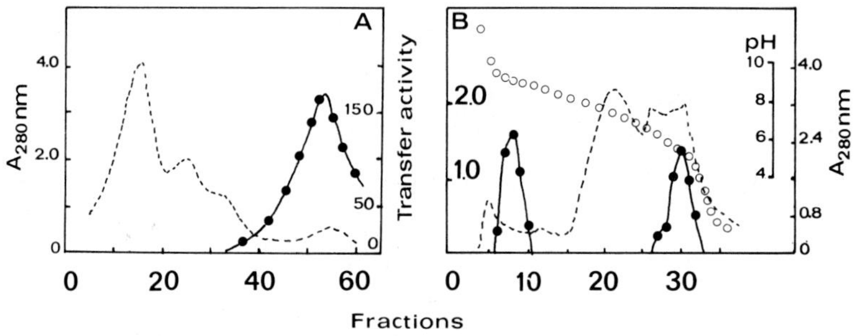

Fig. 1. Isolation of phospholipid transfer proteins. (A) The first isolation of a phospholipid transfer protein from an animal cytosol. Soluble proteins from beef heart were chromatographed on a Sephadex G100 column. The discontinuous line indicates absorbance at 280 nm. Closed circles indicate the transfer activity expressed as the increase in the specific radioactivity of microsomal phospholipids (microsome-mitochondria assay, performed in the presence of 1 mg of protein of each fraction) (in dpm/mg phospholipid). Reproduced from [11] by permission of the authors and the Federation of European Biochemical Societies. (B) Isolation of basic and acidic phospholipid transfer proteins in rat-liver cytosol by isoelectric focussing. The transfer activity (closed circles), determined in liposome-mitochondria assay is expressed in nmol of PC transferred per min. Reproduced from [27] by permission of the authors and the American Oil Chemists Society.

After pH 5.1 treatment, they used ammonium sulphate precipitation, hydroxylapatite adsorption-desorption and Sephadex G100 chromatography (Fig. 1A). The activity of the transfer proteins was determined by following the exchange of [^{32}P]phospholipids between microsomes and mitochondria. This isolation opened a new field of investigation into this original category of proteins.

2. *Methods for the determination of transfer activities*

Although the first assays were done on intracellular membranes, the use of artificial emulsions of lipids rapidly proved to be of great interest.

(a) Transfer between natural membranes

The classical transfer system comprised microsomes and mitochondria (the labelled fraction being either the former or the latter) which were incubated for 5 to 60 min at 37°C (pH 7 to 8) in the presence of transfer protein. The fractions were then separated by centrifugation and their lipids were extracted. The percent of label recovered in the initially non-radioactive membrane indicated the extent of the transfer. The protein-mediated transfer was thus calculated by subtracting from this value, the percent of transfer obtained without any addition of cytosolic protein. This indicated the amount of phospholipid transferred by the protein and allowed the definition of units of activity. To determine the extent of cross contamination between fractions, microsomes labelled from [^{3}H]leucine were incubated with mitochondria in the presence or absence of transfer protein. Since [^{3}H]leucine-labelled compounds were not exchanged between these organelles, the percent of

TABLE 1

Distribution of phospholipid transfer proteins in living cells

Origin	Assay	Refs.
Rat liver	[a]	5–8, 17, 18
	[b]	19, 20, 21, 22
	[c]	21, 23, 24
	microsomes—calcium loaded microsomes	12
	rough or smooth microsomes—mitochondria	13
	microsomes—inner mitochondrial membranes	14, 16
	outer—inner mitochondrial membranes	15, 16
	spin-labelled liposomes—mitochondria	25
	liposomes—erythrocyte ghosts	21
Rat small intestinal smooth muscle	[a]	26
Rat intestine	[b]	27
Rat brain	[a]	28
	[c]	29
	liposomes—myelin	30
Rat hepatoma	[a]	31
Beef heart	[a]	11
	[b] and [d]	32
	[c] and [e]	33
Beef liver	[a,c] and [d]	34
	liposomes—fibroblasts	35
	liposomes—monolayer	34
	[e]	36
Beef retina	liposomes—retinal rod outer segments	37
Beef brain	[b]	38, 39
Guinea pig brain	[a]	40
Calf liver	[c]	41,42
Sheep lung	[a]	43
	microsomes—lamellar bodies	43
Squirrel monkey	plasma membrane or high density lipoproteins	44, 45
	high or low density lipoproteins	
Potato tuber	[a]	46
Castor bean	[a]	47
	[b]	48–50
	[c]	51
	[d]	52, 53
Maize, cauliflower	[b]	54, 55
Jerusalem artichoke		
Spinach and pea leaves	[b]	56
	liposomes—chloroplasts	56
Bacillus subtilis	protoplasts—mesosomes	57, 58
Saccharomyces cerevisiae	[a]	59
Rhodopseudomonas sphaeroides	liposomes—intracytoplasmic membranes	60

[a] Microsomes—mitochondria.
[b] Liposomes—mitochondria.
[c] Liposomes—microsomes.
[d] Liposomes—liposomes.
[e] Liposomes—multilamellar vesicles.

^{3}H-label recovered in the mitochondria was considered to have arisen by the co-sedimentation of labelled microsomes with mitochondria.

Other exchange systems involving natural membranes have been studied: endoplasmic reticulum vesicles and mitochondria [12,13], microsomes and inner or outer mitochondrial membranes [14–16], calcium-loaded microsomes and microsomes or plasmalemma [12] etc. (Table 1).

(b) Transfer between artificial and natural membranes

It rapidly became necessary to use membranes of controlled lipid composition. Liposomes, obtained by ultrasonic irradiation of a mixture of [^{32}P]phosphatidylcholine and [^{14}C]triolein, were incubated with rat liver mitochondria and a pH 5.1 post-mitochondrial supernatant [19]. The mitochondrial fraction was collected by centrifugation, when it was found that an active transfer of phosphatidylcholine (PC) occurred. The incorporation of [^{14}C]triolein, a non-exchangeable marker, into the liposomes, provided a simple means to check that the co-sedimentation of liposomes with mitochondria was not appreciable. Following these successful experiments, other models were found, particularly the liposome-microsome exchange assay [34]. Liposomes, made from PC and [7α-^{3}H]cholesteryl-oleate as a non-exchangeable marker, were incubated with microsomes containing [^{14}C]phosphatidylcholine in the presence or absence of beef liver protein. Microsomes were collected by adjusting the pH to 5.1. The increase in the $^{14}C/^{3}H$ ratio of the liposomal PC indicated the extent of the transfer. Similar experiments have been done with plant cytosol proteins (Fig. 2).

Other membranes were assayed with liposomes: inner mitochondrial membranes, erythrocyte ghosts, fibroblasts, etc. In some experiments [61,62], spin-labelled phospholipids were incorporated into liposomes instead of radiolabelled ones.

(c) Transfer between liposomes

To eliminate the influence of any other membrane components, exchange experiments were done between different kinds of liposomes made from pure phospholipids. Several procedures were used to separate the liposomes after incubation. Ehnohlm and Zilversmit [32] incorporated Forssman antigen into liposomes, incubated these "sensitized" liposomes with normal ones and then collected the antigen-containing liposomes by immuno-precipitation with anti-Forssman antibody. Hellings et al. [63] separated donor liposomes containing sufficient amounts of acidic phospholipids (9 mol%) by binding to a DEAE-cellulose column. The acceptor liposomes, poor in acidic phospholipid, were not retained by the column. Sasaki and Sakagami [64] introduced a glycolipid in a population of liposomes and, after incubation with standard liposomes, collected the sensitized liposomes after agglutination by concanavalin A. Agglutination of liposomes by lectins was also used by other authors [53,65]. De Cuyper et al. [66] separated vesicles by free-flow electrophoresis.

The liposomes used in the exchange assays are usually unilamellar vesicles obtained after sonication. However, large multilamellar vesicles can be prepared by the dispersion of lipids in buffer by hand-shaking [67–69]. The major advantage of

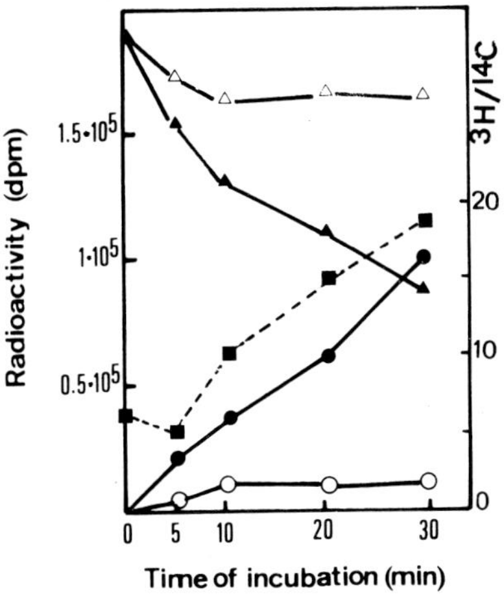

Fig. 2. Influence of incubation time on the transfer of phosphatidylcholine from liposomes to mitochondria in the presence of castor-bean cytosol proteins. Liposomes made from [^{3}H]phosphatidylcholine and cholesteryl [1-^{14}C]oleate were incubated at 30°C with mitochondria in the presence or absence of castor-bean cytosol proteins (11 mg). In the figure are shown (in dpm) the ^{3}H radioactivities recovered in the mitochondrial phosphatidylcholine in the presence (●) and in the absence (○) of cytosol proteins and that remaining in the supernatant in the presence (▲) or in the absence (△) of cytosol proteins. The ^{3}H/^{14}C ratio measured in mitochondria is indicated in the figure (■). (Douady, unpublished.)

these large vesicles is their homogeneous size, contrasting with the heterogeneity of the liposome population. Furthermore, these vesicles interact very little with membranes, whereas fusion and sticking of small-sized liposomes to membranes seem to occur in unilamellar preparations [67–69].

NMR and ESR spectroscopy were used to follow the movement of phospholipids without re-separation of the donor and acceptor vesicles. Barsukov et al. [70] employed NMR techniques using paramagnetic probe ions to study the movement of phospholipids between liposomes. Devaux et al. [61] and Machida and Ohnishi [62] used ESR spectroscopy to follow continuously the transfer of 2-acyl spin-labelled PC from vesicles containing this phospholipid to unlabelled ones.

3. Distribution in living cells

The transfer proteins are universally distributed among eukaryotic cells (Table 1). To the best of our knowledge the only indications we have about their existence in prokaryotic cells are from work on *Bacillus subtilis* [57,58] and on a facultatively photosynthetic bacterium, *Rhodopseudomonas sphaeroides* [60].

(a) Animal cells

Four tissues have been intensively studied: rat liver, beef liver, brain and heart. Few proteins have been purified to homogeneity and characterized.

(i) Beef tissues
One of the best purified phospholipid transfer proteins (PL-TP) was isolated from beef liver [34]. A 2680-fold purification was achieved by different steps: DEAE cellulose and carboxymethyl cellulose chromatography, and gel filtration on Sephadex G50. The purified protein was highly specific for phosphatidylcholine (PC). Only one band was found after SDS-polyacrylamide gel electrophoresis, immunoelectrophoresis or isoelectric focussing. The activity was stable for months when the protein was stored at $-20°C$ in 50% glycerol. The stability of this protein (PC-TP) facilitated the use of the material for several experiments. Crain and Zilversmit [36] have also recently isolated, by carboxymethyl cellulose and octylagarose column chromatography, highly purified proteins from beef liver which have the remarkable property of accelerating the transfer of almost all phospholipids. These non-specific phospholipid transfer proteins (nsPL-TP) are of great potential interest for studies on lipid asymmetry in membranes.

After a first purification [11], Ehnohlm and Zilversmit [32] isolated from beef heart two highly purified proteins, using Sephadex G75 filtration and isoelectric focussing. These proteins differed in their isoelectric points (pIs) (5.5 and 4.7, respectively) and were able to transfer mainly PC between liposomes. Additional purification was achieved by Johnson and Zilversmit [71] using gel filtration and carboxymethyl cellulose chromatography. They obtained a 210-fold purified fraction, which stimulated PC exchange between liposomes and mitochondria. From the same tissue, Dicorleto et al. [33] using a combination of phenyl Sepharose, Sephadex and carboxymethyl cellulose column chromatography, succeeded in purifying two proteins. Both had similar M_r-values although they differ in their isoelectric points. These proteins mediated the transfer of phosphatidylinositol (PI) and, to a lesser extent, of PC, between multilamellar and unilamellar vesicles. These proteins will be designated as PI-TP.

Since beef brain cytosol highly stimulated the exchange of PI [38], it was plausible to suggest the presence of one or several PI-TPs. Two such proteins (I and II) differing in their isoelectric points have been effectively isolated by Helmkamp et al. [38] after DEAE-cellulose and Sephadex chromatography and isoelectric focussing. They showed a marked preference for transfer of PI between microsomes and liposomes and exhibited striking similarities in M_r-values, phospholipid specificity, amino acid composition and immunochemical properties. Beef brain cytosol was also able to transfer phospholipids to myelin [39].

A phosphatidylcholine transfer protein (PC-TP) was isolated from the cytosol of bovine retina which was more active with retinal-rod outer segments than with mitochondria [37].

(ii) Rat tissues

Though first detected in rat liver cytosol, phospholipid transfer proteins have been isolated from this tissue only 8 years after their detection. Independent experiments have led to the discovery of acidic and basic proteins in rat liver cytosol [22,23] (Fig. 1B). The basic protein, when highly purified (140-fold [22] and 7000-fold [23]) had a pI of 8.4. It catalyzed specifically the transfer of PC from liposomes to mitochondria [22] or from microsomes to liposomes [23]. This basic PC-TP is responsible for 50% of the PC-transfer activity in rat liver. It has been recently purified following a new procedure and its biochemical and immunological properties compared to that of acidic beef liver PC-TP [24]. Two other basic proteins of low M_r-value, CM1 and CM2, are able to transfer phosphatidylethanolamine (PE) from liposomes to mitochondria or erythrocyte ghosts. An 876-fold increase in their purification was obtained after gel filtration, ion exchange chromatography, ampholyte displacement chromatography and heat treatment [21,72]. Interestingly, when the authors examined the lipid specificity of these proteins, they found that they were able to transfer PE, PI, PC, sphingomyelin and cholesterol. A similar protein was isolated from rat liver after DEAE-cellulose, Sephadex G50 and hydroxylapatite chromatography. A 1450-fold increase in the specific activity was noted, with a yield as high as 50% [73]. The significance of these non-specific proteins will be discussed later. They are probably identical to the low-M_r proteins able to transfer phosphatidylserine (PS) from liposomes to mitochondria [74].

Basic proteins were first discovered in rat intestine by Lutton and Zilversmit [27]. These proteins were responsible for 65% of the transfer activity, the remaining activity being associated with acidic proteins.

Mitochondria and microsomes from rat hepatomas, unlike the organelles from normal liver, contain significant amounts of sphingomyelin and diphosphatidylglycerol (DPG), respectively. This led to the isolation of transfer proteins from hepatomas. A highly purified protein, able to transfer not only sphingomyelin, but also any microsomal phospholipid from microsomes to mitochondria was obtained. In contrast to the nsPL-TP from rat liver, this universal nsPL-TP was acidic (pI = 5.2) [31].

Dipalmitoylphosphatidylcholine and phosphatidylglycerol (PG) are the major components of mammalian lung surfactant which lowers surface tension (Chapters 1 and 6). This phospholipid is initially stored in organelles of alveolar type II epithelial cells, named lamellar bodies, which are unable to synthesize this component. It was suggested that a transfer of PC from the endoplasmic reticulum to these organelles occurred, mediated by transfer proteins. This hypothesis was validated by the detection of transfer-protein activity in rat [75,76] and rabbit [77] lung, and by the isolation of proteins from sheep lung [43]. Two sheep-lung proteins were found, one being similar to PC-TP from beef liver, the other having a neutral isoionic point. Transfer of PG can also be catalysed by soluble proteins from sheep lung [78] and by a specific protein from rat lung [79]. The supernatant from rat lung type II cells only contains a non-specific transfer protein [80] whereas in whole rat lung cytosol, in addition to this non-specific protein, two specific ones are present, transferring PC or PG [79].

Since synaptosomal plasma membranes cannot synthesize their phospholipids, it has been suggested that phospholipid transfer proteins carry phospholipids from endoplasmic reticulum to these membranes. Two acidic proteins were effectively isolated from synaptosome and myelin fractions of rat brain. Two major proteins were released from synaptosomal membranes by diluted phosphate buffer. These proteins, identical to those isolated from total brain cytosol, stimulated PC transfer less effectively than that of PI [29]. Rat brain cytosol stimulated the transfer of phospholipids to myelin [30].

(iii) Human plasma
A protein with an M_r-value of 100000, facilitating PC transfer from liposomes to liver mitochondria, has been isolated from lipoprotein-free human plasma. The presence of this protein in plasma may help the exchange of PC between lipoproteins and erythrocytes [81]. Beef-liver PC-TP was also used to transfer PC from liposomes to the very low density lipoproteins of human plasma [82]. Plasma also contains cholesteryl-ester exchange protein, transferring cholesteryl esters between lipoproteins [83].

(b) Plants and micro-organisms

Phospholipid exchange activity was detected first between organelles isolated from cauliflower florets and potato tubers [9]. A cytosol of the latter has yielded proteins able to catalyse the exchange of phospholipids between microsomes and mitochondria [46]. A gel filtration step was necessary for the detection of the active proteins. The presence of phospholipid transfer proteins was thereafter demonstrated in other plant tissues: cauliflower florets and Jerusalem artichoke tubers [54]. More active proteins were then discovered in plant tissues with high metabolic activity: germinating castor bean endosperm and maize seedlings. Castor bean endosperm prepared from 4-day-old seedlings contains active proteins, stimulating the transfer of phospholipids, essentially PC, from microsomes to mitochondria [47] or from liposomes to mitochondria [48–50]. Phosphatidylethanolamine exchange was also catalyzed by castor bean proteins [49]. After separation by Sephacryl chromatography, castor bean active proteins exhibited different M_r-values [52]. These proteins were PC-specific. 3-day-old maize seedlings also contain active proteins transferring PC from liposomes to mitochondria [55] or to other liposomes [53]. The major part of the transfer activity of the maize cytosol is associated with a basic phospholipid transfer protein which was purified to homogeneity [55]. The maize protein, which is the first to be highly purified from a plant tissue, has a pI of 8.8 ± 0.2, an apparent M_r-value of 20000 (as determined by SDS electrophoresis) and transfers PC, PI and PE.

All these plant tissues are non-photosynthetic. The first studies on chlorophyll-containing tissues were done on spinach and pea leaves. Cytosols prepared from this material catalyzed the transfer of PC between liposomes and mitochondria or chloroplasts [56,84]. However Murphy and Kuhn [85] concluded that spinach leaves lack phospholipid transfer proteins. The presence of interfering compounds explains this discrepancy, as demonstrated by Julienne et al. [86] who partially purified a phosphatidylcholine transfer protein by gel filtration. Two major transfer proteins seem to be present in spinach leaf cytosol with low and high pI (Julienne, unpublished).

Yeast cell cytosol was shown to contain active proteins stimulating the exchange of phospholipids between microsomes and mitochondria isolated from yeast or rat liver. PI and PC were the major phospholipids exchanged [59].

Lipid movements were first detected in *Bacillus subtilis* between microsomes and protoplasts [57]. Active proteins were then isolated from young *Bacillus* cells [58]. Another cytosol prepared from a prokaryotic cell, *Rhodopseudomonas sphaeroides*, mediated the transfer of phospholipids from unilamellar liposomes to intra-cytoplasmic membrane vesicles.

In conclusion, the presence of phospholipid transfer proteins has been established in almost all tissues or cells investigated. However, a non-catalytic transfer of phospholipids was demonstrated by ESR among microsomes isolated from *Tetrahymena pyriformis* cells [87]. This movement, occurring in the absence of any transfer proteins, was faster with microsomes isolated from cells grown at lower temperatures. Transfer proteins were not detected in these cells though a spontaneous lipid exchange between liposomes was observed, but in a time-scale of days [88].

4. Biochemical properties

As indicated above, PC-TP from beef liver has an M_r-value of 22000 (corrected to 28000) and a pI of 5.8. Other phospholipid transfer proteins differ from this one by their apparent M_r and their isoionic point (Table 2). Two major groups of phospholipid transfer proteins have been distinguished: acidic and basic proteins.

(a) Isoelectric point, M_r*-value and amino acid composition*

The first determination was made by isoelectric focussing on beef-liver PC-TP [34]. Other determinations followed, indicating that transfer proteins from different sources also exhibit a pI around 5.0. These acidic proteins were specific for PC or for PC and PI. An exception was found for rat hepatoma protein, a nsPL-TP with a pI of 5.2. However, a neutral protein specific for PC was isolated from sheep lung.

Basic proteins, first isolated from rat intestine [27], have a pI of 8 to 9. A PC-specific transfer protein of pI 8.4 was highly purified from rat liver. From the same tissue, a group of two proteins of high pI showed a transferring capacity towards various lipids. The highest pIs (9.5 and 9.75) were attributed to beef liver nsPL-TP. It is interesting to note that no special correlation exists between the isoionic point and other properties, like specificity. It was also found that basic transfer proteins are not present in all tissues; for instance, beef heart does not contain them.

Transfer proteins have M_r-values varying from 11000 to 32000. The majority of

TABLE 2

Properties of phosphoplipid transfer proteins

Origin	Isoelectric point		M_r-value		Specificity	Purification factor	Refs.
(a) *Acidic proteins*							
Beef liver		5.8	21320 [a]		PC	2680	34
			22000 [b]				
			23000 [c]				
Beef liver			28000 [c]		PC		24
Beef liver			24681 [a]		PC		90
Beef heart	Protein 1	4.7	21000 [c]		PC, SM	179	32
	Protein 2	5.5	25900 [c]		PC, SM	295	32
Beef heart	Protein I	5.3	23500 [b]	33500 [c]	PI>PC>SM	2008	33
	Protein II	5.6	23500 [b]	33500 [c]	PI>PC>SM	2460	33
Beef brain	Protein 1	5.2	29000 [a]	32500 [c]	PI>PC	508	38
	Protein 2	5.5	30000 [a]	32800 [c]	PI>PC	426	38
Calf liver					PC	2000	41
Rat intestine	4.5/5.3				only PC was studied		27
Rat liver	Protein I	5.1			PI, PC	348	23
	Protein II	5.3			PI, PC	556	
Rat liver	Protein I		13200 [b]		PS, PC	539	74
	Protein II		12400 [b]		PS, PC	394	74
Rat hepatoma		5.2	11168 [a]	11200 [b]	PC, PE, PS, PI, SM	118	31
Rat brain		5.02			PI>PC		29
		5.34			PI>PC		29

Sheep lung	Protein I	5.8	22600 [b]	22000 [c]		PC	152	43
	Protein II	7.1	20600 [b]	21000 [c]		PC	162	43
Potato tuber			22000 [b]			PC, PE, PI	4	46
Castor bean			72400 [b]		30300 [b]	PC	103	52
			22100 [b]	15800 [b]	13300 [b]			
(2) *Basic proteins*								
Rat intestine		8–9				only PC was studied		27
Rat liver		8.4	18700 [c]	17000 [b]		PC	140	22
Rat liver		8.4	16700 [b]	15800 [c]		PC	7410	23
			15023 [a]					
Rat liver		8.4	28000 [a]			PC	5300	24
Rat liver	CM1	8.3	13500 [b]	12500 [c]		PC, PE, PI, SM, cholesterol	876	21
		to						
	CM2	8.7						
Rat liver	CM2		12400 [a]			PC, PE, PI, SM,	876	72
Rat liver			14800 [c]			cholesterol	1540	73
Beef liver	CMI	9.55	14500 [c]	13600 [b]		all phospholipids	1270	36
	CMII	9.75	14500 [c]	13600 [b]		except DPG		
						Also SM and cholesterol		
Rat lung			14000			PG	140	79
Rat lung						PC, PG, PE		79
Maize seeds		8.8	20000 [c]			PC, PE, PI	125	55
			14058 [a]					

[a] Amino acid analysis.
[b] Gel filtration.
[c] SDS-gel electrophoresis.
SM, sphingomyelin.

TABLE 3

Amino acid composition of phospholipid transfer proteins (mol%)

	Asp	Thr	Ser	Glu	Pro	Gly	Ala	1/2 cyst	Val	Met	Ileu	Leu	Tyr	Phen	Lys	His	Arg	Trp
PC-TP (Beef liver) [34]	8.4	2.6	5.8	15.2	5.3	8.4	7.4	1.0	8.4	2.1	3.1	7.9	4.7	4.2	8.4	1.5	4.2	1.0
nsPL-TP (Beef liver) [36]	12.6	4.0	3.7	10.6	3.2	11.8	7.3	2.9	6.5	4.0	4.1	9.0	–	5.7	13.6	0.08	0.15	0.8
nsPL-TP (Rat liver) [72]	10.4	3.7	6.2	11.1	3.6	11.8	9.1	1.7	4.6	3.8	4.4	8.8	–	5.1	15.7	–	–	–
nsPL-TP (Rat liver [73]	8.5	4.8	8.4	11.7	3.7	13.9	8.0	0.9	4.9	3.1	4.2	7.6	1.0	4.1	10.9	1.8	1.9	0.5
nsPL-TP (Maize) [55]	9.1	5.8	13.3	4.2	4.9	11.1	15.5	8.3	5.1	1.4	5.3	4.3	2.1	0.7	3.1	0.7	5.1	nd

nd, not determined.

these proteins have an M_r-value around 22000. This value was initially attributed to PC-TP from beef liver, but it was re-investigated and found to be M_r 28000. The highest value—M_r 72400—was attributed to one of the five proteins transporting PC in castor bean cytosol. The other proteins from this tissue seem to be constituted of elementary peptide chains of about M_r 6000 [52]. Although acidic and basic proteins have similar M_r-values, the smallest proteins are the nsPL-TP (around M_r 11000 for rat hepatoma [31], M_r 13000 for rat liver [21], and M_r 14000 for beef liver [36]).

From Table 3 it may be noted that non-specific transfer proteins [36,72] contain high proportions of lysine, aspartic acid, asparagine, and glycine, whereas histidine, arginine, tryptophan, and tyrosine are absent or present in low amounts. However, the non-specific protein from rat hepatomas contains these four amino acids [31].

The ratio of acidic to basic amino acids roughly reflects the difference in pI. Beef-liver PC-TP [34] and rat hepatoma nsPL-TP [31] have a ratio of 2, whereas rat-liver PC-TP [23] has a ratio of 1. However, purified rat-liver PC-TP [24] (basic protein) has the same ratio as beef-liver PC-TP. The average hydrophobicity of PC-TP from beef liver has been calculated and found to be high; this may explain why the highly purified protein aggregates in concentrated solutions.

The first determination of the primary structure of transfer protein was made by Moonen et al. [89] for beef-liver PC-TP. The amino acid sequence of the hydrophobic binding site was established. These authors have identified the blocked N-terminal residue and have determined the sequence of the first 122 of the 244 amino acid residues of beef liver protein. Akeroyd et al. [90] have recently succeeded in establishing the complete amino acid sequence of beef-liver PC-TP, including the location of the two disulphide bridges.

(b) Molecular specificity

The observation that cytosols of various sources mediated the exchange of the major phospholipids, PC, PE, and PI [4], may be explained by the presence of monospecific transfer proteins in these extracts. The successful isolation of monospecific transfer proteins (highly purified protein from beef liver [34] or rat liver [23,24]; partially purified protein from sheep lung [43] and castor bean [52]) provided a first demonstration. Highly purified PI-TPs were obtained from beef heart [33], beef brain [38], rat liver [23] and rat brain [29], but these proteins also mediated (to a lesser extent) the movement of PC or sphingomyelin. No proteins strictly specific for PI have been isolated up to now. Looking for PE exchange, several authors have found such activity in soluble proteins prepared from various tissues, including those of plants [49]. The isolation of protein able to transfer this phospholipid led to the discovery in rat liver of non-specific proteins mediating the movement of PC, PI, sphingomyelin and cholesterol, in addition to PE. Similar proteins accelerating the transfer of the same phospholipids and also phosphatidylserine (PS) were found in rat hepatomas [31]. The isolation in large amounts of highly purified nsPL-TP from beef liver, acting on all phospholipids (except DPG), cholesterol and sphingomyelin, will allow for a study of its properties. This protein is the first one demonstrated to transfer PG and phosphatidic acid (PA). As far as we know, no protein able to catalyze a transfer of DPG has been isolated. Also, no protein exhibiting a specificity for molecular species of phospholipids (comprising saturated or unsaturated acyl-chains) has been isolated, although the presence of such proteins in rat liver cytosol has been postulated [91].

(c) Specificity for membranes

Are these transfer proteins specific for certain natural membranes? At present, the answer is no. It is true that, at first, PC-TP from beef liver appeared unable to react with intact erythrocytes [34], but recent experiments have revealed that with high concentrations of transfer proteins, protein-mediated transfers are observed with these cells [92]. PL-TPs appear able to function with a large variety of intracellular membranes. It has also been reported that rat liver cytosol accelerates the transfer of PC between liposomes and plant mitochondria [51] (Kader, unpublished experiments).

(d) Immunological properties

Immunochemical techniques have been used to specifically inhibit the activity of transfer proteins. Antisera against PC-TP from beef liver [34,93] or PI-TP [94,95] from beef brain were raised in rabbits. Such antisera specifically inhibit the activity of these proteins, which can thus be detected in crude cytosols. With anti-PI-TP it was shown that PI-transfer proteins in various cytosols have common antigenic determinants. Also, antisera against PC-TP from rat liver have been used to determine the contribution of this specific protein to the bulk of PC transfer activity [24]. It was noted that anti-beef-liver PC-TP did not cross-react with rat-liver PC-TP.

5. Mode of action

How do phospholipid transfer proteins act? The finding that highly purified PC-TP from beef liver contains 1 mol of bound PC [2,96] led to the hypothesis that this bound PC was exchanged with membrane PC. It was thus essential to examine whether this protein is able to carry PC from one membrane to another.

(a) Phospholipid transfer proteins as carriers

Two different approaches have been considered.

(i) Phospholipid monolayers

The elegant experiments of Demel et al. [97] consisted of introducing PC-TP from beef liver into a medium overlaid with a ^{14}C-labelled C_{16}–$C_{18:1}$ PC monolayer. A rapid decrease in the radioactivity of the monolayer followed the injection of the transfer protein. This indicated that the protein acts as a PC carrier. Evidence was also obtained by following the transfer of PC from one monolayer to another or from a monolayer to liposomes. These experiments were confirmed later with PI-TP from beef brain [98]. PI molecules were carried more effectively than PC (Fig. 3).

(ii) Binding experiments

To function as a carrier PL-TPs must bind and release phospholipids. The binding of phospholipids to transfer proteins was independently shown by Kamp et al. [96] and Johnson and Zilversmit [71], using, respectively, beef-liver and -heart proteins. Beef-liver PC-TP was incubated with [^{14}C]PC liposomes and then separated from liposomes by gel filtration or electrophoresis on polyacrylamide gel. Beef-heart protein, after incubation with liposomes, was recovered by isoelectric focussing. By these methods, complexes between PC and the exchange protein were obtained. Binding of PI to bovine-brain PI-TP was demonstrated by Demel et al. [98] using the monolayer technique. But no PI-protein complex was isolated. The release of PC from the PC-transfer protein complex was observed after incubation of this complex with liposomes [94]. Using ESR spectrometry, it was found that spin-labelled PC was incorporated into PC-TP from beef liver [61,62,99]. The release of spin-labelled PC was observed when the complex PC protein was incubated with vesicles of phosphatidic acid or lysophosphatidylcholine [99] or with vesicles of PC [62]. A binding of PS was observed with low-M_r transfer proteins from rat liver [74]. These binding properties were used to purify beef-liver PC-TP by affinity chromatography [100].

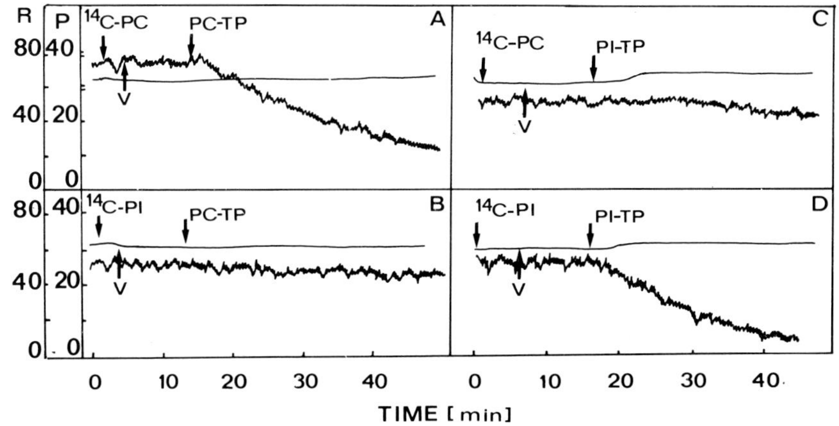

Fig. 3. Transfer of phospholipids from monolayers to vesicles. Monolayers A and C contain [^{14}C]PC and unlabelled PI. Monolayers B and D contain [^{14}C]PI and unlabelled PC. PC-PI (98 : 2 mol%) vesicles (V) and beef-liver PC-TP (in A and B) or beef-brain PI-TP (in C and D) were injected into the subphase (arrows). The surface pressure (P) (thin line) is indicated in dynes/cm. The decrease in the surface radioactivity (R, expressed in arbitrary units) indicates the transfer of labelled phospholipid from monolayer to vesicles. Reproduced from [98] with the authors' permission.

(b) Interactions between phospholipids and phospholipid transfer proteins

What is the nature of these interactions? The high hydrophobicity of beef-liver PC-TP suggests an important role of hydrophobic binding. Kamp et al. [96] displaced [^{14}C]PC molecules bound to the complex between [^{14}C]PC and beef-liver PC-TP by using detergents like deoxycholate (0.1%, w/v) or organic solvents like isobutanol. This displacement rendered the PC molecules susceptible to phospholipase attack [101]. Removal from the complex of [^{14}C]PC was observed for higher concentrations of deoxycholate (0.42%, w/v). Two conclusions could be drawn from these experiments: (*i*) Hydrophobic interactions play an important role in the formation of the complex; (*ii*) PC is embedded in the exchange protein, since phospholipase digestion required pre-treatment of the complex by low concentrations of detergent.

This was confirmed by experiments in which the acyl-chains of PC molecules were modified. Kamp et al. [102] found that the protein-mediated transfer of labelled analogues of PC was partially inhibited when the acyl-chains of PC molecules were saturated or contained D-stereoisomers. In addition, lyso-PC was not transferred. Schulze et al. [91], using rat-liver cytosol, found that the protein-mediated exchange of different molecular species of PC between liposomes and mitochondria was more active with unsaturated than with saturated PC molecules. Using the same cytosol, but the microsome-mitochondria assay, Wirtz et al. [103] have not observed significant changes in the extent of transfer when the acyl chain-composition was varied. Helmkamp [104] noted that the beef brain PI-TP was more active in liposome-microsome assays, when liposomes of increasing degree of unsaturation were used. It seems that the hydrocarbon fluidity of the membrane controls the activity of this transfer protein.

In conclusion, the hydrophobic interactions, although important, do not seem to be highly specific with respect to the acyl moiety of PC molecules.

Electrostatic interactions also play an important role in the binding process. Several arguments are available:

(*i*) PL-TPs with varying specificities towards phospholipids have been detected.

(*ii*) Important changes in ionic strength inhibit phospholipid transfer activity mediated by beef-heart protein [71] and the binding of PS to rat liver protein is highly sensitive to ionic strength [74].

(*iii*) The introduction of cations inhibits the activity of PC-TP from beef liver between liposomes and a monolayer [105].

Which moiety of the PC molecule then controls the electrostatic interactions with the protein? The introduction of analogues of PC into the exchange assay comprising liposomes, mitochondria and beef heart did not inhibit the transfer process [71]. This may be due to the high binding of PC to the transfer protein. However, when labelled analogues of PC were introduced into donor liposomes, variations in the extent of transfer of these analogues were observed in the presence of beef liver PC-TP [102]. With C_{16}–$C_{18:1}$ PC as the standard PC molecule, the transfer was partially inhibited or suppressed when the distance between phosphorus and nitrogen varied and when a methyl group on the quaternary nitrogen was removed or substituted by an ethyl or propyl group. These results clearly show that the binding site for the transfer protein interacts specifically with the phosphocholine group. It was also noted that PC with a spin-label in the polar group was not transferable [61].

In beef-liver PC-TP, PC molecules appear embedded in a cavity on the transfer proteins, since these molecules are not attacked by phospholipases unless a pre-treatment with detergents has been performed. An incorporation of spin-labelled PC into beef-liver PC-TP has been independently observed by Devaux et al. [61] and by Machida and Ohnishi [62]. The ESR spectrum of the protein-phospholipid complex showed a strong immobilization of the spin label. The nitroxide group appeared inaccessible to ascorbate [61].

Machida and Ohnishi [62] also showed an immobilization of phosphatidyltempocholine incorporated in beef-liver PC-TP. These observations confirm that PC molecules are buried in a cavity of the transfer protein and thus protected from the aqueous medium. The depth of this crevice is not yet known. Dicorleto et al. [33], studying the effect of sulphydryl-specific reagents (maleimides) found that the action of these reagents needed the presence of membranes. This suggests that the site of sulphydryl groups is only exposed during the interaction of the protein with membranes (Fig. 4).

Decisive progress in our knowledge of PC binding to the transfer protein has come from the work of Moonen et al. [106]. These authors incorporated PC with a photosensitive group into beef-liver PC-TP. The PC-protein complex was then isolated and partly digested by a protease. The 2-acyl chain of the PC molecule was recovered in a particular segment of a protease peptide of about 65 residues. The sequence of this peptide, determined for the first 38 residues, comprised an extremely hydrophobic group of apolar amino acids: Gly–Ser–Lys–Val–Phe–Met–Tyr–Tyr–. A β-sheet structure was predicted for this hydrophobic segment. As indicated above, this sequential analysis of amino acids has been recently carried out, identifying all of the polypeptide chain of PC-TP.

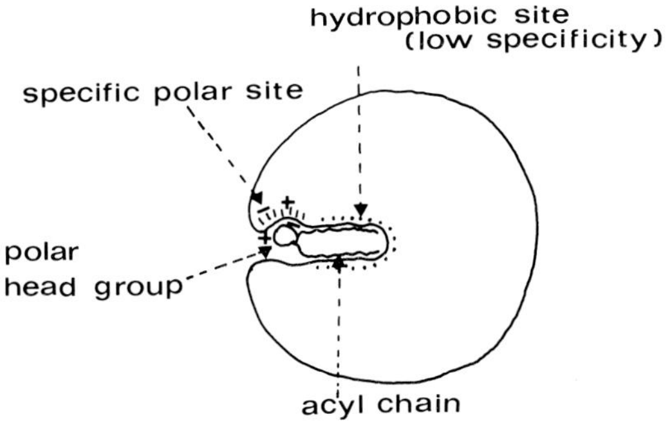

Fig. 4. Possible organization of the complex between beef-liver PC-TP and the transported phosphatidylcholine molecule.

(c) Net transfer

In the first studies on intermembrane movements of phospholipids, it appeared that exchange processes occurred. The discovery of proteins catalyzing this movement led the authors to name these proteins phospholipid exchange proteins (PLEP) [2,5]. The membrane lipid pools remained stable when the donor and acceptor membranes were incubated with transfer proteins. This suggested a one-for-one exchange. Moreover, the fact that beef-liver PC-TP contains 1 mol of PC per mol of protein suggested that this PC molecule might be exchanged with PC extracted from the membrane. However, several findings introduced three sets of arguments against a one-for-one exchange and in favour of a net transfer.

(i) Transfer proteins are able to insert PI or PC into membranes deficient in these phospholipids
Kagawa et al. [107] showed that beef-heart protein was able to introduce PC into vesicles containing all the components needed for [γ-^{32}P]ATP inorganic phosphate exchange, except PC. The protein mediated the incorporation of PC into the initially inactive vesicles and restored the activity. This work demonstrated for the first time that a net transfer of PC was mediated by proteins which can thus serve as tools for modifying membrane lipid composition and thus membrane activity. Similar experiments were conducted on *Micrococcus lysodeikticus* protoplasts [108,109]. Rat liver cytosols were able to substitute the acidic phospholipids of the protoplasts by PC, which is absent from these membranes. A net transfer of PI was also observed from microsomes to liposomes made from pure PC, in the presence of beef brain protein [110,111]. A similar transfer of PI was observed with rat liver or beef brain proteins [20,98], from monolayers to liposomes [98] or between vesicles [70]. Kasper and Helmkamp [65] showed that bovine brain PL-TP catalyzed a net transfer of PC between two populations of single bilayer vesicles. Dicorleto et al. [33] observed that purified beef heart proteins mediated a net transfer of PI between unilamellar vesicles made from PE, PC, PI and multilamellar vesicles containing PC, PE and DPG. However, in all of these experiments, it was not established whether these proteins, after having released their bound phospholipid into the membrane, remained devoid of any phospholipid (net transfer) or charged with another type of phospholipid (replacement).

(ii) Transfer proteins are able to leave the membrane devoid of any lipid, after the transfer process
Evidence in favour of a net transfer was recently given by Wirtz et al. [99] using a 2-stearoyl spin-labelled PC bound to beef-liver PC-TP. This labelled PC was released when micelles of lyso-PC or liposomes of PA were incubated with the phospholipid-protein complex. This indicates that spin-labelled PC was inserted into the membranes lacking this phospholipid and that the protein was not re-charged with lyso-PC or PA from the micelles. A similar insertion of spin-labelled PC, transferred from donor liposomes into unlabelled acceptor vesicles made from PE and PA, was

also catalyzed by the protein. This experiment demonstrated that the protein, which transferred only PC, released PC into the membrane and then left the membrane interface without a bound phospholipid. Kamp et al. [96] also observed that beef-liver PC-TP, depleted of PC by detergents, retained its activity. This protein also released PC into vesicles made from pure dimethylphosphatidylethanolamine which could not be carried by this protein [102].

In the experiment of Wirtz et al. [99], the transfer proceeded until the acceptor vesicles contained 2 mol% of PC. It was calculated that 20% of the spin-labelled PC was transferred under these conditions. Protein-mediated net transfer stopped when donor liposomes were depleted of about 20% of their initial PC content. An exchange process gradually replaced net transfer until an equilibrium concentration, governed by the nature of the interface, was reached. A release of spin-labelled PC from PC-PL-TP complex to receptor-rich membranes from *Torpedo marmorata* has also been observed [112]. Only a partial release of PC was noted when Machida and Ohnishi [62] added vesicles of pure PS to a PC-phospholipid transfer protein complex. It may be assumed that the PA interface competes with PC for the lipid binding site more actively than does the PC interface. No release was observed with pure PE vesicles when beef-liver PC-TP was used [102].

(iii) Transfer proteins are able to catalyze a net mass transfer

The first demonstration of a net transfer of phospholipid mass was made by Crain and Zilversmit [113] using non-specific PL-TP. When liposomes made from PC were incubated with mitochondria devoid of their outer membranes in the presence of the non-specific protein, a high increase in the amounts of PC and total phospholipids was noted in the mitochondrial pellets. This non-specific protein, isolated from beef liver, was also able to catalyze a net transfer of PC and PI from multilamellar vesicles to human high-density lipoprotein, whereas PC-TP from beef liver was unable to stimulate this transfer but catalyzed a phospholipid exchange.

The fact that the proteins considered in the present review can catalyze a true net transfer led to the novel generic name "phospholipid transfer proteins" rather than "phospholipid exchange proteins" previously used.

(d) Control of phospholipid transfer activity by membrane properties

PC binds to beef-liver PC-TP with hydrophobic and electrostatic interactions. It is reasonable to think that the surface properties of the membrane also play a role in the process, controlling the release of bound PC into the membrane and the extraction of another PC molecule from the membrane. As will be described in the next section, only the phospholipids present in the outer monolayer of a membrane are involved in the transfer process.

The influence of surface charge, modulated by the insertion of acidic phospholipids, was investigated first. The introduction of increasing amounts of PA or PI into "donor" PC liposomes diminished and finally suppressed beef-liver PC-TP-mediated transfer of PC between donor and acceptor liposomes [63]. Similar results were obtained between liposomes and mitochondria [105] and also in beef-brain proteins between microsomes and liposomes [110]. Van den Besselaar et al. [114] studying the kinetics of the reaction, proposed that the association of the protein with the donor liposome increases when the acidic phospholipid content of the liposomes is augmented. They observed that the protein was firmly associated with negatively charged interfaces and was less easily dissociated from the membrane. Similar conclusions were reached by Helmkamp et al. [115] with PI-TP from beef brain, suggesting a "ping-pong" mechanism for phospholipid transfer.

Inhibitory effects were also observed when PA (conferring a negative charge) or stearylamine (giving a positive charge) were introduced into liposomes incubated with mitochondria [105]. The optimal transfer needed a slightly positive charge. The neutralization of the negative surface charges by cations like Mg^{2+} restored the transfer activity [105]. However, this effect was limited to low ionic concentrations. An inhibitory effect of PS on PC transfer was demonstrated by ESR spectrometry using spin-labelled liposomes as donors, beef-liver PC-TP, and acceptor liposomes made from PC and PS [62]. Mg^{2+} and Ca^{2+} restored the transfer activity.

The relationship between the transfer activity and the membrane charge turned out to be a matter of controversy when the experiments of Dicorleto et al. [69] confirmed an observation of Zilversmit and Hughes [3]. Dicorleto et al. [69] found that transfer of PC between unilamellar liposomes and mitochondria or multilamellar vesicles, mediated by beef-liver or -heart proteins, was stimulated by the introduction of acidic phospholipids into liposomes. It is difficult to explain this discrepancy. An inhibition of PC transfer from multilamellar vesicles to liposomes was also observed at levels of liposomal PI greater than 15% when beef-liver protein was used [69].

Wirtz et al. [116] developed a kinetic model for the latter assay. They found that the apparent dissociation constant of a protein-vesicle complex decreased when the PA content of multilamellar vesicles was increased. The transfer protein was bound more strongly to vesicles of higher PA content. Similar results were obtained with fluorimetric titration [117].

Beef-brain PI-TP was found to react differently to changes of liposomal lipid composition [118]. PI- or PC-mediated transfer from liposomes to microsomes, inhibited by the incorporation of PI into liposomes, was unaffected by PA, PS or PG, whereas stearylamine inhibited the transfer. Interestingly, PE stimulated the transfer and sphingomyelin exerted an effect dependent on its concentration. These experiments confirmed that transfer proteins of various origins differ not only in their biochemical properties but also in their interaction with membrane interfaces.

All these data indicate that the membrane lipid composition has a marked effect on the transfer process. Also modifications of the acyl chains of liposomal PC molecules influence the transfer of PC from donor liposomes to acceptors in the presence of beef-liver PC-TP. Only 1% of ^{14}C-labelled PC was transferred from di-C_{16}-PC liposomes, whereas 26% was transferred from similarly labelled C_{16}–$C_{18:1}$ PC liposomes [102].

The importance of the phospholipid composition of the membranes was underlined by experiments using a complex between PC and beef-liver PC-specific protein. No PC was transferred when the complex was incubated with liposomes made from pure PE [102]. The same experiment, repeated with spin-labelled PC, revealed that PC molecules were transferred to liposomes containing PE and PA (81 : 19 mol%) [99]. When vesicles of pure PS were incubated with the complex formed by spin-labelled PC and beef-liver PC-TP, only a partial release of PC was observed [62].

Not only the lipid composition but also the membrane curvature influences the transfer activity of beef-liver PC-TP [119]. The transfer rate, determined by ESR spectrometry, was 100 times higher among small sonicated liposomes than between liposomes and large multilamellar vesicles. This effect of membrane curvature was also demonstrated in experiments indicating that the protein-mediated PC transfer from liposomes to spiculated erythrocyte ghosts was four times higher than that found with cup-shaped ghosts. An explanation for these results may lie in the differences in lipid packing in the outer layers of the two types of artificial membranes. Interestingly, Dicorleto and Zilversmit [67] have observed that multilamellar vesicles made from PC did not transfer PC to pure PC-liposomes; an addition of acidic phospholipids was needed to induce the transfer process. It was suggested that the formation and disruption of the protein-membrane complex was 50 to 100 times slower with liposomes than with PC-PA vesicles [116].

In conclusion, membrane properties (electric charge, lipid composition, membrane curvature) profoundly affect the transfer process. These properties govern the activity of the transfer proteins and control the relative contribution of the net transfer process as compared to the exchange process.

(e) Different steps of the exchange process

The different steps of the protein-mediated release and extraction of phospholipid from a membrane may be described as follows (Fig. 5).

(i) Binding of phospholipid to the protein

The phospholipid (PC) is embedded in a crevice. The acyl chains are bound to a non-specific hydrophobic site, whereas the choline head group is associated with a specific site. It is not known if this specific group is exposed to the medium when the protein is free in an aqueous environment but it may be shielded from the medium and unmasked when the protein forms a complex with the membrane. It is not known whether nsPL-TP have one multispecific site or several sites, and one or several crevices for phospholipid binding.

(ii) Formation of a collision complex between the proteins and the membrane

This formation is influenced by the surface properties of the membrane. A conformational change in the protein incubated with lipid interfaces was observed by measurement of fluorescence and circular dichroism [117]. When the interface is highly charged negatively, the protein is irreversibly bound to the membrane and the process stops. Only the outer monolayer of the membrane is involved in the process.

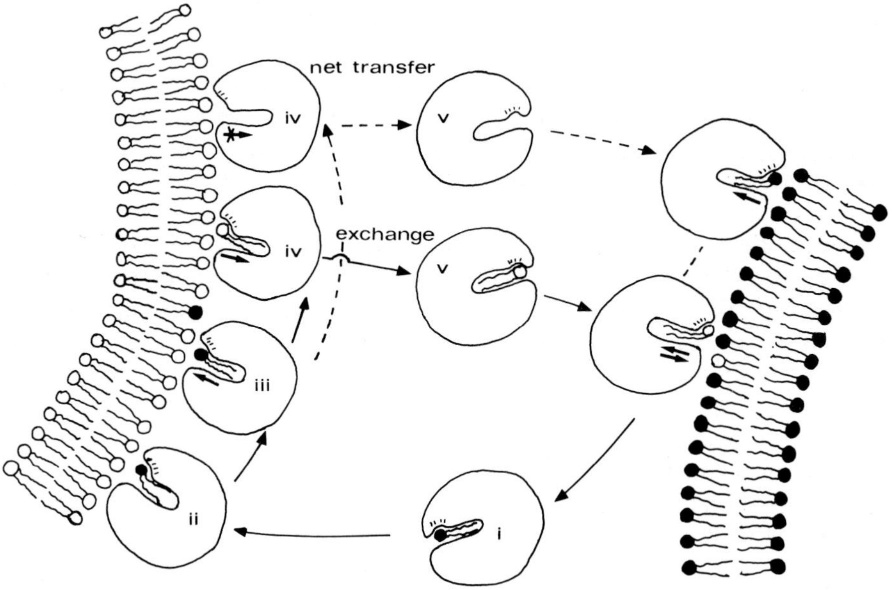

Fig. 5. Hypothetical scheme indicating the probable sequence of events in a transfer process mediated by phospholipid transfer proteins. This process may lead to a replacement (exchange) of the phospholipids of the membrane by those of the other membrane or to a net transfer of phospholipid molecules from one membrane to the other. Only the outer monolayers are involved in the process. The different steps are explained in the text.

(iii) Release of phospholipid

Again, properties of the membrane play a critical role in this step. This release seems to be influenced by the membrane curvature. A release of a phospholipid can occur even if the membrane normally lacks this phospholipid.

(iv) Detachment of phospholipid from the membrane

This step is not obligatory since net transfer has been observed in certain conditions.

(v) Detachment of the protein with or without bound phospholipid

When net transfer occurs, the protein leaves the membrane devoid of phospholipid.

It appears that all these steps are independent of each other.

6. Phospholipid transfer proteins as tools for membrane research

Since these proteins are able to extract a phospholipid from a membrane or to release it into a membrane lacking this phospholipid, they can be used as tools for studying the location of phospholipids within a membrane or for modifying the lipid composition of a membrane.

(a) Asymmetric distribution and transbilayer movement of lipids

The localization of phospholipids within membranes has been studied by several probing methods such as digestion by phospholipases, labelling with chemical reagents and ESR or NMR techniques, all using non-permeating reagents (see reviews [120–123]). PL-TPs, with a minimal M_r-value of approx. 13000, are assumed to be non-permeating and thus can be used as membrane probes.

Beef heart protein was the first used to determine the extent of the exchangeable lipid pool of liposomes [124]. When $[^{32}P]$PC liposomes were incubated with unlabelled mitochondria and the protein, a loss of ^{32}P label occurred. The reaction stopped when about 35% of the labelled PC remained in the liposomes. This experiment showed that a portion of the PC pool, representing about 65% of the total PC, located in the outer monolayer was exchangeable. A similar value was published by Rothman and Dawidowicz, using calf-liver protein to mediate PC transfer from liposomes to erythrocyte ghosts [42]. Other techniques (NMR studies, spin-label, phospholipase digestion [120–123]) have shown that the outer monolayer of the erythrocyte membrane contained twice as much PC as the inner monolayer. These experiments opened up a series of studies on lipid asymmetry and transbilayer lipid movement within membranes. The studies concerned several types of membranes including artificial and natural ones.

(i) Liposomes

The findings of Johnson et al. [124] were confirmed by Dawidowicz and Rothman [125] working on phospholipid vesicles of different density, by Dicorleto and Zilversmit [68] studying large unilamellar vesicles, dialysed cholate vesicles and cytochrome oxidase vesicles, by Machida and Ohnishi [62] using ESR spectrometry with spin-labelled PC-containing liposomes, and by Sandra and Pagano following PC or PE transfer from liposomes to hamster fibroblasts [35] or to mouse phagosomes [126]. All these experiments agree on the presence of an exchangeable pool of about 65 to 70% of the total PC pool. Recent investigations on sonicated PC-DPG-PI vesicles have revealed that 70% of PI molecules are accessible to beef-heart protein [127]. A similar proportion of the pool of PI was also accessible to phospholipase C attack. The transbilayer movement of phospholipids is very slow in these vesicles (half-time of days) as determined by transfer proteins [103,127] or a combination of transfer protein and NMR [128]. However, the protein-mediated introduction of dioleoyl phosphatidyl$[^{13}N\text{-}Me_3]$choline into dimyristoylphosphatidylcholine vesicles provoked the induction of a rapid transbilayer movement (half-time of less than 12 h) [129]. Also, the induction of bilayer to non-bilayer transitions by temperature changes led to an increase in the exchangeability of the PC pool in vesicles, suggesting a rapid transbilayer movement [130].

(ii) Erythrocytes

It is well known that the distribution of phospholipids is asymmetric in erythrocyte membranes, as shown by chemical techniques or phospholipase digestion [120–123] (see also Chapter 1).

This was confirmed by phospholipid transfer proteins. Since intact erythrocytes exchanged PC with liposomes too slowly, released erythrocyte ghosts and "inside-out" vesicles were used [131]. About 75% of the PC of ghosts but only 33% of "inside-out" vesicles was exchangeable. Half-times for the equilibration of the lipidic pools of the outer and inner leaflets were 2.3 h for ghosts and 5.3 h for "inside-out" vesicles.

Recent studies have shown that intact erythrocytes are able to exchange their PC with rat-liver microsomes in the presence of specific rat- or beef-liver proteins [92] or with unilamellar liposomes with nsPL-TP from beef liver [132]. This success allowed a direct determination of the exchangeability of PC in intact erythrocytes. About 75% and 60% of the total PC for human and rat erythrocytes, respectively, were available for transfer [92,132] (Fig. 6). The transbilayer movement of PC was slow (half-time approx. 7 h in rat erythrocytes [92,132]). In conjunction with these results, spin-labelled probes show the absence of rapid transbilayer movement in human erythrocytes [133], while a slow movement was observed using phospholipase digestion [134,135]. This slow transbilayer movement may be responsible for the maintenance of transmembrane asymmetry.

(iii) Mitochondria

Transfer protein has been used to incorporate spin-labelled PC into the outer monolayer of the inner mitochondrial membrane. It was also found that the rate of transbilayer transition was very slow (half-time > 24 h) [25].

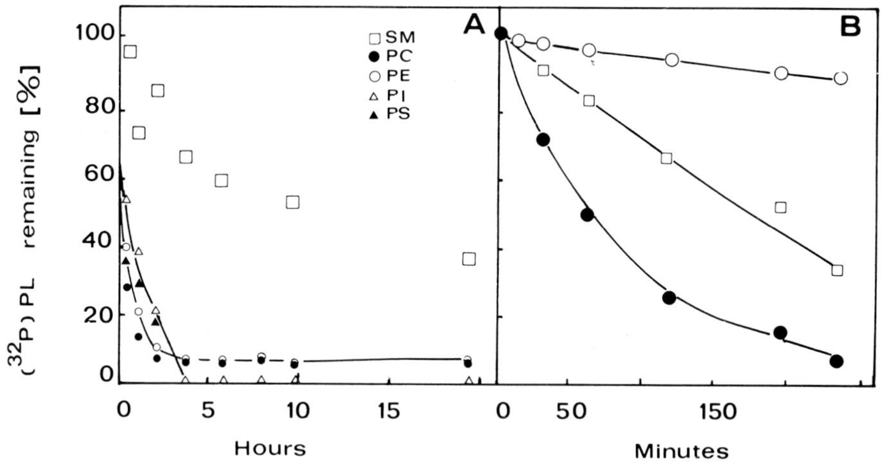

Fig. 6. Extensive transfer of membrane phospholipids mediated by phospholipid transfer proteins. (A) [^{32}P]Phospholipid-containing microsomes were incubated with rat liver mitochondria and rat-liver nsPL-TP (PL, phospholipid; SM, sphingomyelin). (B) [^{32}P]Phospholipid-containing erythrocytes were incubated with unilamellar vesicles and beef-liver nsPL-TP. Reproduced from [136] (A) and [132] (B) with the permission of the authors and publishers.

(iv) Microsomes

When microsomal fractions were studied for exchangeability of their lipid constituents, completely different results were obtained [136]. When rat liver microsomes—impermeable to EDTA—were incubated with an excess of mitochondria and proteins from different sources, a rapid exchange of phospholipids occurred, the exchange being nearly complete in about 2 h. This evolution was followed particularly for PC, using beef-liver PC-TP, and for almost all the phospholipids, using rat-liver nsPL-TP. Independent studies [137] have also established that beef-liver PC-TP is able to mediate the almost complete replacement of rat-liver microsomal PC by the egg-PC of liposomes. Microsomal PC was also fully exchangeable with lipoproteins [138] and exchangeable up to 90% with liposomes [139]. Up to 80% of rat liver microsomal PI was exchanged within 1 h in the presence of beef-liver PC-TP and liposomes [139–140] (Fig. 6).

Using phosphatidyl [N-^{13}C-Me_3]choline and [^{13}C]NMR, De Kruijff et al. [141] observed that 40% of rat sarcoplasmic PC was located in the outer monolayer. Since 80% of the total PC pool was exchangeable with beef-liver PC-TP, it was suggested that a rapid transbilayer movement of PC occurred in these membranes.

This extensive exchangeability of microsomal phospholipids did not lead to a clear conclusion about their location in the membrane (see also Chapter 1). Phospholipase digestion techniques gave conflicting results, pointing to a localization of PE and PI on the outside of the microsomal membranes, with PC equally distributed between the two layers [122], or to a symmetric distribution of phospholipids within the membrane [123].

A rapid transbilayer movement of phospholipids in microsomal membranes was suggested by the experiments with transfer proteins [122–125,136,137]. This extensive "flip-flop" of phospholipids may depend on a membrane-protein-catalyzed mechanism facilitated by non-bilayer structures within the membranes [142]. However, the precise mechanism is still unknown. A rapid transverse movement of phospholipids was also suggested in brush-border membranes from rabbit small intestine, using beef-liver PC-TP [143].

(v) Microorganisms

Little information is available concerning these cells. Rat liver cytosol was used to study the location of phospholipids in the protoplasmic membrane of *Micrococcus lysodeikticus* [108]. It was concluded that DPG was distributed almost equally between the two layers whereas PG and PI were located in the outer and inner layers, respectively. Treatment of the protoplasts by phospholipases revealed a similar distribution. The phospholipid composition of the outer layer of the membrane of influenza virus was different from that of the inner leaflet, as determined by the use of calf-liver or beef-heart protein or by phospholipase attack [41]. The outer surface is enriched in PC and PI whereas PE and PS are equally distributed; sphingomyelin appears to be localized on the inner side of the membrane. The transmembrane movement of phospholipids was found to be very slow (half-time of several days).

(b) Manipulation of the phospholipid composition

It has been tempting to use transfer proteins to modify the phospholipid composition of the outer monolayer of natural membranes, and then to examine the effect of this change on membrane properties. This was done on protoplasts of *Micrococcus lysodeikticus* which lack PC [109]. The replacement of one half of the endogenous phospholipids by PC, mediated by rat liver proteins, provoked changes in enzymatic activities and a restoration of the permeability barrier of the membrane.

In conclusion, in the last 5 years, PL-TPs have been shown to be useful tools with which to study the mobility of the lipid components of biological membranes [121,122]. These proteins are mild, non-permeating reagents with no lytic activity. It is important to know whether they disturb membrane structure, but since a partial penetration of these proteins into the phospholipid bilayer is not excluded [122], a definite conclusion cannot be drawn. A protein-mediated net transfer of phospholipids, by modifying the lipid composition of the membrane, may also disturb the initial arrangement of the membrane constituents. However, in spite of these limitations, it may be predicted that the use of phospholipid transfer proteins as membrane probes will be further developed in the near future. In particular, studies on the lipid dependence of enzymes could be developed by using PL-TP. Recent work by Crain and Zilversmit [144] showed that the activity of glucose-6-phosphatase is modified when the lipid composition of microsomal membranes is manipulated by non-specific PL-TP.

7. Physiological role

Although transfer proteins seem universally distributed within eukaryotic cells and have also been found in two prokaryotic cells, their physiological role has not been clearly demonstrated. The discovery of the intermembrane exchange of phospholipids and of phospholipid transfer proteins arose from the concept of intracellular co-operation in lipid biosynthesis for the whole membrane network [5–9]. The first experiments were based on the inability of mitochondria to form their own phospholipids whereas the endoplasmic reticulum was highly active in this biosynthesis [1–4]. The discovery of an exchange of phospholipids between mitochondria and endoplasmic reticulum (microsomes), mediated by cytosolic proteins, led to the hypothesis that these proteins participated in the biogenesis of membranes by inserting newly formed phospholipids into membranes unable to synthesize these components (Fig. 7).

Labelling experiments in vivo showed a sequence of lipid labelling, first in the microsomal and then in mitochondrial fractions [9,145–147]. These studies gave only indirect evidence. All the other arguments are based on experiments in vitro. One major criticism of this theory would be that these proteins only catalyze an exchange process. In this case, they would participate only in a renewal of phospholipid molecules, mediating the replacement of one type of phospholipid by another. This

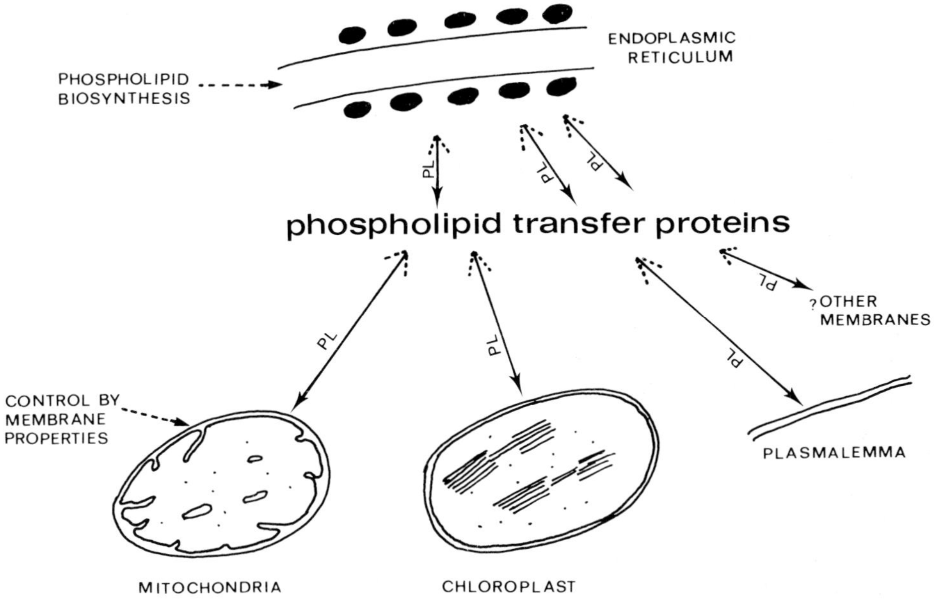

Fig. 7. Postulated participation of phospholipid transfer proteins in membrane biogenesis (PL, phospholipid.)

replacement could participate in the distribution of different types of phospholipid within the various membranes of the cell. In plant cells, such PC-TP could, for instance, remove PC molecules containing $C_{18:1}$ fatty acid from the chloroplast envelope, transfer these phospholipids to the endoplasmic reticulum where the desaturation of $C_{18:1}$ to $C_{18:2}$ would occur and then bring back PC molecules containing $C_{18:2}$ fatty acid to the chloroplast envelope [52,56,148,149]. Since phospholipid transfer proteins only interact with the outer monolayer of membranes, it has been proposed that they may play a major role in the origin of the asymmetric distribution of lipids in membranes, for instance in erythrocytes [134,135].

The extensive exchangeability of the microsomal lipid pool may help the transfer process. The presence of non-specific PL-TPs, able to mediate the movement of almost all the phospholipids, is of great interest for the evaluation of the physiological significance of these proteins. It is now clear that in certain conditions a net transfer process occurs [99]. The recent demonstration that non-specific phospholipid transfer proteins from beef liver catalyze a net mass transfer of phospholipids [113] reinforces the concept of a direct participation of these proteins in membrane biogenesis. Although it is not known at present what factors govern the magnitude of the net transfer or the exchange processes, membrane properties probably play an essential role. Correlations between transfer activity, acyl-chain unsaturation, temperature and membrane fluidity have been demonstrated with PL-TPs from beef liver [150] and beef brain [151]. It has also been shown that ions strongly interfere in vivo with the activity of the transfer proteins [62]. Membrane proteins may be involved, since the PE transfer from vesicles to hamster fibroblasts decreased when these cells were pretreated by trypsin [35]. However, other mechanisms for membrane biogenesis may exist. The concept of "membrane flow" [152] implies a transfer of intact segments of membrane rather than transfer of individual components.

An effective physiological role of PC-TPs may imply that the levels of transfer activity vary with the intensity of membrane biogenesis. Such a modulation of transfer activity was observed during the biogenesis of new membranes in developing rat brain [153], mouse lung [75], and in castor-bean endosperm (Kader, unpublished). However, no significant variation of transfer activity was noted in rat intestine [27]. This relationship was investigated in tumour cells which exhibit an abnormal composition of membrane phospholipids [31]. In particular, mitochondria from rat hepatoma, in contrast to those from rat liver, contain sphingomyelin. The isolation of a universal lipid exchange protein, transferring sphingomyelin, PC, PI and PS between microsomes and mitochondria, suggested that this protein is responsible for the "lipid de-differentiation" of the hepatoma membranes [31]. This was the first attempt to demonstrate that the lipid composition of a membrane can be governed by phospholipid transfer proteins. The relationship between phospholipid metabolism and transfer activity was recently studied in three Morris hepatomas [24]. It was found that the cytosol prepared from a fast-growing hepatoma containing PC- and PI-rich mitochondria exhibited higher PC and PI transfer activities than those observed in other hepatomas. The almost complete absence of PE transfer activity in these three hepatomas as compared to normal liver was attributed to low levels of the non-specific PL-TP. Teerlink et al. [154] developed a double-antibody radioimmunoassay to determine the levels of PC-TPs in the cytosols prepared from normal rat liver or Morris hepatomas. The levels observed are lower in one hepatoma than in the others and in the normal liver. A discrepancy between these values and the results obtained by the immuno-titration technique suggested that inhibited forms of PC-TPs may exist.

An important role has been attributed to PI transfer proteins from the brain and it has been suggested that they transfer PI from the endoplasmic reticulum to the synaptosomal membrane [155]. This transfer restores the pool of PI of the latter membrane which is degraded in response to a stimulus [29] (see Chapter 7). It was noted that the nerve endings of the neuron, where rapid degradation of PI occurred, were rich in PI-TP.

The modulation of transfer activity may be due to controlling factors or to the turnover of the protein. We suggest that phospholipid transfer proteins are synthesized at various rates, depending on the intensity of membrane biogenesis. The radioimmunoassay technique may be of great help in determining these rates. It may be predicted that in young cells, with active membrane formation, the transfer proteins are more abundant than in adult cells where only renewal and slight membrane biogenesis occur. Studies on the biosynthesis of PL-TPs, including the isolation of the RNA messenger coding for them, are necessary to check this hypothesis.

8. Conclusions

Considerable progress has been made in the short period of time since the discovery of PL-TPs. Highly purified proteins, mono-specific or non-specific, are now available. Their properties have been explored, revealing some interesting features, including a relatively high hydrophobicity of the binding site for phospholipids and the likely presence of a crevice protecting the lipid against the hydrophilic environment. The primary structure of beef-liver transfer protein has now been partially elucidated. The mode of action of these proteins has been carefully analyzed, revealing the major influence of the surface properties on the transfer process. The different steps of the process have been described and found to be independent of each other. Recent evidence in favour of a net transfer reinforces the postulated role of these proteins as carriers of phospholipids from the sites of biosynthesis to membranes being formed. A participation of transfer proteins in the control of the lipid composition of the membrane has also been deduced from studies on tumour cells. Finally, the use of PL-TPs as mild membrane probes has been actively developed.

Several points remain unanswered concerning the mode of action of these proteins, e.g. the factors controlling net transfer, the molecular specificity of the proteins and their biogenesis during the life of the cell. It will be of interest to examine if a perturbation of their biogenesis or of their activity can disturb the normal behaviour of cells.

Acknowledgements

The authors are much indebted to Dr. K.W.A. Wirtz (Laboratory of Biochemistry, State University of Utrecht, The Netherlands) for his continuous encouragement and help. We thank Dr. D.B. Zilversmit (Cornell University, Ithaca, USA), Prof. L.L.M. van Deenen (Laboratory of Biochemistry, State University of Utrecht, The Netherlands) and Dr. P.F. Devaux (Institut de Biologie Physico-Chimique de Paris) for their stimulating interest. We are grateful to Dr. A. Kovoor (Université de Paris VII) for critical reading of the manuscript. We thank Mrs. M.F. Laforge for preparing the manuscript and the illustrations.

References

1 Dawson, R.M.C. (1973) Sub-Cell. Biochem. 2, 69–89.
2 Wirtz, K.W.A. (1974) Biochim. Biophys. Acta 344, 95–117.
3 Zilversmit, D.B. and Hughes, M.E. (1976) in Methods in Membrane Biology (Korn, E.D., ed.), Vol. 7, pp. 211–259, Plenum, New York.
4 Kader, J.C. (1977) in Cell Surface Reviews (Poste, G. and Nicolson, G.L., eds.), Vol. 3, pp. 127–204, Elsevier, Amsterdam.
5 Wirtz, K.W.A. and Zilversmit, D.B. (1968) J. Biol. Chem. 243, 3596–3602.
6 Kadenbach, B. (1968) in Biochemical Aspects of the Biogenesis of Mitochondria (Papa, S., Quagliariello, E., Slater, E.C. and Tager, J.M., eds.), pp. 415–429, Adriatica, Bari.

7 McMurray, W.C. and Dawson, R.M.C. (1969) Biochem. J. 112, 91–108.
8 Akiyama, M. and Sakagami, T. (1969) Biochim. Biophys. Acta 187, 105–112.
9 Abdelkader, A.B. and Mazliak, P. (1970) Eur. J. Biochem. 15, 250–262.
10 Wirtz, K.W.A. and Zilversmit, D.B. (1969) Biochim. Biophys. Acta 193, 105–116.
11 Wirtz, K.W.A. and Zilversmit, D.B. (1970) FEBS Lett. 7, 44–46.
12 Kamath, S.A. and Rubin, E. (1973) Arch. Biochem. Biophys. 158, 312–322.
13 Taniguchi, M. and Sakagami, T. (1975) J. Biochem. (Tokyo) 77, 1245–1248.
14 Blok, M.C., Wirtz, K.W.A. and Scherphof, G.L. (1971) Biochim. Biophys. Acta 233, 61–75.
15 Sauner, M.T. and Levy, M. (1971) J. Lipid Res. 12, 71–75.
16 Wojtczak, L., Barabska, J., Zborowski, J. and Drahota, Z. (1971) Biochim. Biophys. Acta 249, 41–52.
17 Strunecka, A. and Zborowski, J. (1975) Comp. Biochem. Physiol. 50B, 55–60.
18 Butler, M.M. and Thompson, W. (1975) Biochim. Biophys. Acta 388, 52–57.
19 Zilversmit, D.B. (1971) J. Lipid Res. 12, 36–42.
20 Zborowski, J. and Wojtczak, L. (1975) FEBS Lett. 51, 317–320.
21 Bloj, B. and Zilversmit, D.B. (1977) J. Biol. Chem. 252, 1613–1619.
22 Lutton, C. and Zilversmit, D.B. (1976) Biochim. Biophys. Acta 441, 370–379.
23 Lumb, R.H., Kloosterman, A.D., Wirtz, K.W.A. and Van Deenen, L.L.M. (1976) Eur. J. Biochem. 69, 15–22.
24 Poorthuis, B.J.H.M., Van der Krift, T.P., Teerlink, T., Akeroyd, R., Hostetler, K.Y. and Wirtz, K.W.A. (1980) Biochim. Biophys. Acta, 600, 376–386.
25 Rousselet, A., Colbeau, A., Vignais, P.M. and Devaux, P.F. (1976) Biochim. Biophys. Acta 426, 372–384.
26 Horiuchi, I. (1973) Sapporo Med. J. 42, 457–464.
27 Lutton, C. and Zilversmit, D.B. (1976) Lipids 11, 16–20.
28 Possmayer, F. (1974) Brain Res. 74, 167–174.
29 Wirtz, K.W.A., Jolles, J., Westerman, J. and Neys, F. (1976) Nature 260, 354–355.
30 Brammer, M.J. (1979) J. Neurochem. 31, 1435–1440.
31 Dyatlovitskaya, E.V., Timofeeva, N.G. and Bergelson, L.D. (1978) Eur. J. Biochem. 82, 463–471.
32 Ehnholm, C. and Zilversmit, D.B. (1973) J. Biol. Chem. 248, 1719–1724.
33 Dicorleto, P.E., Warach, J.B. and Zilversmit, D.B. (1979) J. Biol. Chem. 254, 7795–7802.
34 Kamp, H.H., Wirtz, K.W.A. and Van Deenen, L.L.M. (1973) Biochim. Biophys. Acta 318, 313–325.
35 Sandra, A. and Pagano, R.E. (1979) J. Biol. Chem. 254, 2244–2249.
36 Crain, R.C. and Zilversmit, D.B. (1980) Biochemistry 19, 1433–1439.
37 Dudley, P.A. and Anderson, R.E. (1978) FEBS Lett. 95, 57–60.
38 Helmkamp Jr., G.M., Harvey, M.S., Wirtz, K.W.A. and Van Deenen, L.L.M. (1974) J. Biol. Chem. 249, 6382–6389.
39 Carey, E.M. and Foster, P.C. (1977) Biochem. Soc. Trans. 5, 1412–1414.
40 Miller, E.K. and Dawson, R.M.C. (1972) Biochem. J. 126, 823–835.
41 Rothman, J.E. and Dawidowicz, E.A. (1975) Biochemistry 14, 2809–2816.
42 Rothman, J.E., Tsai, D.K., Dawidowicz, E.A. and Lenard, J. (1976) Biochemistry 15, 2361–2370.
43 Robinson, M.E., Wu, L.N.Y., Brumley, G.W. and Lumb, R.H. (1978) FEBS Lett. 87, 41–44.
44 Illingworth, D.R. and Portman, O.W. (1972) J. Lipid Res. 13, 220–227.
45 Illingworth, D.R., Portman, O.W., Robertson, A.L. and Mayar, W.A. (1973) Biochim. Biophys. Acta 306, 422–436.
46 Kader, J.C. (1975) Biochim. Biophys. Acta 380, 31–44.
47 Yamada, M., Tanaka, T., Kader, J.C. and Mazliak, P. (1978) Plant Cell Physiol. 19, 173–176.
48 Douady, D., Kader, J.C. and Mazliak, P. (1978) in Plant Mitochondria (Ducet, G. and Lance, C., eds.), pp. 357–364, Elsevier, Amsterdam.
49 Boussange, J., Douady, D. and Kader, J.C. (1980) Plant Physiol. 65, 335–358.
50 Douady, D., Kader, J.C. and Mazliak, P. (1980) Plant Sci. Lett. 17, 295–301.
51 Tanaka, T. and Yamada, M. (1979) Plant Cell Physiol. 20, 533–542.
52 Yamada, M., Tanaka, T. and Ohnishi, J. (1980) in Recent Advances in the Biogenesis and Function of Plant Lipids (Mazliak, P., Benveniste, P., Costes, C. and Douce, R., eds.) Elsevier, Amsterdam, pp. 161–168.

53 Guerbette, F., Douady, D., Grosbois, M. and Kader, J.C. (1981) Physiol. Vég., 19, 467–472.
54 Douady, D., Kader, J.C. and Mazliak, P. (1978) Phytochemistry 17, 793–794.
55 Douady, D., Grosbois, M., Guerbette, F. and Kader, J.C. (1982) Biochim. Biophys. Acta, 710, 143–153.
56 Julienne, M. and Kader, J.C. (1981) C.R. Hebd. Séances Acad. Sci. Sér. D 292, 255–258.
57 Bureau, G. and Mazliak, P. (1974) FEBS Lett. 39, 332–336.
58 Bureau, G., Kader, J.C. and Mazliak, P. (1976) C.R. Hebd. Séances Acad. Sci. Sér. D 282, 119–122.
59 Cobon, G.S., Crowfoot, P.D., Murphy, M. and Linnane, A.W. (1976) Biochim. Biophys. Acta 441, 255–259.
60 Cohen, L.K., Lueking, D.R. and Kaplan, S. (1979) J. Biol. Chem. 254, 721–728.
61 Devaux, P.F., Moonen, P., Bienvenue, A. and Wirtz, K.W.A. (1977) Proc. Natl. Acad. Sci. USA 74, 1807–1810.
62 Machida, K. and Ohnishi, S. (1978) Biochim. Biophys. Acta 507, 156–164.
63 Hellings, J.A., Kamp, H.H., Wirtz, K.W.A. and Van Deenen, L.L.M. (1974) Eur. J. Biochem. 47, 601–605.
64 Sasaki, T. and Sakagami, T. (1978) Biochim. Biophys. Acta 512, 461–471.
65 Kasper, A.M. and Helmkamp Jr., G.M. (1981) Biochim. Biophys. Acta 664, 22–32.
66 De Cuyper, M., Joniau, M. and Dangreau, H. (1980) Biochem. Biophys. Res. Commun. 95, 1224–1230.
67 Dicorleto, P.E. and Zilversmit, D.B. (1977) Biochemistry 16, 2145–2150.
68 Dicorleto, P.E. and Zilversmit, D.B. (1979) Biochim. Biophys. Acta 552, 114–119.
69 Dicorleto, P.E., Fakharzadeh, F.F., Searles, L.L. and Zilversmit, D.B. (1977) Biochim. Biophys. Acta 468, 296–304.
70 Barsukov, L.I., Shapiro, Y.E., Viktorov, A.V., Volkova, V.I., Bystrov, V.F. and Bergelson, L.D. (1975) Chem. Phys. Lipids 14, 211–226.
71 Johnson, L.W. and Zilversmit, D.B. (1975) Biochim. Biophys. Acta 375, 165–175.
72 Bloj, B., Hughes, M.E., Wilson, D.B. and Zilversmit, D.B. (1978) FEBS Lett. 96, 87–89.
73 Poorthuis, B.J.H.M., Glatz, J.E.C., Akeroyd, R. and Wirtz, K.W.A. (1981) Biochim. Biophys. Acta 665, 256–261.
74 Baranska, J. and Grabarek, Z. (1979) FEBS Lett. 104, 253–257.
75 Engle, M.J., Van Golde, L.M.G. and Wirtz, K.W.A. (1978) FEBS Lett. 86, 277–281.
76 Koumanov, K., Neitcheva, T., Boyanov, A. and Georgiev, G. (1978) Bull. Eur. Physiopathol. Resp. 14, 375–381.
77 Spalding, J.W. and Hook, G.E.R. (1979) Lipids 14, 606–613.
78 Whitlow, C.D., Pool, G.L., Brumley, G.W. and Lumb, R.H. (1980) FEBS Lett. 113, 221–224.
79 Van Golde, L.M.G., Oldenborg, V., Post, M., Batenburg, J.J., Poorthuis, B.J.H.M. and Wirtz, K.W.A. (1980) J. Biol. Chem. 255, 6011–6013.
80 Post, M., Batenburg, J.J., Schuurmans, E.A.J.M. and Van Golde, L.M.G. (1980) Biochim. Biophys. Acta 620, 317–321.
81 Brewster, M.E., Ihm, J., Brainard, J.R. and Harmony, J.A.K. (1978) Biochim. Biophys. Acta 529, 147–159.
82 Jackson, R.L., Wilson, D. and Glueck, C.J. (1979) Biochim. Biophys. Acta 557, 79–85.
83 Pattnaik, N.M. and Zilversmit, D.B. (1979) J. Biol. Chem. 254, 2782–2786.
84 Tanaka, T., Ohnishi, J.I. and Yamada, M. (1980) Biochem. Biophys. Res. Commun. 96, 394–399.
85 Murphy, D.J. and Kuhn, D.N. (1981) Biochem. J. 194, 257–264.
86 Julienne, M., Vergnolle, C. and Kader, J.C. (1981) Biochem. J. 197, 763–764.
87 Iida, H., Maeda, T., Ohki, K., Nozawa, Y. and Ohnishi, S. (1978) Biochim. Biophys. Acta 508, 55–64.
88 Martin, F.J. and MacDonald, R.C. (1976) Biochemistry 15, 321–327.
89 Moonen, P., Akeroyd, R., Westerman, J., Puyk, W.C., Smits, P. and Wirtz, K.W.A. (1980) Eur. J. Biochem. 106, 279–290.
90 Akeroyd, R., Moonen, P., Westerman, J., Puyk, W.C. and Wirtz, K.W.A. (1981) Eur. J. Biochem. 114, 385–391.

91 Schulze, G., Jung, K., Kunze, D. and Egger, E. (1977) FEBS Lett. 74, 220–224.
92 Van Meer, G., Poorthuis, B.J.H.M., Op den Kamp, J.A.F. and Van Deenen, L.L.M. (1980) Eur. J. Biochem. 103, 283–288.
93 Harvey, M.S., Wirtz, K.W.A., Kamp, H.H., Zegers, B.J.M. and Van Deenen, L.L.M. (1973) Biochim. Biophys. Acta 323, 234–239.
94 Wirtz, K.W.A., Helmkamp Jr., G.M. and Demel, R.A. (1978) in Protides of the Biological Fluids (Peeters, H., ed.), pp. 25–32, Pergamon, Oxford.
95 Helmkamp Jr., G.M., Nelemans, S.A. and Wirtz, K.W.A. (1976) Biochim. Biophys. Acta 424, 168–182.
96 Kamp, H.H., Wirtz, K.W.A. and Van Deenen, L.L.M. (1975) Biochim. Biophys. Acta 398, 401–414.
97 Demel, R.A., Wirtz, K.W.A., Kamp, H.H., Geurts Van Kessel, W.S.M. and Van Deenen, L.L.M. (1973) Nature New Biol. 246, 102–105.
98 Demel, R.A., Kalsbeek, R., Wirtz, K.W.A. and Van Deenen, L.L.M. (1977) Biochim. Biophys. Acta 466, 10–22.
99 Wirtz, K.W.A., Devaux, P.F. and Bienvenue, A. (1980) Biochemistry 19, 3395–3399.
100 Barsukov, L.I., Dam, C.W., Bergelson, L.D., Muzja, G.I. and Wirtz, K.W.A. (1978) Biochim. Biophys. Acta 513, 198–204.
101 Kamp, H.H., Sprengers, E.D., Westerman, J., Wirtz, K.W.A. and Van Deenen, L.L.M. (1975) Biochim. Biophys. Acta 398, 415–423.
102 Kamp, H.H., Wirtz, K.W.A., Baer, P.R., Slotboom, A.J., Rosenthal, A.F., Paltauf, F. and Van Deenen, L.L.M. (1977) Biochemistry 16, 1310–1316.
103 Wirtz, K.W.A., Van Golde, L.M.G. and Van Deenen, L.L.M. (1970) Biochim. Biophys. Acta 218, 176–179.
104 Helmkamp Jr., G.M. (1980) Biochemistry 19, 2050–2056.
105 Wirtz, K.W.A., Geurts Van Kessel, W.S.M., Kamp, H.H. and Demel, R.A. (1976) Eur. J. Biochem. 61, 515–523.
106 Moonen, P., Haagsman, H.P., Van Deenen, L.L.M. and Wirtz, K.W.A. (1979) Eur. J. Biochem. 99, 439–445.
107 Kagawa, Y., Johnson, L.W. and Racker, E. (1973) Biochem. Biophys. Res. Commun. 50, 245–251.
108 Barsukov, L.I., Kulikov, V.I. and Bergelson, L.D. (1976) Biochem. Biophys. Res. Commun. 71, 704–711.
109 Barsukov, L.I., Simakova, I.M., Tikhonova, G.V., Ostrovskii, D.N. and Bergelson, L.D. (1978) Europ. J. Biochem. 90, 331–336.
110 Harvey, M.S., Helmkamp Jr., G.M., Wirtz, K.W.A. and Van Deenen, L.L.M. (1974) FEBS Lett. 46, 260–262.
111 Zborowski, J. (1979) FEBS Lett. 107, 30–32.
112 Rousselet, A., Devaux, P.F. and Wirtz, K.W.A. (1979) Biochem. Biophys. Res. Commun. 90, 871–877.
113 Crain, R.C. and Zilversmit, D.B. (1980) Biochim. Biophys. Acta 620, 37–48.
114 Van den Besselaar, A.M.H.P., Helmkamp Jr., G.M. and Wirtz, K.W.A. Biochemistry 9, 1852–1858.
115 Helmkamp Jr., G.M., Wirtz, K.W.A. and Van Deenen, L.L.M. (1976) Biochim. Biophys. Acta 174, 592–602.
116 Wirtz, K.W.A., Vriend, G. and Westerman, J. (1979) Eur. J. Biochem. 94, 215–221.
117 Wirtz, K.W.A. and Moonen, P. (1977) Eur. J. Biochem. 77, 437–443.
118 Helmkamp Jr., G.M. (1980) Biochim. Biophys. Acta 595, 222–234.
119 Machida, K. and Ohnishi, S. (1980) Biochim. Biophys. Acta 596, 201–209.
120 Rothman, J.E. and Lenard, J. (1977) Science 195, 743–753.
121 Bergelson, L.D. and Barsukov, L.I. (1977) Science 197, 224–230.
122 Zilversmit, D.B. (1978) Ann. N.Y. Acad. Sci. 308, 149–163.
123 Op den Kamp, J.A.F. (1979) Annu. Rev. Biochem. 48, 47–71.
124 Johnson, L.D., Hughes, M.E. and Zilversmit, D.B. (1975) Biochim. Biophys. Acta 375, 176–185.
125 Dawidowicz, E.A. and Rothman, J.E. (1976) Biochim. Biophys. Acta 455, 621–630.

126 Sandra, A. and Pagano, R.E. (1978) Biochemistry 17, 332–338.
127 Low, M.G. and Zilversmit, D.B. (1980) Biochim. Biophys. Acta 596, 223–234.
128 Shaw, J.M., Hutton, W.C., Lentz, B.R. and Thompson, T.E. (1977) Biochemistry 16, 4156–4163.
129 De Kruijff, B. and Wirtz, K.W.A. (1977) Biochim. Biophys. Acta 468, 318–326.
130 Noordam, P.C., Van Echteld, C.J.A., De Kruijff, B. and De Gier, J. (1981) Biochim. Biophys. Acta 646, 483–487.
131 Bloj, B. and Zilversmit, D.B. (1976) Biochemistry 15, 1277–1283.
132 Crain, R.C. and Zilversmit, D.B. (1980) Biochemistry 19, 1440–1447.
133 Rousselet, A., Guthmann, C., Matricon, J., Bienvenue, A. and Devaux, P.F. (1976) Biochim. Biophys. Acta 426, 357–371.
134 Renooij, M., Van Golde, L.M.G., Zwaal, R.F.A. and Van Deenen, L.L.M. (1976) Eur. J. Biochem. 61, 53–58.
135 Renooij, W. and Golde, L.M.G. (1976) FEBS Lett. 71, 321–324.
136 Zilversmit, D.B. and Hughes, M.E. (1977) Biochim. Biophys. Acta 469, 99–110.
137 Van den Besselaar, A.M.H.P., De Kruijff, B., Van den Bosch, H. and Van Deenen, L.L.M. (1978) Biochim. Biophys. Acta 510, 242–255.
138 Jackson, R.L., Westerman, J. and Wirtz, K.W.A. (1978) FEBS Lett. 94, 38–42.
139 Brophy, P.J., Van den Besselaar, A.M.H.P. and Wirtz, K.W.A. (1978) Biochem. Soc. Trans. 6, 280–281.
140 Brophy, P.J., Burbach, P., Nelemans, A.D.S., Westerman, J., Wirtz, K.W.A. and Van Deenen, L.L.M. (1978) Biochem. J. 174, 413–420.
141 De Kruijff, B., Van den Besselaar, A.M.H.P., Van den Bosch, H. and Van Deenen, L.L.M. (1979) Biochim. Biophys. Acta 555, 181–192.
142 Cullis, P.R. and De Kruijff, B. (1979) Biochim. Biophys. Acta 559, 399–420.
143 Barsukov, L.I., Hauser, H., Hasselbach, H.J. and Semenza, G. (1980) FEBS Lett. 115, 189–192.
144 Crain, R.C. and Zilversmit, D.B. (1981) Biochemistry 20, 5320–5326.
145 Wirtz, K.W.A. and Zilversmit, D.B. (1969) Biochim. Biophys. Acta 187, 468–476.
146 Eggens, I., Valtersson, C., Dallner, G. and Ernster, L. (1979) Biochem. Biophys. Res. Commun. 91, 709–714.
147 Lord, J.M. (1976) Plant Physiol. 57, 218–223.
148 Trémolières, A., Dubacq, J.P., Drapier, D., Muller, M. and Mazliak, P. (1980) FEBS Lett. 114, 135–138.
149 Roughan, P.G., Holland, R. and Slack, C.R. (1979) Biochem. J. 184, 193–202.
150 Helmkamp Jr., G.M. (1980) Biochemistry 19, 2050–2056.
151 Kasper, A.M. and Helmkamp Jr., G.M. (1981) Biochemistry 20, 146–151.
152 Morré, D.J. (1977) in Cell Surface Reviews (Poste, G. and Nicolson, G.L., eds.), Vol. 4, pp. 1–83, Elsevier, Amsterdam.
153 Brophy, P.J. and Aitken, P.J. (1979) J. Neurochem. 33, 355–356.
154 Teerlink, T., Poorthuis, B.J.H.M., Van der Krift, T.P. and Wirtz, K.W.A. (1981) Biochim. Biophys. Acta 665, 74–80.
155 Lapetina, E.G. and Miller, R.H. (1973) FEBS Lett. 31, 1–10.

CHAPTER 9

Phospholipases

HENK VAN DEN BOSCH

Laboratory of Biochemistry, State University of Utrecht, Padualaan 8, NL-3584 CH Utrecht, The Netherlands

1. Introduction

When the previous volume on lipid metabolism in this series appeared in 1970, phospholipases were dealt with in only a few pages as part of a general contribution on phospholipid metabolism by Thompson [1]. At that time the occurrence of phospholipase A in many different cell types began to be firmly established through the use of radioactive phospholipid substrates. Numerous publications have appeared in the last decade to extend the initial observations. Rapid progress has been made in the elucidation of the primary structure of pancreatic and venom phospholipases A_2 and detailed information on the mechanism of action of these enzymes is now available [2] (see Chapter 10). Many phospholipases from sources other than venoms and pancreatic tissue have now been purified. Consequently, this volume contains two chapters on phospholipases.

In principle any ester linkage in glycerophospholipids is susceptible to enzymic hydrolysis. The enzymes involved in this hydrolytic cleavage and their sites of attack are indicated for phosphatidylcholine in Fig. 1. Hydrolysis of fatty acylester bonds is catalyzed by phospholipases A. It is now clear that different enzymic activities exist, removing fatty acids from either the *sn*-1- or *sn*-2-position of the glycerol moiety. To differentiate between these different positional specificities the terms phospholipase A_1 (EC 3.1.1.32) and phospholipase A_2 (EC 3.1.1.4) have been proposed [3]. According to this nomenclature phospholipases A_1 and A_2 should produce equimolar amounts of free fatty acid and 2-acyl lysophosphatidylcholine * or 1-acyl lysophosphatidylcholine, respectively. Hydrolysis of the acyl ester bond in lysophosphatidylcholines is catalyzed by lysophospholipases (EC 3.1.1.5). The Enzyme Commission of the International Union of Biochemistry has used the term phospholipase B as a synonym for lysophospholipase. In the older literature phospholipase B has also been used for enzyme preparations catalyzing the complete deacylation of diacylglycerophospholipids. For some time the prevailing opinion has been that these crude preparations contained either phospholipase A_1 or A_2 and a

* The IUPAC-IUB Commission (e.g. Biochem. J. 171 (1978) 21–35) would call this compound 1-lysophosphatidylcholine, but giving the position of the acyl group is less likely to be misunderstood.

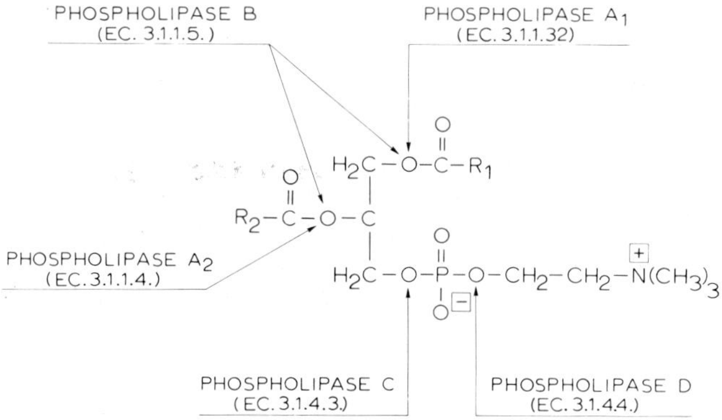

Fig. 1. Sites of attack of phospholipases on phosphatidylcholine.

lysophospholipase which converted the initially produced lysophospholipid to the fully deacylated compound. In recent years, however, several highly purified enzymes have been obtained which indeed appear to be capable of removing both acyl chains from a diacylglycerophospholipid. Such phospholipases B are always active towards lysophosphatidylcholines and in this sense also have lysophospholipase activity, but not all lysophospholipases purified so far are able to attack diacylglycerophospholipids (cf. Sections 2b and 4b). The nomenclature remains somewhat confusing, therefore, especially since in vitro the apparent specificity appears to depend on environmental conditions (e.g. presence or absence of detergents) and in vivo the functioning of the enzymes is not always known. Hydrolysis of phosphodiester bonds in phosphatidylcholine (Fig. 1) is catalyzed by phospholipase C (EC 3.1.4.3) to yield 1,2-diacylglycerol and phosphocholine or by phospholipase D (EC 3.1.4.4) to give phosphatidic acid and choline.

This review deals with intracellular phospholipases A and B and lysophospholipases and their involvement in phospholipid metabolism. At present, studies on pancreatic and venom phospholipase A_2 have reached a much higher level of sophistication and these will be discussed by de Haas et al. in Chapter 10 of this volume. In addition, phospholipases C and D are discussed in general terms, excluding phospholipase C-type enzymes acting on sphingomyelin (see Chapter 4 by Barenholz and Gatt) and phosphatidylinositol (see Chapter 7 by Hawthorne).

2. *Phospholipases A_1*

(a) Occurrence and assay

Phospholipase A_1 activities, i.e. lipolytic enzymes that remove the fatty acid from the 1-position of diacylglycerophospholipids, have been found in both prokaryotic [4–7] and eukaryotic cells [3,8–14]. These references constitute only a few selected examples out of many published papers to indicate the widespread occurrence of this type of lipolytic activity. More detailed compilations on the occurrence of phospholipase A_1 activities can be found in several reviews [15–17] and a monograph devoted entirely to lipolytic enzymes [18]. Within eukaryotic cells the enzyme does not appear to be localized at a single site. Thus, in rat liver phospholipase A_1 activity has been reported to be a true constituent of the plasma membrane [19–22], microsomes [20,21,23,24] and Golgi membranes [21]. In addition, soluble phospholipases A_1 have been described in lysosomes [20,25] and the cytoplasm [23,26]. Remarkably, the soluble phospholipase A_1 enzymes were either not affected [26] or inhibited by addition of Ca^{2+} [20,25]. All other phospholipases A_1 in rat liver are stimulated by the presence of this bivalent cation. This and the difference in pH optimum, acidic for the soluble enzymes and alkaline for the membrane-bound enzymes, strongly suggest the presence of different protein entities with phospholipase A_1 activities in liver tissue. It would be interesting to know whether the alkaline phospholipase A_1 activities found in various subcellular membranes are due to a single protein entity which is present in the different membranes or whether each membrane possesses a different protein with phospholipase A_1 activity. Such questions can only be answered definitively by purification of the enzymes from different subcellular sources. Although today several phospholipases A_1 have been obtained in highly purified form (see Section 2b) the approach of purifying the enzyme from different subcellular fractions of a given tissue has not yet been undertaken.

Purification of an enzyme can only be successfully attempted if a rapid assay method is available. In the case of phospholipase A_1, as is general with intracellular phospholipases A, the assay method should also be sensitive because of the low activity of these enzymes in crude subcellular fractions. Usually, this activity does not exceed a few nmol of substrate hydrolyzed per min and per mg of protein. Consequently, most assays are based on the release of fatty acids from radioactive substrates. It should be realized that the release of a labelled fatty acid from the *sn*-1-position of a diacylglycerophospholipid per se does not prove the presence of a phospholipase A_1. The action of a phospholipase A_2 in combination with a lysophospholipase would give the same result. In initial experiments with crude subcellular fractions the stoichiometry of fatty acid and 2-acyl lysophospholipid formation should be established. Doubly labelled substrates are very useful to reach this goal [8,27]. When single-labelled substrates are used conditions should be worked out so as to minimize the lysophospholipase activity as much as possible. This has often been done by addition of deoxycholate [8,27], although there is some danger in using detergents to inhibit the lysophospholipase activity (cf. Section 2b). Once the lipolytic enzyme has been identified as catalyzing a phospholipase A_1 reaction by careful analysis of product formation using thin-layer chromatography, a more rapid assay method is highly desirable. To circumvent tedious thin-layer chromatographic procedures, methods have been developed in which one of the radioactive reaction products, either lysophospholipid or free fatty acid, is extracted selectively. Extraction of the lysophospholipid was used by Scandella and Kornberg [4] in the purification of a membrane-bound phospholipase A_1 from *Escherichia coli*. This assay is based on the solubility of lysophosphatidylglycerol in the upper water–methanol phase of a lipid extraction according to Bligh and Dyer [28]. Selective extraction of released radioactive fatty acid was applied in the purification of an enzyme with phospholipase A_1 activity from bovine pancreas [8]. The fatty acid was extracted by a modified Dole procedure. Care should be taken to remove labelled substrate from the heptane phase before quantitation of the extracted fatty acid [29].

(b) Purified enzymes and properties

Table 1 lists the purified enzymes with phospholipase A_1 activities. Highly purified lipases, known to hydrolyze not only triacylglycerols but also diacylphospholipids at the primary hydroxyl acylester bond [30,31] are excluded from this table. The possible relationship between intracellular phospholipases A_1 and lipases has recently been discussed in more detail [18,32].

Scandella and Kornberg [4] were the first to purify a phospholipase A_1 from the membranes of *E. coli* B. The enzyme was solubilized with SDS solutions saturated with *n*-butanol and purified about 5000-fold. The nearly homogeneous enzyme showed an M_r-value of 29000 and was optimally active at pH 8.4. Ca^{2+} stimulated enzymic activity. Triolein was not hydrolyzed, thus distinguishing the catalytic capacity of the enzyme from the phospholipase A_1 activity of pancreatic and fungal lipases. 1-Acyl lysophosphatidylethanolamine was degraded about two-fold faster than phosphatidylethanolamine. These rate comparisons were made in the presence of Triton X-100. This detergent had no influence on the V_{max} of phosphatidylethanolamine and phosphatidylglycerol hydrolysis, but was shown to stimulate diphosphatidylglycerol hydrolysis about 100-fold and to increase the apparent K_m for phosphatidylglycerol about 40-fold. These findings stress the diverse effects that detergents can exert on the kinetic constants of lipolytic enzymes. It is obvious from the above examples that such effects can vary greatly with the lipid substrate. The

TABLE 1

Purified enzymes with phospholipase A_1 activity

Source	Authors	Ref.
Escherichia coli B	Scandella and Kornberg	4
Escherichia coli K-12	Nishijima et al.	5
Mycobacterium phlei	Nishijima et al.	6
Bacillus megaterium	Raybin et al.	7
Penicillium notatum	Kawasaki et al.	33
Bovine pancreas	Van den Bosch et al.	8
Human brain	Woelk et al.	9

influence of Triton X-100 on the hydrolysis of 1-acyl lysophosphatidylethanolamine was not studied. At present, the enzyme can therefore best be denoted as a phospholipase A_1 with lysophospholipase activity. However, if the general rule that lysophospholipases are inhibited by detergents [15] also applies to this lipolytic enzyme from *E. coli,* the hydrolysis of lysophosphatidylcholine in the absence of Triton X-100 might exceed that of phosphatidylethanolamine by more than the factor of two observed in the presence of Triton X-100. In that case the enzyme would more appropriately be named a lysophospholipase with phospholipase A_1 activity. An enzyme with these properties has been isolated in highly purified form from bovine pancreas [8]. The brief discussion of this enzyme will serve as another example of the large influences detergents can have on the velocity of enzymic reactions with different phospholipids as substrate. Although these effects are generally recognized and accepted it should be realized that by exerting these influences detergents may completely change the apparent specificity of a lipolytic enzyme (Fig. 2). The bovine pancreatic enzyme showed low activity with diacylphospholipids in the absence of detergents. Addition of deoxycholate, however, stimulated the hydrolysis about 25-fold. Under these conditions the enzyme produced equimolar amounts of free fatty acid and 2-acyl lysophospholipid, indicative of phospholipase A_1 activity. Diacylphospholipids were hydrolyzed about 5 times faster than 1-acyl lysophospholipid and about 50 times faster than 2-acyl lysophospholipid. With diacylphospholipid the reaction stopped at the level of lysophospholipid formation because the amounts of deoxycholate used to stimulate optimally phospholipase A_1 activity appeared to inhibit lysophospholipase activity by more than 95%. In the absence of deoxycholate the phospholipase A_1 activity was

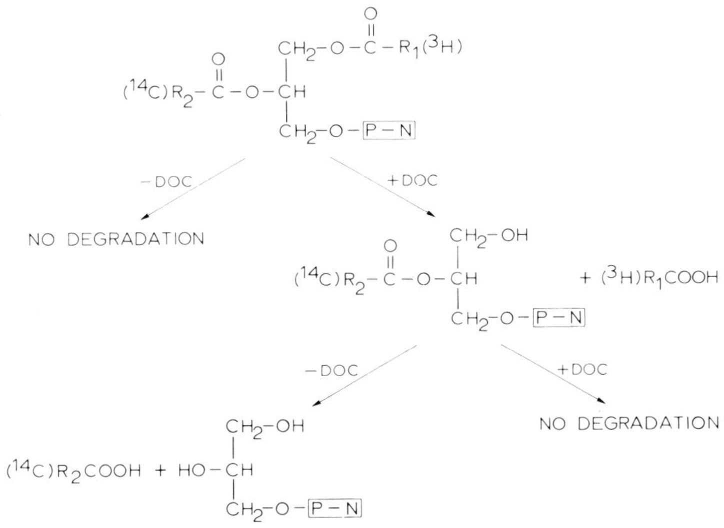

Fig. 2. Change of apparent specificity of a lipolytic enzyme from bovine pancreas dependent upon deoxycholate concentration.

reduced 25-fold and the lysophospholipase activity was fully expressed. This resulted in a ratio of about 200 for the rate of hydrolysis of 1-acyl lysophospholipid versus that of diacylphospholipid. In the absence of detergent the enzyme therefore acts almost exclusively as a lysophospholipase. At intermediate levels of deoxycholate, both the phospholipase A_1 activity and the lysophospholipase activity towards the initially produced 2-acyl lysophospholipid were expressed. Under these conditions the single protein (M_r 60000) exhibited phospholipase B activity, i.e. catalyzed the complete deacylation of diacylphospholipids.

The classical example of a phospholipase B is that of the fungus of *Penicillium notatum*. Kawasaki and Saito [34] purified this enzyme 2300-fold. The ratio of 1-acyl lysophosphatidylcholine hydrolysis to phosphatidylcholine hydrolysis in the absence of detergents was about 100 and this ratio remained essentially constant over the purification procedure, strongly suggesting that one enzyme attacked both diacyl- and monoacyl phosphatidylcholine. The enzyme, optimally active at pH 4.0, had an M_r-value of 116000 and an isoelectric point of 4.0. Subsequent studies [33] showed that the ratio of 100:1 for monoacyl-hydrolase to diacyl-hydrolase activity in the absence of Triton X-100, changed to 1:1 in the presence of this detergent. The apparent K_m for egg phosphatidylcholine increased 16-fold by addition of Triton X-100 (cf. *E. coli* phospholipase A_1). In agreement with what had been found for the pancreatic phospholipase A_1/B [8] the *P. notatum* enzyme was inhibited by diisopropyl fluorophosphate [33], suggesting that these enzymes belong to the class of serine hydrolases. When the *P. notatum* enzyme was incubated with phosphatidylcholine in the presence of Triton X-100 it was possible to detect some accumulation of lysophosphatidylcholine and hence to determine the initial site of attack. These studies revealed an important difference in the mode of action of the fungal and mammalian enzyme. While the mammalian enzyme attacked diacylphospholipids initially at the *sn*-1-position [8], the fungal enzyme preferentially removed first the acyl chain from the *sn*-2-position [33,35]. This proposed sequence is corroborated by the finding that 1-*O*-alk-1′-enyl-2-acyl- and 1-*O*-alkyl-2-acyl-*sn*-glycero-3-phosphocholine appear to be deacylated by the enzyme [36]. However, when 1-acyl- and 2-acyl lysophosphatidylcholines were used as substrates, the enzyme showed a preference for the *sn*-1-position with the 1-acyl isomer being hydrolyzed 15 times faster [36].

Quite recently, two improved methods for the purification of the *P. notatum* enzyme have been reported. Purification by hydrophobic chromatography on palmitoyl cellulose [37] yielded a preparation with a specific lysophospholipase activity of 3430 U/mg*, comparing favourably with the initially reported 2730 U/mg [33]. Even better results were obtained by Saito and collaborators [38], who applied phosphatidylserine-AH Sepharose 4B affinity chromatography and obtained preparations with a specific activity of over 5000 U/mg. The purified preparation gave one band in SDS disc gels in the absence of β-mercaptoethanol with an

* All enzyme units referred to in this paper were recalculated to μmol/min.

apparent M_r of 90000 (in contrast to the 116000 found earlier by gel-filtration). In the presence of β-mercaptoethanol three additional bands were detected in the gels with apparent M_r-values of 68000, 38000 and 33000. This phenomenon was not found when the enzyme was extracted from cells in the presence of the protease inhibitor phenylmethylsulphonylfluoride at concentrations which apparently did not inactivate the phospholipase B. These results suggest that endogenous proteases can modify the covalent structure of the enzyme while leaving the tertiary structure intact, presumably through disulphide bridges. Disruption of these bridges by β-mercaptoethanol in the presence of SDS leads the protein to dissociate into smaller peptides. The modification by endogenous proteases had no effect on the lysophospholipase activity of the enzyme, but markedly reduced its phospholipase B activity (4-fold in the presence of Triton X-100 and 9-fold in the absence of this detergent). As pointed out by the authors [38] the earlier reported ratios of monoacyl- to diacylhydrolase activities should not be regarded as absolute values in view of these differential effects of endogenous proteases on the two catalytic activities of the enzyme.

Nojima and coworkers [5] solubilized and purified an enzyme from *E. coli* K-12 which most likely represents the K-12 analog of the *E. coli* B enzyme isolated by Scandella and Kornberg. The M_r-value of 28000, the pH optimum of 8.0 and the Ca^{2+}-requirement resemble those of the *E. coli* B enzyme. Both enzymes were routinely assayed in the presence of 0.05% (w/v) Triton X-100. While the *E. coli* B enzyme produced only 2-acyl lysophospholipids, the *E. coli* K-12 enzyme formed both positional isomers under these conditions. This indicates that the initial attack can be at either the *sn*-1-position or the *sn*-2-position. At the level of lysophospholipids the 1-acyl isomer of lysophosphatidylethanolamine was hydrolyzed 5 times faster than the 2-acyl isomer. The enzyme therefore also exhibits phospholipase B activity [5].

Nojima's group also purified a lipolytic enzyme from the membranes of *Mycobacterium phlei* [6]. This enzyme had an apparent M_r-value of 45000 and was optimally active at pH 8.0 when assayed with neutral phospholipids. Acidic phospholipids were hydrolyzed best at pH 4.0. 1-Acyl- and 2-acyl-lysophosphatidylethanolamine were deacylated equally well at a rate about twice that observed with phosphatidylethanolamine at the same substrate concentration. When the enzyme was incubated with phosphatidylethanolamine some lysophosphatidylethanolamine accumulated and this was found to be almost exclusively the 2-acyl isomer. Thus, this enzyme resembled the *E. coli* B enzyme in its initial site of attack of diacylphospholipids. On the other hand the enzyme from *M. phlei* shared with the *E. coli* K-12 enzyme the property that the reaction does not stop at the lysophospholipid level. It can thus also be classified as a phospholipase B.

Certainly the most active and perhaps the most specific phospholipase A_1 has been obtained from *Bacillus megaterium* spores by Raybin et al. [7]. Phosphatidylglycerol was hydrolyzed optimally around pH 6.0 with a specific activity of 1560 U/mg. The enzyme required negatively charged substrate or substrate-detergent complexes. Tributyrin, even in the presence of anionic detergents, was degraded at a rate less than 0.2% that of phospholipids. The lysophospholipids produced were shown to have the 2-acyl configuration, i.e. the enzyme acted as a phospholipase A_1. Lysophospholipids were only tested in the presence of detergents and at low substrate concentration. It cannot be excluded, therefore, that the enzyme would exert some phospholipase B activity in the absence of detergents. Similar considerations hold for a partially purified phospholipase A_1 obtained from acetone-dried powders of human brain [9]. This enzyme (M_r 75000; pH optimum 4.0) specifically released fatty acids from the 1-position of phospholipids in the presence of Triton X-100 and taurocholate, albeit at a low rate of 0.04 U/mg.

Summarizing, it can be stated that the phospholipases A_1 discussed in this section do not show the high degree of specificity observed with pancreatic and venom phospholipases A_2. Only the *E. coli* B and K-12 enzymes require Ca^{2+}, while the others do not. The relative rates for hydrolysis of diacyl- and monoacylphospholipids are subject to large variation depending on the nature and concentration of added detergents. It seems justified to predict that for all enzymes of Table 1 conditions can be found to have them act in vitro as phospholipases B. Unfortunately, it cannot be deduced from experiments in vitro what activity the enzymes exert in vivo, i.e. phospholipase A_1, lysophospholipase or phospholipase B activity. Neither can it be stated which intracellular factors, if any, take over the in vitro modulation of activity by detergents.

3. *Phospholipases A_2*

(a) Occurrence and assay

Phospholipases A_2 are most abundant in the venom of snakes, bees and scorpions [2]. In mammals the enzyme occurs in highest amounts in pancreatic secretions. These enzymes and their assay methods are discussed in the next chapter [2]. Phospholipase A_2 activity has been found in almost any cell that has been investigated for its presence, both prokaryotic [39] and eukaryotic [11,15,20–25,27,40–46]. Within the hepatocyte, phospholipase A_2 activity is most easily demonstrated in mitochondria [47–49] since this organelle does not contain phospholipase A_1 [48] or lysophospholipase [47] to any appreciable extent. In other subcellular fractions of liver cells phospholipase A_2 activity has always been found in conjunction with phospholipase A_1 activity. Since none of the phospholipase A_2 activities from these subcellular fractions, except the one from mitochondria [49], has been purified, there is only circumstantial evidence but no firm proof that separate phospholipase A_2 enzymes are present. The possibility that a single phospholipase B-type enzyme accounts for the apparent A_1- and A_2-activities cannot be completely disregarded. With these reservations in mind phospholipase A_2 activities have been described for hepatic plasma membranes [19–22,50], microsomes [20,21,23,24], Golgi membranes [21] and lysosomes [20,25]. With the exception of the lysosomal activity, all phospholipase A_2 activities had an alkaline pH optimum between pH 8.0 and 9.5 and were stimulated by Ca^{2+} ions. The lysosomal enzyme was optimally active at pH 4.0–5.0 and was inhibited rather than stimulated by Ca^{2+}. As mentioned above, the phospholipase A_2 from rat liver mitochondria has been solubilized from the membranes (the enzyme seems to be present in both inner and outer membrane) and partially purified [49]. The 160-fold purified preparation appeared in the void volume fractions of a Sephadex G-200 column, most likely as a result of protein aggregation. This behaviour made it impossible to investigate whether inner and outer mitochondrial membranes contain identical or different phospholipase A_2 species. Phospholipase A_2 has also been detected in mitochondria from myocardial tissue [12].

Again, the activity of intracellular phospholipases A_2 in crude subcellular fractions is at best in the order of a few nmol/min/mg protein. The continuous titration of released fatty acids, the method of choice for venom and pancreatic phospholipases A_2 [2], is not sensitive enough to detect the intracellular phospholipases A_2. Most commonly, the enzyme is assayed by using radioactively labelled phospholipids. Thus, the methods have all the drawbacks and pitfalls discussed for phospholipase A_1 assays in Section 2a. The potential for a continuous assay with sufficient sensitivity has recently been demonstrated [29,51]. The method utilizes a substrate analogue in which the acyl group is attached to the backbone of the molecule in thioester rather than oxyester linkage. During hydrolysis thiol groups are released which can be detected continuously by carrying out the reaction in a spectrophotometer cuvette in the presence of chromogenic thiol reagents. The principle was introduced by using monoacyl phospholipids as substrates for lysophospholipases (cf. Section 4a). The advantage of having optically clear solutions of monoacylphospholipid micelles is obvious. Volwerk et al. [52] have recently synthesized short-chain phosphatidylcholines with acylthioester bonds that proved useful in studying monomer kinetics of pancreatic phospholipase A_2.

(b) Purified enzymes and properties

The purified phospholipases A_2 from sources other than venoms and pancreas are listed in Table 2.

Rahman et al. [41] obtained a phospholipase A_2 in soluble form from rat spleen homogenates by sonication. This suggests that the enzyme is not firmly bound to membrane structures. The purified enzyme showed specificity for the *sn*-2-position of phosphatidylethanolamine in a reaction with a requirement for Ca^{2+} ions. The enzyme had a pH optimum at 7.0, an M_r-value of 15000 and an isoelectric point of pH 7.4.

Starting from a lyophilized powder, obtained by therapeutic bronchoalveolar lavage of patients with alveolar proteinosis, Sahu and Lynn [42] solubilized an active phospholipase A_2 by delipidation of the powder. The cellular source for this enzyme is unknown. The M_r-value of the purified enzyme was estimated by gel filtration and SDS-polyacrylamide gel electrophoresis. Both methods yielded an M_r of 75000, suggesting that the enzyme consisted of a single polypeptide chain despite its

TABLE 2

Purified phospholipases A_2

Source	Authors	Ref.
Rat spleen	Rahman et al.	41
Human pulmonary secretions	Sahu and Lynn	42
Rabbit polymorphonuclear leukocytes	Elsbach et al.	43
Sheep erythrocytes	Kramer et al.	44
Rabbit platelets	Kannagi and Koizumi	45
Human platelets	Apitz-Castro et al.	46
Rat ascites hepatoma	Natori et al.	58

relatively high M_r-value in comparison to venom and pancreatic phospholipases A_2. The enzyme preparation contained alanine as a single *N*-terminal amino acid. Interestingly, the same *N*-terminal has been found in phospholipase A_2 from the pancreas of cow, pig and horse [53].

Elsbach and coworkers [43] showed that a membrane-associated phospholipase A_2 from rabbit polymorphonuclear leukocytes could be solubilized in good yield by treatment of homogenized or intact cells with 0.16 N sulphuric acid. The enzyme was purified 8000-fold from this extract and gave a preparation with a specific activity of 5 U/mg when tested with autoclaved *E. coli* containing labelled phospholipids. This assay uses rather low substrate concentrations and it is claimed that purified venom phospholipases A_2 give specific activities comparable to those found for the leucocyte phospholipase A_2. The enzyme showed an absolute requirement for Ca^{2+} ions and an estimated M_r-value of 14000. These properties are shared with pancreatic phospholipases A_2.

The question whether erythrocytes contain a phospholipase A_2 has been one of considerable debate [15,54]. The group of Zahler has recently purified such an enzyme from sheep erythrocytes. The enzyme was solubilized from ghost membranes with SDS, which was then replaced by cholate in a gel filtration step. While in cholate buffer, the enzyme was absorbed to an affinity column in the presence of Ca^{2+} and eluted with buffer containing EDTA according to the principle outlined by Rock and Snyder [55]. This behaviour of the enzyme on the affinity column is good proof that the enzyme requires Ca^{2+} for enzyme-substrate complex formation. Estimations by gel filtration and SDS-gel electrophoresis gave values of M_r 12000 and M_r 18500, respectively. The enzyme was shown to release preferentially polyunsaturated fatty acids from both phosphatidylcholine and phosphatidylethanolamine [56]. Treatment of intact cells or ghosts with proteases indicated that the membrane-associated phospholipase A_2 is oriented towards the exterior of the cell [57].

Two successful attempts to purify phospholipase A_2 from another type of blood cell, i.e. platelets, have recently been reported. Kannagi and Koizumi [45] started from rabbit platelets and extracted the membrane-bound enzyme in buffers with high salt but no detergent. Applying classical protein purification techniques, i.e. gel filtration and ion-exchange chromatography, yielded a not yet homogeneous, but 1020-fold purified preparation. The phospholipase A_2 in this preparation had a value of M_r 12000 as judged from gel chromatography. Apitz-Castro et al. [46] achieved a 1300-fold enriched phospholipase A_2 from human platelets in a two-step procedure, i.e. solubilization from homogenates with sulphuric acid followed by affinity chromatography. The enzyme gave one band upon electrophoresis in polyacrylamide gradients. Both gel filtration and gel electrophoresis gave an estimated value of M_r 44000. It is somewhat puzzling at present that the M_r-value of the human platelet enzyme is so at variance with that of rabbit platelets. Both enzymes were specific for the *sn*-2-position, optimally active at pH 9.0–9.5 and showed an absolute requirement for Ca^{2+}.

A phospholipase A_2, presumably located in the plasma membrane, was extracted with cholate from the particulate fraction of rat ascites hepatoma cells by Natori et al. [59]. Using ammonium sulphate precipitation, gel filtration and ion-exchange chromatography on DEAE- and CM-cellulose in the presence of detergents, either cholate, sodium dodecylsulphate or Triton X-100, a 13000-fold enriched fraction was obtained with a yield of 34%. These data illustrate adequately the enormous efforts that have to be made to obtain intracellular phospholipases A_2 in homogeneous form. For comparison, pancreatic phospholipase A_2 is obtained homogeneously after a 210-fold enrichment over crude homogenates [59]. The ascites enzyme, despite its high purification factor, was not yet homogeneous in that three bands were still seen in SDS-gels. Its M_r-value was not determined. Several interesting properties of the purified preparation were reported. Remarkably, the purified enzyme lost only 25% of its activity during a 5-min heat treatment at 95°C, whereas the enzyme when still in membrane-bound form was almost completely inactivated during similar treatment at temperatures above 70°C. Purified enzyme required Ca^{2+} and attacked phosphatidylethanolamine exclusively at the *sn*-2-position. 2-Acyl lysophosphatidylethanolamine, but not its 1-acyl isomer, was also hydrolyzed. This specificity for hydrolysis of the acylester bond in β-position to the phosphate ionization is consistent with observations made for venom and pancreatic phospholipase A_2 [59,60]. Characteristically, the enzyme hydrolyzed phosphatidylethanolamine and to a lesser degree phosphatidylglycerol, but phosphatidylcholine and cardiolipin were resistant. This substrate selectivity is in line with, as yet unexplained, observations that have been reported repeatedly, namely that intracellular phospholipases A are much more active with phosphatidylethanolamine than with phosphatidylcholine as substrate (e.g. [23,47–49].

(c) Regulatory aspects

In a simplified way membrane-associated phospholipases can be considered as enzymes floating in a sea of substrate and the question of regulation of phospholipase activity seems highly pertinent. Unfortunately, very little is known as yet about the molecular mechanisms by which the activity of membrane-associated phospholipases is controlled. Most of the research in this area has been directed towards the regulation of phospholipase A_2 activity in view of the role ascribed to this enzyme in the release of arachidonic acid for endoperoxide and prostaglandin formation (cf. Section 5b). As a result of these studies several models for phospholipase A_2 regulation have been put forward in recent years. Even though none of these has been firmly established it is felt appropriate to discuss the available evidence briefly. A more elaborate account of this subject was given recently [32]. By analogy with other enzymes, where the regulation of activity is better known, the main models for phospholipase A_2 regulation involve zymogen to active enzyme conversion, availability of Ca^{2+} ions or interaction with regulatory proteins.

(i) Regulation of phospholipase A_2 activity by zymogen-active enzyme conversion
This model is based on the existence of an inactive zymogen of pancreatic phospholipase A_2. This zymogen is converted into active enzyme by a trypsin-catalyzed removal of a heptapeptide from the N-terminus of the single polypeptide chain [2]. The conversion of a zymogen into an active phospholipase A_2 represents a seemingly irreversible modulation of enzymic activity. The meaning of such a mechanism is easily understood for digestive enzymes, but is more difficult to envisage for intracellular, membrane-associated phospholipases A_2. For these enzymes reversible regulation would seem much more appropriate. On the other hand, a number of reports have described enhanced phospholipase A_2 activity upon treatment of isolated membranes or intact cells with proteolytic enzymes. Trypsin treatment led to a several-fold increased phospholipase A activity in various rat tissues [61], lysates of human erythrocytes [62] and rat plasma [63]. However, it has later been reported that the activation of rat plasma phospholipase A was due to an activating factor in the crude preparation of trypsin which was different from trypsin itself [64] and also the activation of the phospholipase A activity in human red cell lysates could not be repeated with pure trypsin [65]. More recently, increased phospholipase activity has been observed upon treatment of human platelets [66] and transformed mouse fibroblasts [67] by either trypsin or thrombin. It is not known, however, in these cases whether proteolysis causes an increased phospholipase activity by conversion of a proenzyme into active enzyme or whether secondary effects such as removal of inhibitory proteins, changes in Ca^{2+}-concentration or alterations in membrane structure are responsible for the effect of the proteolytic enzyme. Feinstein et al. [68] have demonstrated that phospholipase activation in platelets, as caused by thrombin or collagen, was blocked when the cells had first been treated with the protease inhibitor phenylmethanesulphonyl fluoride. This suggested that an endogenous serine-protease is involved somehow in the activation of the phospholipase A. It is obvious that the exact molecular mechanism underlying these proteolytic activations remains to be elucidated.

(ii) Regulation of phospholipase A_2 activity by availability of Ca^{2+} ions
Membrane-associated phospholipases A_2 require Ca^{2+} for their activity and recent evidence suggests that at least in platelets the phospholipase A_2 could be regulated by the extent of saturation of the enzyme with Ca^{2+} ions. Several authors have

shown that addition of the Ca^{2+}-ionophore A23187 to platelets led to the sudden release of arachidonate from platelet phospholipids [68–72] in much the same way as caused by thrombin or collagen. The experiments allow for a simple model in which the activity of the platelet phospholipase A_2 is regulated by free Ca^{2+} in the cytoplasm [71]. In this model the addition of ionophore or thrombin would result in increased cytoplasmic Ca^{2+} levels by releasing Ca^{2+} from an internal Ca^{2+} store and hence in increased phospholipase activity. Since the free cytoplasmic Ca^{2+} level is supposedly controlled by cyclic-AMP [73], the repeatedly reported finding that platelet phospholipase A_2 is inhibited by dibutyryl-cAMP [69,71,74,75] or agents which increase intracellular cAMP levels [68,74,75] can be fitted into this model. Cyclic-AMP causes a lowering of the cytoplasmic free Ca^{2+} by stimulating the storage of Ca^{2+} [73] and thus would lead to an inhibition of the ionophore- or thrombin-induced phospholipase A_2 activity. It seems unlikely, however, that the free cytoplasmic Ca^{2+} level as controlled by cAMP constitutes the sole factor which regulates platelet phospholipase A_2 activity. The above model assumes saturation of existing phospholipase A_2 molecules with Ca^{2+} as the common final step in thrombin- or collagen- and ionophore-induced phospholipase A_2 activation, which may well be an oversimplification. Feinstein et al. [68] have provided evidence to suggest that ionophore-induced stimulation would be due to Ca^{2+} saturation of existing phospholipase A_2 molecules. In contrast, collagen or thrombin, in addition to being able to increase cytoplasmic free Ca^{2+}, could cause the conversion of some inactive form of the phospholipase A_2 to an active form in a process involving an endogenous serine-protease. Certainly, the model of phospholipase A_2 regulation by availability of Ca^{2+} ions should not be generalized. Frei and Zahler [57] have studied the Ca^{2+} requirement of the phospholipase A_2 in washed sheep erythrocyte membranes. The enzyme was found inactive at Ca^{2+} concentration below 10^{-5} M, increased sharply above $5 \cdot 10^{-5}$ M and reached a plateau value at 0.5 mM Ca^{2+}. Since this phospholipase appeared to be oriented towards the exterior of the red cell [57] and the plasma concentration of Ca^{2+} is 1.5 mM these data seem to exclude a regulation of phospholipase A_2 activity by the availability of Ca^{2+} ions.

(iii) Regulation of phospholipase A_2 activity by interaction with regulatory proteins

This hypothetic model for phospholipase A_2 regulation can be put forward by analogy with proteolytic enzymes, where non-enzymic inhibitory proteins are well known [76]. The evidence for this type of regulation of phospholipases is still scanty. It might well be that the stimulated phospholipase A_2 activity observed after proteolysis, as discussed above, is due to proteolytic removal or modification of inhibitory proteins rather than to a direct modification of a proenzyme by the protease. Studies on prostaglandin release, however, have recently provided more direct evidence to suggest that inhibitory proteins for phospholipase A_2 exist. Anti-inflammatory corticosteroids have been suggested to interfere with prostaglandin production not by affecting the cyclooxygenase but by reducing the release of arachidonate substrate from intracellular phospholipids [77–80]. This inhibition of arachidonate release by corticosteroids appears to require RNA and protein synthesis [80,81]. These observations have led to the suggestion that corticosteroids induce the synthesis of a protein factor which inhibits phospholipase A_2 activity. Indeed, the perfusate of dexamethasone-treated lungs, in contrast to that of control lungs, was shown to contain a phospholipase A_2 inhibitor [81]. Recently, Hirata et al. [82] have provided additional evidence for the protein nature of such a factor. In response to glucocorticoids rabbit peritoneal neutrophils showed a decrease in phospholipase A_2 activity. Inhibitors of RNA and protein synthesis suppressed this inhibitory effect of glucocorticoids on phospholipase A_2 activity. In line with these results the membranes of glucocorticoid-treated cells, after solubilization and Sephadex G-200 filtration, were found to contain enhanced levels of material which inhibited pancreatic phospholipase A_2. This material had an apparent M_r-value of about 40000 and its protein nature was further deduced from the finding that glucocorticoid-treated cells after pronase digestion contained hardly any of the inhibitory material. Such cells retained full ionophore-induced phospholipase A_2 activity, suggesting a different localization of the phospholipase A_2 and its inhibitory protein in the plane of the membrane.

While the evidence for the occurrence of inhibitory proteins of phospholipase A_2 is thus starting to accumulate, indications for stimulatory proteins or peptides are still less direct. Nevertheless, Nijkamp et al. [83], in their attempts to purify and characterize rabbit aorta-contracting substance-releasing factor from anaphylactic lungs, have suggested that this material is a peptide consisting of less than 10 amino acids and demonstrated that it stimulated phospholipase A_2 activity of perfused lungs. It has recently also been demonstrated that chemotactic peptides enhance the release of arachidonic acid from phospholipids in rabbit neutrophils [84]. Also prostaglandin production in transformed mouse fibroblasts as stimulated by thrombin and bradykinin required protein synthesis to become expressed [85,86]. Since these stimuli did not affect the cyclooxygenase it was presumed that stimulated prostaglandin formation was due to enhanced phospholipase activity. Whether this increase in phospholipase A_2 activity is due to synthesis of new enzyme or to synthesis of a stimulatory protein remains to be determined.

By analogy, it should be noted that stimulation of lipolytic activities by non-enzymic proteins is well documented. Lipoprotein lipases are stimulated by apolipoprotein C-II [87,88] and lecithin-cholesterol acyltransferase requires apolipoprotein A-I for full activity [89,90]. In addition, several lysosomal hydrolases acting on complex glycolipids appear to be activated by non-enzymic proteins [91] and the activity of pancreatic lipase is greatly influenced by the presence of co-lipase [92]. Even in these cases, however, the detailed mechanism by which the activator protein exerts its action is not always understood and seems to be different in the various cases. Much effort will be required before the possible regulation of membrane-associated phospholipase A_2 by non-enzymic inhibitory and stimulatory proteins is fully unravelled.

4. *Lysophospholipases*

(a) Occurrence and assay

Lysophospholipases, defined here in general terms as enzymes that catalyze the hydrolysis of acylester bonds in lysophospholipids, occur widespread in nature. Not infrequently the presence of the enzyme was inferred from the observation that the breakdown of a diacylphospholipid proceeded at least partially to the completely deacylated product. Invariably, when such enzyme preparations were then assayed with a lysophospholipid, hydrolysis of this substrate was observed. It should be recalled that these observations do not necessarily prove the existence of a separate lysophospholipase, as enzymes (phospholipases B) are now known to exist which themselves catalyze the complete deacylation of diacylphospholipids (compare Section 2b). On the other hand, lipolytic enzymes active towards lysophospholipids but not diacylphospholipids are known as well. With these thoughts in mind, lysophospholipase activities have been detected in both prokaryotic and eukaryotic microorganisms and in almost all other eukaryotic cells that have been assayed for the presence of this enzyme. For example, in microorganisms, lysophospholipases have been reported in *E. coli* [93–96], *M. phlei* [97], *Saccharomyces cerevisiae* [98,99], *Tetrahymena pyriformis* [100], *Dictyostelium discoideum* [101], *Acanthamoeba castellanii* [102], *Neurospora crassa* [103], *Penicillium notatum* [104,105] and *Mycoplasma laidlawii* [106]. In addition, the enzyme has been found in insects [13], plants [107], fish [108] and mammalian tissues (see [15] and [109] for reviews). In a comparative study of rat tissues Marples and Thompson [110] detected high levels of lysophospholipase in intestine, lung, spleen, liver and pancreas and lower levels in muscle, kidney, testes, brain and blood.

Studies on the subcellular distribution of lysophospholipase activity in mammalian cells have not provided a uniform picture. In brain [111,112], heart [12] and adrenal medulla [113] most of the activity was found associated with the microsomal fraction. By contrast, the bulk of the lysophospholipase activity in rat liver [23,114,115], spleen [116] and lung [117] was recovered in the $100000 \times g$ supernatant. Even though in rat liver most of the activity appears soluble, considerable activity can still be detected in microsomes [118,119]. Mitochondria [47], plasma membranes [50] and even lysosomes [120] from rat liver do not seem to contain any significant amounts of lysophospholipase. On the other hand, lysosomal preparations have been demonstrated to catalyze the complete deacylation of both diacyl- and monoacylphospholipids [121,122], but it remains to be seen whether this is due to a combined action of phospholipases A_1 and A_2, to a single phospholipase B-type enzyme or to the consecutive action of either phospholipase A_1 or A_2 and a separate lysophospholipase.

As in rat liver, the presence of lysophospholipases is usually not restricted to a single subcellular site in eukaryotic cells. Leibovitz and Gatt [112] reported the microsomal fraction of rat brain to have the highest specific lysophospholipase activity, but found activity in the mitochondrial and cytosol fraction as well. Similar findings were reported for rat lung, intestine and kidney [119]. In brain, the particulate and soluble lysophospholipase activity exhibited quite different kinetic properties [123], suggesting that the kinetic properties were influenced by the physical form in which a single enzyme became expressed (i.e. membrane-associated or soluble) or that in fact different protein entities with lysophospholipase activity occurred. The latter was shown to be the case in bovine liver. From this tissue two proteins with lysophospholipase activity could be easily extracted and separated on DEAE-Sephadex columns [124]. These enzymes were provisionally denoted lysophospholipase I and lysophospholipase II (see next section for properties). In subsequent cell-fractionation studies the total lysophospholipase activity of the individual subcellular fractions was separated into a lysophospholipase I and II contribution, thus allowing a determination of the subcellular distribution of each individual enzyme, rather than of total lysophospholipase activity [125]. Lysophospholipase I appeared to be a soluble enzyme with a bimodal distribution. Highest relative specific activities were observed in the mitochondrial and cytosolic fractions. Evidence was provided to show that the mitochondrial enzyme is present in the matrix fraction. No differences between the mitochondrial and cytosolic form of lysophospholipase I have become apparent. The lysophospholipase II occurred in membrane-associated form with highest relative specific activity in the microsomal fraction. These data clearly demonstrate that the lysophospholipase activity of different subcellular fractions from bovine liver can be ascribed to different protein entities. The possibility of lysophospholipase I being an artifactual hydrolytic product of the larger lysophospholipase II was excluded [125,126].

Some properties of lysophospholipases can be deduced readily from crude or only partially purified enzyme preparations. Such studies have shown that lysophospholipases do not require Ca^{2+} for activity and that they are almost universally inhibited by such different kinds of detergents as Tween-80, saponin, cholate, Triton X-100, deoxycholate and hexadecyltrimethylammonium bromide and cetyltrimethylpyridinium bromide. As pointed out by Brockerhoff and Jensen [109], these results suggest that it is the substrate rather than the enzyme that is affected by the detergent. Presumably, the lysophospholipid substrate is incorporated into mixed micelles of substrate and detergent in such a way that it becomes inaccessible to the enzyme. It follows from these considerations that in crude preparations inhibition of a lysophospholipase activity and stimulation of e.g. a phospholipase A_1 activity by a given detergent constitutes poor evidence for the existence of two different enzymes. Differential effects of e.g. deoxycholate on the phospholipase A_1 and lysophospholipase activity of a single purified protein, because of a differential effect of the detergent on the different substrates used in the assay of these activities, illustrate this note of caution (cf. Section 2b).

Using rat brain microsomes as the enzyme source Gatt and co-workers [111,123] analyzed the irregular kinetic behaviour of the enzyme in response to variation in substrate concentration. These investigators arrived at the conclusion that the enzyme would only be active on substrate monomers with substrate micelles being inhibitory. In view of the occurrence of both 1-acyl and 2-acyl lysophospholipids in mammalian tissues we have investigated whether both isomers could be deacylated by lysophospholipases. A crude enzyme preparation, i.e. the $100000 \times g$ supernatant of rat liver, hydrolyzed 1-acyl-*sn*-glycero-3-phosphocholine twice as fast as the isomeric 2-acyl derivative [115]. In addition, monoacyl derivatives of *sn*-glycero-1-phosphocholine and *sn*-glycero-2-phosphocholine were deacylated. These data suggested a lack of both positional and stereo-specificity for lysophospholipases. The degradation of 1-acyl propanediol-3-phosphocholine at a rate comparable to that of 1-acyl lysophosphatidylcholine indicated further that the presence of a free hydroxyl group in the substrate is not mandatory for lysophospholipase action. However, since other lipolytic or esterolytic enzymes might be present in the $100000 \times g$ supernatant, more detailed studies on the substrate specificity of lysophospholipase required the availability of homogeneous enzymes (see next section).

Even though the activity of intracellular lysophospholipases is usually higher than that of phospholipases A_1 and A_2, as can be deduced from the fact that often little accumulation of lysophospholipid is observed during degradation of diacylphospholipid, the rate of fatty acid production in crude systems is still too small to be determined by continuous titration. Discontinuous assay procedures have therefore been used in most cases. After the reactions have been terminated, the lysophospholipase activity can be determined by classical methods from an analysis of the free fatty acids or the glycerophosphoester produced during the incubation. Mostly, however, lysophospholipase activity has been determined in recent years by employing radioactive lysophospholipids, especially those having a radioactive fatty acid. To circumvent time-consuming thin-layer chromatography, modified Dole-extraction procedures have been developed [127,128]. In these methods the reaction is terminated by addition of a mixture of hexane-isopropanol-sulphuric acid which

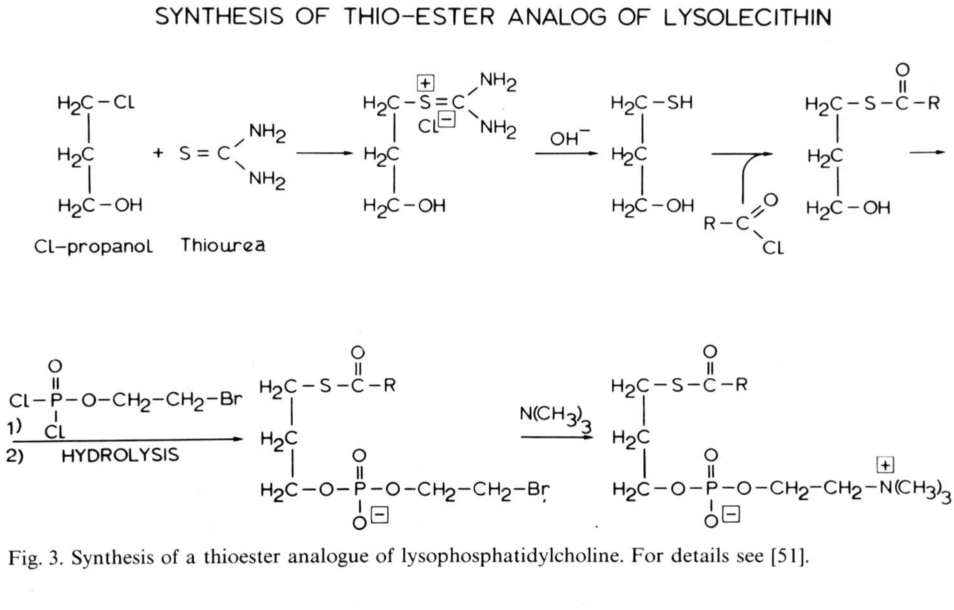

Fig. 3. Synthesis of a thioester analogue of lysophosphatidylcholine. For details see [51].

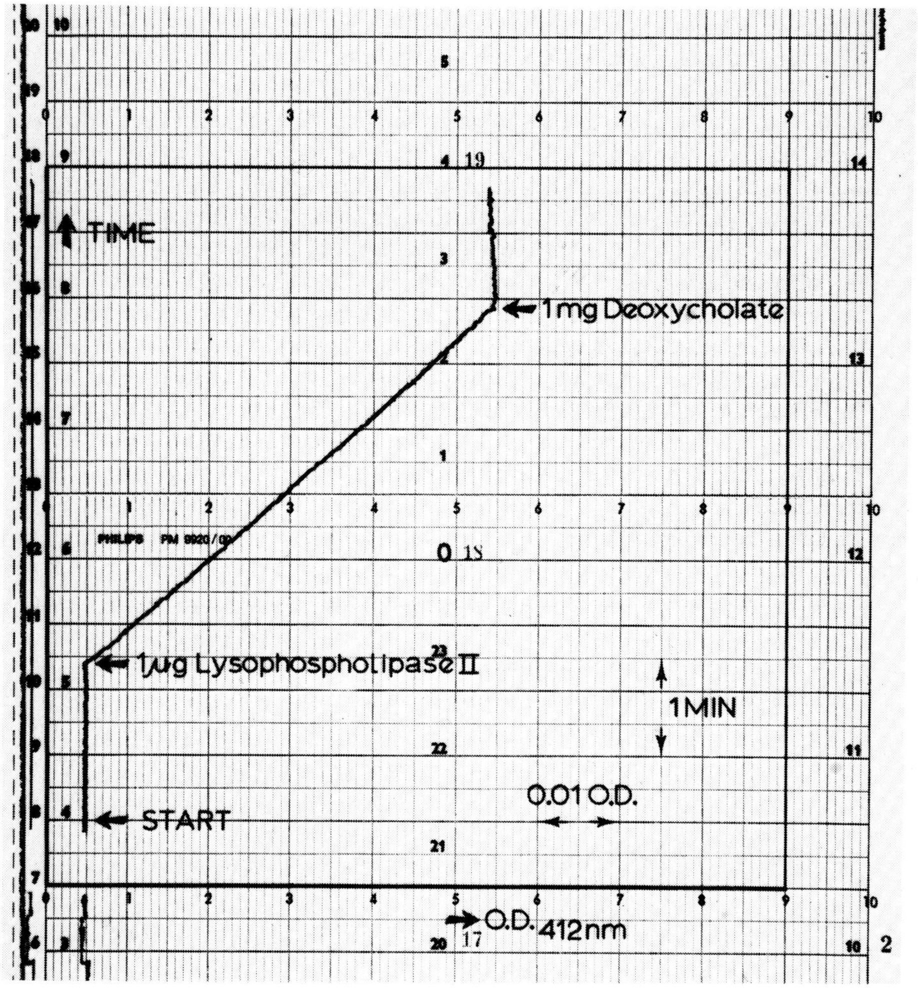

Fig. 4. Continuous spectrophotometric assay of lysophospholipase. The method uses a substrate analogue with an acylthioester bond in the presence of 5,5′-dithiobis-(2-nitrobenzoic acid). Upon addition of enzyme the increase in absorbance at 412 nm is recorded. The slope corresponds to an enzymatic activity of 1.1 nmol/min. (Reproduced with permission of Bioorganic Chemistry.)

extracts the fatty acid into the upper heptane phase and leaves the lysophospholipid substrate in the isopropanol-water phase, provided silica gel is also added to prevent lysophospholipid from partitioning partly into the heptane phase [29,115]. Although this is a rather fast method it is still a fixed-time assay with all the disadvantages inherent in discontinuous assays. Recently, an assay procedure was developed which allowed continuous measurement of enzymatic activity in a spectrophotometer. The method used a substrate analogue, i.e. 3-palmitoylthio-propane-1-phosphocholine, in which the acyl chain is esterified in a thioester rather than an oxyester linkage to the backbone (Fig. 3). Thus, as shown in Fig. 4, upon addition of enzyme, thiol groups are released which can be detected continuously in the presence of thiol reagents such as e.g. 5,5′-dithiobis-(2-nitrobenzoic acid). Also shown in Fig. 4 is the stability of the substrate in the absence of enzyme and the previously mentioned inhibitory effect of detergents on lysophospholipase activity. This assay method has proved to be useful in the determination of both soluble and membrane-bound lysophospholipases [129] and during the purification of these enzymes [130].

(b) Purified enzymes and properties

Despite the widespread occurrence of lysophospholipases relatively few attempts have been undertaken to purify this type of lipolytic activity to homogeneity for studies of its substrate specificity. Some proteins have been purified which hydrolyse both diacylphospholipids and monoacylphospholipids (cf. Section 2b), but these will not be reconsidered here. The lysophospholipases which have (virtually) no activity towards diacylphospholipids are listed in Table 3.

De Jong et al. [124] solubilized the lysophospholipase activity from bovine liver by treatment of homogenates with *n*-butanol. When the extract was chromatographed on DEAE-cellulose columns two protein peaks with lysophospholipase activity emerged, well separated, from the column. According to this appearance in the eluate these were denoted lysophospholipase I and lysophospholipase II. Lysophospholipase I required a 3600-fold enrichment to obtain a homogeneous enzyme. It had a specific activity of 1.5 U/mg and a value of about M_r 25000 both on Sephadex G-100 columns and SDS-polyacrylamide disc gel electrophoresis, indicating that it consisted of a single polypeptide chain. The enzyme showed a broad pH optimum between pH 6 and 8 and had an isoelectric point of pH 5.2. Lysophospholipase II was obtained in homogeneous form after a 770-fold purification. The single polypeptide chain had an M_r-value of about 60000 and an isoelectric point of pH 4.5. This enzyme was optimally active at pH 8.5 and then showed a specific activity of about 1.5 U/mg. Neither enzyme required Ca^{2+} and both were inhibited by sodium deoxycholate and Triton X-100, although enzyme II appeared to be more sensitive to these detergents. Also, enzyme I was more resistant to serine reagents such as diisopropylfluorophosphate and bis-*p*-nitrophenylphosphate. The

TABLE 3

Purified lysophospholipases

Source	Authors	Ref.
Bovine liver	De Jong et al.	124
Escherichia coli	Doi and Nojima	96
Vibrio parahaemolyticus	Misaki and Matsumoto	131
Rat lung	Brumley and Van den Bosch	117

latter reagents completely and stoichiometrically inhibited lysophospholipase II [132]. By contrast, lysophospholipase I was more sensitive to SH-reagents [124]. Both enzymes possessed general esterolytic properties, in that *p*-nitrophenylacetate and tributyrin were hydrolyzed as well. Long-chain diacylphospholipids, if attacked at all, were hydrolyzed at rates less than 2% of those observed with palmitoyl lysophosphatidylcholine. However, when enzyme II was assayed with short-chain phosphatidylcholines, an almost stoichiometric production of 2-acyl lysophosphatidylcholine and free fatty acid was seen, indicative of phospholipase A_1 activity towards these unnatural substrates [133]. Studies with a series of 1-acyl lysophosphatidylcholines demonstrated that the enzyme had virtually no activity with substrates having acyl residues of 8 or less C-atoms. The purified enzyme was fully active on substrate monomers and micelles (Fig. 5), with no abrupt changes in the activity vs. substrate concentration profile at the critical micellar concentration of the substrates [133]. Thus, the suggestion that a brain microsomal lysophospholipase was only active with substrate monomers and inhibited by substrate micelles [111,123] was not confirmed for the purified bovine liver enzyme. Although the stoichiometric inhibition of lysophospholipase II by diisopropylfluorophosphate suggested this enzyme to be a serine-hydrolase and the enzyme was shown to act by an acyl-cleavage mechanism [132], a covalent acyl-enzyme intermediate could not be isolated.

A lysophospholipase from *E. coli,* not active towards diacylphospholipid was purified about 1500-fold, to near homogeneity, by Doi and Nojima [96]. The enzyme was optimally active between pH 8 and 10 with a specific activity of about 2.6 U/mg and had a value of about M_r 39500. Hydrolysis of 1-acylglycerol proceeded at rates similar to those for 1-acyl lysophosphatidylethanolamine. The latter was degraded about 3 times faster than the 2-acyl isomer. No hydrolysis of *p*-nitrophenylacetate was observed, but several other properties were similar to those

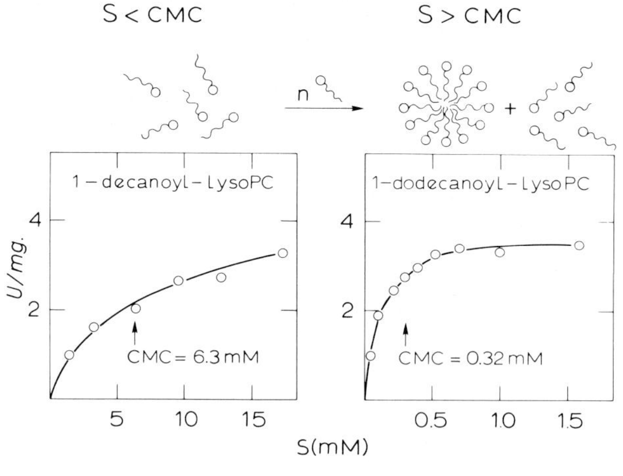

Fig. 5. Activity of bovine liver lysophospholipase II on monomeric and micellar lysophosphatidylcholine.

of bovine liver lysophospholipase. Among these are: inhibition by diisopropylfluorophosphate and detergents and activity towards both monomolecular and micellar forms of lysophospholipids [96].

A homogeneous lysophospholipase, obtained after a 592-fold purification from *Vibrio parahaemolyticus* by Misaki and Matsumoto [131] was 3 times more active with 2-acyl lysophosphatidylethanolamine than with the 1-acyl isomer. At its pH optimum of 10 the enzyme hydrolyzed 1-acyl lysophosphatidylcholine with a specific activity of 48 U/mg. Monoacylglycerol and tributyrin were not attacked. The enzyme did not require bivalent cations, was unaffected by SH-reagents and was completely inhibited by diisopropylfluorophosphate. Isoelectric focussing indicated an isoelectric point of pH 3.6. The M_r-value of 89000 as estimated from gel filtration and of 15000 on SDS-polyacrylamide disc electrophoresis suggested the enzyme to be a hexamer. Enzymatically active fractions with M_r-values of 45000 and 29000, presumably representing trimeric and dimeric forms, were found during Sephadex filtration at pH 3. Also a 20-fold purified lysophospholipase from rat brain gave an estimated M_r-value of about 15000–20000 [112].

Brumley and Van den Bosch [117] purified a lysophospholipase from rat lung $100000 \times g$ supernatant. The enzyme had a pH optimum of 6.5 and a specific activity of 2.1 U/mg. The M_r-value as estimated from gel filtration was 50000 and diisopropylfluorophosphate inhibited enzymic activity completely [134]. This enzyme hydrolyzes neither *p*-nitrophenylacetate nor monoacylglycerol [135]. Apart from its capacity to hydrolyze lysophosphatidylcholine, this enzyme was able to catalyze a transacylation between two molecules of lysophosphatidylcholine to yield phosphatidylcholine and glycerophosphocholine.

In summary, lysophospholipases appear to comprise a rather heterogeneous group of enzymes. They vary widely in their M_r-value and substrate specificity. In general, they appear to be lipolytic esterases, which hydrolyze acylester bonds in rather hydrated lipidic substrates, either monomers or micelles. Consequently, some enzymes (e.g. bovine liver lysophospholipase II) are active towards short-chain but not long-chain diacylphospholipids. Other lysophospholipases hydrolyze monoacylglycerol (e.g. *E. coli*) and may in fact be identical to monoglyceride lipase or acyl-CoA hydrolase. Paradoxically then, certain enzymes that have been classified as lysophospholipases do not even show an absolute requirement for a phosphate group in their substrate. The enzymes from *V. parahaemolyticus* and rat lung, which hydrolyze neither monoacylglycerol nor diacylphospholipid, are presently the two enzymes which exhibit the greatest specificity for lysophospholipids. The substrate preference of some other lysophospholipases extends apparently somewhat further into the hydrophobic region, allowing these enzymes to act under certain conditions on long-chain diacylphospholipids. Such enzymes are currently denoted as phospholipases B (Section 2b).

5. Functions of phospholipases A and lysophospholipases

Over the years many different functions have been ascribed to intracellular phospholipases depending on the cell type or subcellular organelle where these enzymes were detected. It is impossible to review all these in this chapter. Instead, the discussion will be confined to a few thoughts on the functions of phospholipases A and lysophospholipases in phenomena that may be of more general significance.

(a) Phospholipid turnover

It is now well-recognized that membrane phospholipids exist in a dynamic flux in which continuous biosynthesis is balanced by degradation. At the same time, it appears to be extremely difficult to formulate compelling reasons for the necessity of this dynamic turnover. A more elaborate discussion to illustrate these points was recently presented [32].

The catabolic part of the turnover of diacylglycerophospholipids is thought to proceed in most cells by a stepwise deacylation catalyzed by the ubiquitous phospholipases A_1 and A_2 and lysophospholipases. Apart from these hydrolytic enzymes, many cells appear to contain phospholipase C-type enzymes which are specific for phosphatidylinositol [136]. It is only recently that evidence for the existence of phospholipase C activity towards the other diacylglycerophospholipids in mammalian tissues has been obtained [137,138]. This enzyme was localized in the soluble fraction of rat liver lysosomes. Evidence was provided to indicate that a soluble, delipidated lysosomal protein fraction contained all the enzymes for the degradation of phosphatidylcholine by two pathways, i.e. complete deacylation to glycerophosphocholine and phosphodiester cleavage to yield phosphocholine and diacylglycerol, which was then further metabolized to monoacylglycerol [137]. This study added the phospholipase C pathway to the well-established deacylation pathway [139] in the lysosomal digestion of all major glycerophospholipids. Previously, the lysosomal phospholipase C was shown to have a marked specificity towards phosphatidylinositol, although a slow hydrolysis of phosphatidylcholine, to an extent comparable to deacylation, was also reported [140]. When microsomal membranes containing labelled phospholipids were incubated with lysosomal extracts, phosphatidylinositol degradation proceeded mainly via the phospholipase C pathway, whereas phosphatidylcholine and phosphatidylethanolamine hydrolysis took place largely via the deacylation pathway, albeit with a considerable accumulation of the lyso-derivatives of these phospholipids [141]. However, lysosomal digestion of autophagocytosed membranes, although most likely contributing to the phenomenon of phospholipid turnover, has to be distinguished from the independent turnover of membrane phospholipids. This is suggested by the different half-lives observed for membrane proteins and membrane lipids [142,143]. Since the phospholipase C is exclusively located within the lysosomes, the available evidence suggests that this part of phospholipid turnover occurs via a stepwise deacylation and without any appreciable accumulation of the intermediary lysophospholipids

[32]. Purified lysophospholipases have been shown to be active on lysophospholipid embedded in model membranes [144] and in microsomal membranes [145]. Evidence has been provided to show that lysophospholipid deacylation is almost linearly dependent on the substrate density in the membranes, suggesting that increased concentrations of membrane-associated lysophospholipids may result in enhanced deacylation rates to keep the concentration of these lytic components within acceptable limits.

The available data indicate that the activity of membrane-associated phospholipases A and that of intracellular lysophospholipases is sufficient to account for the half-lives of the major membrane phospholipids ([32] and refs. therein).

A large body of evidence is now available to sustain the conclusion that synthesis de novo of phosphatidylcholine, -ethanolamine and -inositol produces primarily the monoenoic and dienoic species of these phospholipids with palmitate at the *sn*-1-position (see [32] and [146] for reviews; also Chapter 1). Yet these lipids are known to contain considerable amounts of stearate at the *sn*-1-position and arachidonate at the *sn*-2-position, which are held to be incorporated largely through deacylation-reacylation mechanisms. This independent turnover of the acyl constituents of phospholipids requires the action of phospholipases A_1 and A_2 on de novo synthesized monoenoic and dienoic species, to provide the 2-acyl and 1-acyl lysophospholipids necessary for acylation with stearate and arachidonate, respectively. Indeed, the lysophosphatidylcholine fraction from rat liver was shown to consist of a mixture of the 1-acyl and 2-acyl isomer [27] and the presence of specific acyl-CoA:lysophospholipid acyltransferases is well documented [147]. The involvement of phospholipases A_1 and A_2 in the remodelling of phosphatidylcholine and -ethanolamine species during incubation of isolated hepatocytes was clearly demonstrated by Kanoh and Åkesson [148]. In line with the notion that monoenoic and dienoic species serve as precursors for tetraenoic species, several-fold shorter half-lives for the former species have been observed in vivo [149–151].

(b) Release of prostaglandin precursors

Another area of lipid physiology in which phospholipases have been implicated in the last decade is in providing arachidonate to the cyclooxygenase for cyclic endoperoxide and subsequent prostaglandin, thromboxane and prostacyclin synthesis. It has become well-accepted that endoperoxide formation is limited by the availability of free arachidonate [152–154]. In view of the preponderant occurrence of this acid at the *sn*-2-position of glycerophospholipids, phospholipase A_2 was proposed to be mainly responsible for arachidonate release that preceded prostaglandin formation. Some of the evidence to support this idea was discussed in Section 3c. Thus, indications for phospholipase A_2 as the enzyme responsible for arachidonate release were reported in experiments with e.g. platelets [68,71,72,74,155–160], mouse BALB/3T3 fibroblasts [161] and a methylcholanthrene-transformed cell line thereof [77,85,86,162], renomedullary interstitial cells [163], spleen slices [164], perfused hearts [165,166] and kidneys [167]. In several cases, when the cells or tissues were prelabelled with different fatty acids, certain stimuli for prostaglandin production elicited a specific release of arachidonate but not of the other fatty acids [156,161,162,165]. This suggested that a phospholipase is activated which either distinguishes different fatty acids or is compartmentalized in a selective way with arachidonoyl phospholipid species. It has also been found that the arachidonate released upon treatment with such specific stimuli as hormones, is more efficiently converted to prostaglandins than arachidonate released by presumably unspecific stimuli such as ischaemia or ionophores [162,165,167]. This lends support to the idea that the phospholipase which selectively releases arachidonate is tightly coupled to the prostaglandin-generating system.

In many studies on stimulated prostaglandin production, indomethacin, an inhibitor of cyclooxygenases, has been used to differentiate between stimulated lipolysis and stimulated cyclooxygenase activity as the cause of increased prostaglandin release. This approach has been very useful to uncouple the processes and to demonstrate stimulated lipolysis in the absence of further metabolism of arachidonate by oxygenation (e.g. [74,159,161,163,164,167]). Recent reports [168–171] suggest that the validity of this approach may depend on the relative concentrations of indomethacin and Ca^{2+}. It was demonstrated that membrane-associated or highly purified phospholipases A_2 from platelets, alveolar macrophage and polymorphonuclear leukocytes were inhibited by indomethacin [168–170]. At 5 mM Ca^{2+} rather high concentrations of the drug were required to give 50% inhibition of the phospholipase A_2, but at 0.5 mM Ca^{2+} inhibition was seen in the nanomolar range of non-steroidal anti-inflammatory agents [170]. Thus, the anti-inflammatory action of indomethacin and its analogues may not be due solely to inhibition of the cyclooxygenase [170,171].

Much research has been directed to the question of which phospholipids actually donate the arachidonate for prostaglandin formation. Early studies with pre-labelled human platelets indicated a major decrease in the labelled arachidonate content of phosphatidylcholine and phosphatidylinositol during stimulated production of arachidonate and its oxygenated metabolites [155–158]. In isolated human platelet membranes [172] and in a recent study on intact human platelets [173] phosphatidylethanolamine was also reported to be a major substrate for the phospholipase A_2. Similarly, in horse [72,75] and rabbit [159,160] platelets, phosphatidylcholine, -ethanolamine and -inositol were found to lose arachidonate upon stimulation with thrombin. By contrast, Bills et al. [156] arrived at the conclusion that phosphatidylethanolamine in human platelets is not a substrate for the stimulated phospholipase A_2. The discrepancy with the results obtained by Broekman et al. [173] might be explained by the use of pre-labelled cells by Bills et al. [156]. In other words, during prelabelling the arachidonate may be incorporated into a phosphatidylethanolamine pool which is not accessible to the phospholipase A_2 following stimulation by thrombin. The disadvantage of using pre-labelled platelets in establishing the relative contribution of the different phospholipid classes to arachidonate release was emphasized by Blackwell and colleagues [159,160]. That compartmentalization may play an important role in the availability of phospholipids for the stimulated platelet phospholipase(s) was demonstrated by Bills et al. [156]. When platelets were pre-labelled with various fatty acids, significant loss of the radioactivity in the phosphatidylcholine fraction was only noticed in cells pre-labelled with arachidonate. In line with the molecular species composition of phosphatidylcholine a 25.6% loss of radioactivity from this phospholipid was accompanied by only a 7.6% loss of phosphorus. At this point it should be pointed out that a loss of arachidonate from phospholipids, even though this fatty acid is present almost exclusively at the *sn*-2-position, and a loss of the phosphorus content of that phospholipid class do not prove the action of a phospholipase A_2. A phospholipase C followed by a diacylglycerol lipase can accomplish what phospholipase A_2 can do alone, namely cause a loss of lipid phosphorus and a release of arachidonate. Evidence for the concerted action of these two enzymes in platelets has recently accumulated. Rittenhouse-Simmons [174] was the first to show generation of up to 30-fold increased levels of diacylglycerol within 5 s of platelet exposure to thrombin. Within this short period only phosphatidylinositol lost radioactivity in cells pre-labelled with arachidonate. Loss of radioactivity from phosphatidylcholine only became apparent after 30 s. These data suggested the action of a phospholipase C on phosphatidylinositol. The presence of this enzyme in platelets was reported by several investigators [174–177] and its specificity for phosphatidylinositol was established [174,177]. Mauco et al. [178], Bell et al. [176], and Rittenhouse-Simmons [180] have shown a diacylglycerol lipase in platelets. The latter enzyme was thought to be responsible for the early arachidonate release following thrombin stimulation of platelets [176,179]. In this view the initial release of arachidonate would be from phosphatidylinositol via the phospholipase C plus diacylglycerol lipase pathway and arachidonate release from other phospholipids by the action of phospholipase A_2 would be a secondary event. It is obvious then that the early studies on the relative contribution of the various phospholipids to arachidonate production have given variable results depending on how long after thrombin addition the cells were analyzed. It should be mentioned, however, that a recent contribution by Broekman et al. [173] has demonstrated lysophosphatidylethanolamine formation upon treatment of platelets with thrombin, at a rate comparable to phosphatidylinositol disappearance. These authors followed changes in the phospholipid composition of unlabelled platelets as early as 5 s after thrombin addition and arrived at the conclusion that both the phospholipase C plus diacylglycerol lipase and phospholipase A_2 pathway contributed to arachidonate release for cyclooxygenase and lipoxygenase activity. Much work will be required to establish these points further and to see whether similar patterns hold for systems other than platelets that are known to give enhanced prostaglandin release in response to various stimuli.

6. *Phospholipases C*

(a) Occurrence and assay

Phospholipases C are defined as enzymes that hydrolyze the glycerophosphate ester bond in a variety of phospholipids with the formation of 1,2-diacylglycerols (or

N-acylsphingosine in the case of sphingomyelin) and a phosphate monoester (see Fig. 1). This type of lipolytic activity was first detected by Macfarlane and Knight [181] in the toxin of *Clostridium welchii* (also named *C. perfringens*). The enzymes appear to be secreted into the culture medium of various *Clostridium* [182], *Bacillus* [183] and *Pseudomonas* [184,185] species and of *Acinetobacter calcoaceticus* [186]. Most of the early investigations on phospholipase C used lyophilized powder or ammonium sulphate precipitates of culture filtrates of *C. perfringens* or *B. cereus.* Invariably, these crude preparations showed a broad substrate specificity with varying degrees of activity towards all phospholipids including sphingomyelin. Slein and Logan [187] were the first to achieve a partial resolution of the phospholipase C activity from *B. cereus* on DEAE-cellulose columns. The first peak degraded phosphatidylcholine and phosphatidylethanolamine, but showed no activity with sphingomyelin or phosphatidylinositol. A second peak attacked sphingomyelin and overlapped partially with a third peak with marked specificity for phosphatidylinositol. Following these observations, Pastan et al. [188] resolved the phospholipase C of *C. perfringens* into two enzymes on Sephadex G-100 columns. One enzyme preferentially hydrolyzed sphingomyelin whereas a second enzyme hydrolyzed both phosphatidylcholine and sphingomyelin, but with a preference for phosphatidylcholine. With these findings in mind it is obvious that the general point of view that substrate specificity can only be studied adequately with highly purified enzymes holds especially for phospholipases C (see next section). As the result of the aforementioned and many subsequent studies it is now recommended to subdivide phospholipases C into three groups of enzymes. Phospholipases C with activity against, in principle, all diacylglycerophospholipids except phosphatidylinositol remain to be noted by the original EC 3.1.4.3 number. A second group of enzymes with absolute specificity for phosphatidylinositol has received the number EC 3.1.4.10, whereas EC 3.1.4.12 has been assigned to a third group of lipolytic phosphodiesterases with specificity for sphingomyelin. The latter two groups of enzymes, occurring in both bacterial and mammalian cells, will not be discussed here, as their description is included in other chapters of this volume [136,189]. This nomenclature is somewhat confused by the fact that bacterial phospholipases C are known that hydrolyze both phosphatidylcholine and sphingomyelin. These enzymes are to be distinguished from the specific sphingomyelinases and, as they are denoted by the number EC 3.1.4.3, will be discussed in Section 6b.

Apart from its presence in bacteria, including *E. coli* [93], phospholipase C (EC 3.1.4.3) has been shown to occur in the marine planktonic alga *Monochrysis lutheri* [190]. Evidence that the enzyme occurs in yeast [191] and plants [192] has also been reported, but these studies have not been followed up. The presence of phospholipase C in mammalian tissues has long been questioned. Original observations on the hydrolysis of phosphatidylcholine and phosphatidylethanolamine by a phospholipase C in crude preparations of mammalian tissues [193,194] have later been ascribed to erroneously interpreted data obtained by using non-specific tests [195]. Using sphingomyelin as a substrate Kanfer et al. [193] partially purified a phospholipase C type enzyme from rat liver which attacked sphingomyelin but showed no activity towards phosphatidylcholine and phosphatidylethanolamine. Instead, phosphatidylcholine appeared to be a potent competitive inhibitor of sphingomyelin hydrolysis. Nevertheless, the finding that crude liver extracts hydrolyzed all three phospholipids left the possibility of a separate phospholipase C with specificity for phosphatidylcholine and -ethanolamine open. This uncertainty persisted for quite some time. Although indications for the presence of such a phospholipase C in liver [193] and brain [196,197] were briefly mentioned, it is only quite recently that a more detailed account of phospholipase C activity in rat liver has been published by Matsuzawa and Hostetler [137]. The enzyme was optimally active at pH 4.4 and was found only in the lysosomal fraction. A soluble, delipidated fraction from lysosomes hydrolyzed not only sphingomyelin and phosphatidylinositol, but also phosphatidylcholine, phosphatidylglycerol, phosphatidylserine and phosphatidylethanolamine, in that order. Differential effects of bivalent cations and EDTA on this phospholipase C and the phosphatidylinositol-specific phospholipase C as reported by Dawson and coworkers [140,141] suggest that different enzymes are involved, although a definitive conclusion has to await further purification of these enzymes. The absolute specificity of mammalian phospholipases C (EC 3.1.4.12) for sphingomyelin [193,199] and the observation that phosphatidylcholine hydrolysis by the liver lysosomal phospholipase C (EC 3.1.4.3) was uninhibited by a 4-fold excess of sphingomyelin [137] provide strong support for the idea that these are different enzymes. Qualitative evidence for the presence of this phospholipase C in a wide variety of rat tissues has since been obtained [138].

Phospholipases C can conveniently be assayed by continuous titration of the released acidic group. In the case of bacterial enzymes the enzymatic activity in crude culture filtrates is usually sufficient for the application of this technique [200]. Alternatively the formation of water-soluble phosphate esters, or radioactivity from appropriately labelled substrates, can be measured after acid precipitation or solvent extraction of the substrate. Care should be taken to use these methods only when a prior identification of the reaction products has unequivocally demonstrated that phospholipase C is the sole lipolytic enzyme operative under the reaction conditions. Obviously, the presence of phospholipase A and lysophospholipase can easily enough disturb the relatively unspecific assays that employ continuous titration or release of water-soluble phosphate. A general assay using nonradioactive substrates was recently described by Krug et al. [201]. In this method *C. perfringens* phospholipase C was incubated with phosphatidylcholine in the presence of Ca^{2+}. The reaction was stopped by addition of EDTA, whereafter alkaline phosphatase was added to liberate inorganic phosphate from phosphocholine produced by phospholipase C. Alkaline phosphatase action was stopped by addition of sodium dodecylsulphate prior to determination of inorganic phosphate. Apart from being a discontinuous assay this procedure has the drawback that it is only applicable to Ca^{2+}-requiring phospholipases C. Kuriola and coworkers [202,203] have advocated the use of *p*-nitrophenylphosphocholine as substrate for phospholipase C. The release of chromogenic *p*-nitrophenol allows for a continuous assay that seems especially useful during enzyme purification. The K_m of *C. perfringens* phospholipase

C is rather high (0.2 M) but assays can be done conveniently at 20 mM substrate. Reaction rates are greatly stimulated in the presence of high concentrations of sorbitol and glycerol [203]. With crude enzyme preparations one should be aware of the possibility that less lipophilic phosphodiesterases than phospholipase C might hydrolyze such a substrate which differs considerably from phosphatidylcholine. This drawback does not seem to be present in the model substrates that have recently been synthesized by Cox et al. [204]. These authors introduced a subtle change in the structure of phosphatidylcholine and phosphatidylethanolamine by substituting the C–O–P-bond to be hydrolyzed by phospholipase C for a C–S–P-bond. Hydrolysis of the thiophosphoester bond allowed for a spectrophotometric assay in the presence of chromogenic thiol reagents.

(b) Purified enzymes and properties

The purified phospholipases C (EC 3.1.4.3) are listed in Table 4. Many investigators have attempted to purify the phospholipase C from *C. perfringens.* After several partially purified preparations had been isolated (e.g. [205–208]), almost homogeneous enzyme with high specific activity was obtained by Takahashi et al. [209], Zwaal et al. [210] and Yamakawa and Ohsaka [211]. Some of the early preparations appeared fairly homogeneous on polyacrylamide disc electrophoresis [206–208], but had rather low specific activities, suggesting that much inactivation had occurred during the purification procedures. Inclusion of glycerol in the buffers appears to highly improve enzyme recoveries [200,209,210], thus allowing nearly homogeneous enzymes to be obtained with a specific activity of about 1600–2000 U/mg protein [209–211]. Takahashi et al. [209] have applied an affinity absorbent by coupling egg yolk lipoprotein to Sepharose 4B to achieve over 80-fold purification in a single step. A more recent procedure, employing ion-exchange chromatography and Sephadex G-100 filtration, suitable for large-scale preparations, yielded enzyme of similar high activity in a reasonable recovery of 15% [211].

TABLE 4

Purified phospholipases C (EC 3.1.4.3)

Source	Authors	Ref.
Clostridium perfringens (*C. welchii*)	Takahashi et al.	209
	Zwaal et al.	210
	Yamakawa and Ohsaka	211
Clostridium novyi	Taguchi and Ikezawa	212
Pseudomonas fluorescens	Doi and Nojima	185
Pseudomonas aureofaciens	Sonoki and Ikezawa	214
Bacillus cereus	Zwaal et al.	200
	Otnaess et al.	216
	Little et al.	217
	Imamura and Horiuti	218

Considerably varying values for the M_r-value of *C. perfringens* phospholipase C have been reported, ranging from 30000 to 90000 (cf. Table III of [211]). The purest preparations gave estimated M_r-values of about 30000 by gel filtration [208,211] and of about 44 000 by the sodium dodecylsulphate electrophoresis technique [207,209,211]. The discrepancy is not easily explained at present and will have to be resolved eventually by amino acid sequence data. An isoelectric point of 5.7 was reported [208] but Takahashi et al. [209] have later resolved the seemingly homogeneous enzyme into four peaks with isoelectric points of 5.2, 5.3, 5.5 and 5.6 by isoelectric focusing. These multiple forms could not be distinguished by immunodiffusion and polyacrylamide electrophoresis, neither in the absence nor in the presence of sodium dodecylsulphate. All four forms hydrolyzed sphingomyelin at a rate of about 70% of that for phosphatidylcholine. It thus appears that *C. perfringens* secretes a phospholipase C (EC 3.1.4.12) that is specific for sphingomyelin [188] and a phospholipase C (EC 3.1.4.3) which hydrolyzes both phosphatidylcholine and sphingomyelin. In addition the latter protein has haemolytic activity [208–210]. As shown by Pastan et al. [188] the sphingomyelin-specific enzyme does not require Ca^{2+} and is in fact completely inhibited by 1 mM Ca^{2+}. This is probably the reason why other workers have not detected this enzyme, since Ca^{2+} was generally included in the assay medium [208–210]. This phosphatidylcholine- and sphingomyelin-hydrolyzing enzyme requires 5–10 mM Ca^{2+} for optimal activity. Of a variety of other bivalent cations only Co^{2+}, Mn^{2+} and Zn^{2+} could substitute for Ca^{2+} to give activities of 30–50% of those observed with Ca^{2+}.

A phosphatidylcholine- and sphingomyelin-hydrolyzing phospholipase was purified 2000-fold from the culture filtrate of *C. novyi* by Taguchi and Ikezawa [212,213]. Insufficient data were provided to judge the purity of this preparation. The purified enzyme hydrolyzed phosphatidylcholine with a specific activity of 95 U/mg protein. Whether this enzyme has a low activity, contains other proteins or inactivated enzyme cannot be deduced from the publications. Purification was carried out in buffers without glycerol, although the stabilizing effect of glycerol in preserving activity of the purified enzyme was demonstrated [212]. The optimal pH for activity, pH 7.0, and the M_r-value of 30000 as estimated from Sephadex filtration agree very well with those values for *C. perfringens* phospholipase C. The isoelectric point of the *C. novyi* enzyme, pH 7.1, is considerably higher, however. In the presence of optimally stimulating amounts of deoxycholate a further 5-fold increase in activity towards phosphatidylcholine was noted by addition of either Ca^{2+} or Mg^{2+}, which were equally effective. Preincubation of the enzyme with EDTA or the Zn^{2+}-chelating agent *o*-phenanthroline completely inhibited activity. The latter was fully restored only by Zn^{2+}, while Ca^{2+}, Co^{2+} and Mn^{2+} were much less effective. Mg^{2+} and Ni^{2+} did not restore activity. These data strongly suggest that the enzyme requires Zn^{2+} for activity. The stimulation observed by Ca^{2+} and Mg^{2+} may be due to effects on substrate emulsions rather than indicating participation in the catalytic process itself [213].

Doi and Nojima [185] obtained a 2500-fold enriched preparation of a phospholipase C from *Ps. fluorescens.* Polyacrylamide disc gel electrophoresis showed one band with phospholipase C activity along with major and minor contaminating bands. The enzyme had a relatively low specific activity of 36 U/mg protein and exhibited the unusual property of being more active towards phosphatidylethanolamine than phosphatidylcholine. However, this may be a property of enzymes from *Pseudomonas* species, as a similar behaviour was reported for a homogeneous enzyme isolated by Sonoki and Ikezawa from *Ps. aureofaciens* [214]. Phosphatidylglycerol, -serine, -inositol and cardiolipin and sphingomyelin were not attacked. The purified enzyme acted optimally at pH 7–8 on phosphatidylcholine with a specific activity of 175 U/mg protein. Phosphatidylethanolamine hydrolysis occurred optimally at pH 8–8.5. Almost complete inhibition was seen in the presence of EDTA and *o*-phenanthroline. Subsequent reactivation was most effective with Zn^{2+} [215]. The M_r-values of this enzyme amounted to 35000 [214] and its isoelectric point was pH 6.4 [215].

The first complete purification of phospholipase C from *B. cereus* was achieved by Zwaal et al. [200], who reported a specific activity of 1010 U/mg protein using an egg-yolk test system. These authors used glycerol-supplemented buffers to prevent inactivation of the enzyme. It has later been found that the presence of Zn^{2+} ions during the purification steps exerts a stabilizing effect also [216]. Using an egg-yolk lipoprotein affinity column in the presence of Zn^{2+} ions, Little et al. [217] succeeded in obtaining highly purified enzyme with a specific activity of about 2900 U/mg in the egg-yolk test at 37°C in an overall yield of 73%. Imamura and Horiuti [218] developed another affinity absorbent, i.e. palmitoyl cellulose, to obtain homogeneous *B. cereus* phospholipase C with similar high specific activity. The enzyme appeared to be adsorbed to the palmitoylated cellulose through a hydrophobic site distinct from the catalytic site since adsorbed enzyme partially retained enzymatic activity. There is general agreement that *B. cereus* phospholipase C consists of a single polypeptide chain with an M_r-value of about 23000 ± 3000 [200,216,218,219]. The enzyme hydrolyzes phosphatidylcholine, -ethanolamine and -serine [200,220,227]. In its action on phosphatidylcholine the enzyme hydrolyzed both monomolecular and micellar substrates, but a clear-cut preference for micellar substrate was deduced from the at least 10-fold increased hydrolysis rates observed upon passing the critical micellar concentration [228]. The native enzyme did not attack phosphatidylinositol and sphingomyelin [200,220]. In this regard it is worth noting that phospholipases C with specificity for either sphingomyelin [225] or phosphatidylinositol [221] have also been purified from *B. cereus*. The sphingomyelin-hydrolyzing enzyme was markedly stimulated by Mg^{2+} and, in agreement with what has been found for the *C. perfringens* sphingomyelinase [188], was completely inhibited by 5 mM Ca^{2+} [225]. The *B. cereus* sphingomyelinase was also inhibited by EDTA, but not by *o*-phenanthroline. The phosphatidylinositol-specific phospholipase C from *B. cereus* is neither inhibited by EDTA nor by *o*-phenanthroline [221]. By contrast, the phosphatidylcholine-hydrolyzing phospholipase C is completely inhibited by both agents and this inactivation can be fully reversed by addition of Zn^{2+} but not by addition of Ca^{2+} [222,223]. Little and Otnaess [223] have determined that the native enzyme contains 2 atoms of zinc per mol of enzyme. Removal of one atom of zinc by EDTA or *o*-phenanthroline yielded an inactive enzyme species which could be activated by Zn^{2+} or Co^{2+}. Prolonged exposure to *o*-phenanthroline removed also the second zinc atom and produced an enzyme species which was only reactivated by Zn^{2+}. Full reactivation was observed when two atoms of zinc were bound per mol of enzyme. These results strongly support the view that *B. cereus* phospholipase C is a zinc metalloenzyme. Interestingly, when *o*-phenanthroline-inactivated enzyme was reactivated by Co^{2+} rather than Zn^{2+}, hydrolysis of sphingomyelin was seen [224].

There has been some dispute concerning the exact isoelectric point of the enzyme, with values ranging from pH 8.1 [219] to pH 6.5 [216] being reported. Recent studies have shown that the isoelectric point changes considerably with the Zn^{2+}-content of the enzyme. A value of pH 6.9 for native enzyme containing 2 atoms of Zn^{2+} per mol of enzyme was found [226].

Both from a structural and mechanistic point of view, the phosphatidylcholine-hydrolyzing phospholipase C from *B. cereus* is the best characterized phospholipase C. The enzyme has been crystallized [226] and its amino acid composition as well as the sequence of the first 25 residues, starting with tryptophan as *N*-terminus, have been reported [220]. Despite earlier reports to the contrary, purified *B. cereus* phospholipase C appears to be extremely stable. The enzyme retains full activity in the presence of 8 M urea [229]. This treatment caused no loss of zinc from the enzyme. Guanidinium chloride, however, caused unfolding of the native enzyme which was accompanied by release of the structural zinc [230]. The zinc-free enzyme irreversibly lost its activity during a pre-incubation for 5 min, at 50°C, whereas native enzyme retained 60% of its activity after such treatment. The structural Zn^{2+} ions would therefore appear to contribute substantially to the general stability of the enzyme. Remarkably, both native and zinc-free enzyme were found to be far less susceptible to irreversible thermal inactivation in the presence of 8 M urea than in its absence. This apparent stabilization of the enzyme by urea remains difficult to explain.

Although the exact mechanism of action of phospholipase C remains to be elucidated, a number of papers have dealt with enzyme modification to obtain insight into the amino acids participating in the catalytic process. As a result of these studies it was concluded that a single carboxyl group [231], two lysine residues [232] and a histidine residue [233] were essential for catalytic activity. Modification of these residues did not impair the enzyme's capacity to bind to a substrate-based affinity gel. Assuming that binding to egg yolk lipoprotein, covalently coupled to agarose, mimics enzyme-substrate binding, it can be concluded that the above-mentioned residues participate in the catalytic process itself and not in substrate binding. Following similar approaches it was suggested that *B. cereus* phospholipase C contains also an arginine residue which is essential for both catalytic activity and substrate binding [234].

7. Phospholipases D

(a) Occurrence and assay

Phospholipase D (EC 3.1.4.4) catalyzes the hydrolysis of the phosphoester bond between phosphatidic acid and the alcoholic moiety of a variety of phospholipids. Heller [236] has recently presented an extensive review on this type of lipolytic activity. The enzyme was first detected by Hanahan and Chaikoff [235] in carrot extracts and has since been found to be extremely widespread in the plant kingdom [237–240]. In a comparative study on the distribution of phospholipase D in developing and mature plants Quarles and Dawson [239] found highest activity in cabbage, cauliflower, celery, carrot, kohlrabi, lettuce and the seeds of marrow, pea and soya bean. Vaskovsky et al. [240] have presented a qualitative comparison of enzyme content in leaves, stalks and roots of some 200 species of higher Far-Eastern plants. Large differences between families, certain species of a given family and leaves, stalks or roots of a given species were encountered. Originally, the phospholipase D was found associated with a plastid fraction (chloroplasts and chromoplasts) of carrot roots, sugar beets and spinach or cabbage leaves [241]. However, by grinding the tissues with sand most of the phospholipase D was apparently solubilized [239]. Clermont and Douce [242] subsequently showed that purified chloroplasts and mitochondria from spinach and maize were devoid of phospholipase D activity. This conclusion for chloroplasts was confirmed by Roughan and Slack [243]. However, these authors found 34% of the total activity associated with a mitochondrial and 23% with a microsomal fraction. Although the supernatant contained the highest percentage of total activity (41%), the highest specific activity was associated with the microsomal pellet. Large variations in the subcellular distribution were found depending on whether the homogenate was prepared in water or in 10 mM Tris buffer, pH 7.5. The content [239,243] and the subcellular distribution of phospholipase D appear to be influenced by the development of the plant tissue. Thus, Heller et al. [244] reported that immature peanut seeds, which contained only 5% of the phospholipase D activity of dry seeds, had most of the enzyme associated with particles, whereas in dry seeds it was exclusively recovered in the soluble fraction. We are thus faced with the problem of not knowing the in situ localization of phospholipase D in plant cells. We do not know whether the soluble enzyme originates from the cytoplasm or arises by solubilization from subcellular membranes during homogenization. Conversely, particulate enzymes may be associated with membrane structures in situ or might stick to these membranes during tissue fractionation. An exact in situ localization of phospholipase D could be very helpful in delineating its function in normal cellular metabolism. Such a function is at present unknown and it has been suggested that phospholipase D might be a structural membrane protein that would only exhibit enzyme properties under certain non-physiological conditions [243], e.g. after tissue disruption or during lipid extraction from plant tissues. Phospholipase D was discovered following the observation that lipid extracts from fresh tissue contained far less choline phospholipids than those from steamed tissues [235] and initiation of phospholipase D activity by extraction of the tissue with methanol has frequently been described. In the absence of any compelling evidence for phospholipase D as a structural protein, it will be discussed rather as an enzyme whose intracellular function has yet to be disclosed.

The presence of phospholipase D is not restricted to higher plants. The enzyme was also identified in the unicellular red alga *Porphyridium cruentum* [245], in the mitochondrial fraction of *Saccharomyces cerevisiae* [246], in a particulate fraction of the slime mould *Physarum polycephalum* [247] and in the culture medium of *Streptomyces hachijoensis* [248]. Phospholipase D-type enzymes specific for cardiolipin have been reported for *Haemophilus parainfluenzae* [249], *E. coli* [250] and *Salmonella typhimurium, Proteus vulgaris* and *Pseudomonas aeruginosa* [251]. These enzymes hydrolyze cardiolipin into phosphatidylglycerol and phosphatidic acid, but do not attack the other major phospholipids found in these bacteria, i.e. phosphatidylethanolamine and -glycerol. Optimal activity with cardiolipin is found at pH 7 in the presence of 10 mM Mg^{2+} [249–251]. Another phospholipase D-type enzyme with specificity for sphingomyelin and lysophosphatidylcholine was found to be secreted into the culture medium of *Corynebacterium pseudotuberculosis* [252]. A 176-fold purification over the culture filtrate to yield an enzyme with a specific activity of 2.65 U/mg protein was reported. The partially purified enzyme with an estimated M_r-value of 90000 displayed optimal activity at pH 7.6–8.0 and was not stimulated by Ca^{2+} [252].

The presence of a phospholipase D-type enzyme in mammalian tissues was first suggested by Dils and Hübscher [253] to explain their findings on the Ca^{2+}-stimulated incorporation of choline into the phospholipids of rat liver microsomes. Although choline release from microsomal phosphatidylcholine could not be detected, Ca^{2+} ions caused a small but significant production of phosphatidic acid. It was proposed that the exchange of bases such as choline, ethanolamine and serine might be a reversal of phospholipase D activity [254]. Support to this idea was lent by the subsequent finding of Yang et al. [255] that a partially purified phospholipase D from cabbage not only hydrolyzed phosphatidylcholine but also catalyzed a transphosphatidylation reaction in which exchange of bases, e.g. choline, took place. Based on kinetic evidence Porcellati et al. [256] suggested the base-exchange reactions to be due to an enzyme different from phospholipase D. However, in contrast to Hübscher [254] a single enzyme was thought to be responsible for the various base-exchange reactions with choline, ethanolamine and serine [256]. A major breakthrough towards the unravelling of the various possibilities came from the recent work of Kanfer and associates. These authors solubilized both base-exchange and phospholipase D activity from a rat brain particulate fraction by treatment with 1% Miranol H2M, an amphoteric detergent [257]. The solubilized preparation produced phosphatidic acid from phosphatidylcholine, indicative of phospholipase D activity. This reaction showed a broad pH optimum with an apparent peak at pH 6. The transphosphatidylating base-exchange showed a much sharper pH profile with an optimum at pH 7.2 [258]. The suggestion from these results that different enzymes might be involved was borne out in subsequent studies when both a serine-base-exchange enzyme devoid of phospholipase D activity [259] and a phospholipase D devoid of any associated base-exchange activity [260] were obtained in partially purified form. The membrane-bound phospholipase D was solubilized from freeze-dried rat brain with 0.8% Miranol H2M and 0.5% cholate, a procedure which left the bulk of the base-exchange enzymes in the particulate fraction. The phospholipase D was purified 240-fold to a specific activity of 2 U/mg. A value of M_r 200000 was estimated from gel filtration experiments. The partially purified enzyme was optimally active at pH 6.0 and it was found that Ca^{2+} was not absolutely required for activity, although a 2-fold stimulation with the optimal concentration of 5 mM Ca^{2+} was observed [258,260]. Diethylether and anionic detergents, such as sodium dodecylsulphate and taurocholate, known to be activators of plant phospholipases D [236], completely inhibited the mammalian phospholipase D [260].

Another phospholipase D-type enzyme from mammalian tissues has been described by Wykle et al. [261–263] and has to be distinguished from the one described in the preceding lines on the basis of its properties. The enzymic activity detected by Wykle and co-workers showed an absolute specificity for 1-alkyl-*sn*-glycero-3-phosphoethanolamine or -choline. The corresponding 1-acyl analogues were not attacked and 1-alkyl compounds in which the 2-position was acylated, albeit only with an acetyl group, were only negligibly hydrolyzed [263]. This lysophospholipase D, specific for ether-linked lysophospholipids, required Mg^{2+} and was inhibited by Ca^{2+}. It appears to be localized in microsomes of rat brain [261], kidney, intestine, lung, testes and liver [262] with highest activity in the latter organ.

Evidence for the presence of phospholipase D in human eosinophils was reported by Kater et al. [264]. The enzyme was purified 162-fold to give a specific activity of 3 U/mg. Isoelectric focussing gave one band with phospholipase D activity (pI about 6.0) along with four contaminating bands. A value of M_r 60000 was deduced from the enzyme's behaviour on Sephadex columns.

Despite the widespread occurrence of phospholipase D in the plant kingdom, it is only recently that homogeneous enzyme preparations have been obtained from peanuts and cabbage (see next section). Most of the properties of plant phospholipases D were deduced from partially purified preparations, especially those from cabbage leaves and peanut seeds.

Davidson and Long [238] prepared a 46-fold enriched protein fraction with a specific activity of 15 U/mg, from Savoy cabbage leaves. They demonstrated that the enzyme lacked stereospecificity. Modifications to further purify the enzyme, though not giving higher specific activities, were reported by Dawson and Hemington [265] and Yang et al. [255]. The latter authors were the first to report the transphosphatidylation capacity of phospholipase D preparations. The ratio of phosphatidylcholine hydrolysis to aminoethanolysis, yielding the transphosphatidylation product phosphatidylethanolamine, remained constant over a 110-fold purification of phospholipase D. This suggested that both transphosphatidylation and hydrolysis were catalyzed by a single enzyme. Support to this hypothesis was lent by the observations that both reactions proceeded optimally between pH 5.5 and 6.0, showed an absolute requirement for Ca^{2+} and were strongly inhibited by 0.1 mM *p*-chloromercuribenzoate. Unexpectedly, neither activity was inhibited by other thiol-reagents such as *N*-ethylmaleimide and iodoacetamide. The actual ratio of hydrolysis to transphosphatidylation depended strongly on the concentration of the alcoholic acceptor compound. Inositol, threonine, glucose and DL-3-glycerophosphate were inactive as acceptors of the phosphatidyl unit. The reaction with glycerol proceeded racemically so that only 50% of the product had the naturally occurring 1,2-diacyl-*sn*-glycero-3-phospho-1′-glycerol configuration [255]. By using circular dichroism as method of analysis Batrakov et al. [266] arrived at the conclusion that the reaction was stereospecific and yielded 100% of the natural stereochemical isomer of phosphatidylglycerol. A more direct approach, i.e. degradation of the phosphatidylglycerol formed in the transphosphatidylation reaction with phospholipase C and enzymic analysis of the glycerophosphate produced, led Joutti and Renkonen [267] to confirm the initial conclusion of Yang et al. [255] that racemic phosphatidylglycerol was produced. The specificity of the transphosphatidylation reaction was further investigated by Dawson, who concluded that the acceptor molecule must contain a primary alcoholic grouping [268].

The transphosphatidylating capacity of phospholipase D preparations has proved to be very useful in the partial synthesis of phospholipids with modified polar headgroups. Systematic studies in this area were conducted by Jezyk and Hughes [269] and Kovatchev and Eibl [270]. The preparation of phosphatidylserine by transphosphatidylation has also been achieved [271].

A constant ratio of hydrolysis to transphosphatidylation was also noticed during a 1000-fold purification of peanut seed phospholipase D [272,274]. It was therefore the prevailing opinion for some time that transphosphatidylation was intrinsic to phospholipase D. The universality of this conclusion was challenged when Saito et al. [273] discovered certain differences between the two reactions catalyzed by phospholipase D preparations from cabbage and when Taki and Kanfer [260] isolated a phospholipase D from rat brain which did not show the transphosphatidylation reaction. In contrast to the results of Yang et al. [255] it was found by Saito et al. [273] that base-exchange and hydrolysis by cabbage phospholipase D preparations showed widely varying pH optima of pH 9.0 and pH 5.6, respectively. Transphosphatidylation required 4 mM Ca^{2+} but hydrolysis required at least 28 mM Ca^{2+} for optimal expression. Differential effects of heat treatment and the drug hemicholinium-3 on the two activities suggested that different enzymes might be involved. Definitive proof to sustain or reject this possibility must await complete purification of the enzyme.

Phospholipase D activity can most conveniently be assayed by measuring the release of the water-soluble alcoholic moiety esterified to the phosphate group. With phosphatidylcholine as substrate methods for the colorimetric determination of choline as its reineckate or enneaiodide have been developed [237]. The sensitivity can be greatly enhanced by using radioactive substrates. Taki and Kanfer [260] have used phosphatidylcholine with a label in the phosphatidate moiety to measure phospholipase D activity by phosphatidic acid production. The latter had to be separated from the substrate by thin-layer chromatography. This method, though time-consuming, has the advantage of unequivocally indicating phospholipase D action. With substrate labelled in the choline part and measuring release of water-soluble radioactivity [244,274] one has to be aware of the possible presence of deacylating enzymes in crude systems yielding water-soluble glycerophosphocholine. This has proved to be disturbing when crude peanut seed phospholipase D was assayed with choline-labelled lysophosphatidylcholine [275]. An enzymatic method for the determination of released choline, using choline oxidase from *Arthrobacter globiformis,* was recently applied by Imamura and Horiuti [276]. Continuous assays for phospholipase D were developed by Allgyer and Wells [277]. Both methods used dihexanoyl phosphatidylcholine as substrate and are based on quantitating the liberation of hydrogen ions from the phosphatidic acid product. A spectrophotometric assay in the presence of a pH indicator measured the disappearance of the basic form of the indicator. The other method measured substrate hydrolysis by a pH stat technique.

(b) Purified enzymes and properties

The purified phospholipases D are listed in Table 5. The phospholipase D from peanut seeds was purified 1170-fold to a specific phosphatidylcholine-hydrolyzing activity of 234 U/mg by Tzur and Shapiro [274]. At this stage the enzyme, despite the high purification factor, appeared to be only 20% pure. Polyacrylamide gel electrophoresis showed that 80% of the protein appeared in bands devoid of enzymic activity. Final purification was achieved by preparative electrophoresis to yield an enzyme which gave a single band in disc gels [278]. Consistent values for specific activity were difficult to obtain due to molecular size transformations and lability of the enzyme in dilute solutions, but exceeded 200 U/mg protein. The purified enzyme isoelectrofocused at pH 4.65 and in spite of its pH optimum of pH 5.6 was unstable at acidic pH values. The amino acid composition was reported and glycine was found to be the single N-terminal amino acid. M_r determinations gave varying values depending on the methods used. Thus, sedimentation equilibrium centrifugation at various pHs and temperatures indicated a minimal value of about M_r 22000. When similar runs were made in the presence of 8 M urea a value of about M_r 48000 was calculated. Similar values were obtained with SDS-disc gel electrophoresis. On

TABLE 5

Purified phospholipases D

Source	Authors	Ref.
Arachis hypogea var. Virginia (peanut seeds)	Tzur and Shapiro	274
	Heller et al.	278
Brassica oleracea (Savoy cabbage)	Allgyer and Wells	277
Streptomyces hachijoensis	Okawa and Yamaguchi	248
Streptomyces chromofuscus	Imamura and Horiuti	276

the other hand, gel filtration yielded an M_r-value of 200000 ± 10000. Ultrafiltration experiments suggested a time-dependent conversion of enzyme species with a value of M_r 200000 to subunits of M_r 20000–25000 [278]. Since the enzyme is greatly stimulated by sodium dodecylsulphate [274], which was therefore routinely included in the assay mixture, it is not known at present which species is actually catalytically active in the assay medium containing phosphatidylcholine and dodecylsulphate (molar ratio 2:1) in addition to 50 mM Ca^{2+} and 50 mM acetate buffer at pH 5.6. Phosphatidylcholine hydrolysis was stimulated by including detergents or ether in the assay medium. Similar effects of ether were noticed with phosphatidylglycerol as substrate. Interestingly, this enzyme also hydrolyzes cardiolipin to give phosphatidic acid and phosphatidylglycerol, provided ether was omitted from the assay medium [279].

The complete purification of the phospholipase D of Savoy cabbage was only recently achieved by Allgyer and Wells [277]. A 680-fold enrichment over a commercial preparation of this enzyme was necessary to obtain a preparation which gave one band on gel electrophoresis in both the presence and absence of dodecylsulphate. The enzyme was routinely assayed with dihexanoylphosphatidylcholine in the absence of detergents. At 30 mM substrate, specific activities in excess of 300 U/mg were measured at pH 7.25. The pH optimum of the purified enzyme depended on the Ca^{2+}-concentration. At 0.5 mM Ca^{2+} the pH optimum was pH 7.25, which shifted to pH 6.0 in the presence of 50 mM Ca^{2+}. Shifts in optimal pH-values for this enzyme were previously reported in less pure preparations. Thus, Quarles and Dawson [280] found a pH optimum of 4.9 when hydrolyzing sonicated phosphatidylcholine and of pH 5.2 when hydrolyzing large aggregates of phosphatidylcholine in the presence of ether. When anionic amphipatic compounds such as phosphatidic acid or dodecylsulphate were included to activate the enzyme, the optimum shifted to about pH 6.5. Determinations by sodium dodecylsulphate disc gel electrophoresis and sedimentation equilibrium ultracentrifugation gave M_r-values of 112500 ± 7500 and 116600 ± 6900 respectively [277]. Preliminary indications were obtained that these M_r estimates are of an associated species. A complex kinetic behaviour of the enzyme was noticed, perhaps related to the apparent multi-subunit structure. With increasing substrate concentration a sharp increase in activity was found at around 4 mM dihexanoylphosphatidylcholine. This apparently critical concentration does not coincide with the critical micellar concentration of about 10 mM for this substrate. No discontinuity at the critical micellar concentration was observed in the substrate-velocity curve [277].

The most active phospholipase D was obtained in homogeneous form by Okawa and Yamaguchi [248] after a 570-fold enrichment from the culture filtrate of *Streptomyces hachijoensis.* The enzyme hydrolyzed phosphatidylethanolamine with a specific activity of 631 U/mg at the optimal pH of 7.5. An M_r-value of only 16000 and an isoelectric point of pH 8.6 were found. The enzyme retained full activity during 24 h storage at 25°C, in buffers with pH 6–8, but lost more than 80% of its activity during similar treatment at pH 4.0. This acid lability was also reported for the peanut phospholipase D [278]. The *S. hachijoensis* phospholipase D was slightly stimulated by Ca^{2+}, and inhibited by EDTA and ionic detergents. Significant stimulation was observed with ether and Triton X-100. The enzyme showed a broad substrate specificity, attacking phosphatidylethanolamine, -choline, -serine, cardiolipin, sphingomyelin and lysophosphatidylcholine.

A second bacterial phospholipase D purified to homogeneity was obtained by Imamura and Horiuti [276] from supernatants of *Streptomyces chromofuscus* cultures. A large part, over 250-fold, of the total 1000-fold purification was achieved by using palmitoylated gauze as hydrophobic absorbent. The final preparation hydrolyzed phosphatidylcholine with a specific activity of 152 U/mg at the optimal pH of 8.0. The isoelectric point of this enzyme was pH 5.1. M_r estimates yielded values of 50000 by gel filtration and 57000 by sodium dodecylsulphate gel electrophoresis. The enzyme adsorbed on palmitoyl-cellulose showed still about 10% of the activity of the free enzyme and was protected against heat inactivation, suggesting that it possessed a hydrophobic site different from the catalytic site. Like the other bacterial phospholipases D and in contrast to the peanut and cabbage enzymes, the *S. chromofuscus* phospholipase D was inhibited by sodium dodecylsulphate. The activity was stimulated ten-fold by Triton X-100 and was further increased nearly two-fold by 1 mM Ca^{2+}, but not by other bivalent metal ions. The enzyme attacked phosphatidylethanolamine, phosphatidylcholine and its lyso-derivative and sphingomyelin.

8. Concluding remarks

The preceding sections testify to the considerable progress that has been made during the last decade in the field of phospholipases and lysophospholipases. 10 years ago none of the enzymes listed in Tables 1 through 5 had been purified to (near) homogeneity. Together, they constitute a total of 28 purified lipolytic enzymes whose main properties I have attempted to review in this chapter. Obviously, what are considered to be the main properties of enzymes are somewhat subjective and certainly not constant in time. I have emphasized in the discussion the properties of apparently homogeneous enzymes to give, hopefully, a rather complete account of current knowledge of phospholipases. As mentioned in the text certain phospholipases, i.e. pancreas and venom phospholipases A_2 and phosphatidylinositol-specific phospholipases C were excluded because they are dealt with in accompanying chapters of this volume. While concentrating on the properties of homogeneous enzymes in a comparative way, occasionally some results obtained with partially purified enzymes were also discussed in the sections on occurrence and assay. It is realized that this was not always done consistently. Obviously, personal views and interests influenced to a large extent the selections which had to be made due to limitations in time and space. An apology is made herewith to those colleagues whose work in the field was omitted or only partially considered. The purification of the lipolytic enzymes discussed and the studies of their properties was undoubtedly shaped by previous findings with partially purified enzymes.

Acknowledgement

I am much indebted to Dr. K.Y. Hostetler for reading the manuscript and for correcting at least the most pertinent violations of the English language.

References

1 Thompson, G.A. (1970) in Comprehensive Biochemistry (Florkin, M. and Stotz, E.H., eds.), Vol. 18, Elsevier, Amsterdam, pp. 157–199.
2 Slotboom, A.J., Verheij, H.M. and De Haas, G.H. this volume, Chapter 10.
3 Van den Bosch, H., Postema, N.M., De Haas, G.H. and Van Deenen, L.L.M. (1965) Biochim. Biophys. Acta 98, 657–659.
4 Scandella, C.J. and Kornberg, A. (1971) Biochemistry 10, 4447–4456.
5 Nishijima, M., Nakaike, S., Tamori, Y. and Nojima, S. (1977) Eur. J. Biochem. 73, 115–124.
6 Nishijima, M., Akamatsu, J. and Nojima, S. (1974) J. Biol. Chem. 249, 5658–5667.
7 Raybin, D.M., Bertsch, L.L. and Kornberg, A. (1972) Biochemistry 11, 1754–1760.
8 Van den Bosch, H., Aarsman, A.J. and Van Deenen, L.L.M. (1974) Biochim. Biophys. Acta 348, 197–209.
9 Woelk, H., Fürniss, H. and Debuch, H. (1972) Hoppe-Seyler's Z. Physiol. Chem. 353, 1111–1119.
10 Lloveras, J., Douste-Blazy, L. and Valdigué, P. (1963) C.R. Acad. Sci. (Paris) 256, 1861–1862.
11 Winkler, H., Smith, A.D., Dubois, F. and Van den Bosch, H. (1967) Biochem. J. 105, 38C–40C.
12 Weglicki, W.B., Waite, M., Sisson, P. and Shohet, S.B. (1971) Biochim. Biophys. Acta 231, 512–519.
13 Kumar, S.S., Millay, R.H. and Bieber, L.L. (1970) Biochemistry 9, 754–759.
14 Gatt, S. (1968) Biochim. Biophys. Acta 159, 304–306.
15 Van den Bosch, H., Van Golde, L.M.G. and Van Deenen, L.L.M. (1972) Annu. Rev. Physiol. 66, 13–145.
16 Van den Bosch, H. (1974) Annu. Rev. Biochem. 43, 243–277.
17 Gatt, S. and Barenholz, Y. (1973) Annu. Rev. Biochem. 42, 61–85.
18 Brockerhoff, H. and Jensen, R.G. (1974) Lipolytic Enzymes, Academic Press, New York, pp. 243–254.
19 Newkirk, J.D. and Waite, M. (1971) Biochim. Biophys. Acta 225, 224–233.
20 Nachbaur, J., Colbeau, A. and Vignais, P.M. (1972) Biochim. Biophys. Acta 274, 426–446.
21 Van Golde, L.M.G., Fleischer, B. and Fleischer, S. (1971) Biochim. Biophys. Acta 249, 318–330.
22 Colard-Torquebiau, O., Bereziat, G. and Polonovski, J. (1975) Biochimie 57, 1221–1227.
23 Waite, M. and Van Deenen, L.L.M. (1967) Biochim. Biophys. Acta 137, 498–517.
24 Lumb, R.H. and Allen, K.F. (1976) Biochim. Biophys. Acta 450, 175–184.
25 Franson, R., Waite, M. and La Via, M. (1971) Biochemistry 10, 1942–1946.
26 Suzuki, Y. and Matsumoto, M. (1978) J. Biochem. 84, 1411–1422.
27 Van den Bosch, H. and Van Deenen, L.L.M. (1965) Biochim. Biophys. Acta 106, 326–337.
28 Bligh, E.G. and Dyer, W.J. (1959) Can. J. Biochem. Physiol. 37, 911–917.
29 Van den Bosch, H. and Aarsman, A.J. (1979) Agents and Actions 9, 382–389.
30 De Haas, G.H., Sarda, L. and Roger, J. (1965) Biochim. Biophys. Acta 106, 638–640.
31 Slotboom, A.J., De Haas, G.H., Bonsen, P.P.M., Burbach-Westerhuis, G.J. and Van Deenen, L.L.M. (1970) Chem. Phys. Lipids 4, 15–29.
32 Van den Bosch, H. (1980) Biochim. Biophys. Acta, 604, 191–246.
33 Kawasaki, N., Sugatani, J. and Saito, K. (1975) J. Biochem. 77, 1233–1244.
34 Kawasaki, N. and Saito, K. (1973) Biochim. Biophys. Acta 296, 426–430.
35 Saito, K. and Kates, M. (1974) Biochim. Biophys. Acta 369, 245–253.
36 Sugatani, J., Kawasaki, N. and Saito, K. (1978) Biochim. Biophys. Acta 529, 29–37.
37 Imamura, S. and Horiuti, Y. (1980) J. Lipid Res. 21, 180–185.

38 Okumura, T., Kimura, S. and Saito, K. (1980) Biochim. Biophys. Acta 617, 264–273.
39 Albright, F.R., White, D.A. and Lennarz, W.J. (1973) J. Biol. Chem. 248, 3968–3977.
40 Brockerhoff, H. and Jensen, R.G. (1974) Lipolytic Enzymes, Academic Press, New York, pp. 235–241.
41 Rahman, Y.E., Cerny, E.A. and Peraino, C. (1973) Biochim. Biophys. Acta 321, 526–535.
42 Sahu, S. and Lynn, W.S. (1977) Biochim. Biophys. Acta 489, 307–317.
43 Elsbach, P., Weiss, J., Franson, R.C., Beckerdite-Quagliata, S., Schneider, A. and Harris, L. (1979) J. Biol. Chem. 254, 11000–11009.
44 Kramer, R.M., Wütrich, C., Bollier, C., Allegrini, P.R. and Zahler, P. (1978) Biochim. Biophys. Acta 507, 381–394.
45 Kannagi, R. and Koizumi, K. (1979) Biochim. Biophys. Acta 556, 423–433.
46 Apitz-Castro, R.J., Mas, M.A., Cruz, M.R. and Jain, M.K. (1979) Biochem. Biophys. Res. Commun. 91, 63–71.
47 Bjørnstad, P. (1960) J. Lipid Res. 7, 612–620.
48 Scherphof, G.L., Waite, M. and Van Deenen, L.L.M. (1966) Biochim. Biophys. Acta 125, 406–409.
49 Waite, M. and Sisson, P. (1971) Biochemistry 10, 2377–2383.
50 Victoria, E.J., Van Golde, L.M.G., Hostetler, K.Y., Scherphof, G.L. and Van Deenen, L.L.M. (1971) Biochim. Biophys. Acta 239, 443–457.
51 Aarsman, A.J., Van Deenen, L.L.M. and Van den Bosch, H. (1976) Bioorg. Chem. 5, 241–253.
52 Volwerk, J.J., Dedieu, A.G.R., Verheij, H.M., Dijkman, R. and De Haas, G.H. (1979) Rec. Trav. Chim. Pays-Bas 98, 214–220.
53 Fleer, E.A.M., Verheij, H.M. and De Haas, G.H. (1978) Eur. J. Biochem. 82, 261–269.
54 Mulder, E. and Van Deenen, L.L.M. (1965) Biochim. Biophys. Acta 106, 348–356.
55 Rock, C.O. and Snyder, F. (1975) J. Biol. Chem. 250, 6564–6566.
56 Jimeno-Abendaño, J. and Zahler, P. (1979) Biochim. Biophys. Acta 573, 266–275.
57 Frei, E. and Zahler, P. (1979) Biochim. Biophys. Acta 550, 450–463.
58 Natori, Y., Nishijima, M., Nojima, S. and Satoh, H. (1980) J. Biochem. 87, 959–967.
59 De Haas, G.H., Postema, N.M., Nieuwenhuizen, W. and Van Deenen, L.L.M. (1968) Biochim. Biophys. Acta 159, 103–117.
60 Van Deenen, L.L.M. and De Haas, G.H. (1963) Biochim. Biophys. Acta 70, 538–553.
61 Ottolenghi, A. (1967) Lipids 2, 303–307.
62 Paysant, M., Bitran, M., Wald, R. and Polonovski, J. (1970) Bull. Soc. Chim. Biol. 52, 1257–1269.
63 Paysant, M., Bitran, M., Etienne, J. and Polonovski, J. (1969) Bull. Soc. Chim. Biol. 51, 863–873.
64 Duchesne, M.J., Etienne, J., Grüber, A. and Polonovski, J. (1972) Biochimie 54, 257–260.
65 Zwaal, R.F.A., Flückiger, R., Moser, S. and Zahler, P. (1974) Biochim. Biophys. Acta 373, 416–424.
66 Pickett, W.C., Jesse, R.L. and Cohen, P. (1976) Biochem. J. 160, 405–408.
67 Hong, S.L., Polsky-Cynkin, R. and Levine, L. (1976) J. Biol. Chem. 251, 776–780.
68 Feinstein, M.B., Becker, E.L. and Fraser, C. (1977) Prostaglandins 14, 1075–1093.
69 Pickett, W.C., Jesse, R.L. and Cohen, P. (1977) Biochim. Biophys. Acta 486, 209–213.
70 Rittenhouse-Simmons, S. and Deykin, D. (1977) J. Clin. Invest. 60, 495–498.
71 Rittenhouse-Simmons, S. and Deykin, D. (1978) Biochim. Biophys. Acta 543, 409–422.
72 Lapetina, E.G., Chandrabose, K.A. and Cuatrecasas, P. (1978) Proc. Natl. Acad. Sci. USA 75, 818–822.
73 Käser-Glanzmann, R., Jakábová, M., George, J.N. and Lüscher, E.F. (1977) Biochim. Biophys. Acta 466, 429–440.
74 Minkes, M., Stanford, N., Chi, M.M., Roth, G.J., Raz, A., Needleman, P. and Majerus, P.W. (1977) J. Clin. Invest. 59, 449–454.
75 Lapetina, E.G., Schmitges, C.J., Chandrabose, K. and Cuatrecasas, P. (1977) Biochem. Biophys. Res. Commun. 76, 828–835.
76 Laskowski, M. and Sealock, R.W. (1971) in The Enzymes (Boyer, P.D., ed.), Vol. 3, Academic Press, New York, pp. 375–473.
77 Hong, S.L. and Levine, L. (1976) Proc. Natl. Acad. Sci. USA 73, 1730–1734.
78 Gryglewski, R.J., Panczenko, B., Korbut, R., Grodzinska, L. and Ocetkiewicz, A. (1975) Prostaglandins 10, 343–355.

79 Blackwell, G.J., Flower, R.J., Nijkamp, F.P. and Vane, J.R. (1978) Br. J. Pharmacol. 62, 79–89.
80 Danow, A. and Assouline, G. (1978) Nature 273, 552–554.
81 Flower, R.J. and Blackwell, G.J. (1979) Nature 278, 456–459.
82 Hirata, F., Schiffmann, E., Venkatasubramanian, K., Salomon, D. and Axelrod, J. (1980) Proc. Natl. Acad. Sci. USA 77, 2533–2536.
83 Nijkamp, F.P., Flower, R.J., Moncada, S. and Vane, J.R. (1976) Nature 263, 479–482.
84 Hirata, F., Corcoran, B.A., Venkatasubramanian, K., Schiffmann, E. and Axelrod, J. (1979) Proc. Natl. Acad. Sci. USA 76, 2640–2643.
85 Levine, L., Pong, S.S., Hong, S.L. and Tam, S. (1977) in Biochemical Aspects of Prostaglandins and Thromboxanes (Kharasch, N. and Fried, J., eds.), Academic Press, New York, pp. 15–38.
86 Pong, S.S., Hong, S.L. and Levine, L. (1977) J. Biol. Chem. 252, 1408–1413.
87 Chung, J. and Scanu, A.M. (1977) J. Biol. Chem. 252, 4204–4209.
88 Bengtsson, G. and Olivecrona, T. (1980) Eur. J. Biochem. 106, 549–556.
89 Fielding, C.J., Shore, V.G. and Fielding, P.E. (1972) Biochem. Biophys. Res. Commun. 46, 1493–1498.
90 Aron, L., Jones, S. and Fielding, C.J. (1978) J. Biol. Chem. 253, 7220–7226.
91 Sandhoff, K. and Conzelmann, E. (1979) Trends Biochem. Sci. 4, 231–233.
92 Borgström, B., Erlanson-Albertsson, E. and Wieloch, T. (1979) J. Lipid Res. 20, 805–816.
93 Proulx, P. and Van Deenen, L.L.M. (1967) Biochim. Biophys. Acta 144, 171–174.
94 Okuyama, H. and Nojima, S. (1968) Biochim. Biophys. Acta 176, 120–124.
95 Albright, F.R., White, D.A. and Lennarz, W.J. (1973) J. Biol. Chem. 248, 3968–3977.
96 Doi, O. and Nojima, S. (1975) J. Biol. Chem. 250, 5208–5214.
97 Ono, Y. and Nojima, S. (1969) Biochim. Biophys. Acta 176, 111–119.
98 Van den Bosch, H., Van der Elzen, H.M. and Van Deenen, L.L.M. (1967) Lipids 2, 279–280.
99 Nurminen, T. and Suomalainen, H. (1970) Biochem. J. 118, 759–763.
100 Thompson, G.A. (1969) J. Protozool. 16, 397–402.
101 Ferber, E., Munder, P.G., Fischer, H. and Gerisch, G. (1970) Eur. J. Biochem. 14, 253–257.
102 Victoria, E.J. and Korn, E.D. (1975) Arch. Biochem. Biophys. 171, 255–258.
103 De Goede, W.G., Samallo, J., Holtrop, M. and Scherphof, G.L. (1976) Biochim. Biophys. Acta 424, 195–203.
104 Dawson, R.M.C. (1958) Biochem. J. 68, 352–357.
105 Beare, J.L. and Kates, M. (1967) Canad. J. Biochem. 45, 101–113.
106 Van Golde, L.M.G., McElhaney, R.N. and Van Deenen, L.L.M. (1971) Biochim. Biophys. Acta 231, 245–249.
107 Bartels, C.T. and Van Deenen, L.L.M. (1966) Biochim. Biophys. Acta 125, 395–397.
108 Yurkovski, M. and Brockerhoff, H. (1965) J. Fish. Res. Bd. Can. 22, 643–652.
109 Brockerhoff, H. and Jensen, R.G. (1974) Lipolytic Enzymes, Academic Press, New York, pp. 254–265.
110 Marples, E.A. and Thompson, R.H.S. (1960) Biochem. J. 74, 123–127.
111 Leibovitz-Ben Gershon, Z., Kobiler, I. and Gatt, S. (1972) J. Biol. Chem. 247, 6840–6847.
112 Leibovitz, Z. and Gatt, S. (1968) Biochim. Biophys. Acta 164, 439–441.
113 Hörtnagl, H., Winkler, H. and Hörtnagl, H. (1969) Eur. J. Biochem. 10, 243–248.
114 Erbland, J.F. and Marinetti, G.V. (1965) Biochim. Biophys. Acta 106, 128–138.
115 Van den Bosch, H., Aarsman, A.J., Slotboom, A.J. and Van Deenen, L.L.M. (1968) Biochim. Biophys. Acta 164, 215–225.
116 Paysant, M., Wald, R. and Polonovski, J. (1968) Bull. Soc. Chim. Biol. 50, 1445–1453.
117 Brumley, G. and Van den Bosch, H. (1977) J. Lipid Res. 18, 523–532.
118 Bjørnstad, P. (1966) Biochim. Biophys. Acta 116, 500–510.
119 Leibovitz-Ben Gershon, Z. and Gatt, S. (1976) Biochem. Biophys. Res. Commun. 69, 592–598.
120 Stoffel, W. and Greten, H. (1967) Z. Physiol. Chem. 348, 1145–1150.
121 Mellors, A. and Tappel, A.L. (1967) J. Lipid Res. 8, 479–485.
122 Fowler, S. and De Duve, C. (1969) J. Biol. Chem. 244, 471–481.
123 Leibovitz-Ben Gershon, Z. and Gatt, S. (1974) J. Biol. Chem. 249, 1525–1529.

124 De Jong, J.G.N., Van den Bosch, H., Rijken, D. and Van Deenen, L.L.M. (1974) Biochim. Biophys. Acta 369, 50–63.
125 Van den Bosch, H. and De Jong, J.G.N. (1975) Biochim. Biophys. Acta 398, 244–257.
126 De Jong, J.G.N., Van den Besselaar, A.M.H.P. and Van den Bosch, H. (1976) Biochim. Biophys. Acta 441, 221–230.
127 Dole, V.P. and Meinerz, H. (1960) J. Biol. Chem. 235, 2595–2599.
128 Ibrahim, S.A. (1967) Biochim. Biophys. Acta 137, 413–419.
129 Aarsman, A.J. and Van den Bosch, H. (1977) FEBS Lett. 79, 317–320.
130 Aarsman, A.J., Hille, J.D.R. and Van den Bosch, H. (1977) Biochim. Biophys. Acta 489, 242–246.
131 Misaki, H. and Matsumoto, M. (1978) J. Biochem. 83, 1395–1405.
132 Aarsman, A.J. and Van den Bosch, H. (1979) Biochim. Biophys. Acta 572, 519–530.
133 De Jong, J.G.N., Dijkman, R. and Van den Bosch, H. (1975) Chem. Phys. Lipids 15, 125–137.
134 Vianen, G.M. and Van den Bosch, H. (1978) Arch. Biochem. Biophys. 190, 373–384.
135 Van Heusden, G.P.H., Reutelingsperger, C.P.M. and Van den Bosch, H. (1981) Biochim. Biophys. Acta, 663, 22–33.
136 Hawthorne, J.N., this volume, Chapter 7.
137 Matsuzawa, Y. and Hostetler, K.Y. (1980) J. Biol. Chem. 255, 646–652.
138 Hostetler, K.Y. and Hall, L.B. (1980) Biochem. Biophys. Res. Commun. 96, 388–393.
139 Waite, M., Griffin, H.D. and Franson, R.C. (1976) in Lysosomes in Biology and Pathology (Dingle, J.T. and Dean, R., eds.), Vol. 4, North-Holland, Amsterdam, pp. 257–305.
140 Irvine, R.F., Hemington, N. and Dawson, R.M.C. (1978) Biochem. J. 176, 475–484.
141 Richards, D.E., Irvine, R.F. and Dawson, R.M.C. (1979) Biochem. J. 182, 599–606.
142 Omura, T., Siekevitz, P. and Palade, G.E. (1967) J. Biol. Chem. 242, 2389–2396.
143 Bailey, E., Taylor, C.B. and Bartley, W. (1967) Biochem. J. 104, 1026–1032.
144 Van den Besselaar, A.M.H.P., Verheijen, J.H. and Van den Bosch, H. (1976) Biochim. Biophys. Acta 431, 75–85.
145 Moonen, H., Trienekens, P. and Van den Bosch, H. (1977) Biochim. Biophys. Acta 489, 423–430.
146 Van Golde, L.M.G. and Van den Bergh, S.G. (1977) in Lipid Metabolism in Mammals (Snyder, F., ed.), Vol. 1, Plenum, New York, pp. 35–149.
147 Lands, W.E.M. and Crawford, C.G. (1976) in The Enzymes of Biological Membranes (Martonosi, A., ed.), Vol. 2, Wiley, London, pp. 3–85.
148 Kanoh, H. and Åkesson, B. (1977) Biochim. Biophys. Acta 486, 511–523.
149 Curstedt, T. (1974) Biochim. Biophys. Acta 369, 196–208.
150 Trewhella, M.A. and Collins, F.D. (1973) Biochim. Biophys. Acta 296, 51–61.
151 Sundler, R. and Åkesson, B. (1975) Biochem. J. 146, 309–315.
152 Lands, W.E.M. and Samuelsson, B. (1968) Biochim. Biophys. Acta 164, 426–429.
153 Vonkeman, H. and Van Dorp, D.A. (1968) Biochim. Biophys. Acta 164, 430–432.
154 Kunze, H. (1970) Biochim. Biophys. Acta 202, 180–183.
155 Bills, T.K., Smith, J.B. and Silver, M.J. (1976) Biochim. Biophys. Acta 424, 303–314.
156 Bills, T.K., Smith, J.B. and Silver, M.J. (1977) J. Clin. Invest. 60, 1–6.
157 Rittenhouse-Simmons, S., Russell, F.A. and Deykin, D. (1977) Biochim. Biophys. Acta 488, 370–380.
158 Schoene, N.W. (1978) in Advances in Prostaglandin and Thromboxane Research (Galli, C., Galli, G. and Porcellati, G., eds.), Vol. 3, Raven Press, New York, pp. 121–126.
159 Blackwell, G.J., Duncombe, W.G., Flower, R.J., Parsons, M.F. and Vane, J.R. (1977) Br. J. Pharmacol. 59, 353–366.
160 Blackwell, G.J. (1978) in Advances in Prostaglandin and Thromboxane Research (Galli, C., Galli, G. and Porcellati, G., eds.), Vol. 3, Raven Press, New York, pp. 137–142.
161 Lindgren, J.A., Claesson, H.E. and Hammerström, S. (1978) in Advances in Prostaglandin and Thromboxane Research (Galli, C., Galli, G. and Porcellati, G., eds.), Vol. 3, Raven Press, New York, pp. 167–174.
162 Hong, S.L. and Deykin, D. (1979) J. Biol. Chem. 254, 11463–11466.
163 Zusman, R.M. and Keiser, H.R. (1977) J. Biol. Chem. 252, 2069–2071.

164 Flower, R.J. and Blackwell, G.J. (1976) Biochem. Pharmacol. 25, 285–291.
165 Hsueh, W., Isakson, P.C. and Needleman, P. (1977) Prostaglandins 13, 1073–1091.
166 Sivakoff, M., Pure, E., Hsueh, W. and Needleman, P. (1979) Fed. Proc. 38, 78–82.
167 Schwartzman, M. and Raz, A. (1979) Biochim. Biophys. Acta 472, 363–369.
168 Kaplan, L., Weiss, J. and Elsbach, P. (1978) Proc. Natl. Acad. Sci. USA 75, 2955–2958.
169 Jesse, R.L. and Franson, R.C. (1979) Biochim. Biophys. Acta 575, 467–470.
170 Franson, R.C., Eisen, D., Jesse, R.L. and Lanni, C. (1980) Biochem. J. 186, 633–636.
171 Kaplan-Harris, L. and Elsbach, P. (1980) Biochim. Biophys. Acta 618, 318–326.
172 Jesse, R.L. and Cohen, P. (1976) Biochem. J. 158, 283–287.
173 Broekman, M.J., Ward, J.W. and Marcus, A.J. (1980) J. Clin. Invest. 66, 275–283.
174 Rittenhouse-Simmons, S. (1979) J. Clin. Invest. 63, 580–587.
175 Mauco, G., Chap, H. and Douste-Blazy, L. (1979) FEBS Lett. 100, 367–370.
176 Bell, R.L., Kennerly, D.A., Stanford, N. and Majerus, P.W. (1979) Proc. Natl. Acad. Sci. USA 76, 3238–3241.
177 Billah, M.M., Lapetina, E.G. and Cuatrecasas, P. (1979) Biochem. Biophys. Res. Commun. 90, 92–98.
178 Mauco, G., Chap, H., Simon, M.F. and Douste-Blazy, L. (1978) Biochimie 60, 653–661.
179 Bell, R.L. and Majerus, P.W. (1980) J. Biol. Chem. 255, 1790–1792.
180 Rittenhouse-Simmons, S. (1980) J. Biol. Chem. 255, 2259–2262.
181 Macfarlane, M.G. and Knight, B.C.J. (1941) Biochem. J. 35, 884–902.
182 Macfarlane, M.G. (1948) Biochem. J. 42, 590–595.
183 Chu, H.P. (1949) J. Gen. Microbiol. 3, 255–273.
184 Kurioka, S. and Liu, P.V. (1967) J. Bacteriol. 93, 670–674.
185 Doi, O. and Nojima, S. (1971) Biochim. Biophys. Acta 248, 234–244.
186 Lehman, V. (1972) Acta Pathol. Microbiol. Scand. Sect. B. 80, 827–834.
187 Slein, M.W. and Logan, G.F. (1965) J. Bacteriol. 90, 69–81.
188 Pastan, I., Macchia, V. and Katzen, R. (1968) J. Biol. Chem. 243, 3750–3755.
189 Barenholz, Y. and Gatt, S., this volume, Chapter 4.
190 Bilinski, E., Antia, N.J. and Lau, Y.C. (1968) Biochim. Biophys. Acta 159, 496–502.
191 Harrison, J.S. and Trevelyan, W.E. (1963) Nature 200, 1189–1190.
192 Kates, M. (1955) Can. J. Biochem. Physiol. 33, 575–579.
193 Kanfer, J.N., Young, O.M., Shapiro, D. and Brady, R.O. (1966) J. Biol. Chem. 241, 1081–1084.
194 Heller, M. and Shapiro, B. (1966) Biochem. J. 98, 763–769.
195 Gatt, S. (1970) Chem. Phys. Lipids 5, 235–249.
196 Ansell, G.B. (1972) Biochem. J. 128, 6P.
197 Williams, D.J., Spanner, S. and Ansell, G.B. (1973) Biochem. Soc. Trans. 1, 466–467.
(ref. 198 has been deleted)
199 Pentchev, P.G., Brady, R.O., Gal, A.E. and Hibbert, S. (1977) Biochim. Biophys. Acta 488, 312–321.
200 Zwaal, R.F.A., Roelofsen, B., Comfurius, P. and Van Deenen, L.L.M. (1971) Biochim. Biophys. Acta 233, 474–479.
201 Krug, E.L., Truesdale, N.J. and Kent, C. (1979) Anal. Biochem. 97, 43–47.
202 Kurioka, S. (1968) J. Biochem. 63, 678–680.
203 Kurioka, S. and Matsuda, M. (1976) Anal. Biochem. 75, 281–289.
204 Cox, J.W., Snyder, W.R. and Horrocks, L.A. (1979) Chem. Phys. Lipids 25, 369–380.
205 Diner, B.A. (1970) Biochim. Biophys. Acta 198, 514–522.
206 Casu, A., Pala, V., Monacelli, R. and Nanni, G. (1971) It. J. Biochem. 20, 166–178.
207 Stahl, W.L. (1973) Arch. Biochem. Biophys. 154, 47–55.
208 Möllby, R. and Wadström, T. (1973) Biochim. Biophys. Acta 321, 569–584.
209 Takahashi, T., Sugahara, T. and Ohsaka, A. (1974) Biochim. Biophys. Acta 351, 155–171.
210 Zwaal, R.F.A., Roelofsen, B., Comfurius, P. and Van Deenen, L.L.M. (1975) Biochim. Biophys. Acta 406, 83–96.
211 Yamakawa, Y. and Ohsaka, A. (1977) J. Biochem. 81, 115–126.

212 Taguchi, R. and Ikezawa, H. (1975) Biochim. Biophys. Acta 409, 75–85.
213 Taguchi, R. and Ikezawa, H. (1977) J. Biochem. 82, 1217–1223.
214 Sonoki, S. and Ikezawa, H. (1975) Biochim. Biophys. Acta 403, 412–424.
215 Sonoki, S. and Ikezawa, H. (1976) J. Biochem. 80, 361–366.
216 Otnaess, A.-B., Prydz, H., Bjørklid, E. and Berre, Å. (1972) Eur. J. Biochem. 27, 238–243.
217 Little, C., Aurebekk, B. and Otnaess, A.-B. (1975) FEBS Lett. 52, 175–179.
218 Imamura, S. and Horiuti, Y. (1979) J. Lipid Res. 20, 519–524.
219 Ottolenghi, A.C. (1969) in Methods in Enzymology (Lowenstein, J.M., ed.), Vol. 14, Academic Press, New York, pp. 188–197.
220 Otnaess, A.-B., Little, C., Sletten, K., Wallin, R., Johansen, S., Flengsrud, R. and Prydz, H. (1977) Eur. J. Biochem. 79, 459–468.
221 Ikezawa, H., Yamanegi, M., Taguchi, R., Miyashita, T. and Ohyabu, T. (1976) Biochim. Biophys. Acta 450, 154–164.
222 Ottolenghi, A.C. (1965) Biochim. Biophys. Acta 106, 510–518.
223 Little, C. and Otnaess, A.-B. (1975) Biochim. Biophys. Acta 391, 326–333.
224 Otnaess, A.-B. (1980) FEBS Lett. 114, 202–204.
225 Ikezawa, H., Mori, M., Ohyabu, T. and Taguchi, R. (1978) Biochim. Biophys. Acta 528, 247–256.
226 Bjørklid, E. and Little, C. (1980) FEBS Lett. 113, 161–163.
227 Roberts, M.F., Otnaess, A.-B., Kensil, C.A. and Dennis, E.A. (1978) J. Biol. Chem. 253, 1252–1257.
228 Little, C. (1977) Acta Chem. Scand. B31, 267–272.
229 Little, C. (1978) Biochem. J. 175, 977–986.
230 Little, C. and Johansen, S. (1979) Biochem. J. 179, 509–514.
231 Little, C. and Aurebekk, B. (1977) Acta Chem. Scand. B31, 273–277.
232 Aurebekk, B. and Little, C. (1977) Biochem. J. 161, 159–165.
233 Little, C. (1977) Biochem. J. 167, 399–404.
234 Aurebekk, B. and Little, C. (1977) Int. J. Biochem. 8, 757–762.
235 Hanahan, D.J. and Chaikoff, I.L. (1947) J. Biol. Chem. 169, 699–705.
236 Heller, M. (1978) in Advances in Lipid Research (Paoletti, R. and Kritchevsky, D., eds.), Vol. 16, Academic Press, New York, pp. 267–326.
237 Kates, M. and Sastry, P.S. (1969) in Methods in Enzymology (Lowenstein, J.M., ed.), Vol. 14, Academic Press, New York, pp. 197–203.
238 Davidson, F.M. and Long, C. (1958) Biochem. J. 69, 458–466.
239 Quarles, R.H. and Dawson, R.M.C. (1969) Biochem. J. 112, 787–794.
240 Vaskovsky, V.E., Gorovoi, P.G. and Suppes, Z.S. (1972) Int. J. Biochem. 3, 647–656.
241 Kates, M. (1954) Can. J. Biochem. Physiol. 32, 571–583.
242 Clermont, H. and Douce, R. (1970) FEBS Lett. 9, 284–286.
243 Roughan, P.G. and Slack, C.R. (1976) Biochim. Biophys. Acta 431, 86–95.
244 Heller, M., Aladjem, E. and Shapiro, B. (1968) Bull. Soc. Chim. Biol. 50, 1395–1408.
245 Antia, N.J., Bilinski, E. and Lau, Y.C. (1970) Can. J. Biochem. 48, 643–648.
246 Grossman, S., Cobley, J., Hogue, P.K., Kearney, E.B. and Singer, T.P. (1973) Arch. Biochem. Biophys. 158, 744–753.
247 Comes, P. and Kleinig, H. (1973) Biochim. Biophys. Acta 316, 13–18.
248 Okawa, Y. and Yamaguchi, T. (1975) J. Biochem. 78, 363–372.
249 Ono, Y. and White, D.C. (1970) J. Bacteriol. 104, 712–718.
250 Cole, R., Benns, G. and Proulx, P. (1974) Biochim. Biophys. Acta 337, 325–332.
251 Cole, R. and Proulx, P. (1976) J. Bacteriol. 124, 1148–1152.
252 Souček, A., Michalec, C. and Součkova, A. (1971) Biochim. Biophys. Acta 227, 116–128.
253 Dils, R.R. and Hübscher, G. (1961) Biochim. Biophys. Acta 46, 505–513.
254 Hübscher, G. (1962) Biochim. Biophys. Acta 57, 555–561.
255 Yang, S.F., Freer, S. and Benson, A.A. (1967) J. Biol. Chem. 242, 477–484.
256 Porcellati, G., Arienti, G., Pirotta, M. and Georgini, D. (1971) J. Neurochem. 18, 1395–1417.
257 Saito, M. and Kanfer, J. (1973) Biochem. Biophys. Res. Commun. 53, 391–398.
258 Saito, M. and Kanfer, J. (1975) Arch. Biochem. Biophys. 169, 318–323.

259 Taki, T. and Kanfer, J. (1978) Biochim. Biophys. Acta 528, 309–317.
260 Taki, T. and Kanfer, J. (1979) J. Biol. Chem. 254, 9761–9765.
261 Wykle, R.L. and Schremmer, J.M. (1974) J. Biol. Chem. 249, 1742–1746.
262 Wykle, R.L., Kraemer, W.F. and Schremmer, J.M. (1977) Arch. Biochem. Biophys. 184, 149–155.
263 Wykle, R.L., Kraemer, W.F. and Schremmer, J.M. (1980) Biochim. Biophys. Acta 619, 58–67.
264 Kater, L.A., Goetz, E.J. and Austen, K.F. (1976) J. Clin. Invest. 57, 1173–1180.
265 Dawson, R.M.C. and Hemington, N. (1967) Biochem. J. 102, 76–86.
266 Batrakov, S.G., Panosayan, A.G.A., Kogan, G.A. and Bergelson, L.D. (1975) Biochem. Biophys. Res. Commun. 65, 755–762.
267 Joutti, L.K. and Renkonen, O. (1976) Chem. Phys. Lipids 17, 264--266.
268 Dawson, R.M.C. (1967) Biochem. J. 102, 205–210.
269 Jezyk, P.F. and Hughes, N.N. (1973) Biochim. Biophys. Acta 296, 24–33.
270 Kovatchev, S. and Eibl, H. (1978) in Advances in Experimental Medicine and Biology (Gatt, S., Freysz, L. and Mandel, P., eds.), Vol. 101, Plenum, New York, pp. 221–225.
271 Comfurius, P. and Zwaal, R.F.A. (1977) Biochim. Biophys. Acta 488, 36–42.
272 Heller, M., Mozes, N. and Maes, E. (1975) in Methods in Enzymology (Lowenstein, J.M., ed.), Vol. 35, part B, Academic Press, New York, pp. 226–232.
273 Saito, M., Bourque, E. and Kanfer, J.N. (1974) Arch. Biochem. Biophys. 164, 420–428.
274 Tzur, R. and Shapiro, B. (1972) Biochim. Biophys. Acta 280, 290–296.
275 Strauss, H., Leibovitz-Ben Gershon, Z. and Heller, M. (1976) Lipids 11, 442–448.
276 Imamura, S. and Horiuti, Y. (1979) J. Biochem. 85, 79–95.
277 Allgyer, T.T. and Wells, M.A. (1979) Biochemistry 18, 5348–5353.
278 Heller, M., Mozes, N., Peri, I. and Maes, E. (1974) Biochim. Biophys. Acta 369, 397–410.
279 Heller, M. and Arad, R. (1970) Biochim. Biophys. Acta 210, 276–286.
280 Quarles, R.H. and Dawson, R.M.C. (1969) Biochem. J. 112, 795–799.

CHAPTER 10

On the mechanism of phospholipase A_2

A.J. SLOTBOOM, H.M. VERHEIJ and G.H. DE HAAS

Laboratory of Biochemistry, State University of Utrecht,
Transitorium III, Padualaan 8,
De Uithof, NL-3584 CH Utrecht,
The Netherlands

1. Introduction

Phospholipase A_2 (PLA *, EC 3.1.1.4) is a ubiquitously occurring esterase in nature catalysing the reaction:

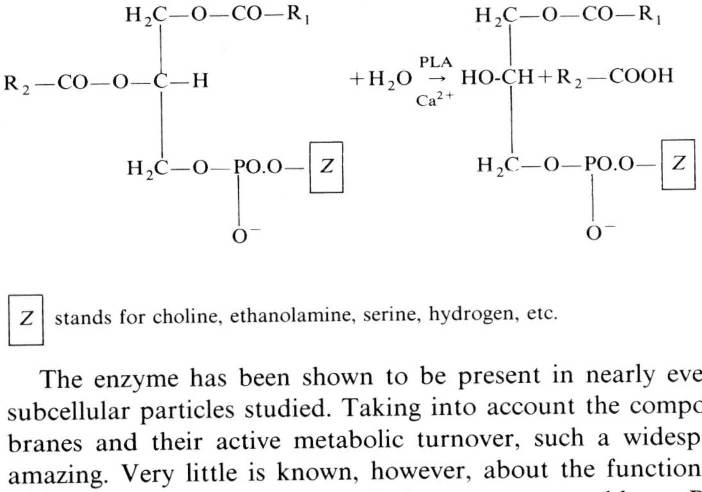

Z stands for choline, ethanolamine, serine, hydrogen, etc.

The enzyme has been shown to be present in nearly every cell and even in all subcellular particles studied. Taking into account the composition of natural membranes and their active metabolic turnover, such a widespread occurrence is not amazing. Very little is known, however, about the function and properties of this endocellular enzyme. This is all the more regrettable as PLA is supposed to be involved as a "trigger" in important processes such as membrane metabolism, haemostasis and blood clotting, prostaglandin synthesis, lung surfactant synthesis, pancreatitis, etc. Our lack of knowledge of these proteins is not due to a vanishing interest in lipolysis; on the contrary, numerous reports in the literature deal with this subject (for recent reviews see [1,1a,1b]). However, because of the low concentration of the enzyme in many tissues and the weak specific activity (at least under the experimental assay conditions used) thorough studies of structure and function are rare.

* List of Abbreviations on pp. 433–434.

Hawthorne/Ansell (eds.) Phospholipids

Fortunately, a few secretory organs are very rich in PLA and important quantities of enzyme have been isolated from mammalian pancreas and the venom glands of snakes and bees. This fact, combined with the high stability of the enzyme and its relatively simple structure, has enabled various investigators to make considerable progress in the elucidation of structure-function relationships of this special class of esterases. Therefore, it seems timely to review our present knowledge of these secretory PLA's in the hope that the results obtained will facilitate similar studies on the less accessible endocellular and often membrane-bound enzymes. *

2. *Purification and assays*

Phospholipase A_2 catalyzes the reaction: diacyl phospholipid → monoacylphospholipid + fatty acid. Among the methods to determine reaction products many applications, advantages and drawbacks have been discussed by Van den Bosch and Aarsman [2]. Although some of these assays are easy to carry out and may be useful to screen a large number of samples for phospholipase activity, comparison of different enzymes is difficult because no absolute activities are obtained.

The liberation of fatty acids is more easily quantitated; the most widely used method utilizes titration in a pH stat. Both purified lecithin and whole egg-yolk have been used, either with or without detergent. Following the reports of Magee et al. [3] and Uthe and Magee [4], deoxycholate has been widely used, although the optimal conditions with respect to Ca^{2+} and deoxycholate concentrations for enzymes from different sources vary widely [5–7]. The non-ionic detergent Triton X-100 as introduced by Salach et al. [8,9] and used, also for kinetic studies, by Dennis and coworkers (see Section 4, "Kinetic data"), has been applied in routine assays in many studies. However, as with deoxycholate, often little attention has been paid to the optimal conditions. In our hands it appeared that every enzyme has its characteristic optimum for Ca^{2+} and Triton X-100 concentration. In conclusion the egg-yolk assay is rapid, cheap with respect to substrate and reproducible with a good sensitivity: specific activities vary between 100 and 5000 $\mu mol \cdot min^{-1} \cdot mg^{-1}$ and the method allows detection and determination of about 0.2 μmol/min (corresponding to about 2 μg down to 40 ng of protein).

As long-chain phospholipids are insoluble in water, their enzymatic hydrolysis can only be accurately measured in the presence of detergents. Synthetic short-chain phospholipids dissolve in water and form true (monomeric) solutions or, at higher concentrations, micelles [10]. Assays based on monomeric substrates and on micellar

* For a more extensive review on this subject the reader is referred to Verheij, H.M., Slotboom, A.J. and de Haas, G.H. (1981), Structure and Function of Phospholipase A_2, in W. Vogt (Ed.), Reviews of Physiology, Biochemistry and Pharmacology, Vol. 91, Springer Verlag, Heidelberg, pp. 91–203. Recently a more detailed review on pancreatic phospholipase A_2 entitled "Pancreatic Phospholipase A_2. A model for lipid protein interactions?" has been published by J.J. Volwerk and G.H. de Haas in O.H. Griffith and P. Jost (Eds.), Molecular Biology of Lipid-Protein Interactions, J. Wiley and Sons, New York, 1982, pp. 69–149.

medium-chain substrates have been used. However, these methods are quite expensive with regard to substrate and can only be justifiably used for special purposes: e.g. for kinetic analysis in the monomeric or micellar substrate region (see also Section 4). In addition, the use of dioctanoyllecithin as a substrate offers an extremely sensitive assay to determine trace amounts of PLA. All phospholipases tested in our laboratory showed a higher activity on this substrate than on any other, including egg-yolk.

Finally a number of specific assays deserve attention. Aarsman et al. [11] introduced the use of thioester substrates. During hydrolysis, thiol groups are released which can be detected spectrophotometrically after reaction with Ellmann's reagent. The introduction of the thiol ester function has been used to study the hydrolysis of monomeric lecithins by porcine pancreatic phospholipase [12] and was found to be about 100-fold more sensitive than titration of liberated fatty acids.

Recently, a sensitive assay of phospholipase using the fluorescent probe 2-parinoyllecithin has been published [12a].

The use of ^{31}P NMR to study hydrolysis was introduced by Henderson et al. [13] and Brasure et al. [14]. This method is based on the difference in chemical shifts of phosphatidylcholine and lysophosphatidylcholine. In an elegant study by Roberts et al. [15], this method was used to simultaneously analyze the hydrolysis of individual phospholipid species in phospholipid mixtures.

Venoms as well as pancreatic tissue contain high amounts (1–10% of all proteins present) of (pro)phospholipase A_2. As these proteins are very stable with respect to heat, variations in pH and denaturing conditions their isolation is relatively simple. Multiple forms of pancreatic phospholipase have been isolated from pancreatin as described by Tsoa et al. [16]. The questionable results obtained with commercial pancreatin argue against its use. Pure preparations of (iso) precursors and activation of these to the corresponding enzymes have been described for pancreatic tissue and juice from: pig [6 and its listed references, 17]; ox and sheep [18]; horse [7] and man [5,5a,b,19]. The isolation of another secretory phospholipase A, from rabbit parotid gland, which is considered to be analogous to venom glands, has been described recently [19a].

As venoms from a great variety of animals can be bought and since there is no need for extensive extraction or homogenisation procedures, these venoms have proved to be popular sources of phospholipase A_2. Yet the elution patterns contain in general more phospholipase A_2 peaks than observed with the pancreatic enzymes. Most purification methods employ a combination of gel filtration and the use of one or more ion exchangers. The more rational order of their application undoubtedly includes first a group separation on a molecular sieve which in general improves the specific activity 2–3-fold and removes small toxins (like direct lytic factor) and most other enzymatic activities from the phospholipase fraction. Subsequent chromatography on an ion exchange column allows separation into the iso-enzymes. Because of the greater capacity of the ion exchange columns the order is frequently reversed.

A number of other interesting methods have been described: (1) precipitation of phospholipase A_2 from aqueous isopropanol with $NdCl_3$ [20]; (2) affinity chromatography with an immobilised substrate analogue [21] which makes use of the fact that only the enzyme-calcium complex of *Crotalus adamanteus* phospholipase binds to the columns. Elution was performed with EDTA, but in our hands a more satisfactory elution takes place by eluting with about 30% organic solvent (acetonitril, dimethylformamide) or 6 M urea (unpublished results); (3) hydrophobic chromatography on phenyl Sepharose CL-4B as described for the removal of traces of phospholipase A_2 from cardiotoxin preparations [22]; (4) affinity chromatography using immobilised antibodies against phospholipase A_2 [23,24]; and (5) the use of concanavalin-Sepharose 4B [25] for the isolation of the carbohydrate-containing bee venom phospholipase.

Phospholipases or phospholipase-containing complexes have been isolated in a pure state and have been characterised from the following venoms: *Agkistrodon halys blomhoffi* [26–28], *Agkistrodon piscivoris* [29], Bee venom [25,30], and wasp venom [30a], *Bitis arientans* [31], *Bitis gabonica* [32], *Bothrops asper* [33,34], *Bothrops atrox, B. jararaca, B. jararacussu, B. neuwiedii* [35], *Bungarus caerulus* [36,37].

From *Bungarus multicinctus* venom several components with weak phospholipase activity and presynaptic activity have been isolated. The β-type toxin apparently contains two chains (M_r 22000 for the covalent complex) based on M_r determinations and amino acid composition of the unreduced toxin [36] and on the sequence analysis [38 and its listed references]. However, there are also studies showing that in addition to the double-chain toxin β-type toxins composed of a single chain ($M_r = 11000$) are present in this venom [39,40]. In addition a non-toxic phospholipase is also present [41,41a].

Further publications dealing with PLA isolation are: *Crotalus adamanteus* [20,42] and *C. atrox* [43,44]. The venom of *C. durissus terrificus* contains the first venom toxin (crotoxin) ever isolated [45; for review see ref. 46]. Depending on the source of the venom, the crotoxin complex contains one or two basic isophospholipases [47]; an acidic non-toxic phospholipase is also present in this venom [48]. *C. scutulatus scutulatus* venom contains a toxic complex very similar in properties to crotoxin [49,50]. From the venom of *C. scutulatus salvanii*, a phospholipase or phospholipase complex ($M_r = 30000$) was isolated with two different amino terminal residues [51].

From the venoms of the following snakes, one or more phospholipases have been isolated: *Enhydrina schistosa* [52]; *Hemachatus haemachatus* [53,54]; *Laticauda semifasciata* [55 and its listed references]; *Micrurus fulvius microgalbineus* [56]; *Naja n. atra* [57]; *N. n. naja* [9,58 and its listed references]; *N. n. kaouthia* (= siamensis) [59,60,214]; *N. n. oxiana* [23]; *N. melanoleuca* [61]; *N. mossambica mossambica* [62,63]; *N. nigricollis* [64,65]. For the venom of *Notechis scutatus scutatus*, the isolation of three isoenzymes including one without phospholipase activity has been described [66–68]. Further purifications have been described for the venoms of: *Oxyuranus scutellatus* [69]; *Parademansia microlepidotus* [70]; *Pseudechis australis* [71,210]; *Pseudechis colletti* [72,210]; *Pseudechis porphyriacus* [71a,210]; *Trimerisurus flavoviridis* [73]; *T. mucrosquamatus* [73a] and *T. okinavensis* [73b].

The isolation of neurotoxic PLA complexes has been reported from the venom of many true vipers. *Vipera ammodytes* contains a complex constituted of a basic phospholipase and an acidic subunit [74–76 and the listed references] and several other toxic as well as non-toxic phospholipases [77]. Phospholipases have also been isolated from the venoms of *Vipera aspis* [78] and *Vipera berus* [79,80]. The venom of *Vipera palestinae* contains one phospholipase. During isolation, this protein is partly converted into a species with different electrophoretic mobility but identical amino acid composition [81]. The venom also contains a neurotoxin which appears to be a 1:1 complex of the acidic phospholipase and a basic polypeptide. The basic component was able to enhance the toxicity of a number of phospholipases isolated from other snake venoms but did not render porcine pancreatic PLA toxic [82].

Finally, from the venom of the elapid *Walterinnesia aegyptia* a pure PLA has been isolated [83].

3. Structural aspects

PLAs isolated from all sources are heat-stable, resistant to denaturing agents and Ca^{2+}-dependent. One may expect, therefore, that there are several similarities in the structural aspects of these enzymes. Because of their low molecular weight, the determination of the amino acid sequence of PLA has become relatively easy and the amino acid sequences of more than 30 "true" PLAs have been determined. In addition, the sequences of a number of homologous proteins like the γ-chain of taipoxin and the B-chain of β-bungarotoxin have been determined. The structures of these proteins are compared in Table 1. It is obvious that all PLA's shown in this table are homologous proteins which have probably developed from a common ancestor. Bee venom PLA [84,85] is not included in Table 1, because its sequence differs too greatly from all other PLAs to allow a homology comparison.

With the exception of the proteins from *Bitis gabonica*, β-bungarotoxin B-chain and taipoxin γ-chain, all PLAs contain 7 disulphide bridges. The disulphide connections of 12 half-cysteine residues were determined for the porcine PLA [86,87], but since a reinvestigation of the sequence showed that this enzyme also contains 14 half-cysteines [88], the disulphide bridge assignment was not totally correct. A second attempt to assign the bridges was made using a low resolution X-ray structure of porcine precursor, but unfortunately two bridges were interchanged [89]. The three-dimensional structure of bovine pancreatic PLA at 1.7 Å resolution revealed the correct pairing beyond any doubt [90,90a]. The disulphide bridges are indicated in Fig. 1. As no attempts have been made to determine the disulphide bridges in snake venom PLA's, we can only assume that they are present at homologous places as in bovine pancreatic PLA. From Table 1 it is obvious that in all *elapidae* and *hydrophidae* PLAs (with the exception of β-bungarotoxin B-chain) the half-cysteine residues are completely conserved *. Hence one must assume that in

* It should be noticed, however, that the alignment of the sequences as shown in Table 1 is also based on the positions of the half-cysteine residues. Because of their highly conserved character they contribute much to this alignment.

these enzymes, the disulphide bridges are connected as in the bovine pancreatic PLA (Fig. 1). As already pointed out by Heinrikson et al. [91], in *viperidae* and *crotalidae* PLA the half-cysteine residues 11 and 77 (Table 1, Fig. 1) are absent. In these enzymes two half-cysteines are found at position 50 and at the C terminus. At these positions, half-cysteines are not present in the PLAs from pancreas, or *elapidae* and *hydrophidae* venoms. Again in the absence of chemical evidence one must assume that these half-cysteines form a disulphide bridge. This assumption has been confirmed recently by the X-ray structure of *Crotalus atrox* phospholipase [91a].

The high number of disulphide bridges contributes to the stability of the enzyme and their correct pairing must be a prerequisite for enzymatic activity. When the disulphide bridges are broken by reduction, the activity is lost and without special precautions the activity is recovered only partly or not at all following reoxidation [92]. Using porcine pancreatic PLA, the authors showed that reduction led to a complete loss of activity. When the reoxidation was carried out in the absence of thiols, only about 35% of the enzymatic activity was recovered. The authors assumed that the relatively low recovery was due to the formation of mismatched disulphide bridges. When the reoxidation was carried out in the presence of cysteine and 0.9 M guanidine chloride to increase the solubility of the reduced protein, 90–95% of the enzymatic activity could be recovered. After purification, this enzyme was indistinguishable from the native enzyme.

When all sequences are compared it appears that 32 amino acids are absolutely conserved. In addition, 29 residues are usually substituted by residues with similar properties with respect to size, charge or hydrophobicity. When only pancreatic and elapid PLA's are compared, these numbers are as high as 36 and 45, respectively. The residues which are absolutely conserved are so because of two major reasons:

TABLE 1 (see also p. 365)

Comparison of amino acid sequences of phospholipases from various sources

Sequences compared are: (1) pig [88]; (2) horse [249; Verheij et al., unpublished results]; (3) ox [250]; (4) iso-pig [251]; (5) man [5b, Verheij et al., unpublished results]; (6-8) *Laticauda semifasciata*, fractions I, III and IV (Nishida et al., unpublished results); (9) *Enhydrina schistosa* [252]; (10) *Notechis scutatus*, notexin [66]; (11) *N. scutatus*, fraction II-5 [67]; (12) *N. scutatus*, fraction II-1 [94]; (13) *Hemachatus haemachatus* [53]; (14-16) *Naja melanoleuca*, fractions DE-I, DE-II and DE III [253,254]; (17-19) *N. mosambica*, fractions CM-I, CM-II and CM-III [62]; (20) *N. nigricollis*, basic (Obidairo et al., unpublished results); (21) *N. n. oxiana* [255]; (22,23) *N. n. kaouthia* [214]; (24) *N. n. atra* [256]; (25–27) *Oxyguranus scutallatus*, α and β chain and the γ chain starting at residue 9 [257]; (28-30) *Bungarus multicinctus*, β-bungarotoxin, A1, A2 and A3 chains [38]; (31) *B. multicinctus*, phospholipase [38a]; (32) *Bitis caudalis* (Viljoen, unpublished results); (33) *B. gabonica* [258]; (34) *Crotalus adamanteus*, fraction α [91]; (35) *C. atrox* [259]; (36) *C. durissus terrificus* [260]; (37) ibid, microheterogeneity [260]; sequences 36 and 37 probably represent the iso-enzymes described by Breithaupt et al. [48]; (38) *Trimeresurus okinavensis* [73b]. Gaps (−) have been introduced in order to obtain alignments of half cysteines and maximal homology. Residues identical to the corresponding residue in porcine pancreatic PLA are indicated with an asterisk. The numbering has been based on horse pancreas PLA; note that gaps introduced in the pancreatic model do not affect the numbering. Note also that the numbers used here do not necessarily correspond to the numbers used in the original publications. The IUPAC one-letter notation for amino acids [262] has been used.

→

```
   1                 10                  20                  30                  40

 1 A L W Q F R S M I K C A I P G S H P L M D F N N Y G C Y C G L G G S G T P V D E L D R C C E
 2 * V * * * * * * * Q * T * * N * K * Y L E * * D * * * * * * * * * * * * * * * * * * A * * Q
 3 * * * * * N G * * * * K * * S * E * * L * * * * * * * * * * * * * * * * * * * D * * * * * Q
 4 * * * * * * * * * * * T * * * * D * * L * * * * * * * * * * * * * * * * * * * * * * * * * *
 5 * V * * * * K * * * * V * * * * D * F L E Y * * * * * * * * * * * * * * * * * * * * K * * Q
 6 N * V * * S N L * Q * N V K * * R A S Y H Y A D * * * * * * A * * * * * * * * * * * * * * K
 7 N * V * * T Y L * Q * * N S * K R A S Y H Y A D * * * * * * A * * * * * * * * * * * * * * K
 8 N * V * * S Y L * Q * * N T * K R A S Y H Y A D * * * * * * A * * * * * * * * * * * * * * K
 9 N * V * * S Y V * T * * N H N R R S S L * Y A D * * * * * * A * * * * * * * * * * * * * * K
10 N * V * * S Y L * Q * * N H * K R * T W H Y M D * * * * * * A * * * * * * * * * * * * * * K
11 N * V * * S Y L * Q * * N H * R R * T R H Y M D * * * * * * W * * * * * * * * * * * * * * K
12 N * V * * S N * * Q * * N H * * R * S L A Y A D * * * * * S A * * * * * * * * * * * * * * K
13 N * Y * * K N * * * * T V * S - R S W W H * A * * * * * * * R * * * * * * * * D * * * * * Q
14 N * Y * * K N * * H * T V * N - R * W W H * A * * * * * * * R * * K * * * * * D * * * * * Q
15 N * Y * * K N * * Q * T V * N - R S W W H * A * * * * * * * R * * * * * * * * D * * * * * Q
16 N * Y * * K N * * H * T V * N - R S W W H * A * * * * * * * R * * * * * * * * D * * * * * Q
17 N * Y * * K N * * H * T V * S - R * W W H * A D * * * * * * R * * K * * A * * D * * * * * Q
18 N * Y * * K N * * H * T V * S - R * W W H * A D * * * * * * R * * K * * A * * D * * * * * Q
19 N * Y * * K N * * H * T V * S - R * W W H * A D * * * * * * R * * K * * * * * D * * * * * Q
20 N * Y * * K N * * H * T V * S - R * W W H * A D * * * * * * R * * K * * * * * D * * * * * Q
21 N * Y * * K N * * * * T V * S - R S W L * * A * * * * * * * R * * * * * * * * D * * * * * Q
22 N * Y * * K N * * Q * T V * N - R S W W * * A D * * * * * * R * * * * * * * * D * * * * * Q
23 N * Y * * K N * * Q * T V * S - R S W W N * A D * * * * * * R * * * * * * * * D * * * * * Q
24 N * Y * * K N * * Q * T V * S - R S W W * * A D * * * * * * R * * * * * * * * D * * * * * Q
25 N * L * * G F * * R * * N R R * R * V W H Y M D * * * * * * K * * * * * * * * D * * * * * Q
26 N * V * * G K * * E * * * R N R R * A L * * M * * * * * * * K * * * * * * * * D * * * * * Q
27 D F E * * S N * * Q * T * * C(G/S)E C L A Y M D * * * * * * P * * * * * * I * D * * * * * K
28 N * I N * M E * * R Y T * * C E K T W G E Y A D * * * * * * A * * * * R * I * A * * * * * Y
29 N * I N * M E * * R Y T * * C E K T W G E Y A D * * * * * * A * * * * R * I * A * * * * * Y
30 N * I N * M E * * R Y T * * C E K T W G E Y A T * * * * * * A * * * * R * I * A * * * * * Y
31 N * Y * * K N * * V * * G T - - R * W I G Y V * * * * * * * A * * * * * * * * * * * * * * Y
32 N * I * * G N * * S - * M T * - K S S L A Y A S * * * * * * W * * K * Q * K * D T * * * * F
33 D * T * * G N * * N - K M - * - Q S V F * Y I Y * * * * * * W * * K * K * I * A T * * * * F
34 S * V * * E T L * M - K V A K R S G * L W Y S A * * * * * * W * * H * R * Q * A T * * * * F
35 S * V * * E T L * M - K * A * R S G * L W Y S A * * * * * * W * * H * L * Q * A T * * * * F
36 H * L * * N K * * * - F Z T R K B A V P F Y A F * * * * * * W * * Z * R * K B A T B * * * F
37 S                                   I                 L
38 H * M * * E T L * M - K * A * R S G V W W Y G S * * * * * * A * * Q * R * Q * P S * * * * F
```

```
         50                  60                  70                  80

 1 T H D N C Y R D A - K N L D S C K F L V D N P Y T E S Y S Y S - C S N T E I T - C N S K N N
 2 V * * * * * T Q * - * E * S * * R * * * * * * * * * * * K F * - * * G * * V * - * S D * * *
 3 * * * * * * K Q * - * K * * * * * V * * * * * * * N N * * * * - * * * * * * * - * S * E * *
 4 * * * * * * * * * - * * * * * * * * * * * * * * * N * * * * * - * * * * * * * - * * * * * *
 5 * * * * * * D Q * - * K * * * * * * * L * * * * * H T * * * * - * * G(B A I T - C S S K B K)
 6 I * * * * * G E * E * - M - G * - Y - - - - * K W T L * T * E S * T D * S P - - * D E * T -
 7 I * * * * * G E * E * - M - G * - Y - - - - * K L T M * N * Y - * G T Q S P * - * D D * T -
 8 I * * * * * G E * E * - M - G * - Y - - - - * K L T M * N * Y - * G T Q S P * - * * D * T -
 9 I * * D * * G E * E * - Q - G * - Y - - - - * K M L M * D * Y - * G S N G P Y - * R N V K K
10 I * * D * * D E * G * - K - G * - * - - - - * K M S A * D * Y - * G E N G P Y - * R N I K K
11 I * * D * * S * * E * - K - G * - S - - - - * K M S A * D * Y - * G E N G P Y - * R N I K K
12 * * * D * * A R * T * - S Y S * - T - - - - * * W T L * * W Q - * I E K T P * - * D * * T -
13 * * * * * * S * * E * - I S G * - R - - - - * * F K T * * * D - * T K G K L * - * K E G * *
14 I * * K * * D E * E * - I S G * - W - - - - * * I K T * T * E S * - Q G T L * - * K D G G -
15 I * * * * * G E * E * - I S G * - W - - - - * * I K T * T * E S * - Q G T L * S * G A N * -
16 I * * * * * G E * E * - I S G * - W - - - - * * I K T * T * D S * - Q G T L * S * G A A * -
17 V * * * * * G E * E * - * - G * - W - - - - * * L T L * K * E - * * Q G K L * - * S G G * *
18 V * * * * * G E * E * - * - G * - W - - - - * * L T L * K * E - * * Q G K L * - * S G G * *
19 V * * * * * E K * G * - M - G * - W - - - - * * F T L * K * K - * * Q G K L * - * S G G * S
20 V * * * * * E K * G * - M - G * - W - - - - * * L T L * K * K - * * Q G K L * - * S G G * S
21 I * * * * * N E * C * - I S G * - W - - - - * * F K T * * * E - * * Q G T L * - * K G D * *
22 V * * * * * D E * E * - I S R * - W - - - - * * F K T * * * E - * * Q G T L * - * K N G * *
23 V * * * * * D E * E * - I S G * - W - - - - * * F K T * * * E - * * Q G T L * - * K G G * *
24 V * * * * * N E * E * - I S G * - W - - - - * * F K T * * * E C S * Q G T L * - * K G G * *
25 V * * E * * G E * V R - R F G * - A - - - - * * W T L * * W K - * Y G K A P * - * * T * T -
26 V * * E * * A E * E * - H - G * - Y - - - - * S L T T * T W E - * R Q V G P Y - * * * * T -
27 * * * E * * A E * G * - * S A * * S V L S E * N N D T * * * E - * N E G Q L * - * * D D N D
28 V * * * * * G * * E * - K H K * - - - - - * * K * S Q * * * K - L T K R T * I - * Y G A A G
29 V * * * * * G * * E * - K H K * - - - - - * * K * S Q * * * K - L T K R T * I - * Y G A A G
30 V * * * * * G * D A A - I R D * - - - - - * * K * S Q * * * K - L T K R T * I - * Y G A A G
31 V * * * * * G E * E * - I P G * - - - - - * * K * K T * * * T - * T K P N L * - * T D A A G
32 V * * C * * G K * D * - - - - * - - - - - S * K M I L * * * K - F H * G N * V - * G D * * -
33 V * * C * * G K M G T - - - - Y - - - - - D T K W T * * N * E - I Q * G G * D - * D E D P -
34 V * * C * * G K * T N - - - - * - - - - - * * K * V * * T * S - E E * G * * V - * G G D D -
35 V * * C * * G K * T D - - - - * - - - - - * * K * V * * T * S - E E * G * * I - * G G D D -
36 V * B C * * G K L A * - - - - * - - - - D T K * W B I * R * S - L K S G Y * * - * G K G T -
37
38 V * * C * * G K V T G - - - - * - - - - - * T K D * F * T * T - E E E G A * S - * G G N(D -
```

```
   90                  100                 110                 120                 130

 1 A C E A F I C N C D R N A A I C F S K A - - P Y N K E H K N L D T K - K Y C
 2 * * * * * * * * * * * * * * * * * * * * - - * * * P * N * * * * S * - R K * A
 3 * * * * * * * * * * * * * * * * * * * V - - * * * * * * * * * * - * - * N *
 4 * * * * * * * * * * * * * * * * * * * * - - * * * * * * * * * * * * - * * *
 5 E * * * * * * * * * * * * * * * * * * * - - * * * * A * * * * * * * - Y S * Q
 6 G * Q G * V * A * * L E * * K * * A R S - - * * * N K N Y * I * * S - * R * K
 7 G * Q R Y V * A * * L E * * K * * A R S - - * * * N K N Y * I * * S - * R * K
 8 G * Q R Y V * A * * L E * * K * * A R S - - * * * N K N Y * I * * S - * R * K
 9 K * N R K V * D * * V A * * E * * A R N - - A * * N A N Y * I * * * - * R * K
10 K * L R * V * D * * V E * * F * * A * * - - * * * N A N W * I * * * - * R * Q
11 K * L R * V * D * * V E * * F * * A * * - - * * * N A N W * I * * * - * R * Q
12 G * Q R * V * D * * A T * * K * * A * * - - * * * * * N Y * I * P * - * R * Q
13 E * A * * V * K * * * L * * * * * A G * - - H * * D N N N Y I * L A - R H * Q
14 K * A * S V * D * * * V * * N * * A R * - - T * * D K N Y * I * F N - A R * Q
15 K * A * S V * D * * * V * * N * * A R * - - T * * D K N Y * I * F N - A R * Q
16 N * A * S V * D * * * V * * N * * A R * - - * * I D K N Y * I * F N - A R * Q
17 K * * * A V * * * * L V * * N * * A G * - - * * I D A N Y * V N L * - E R * Q
18 K * A * A V * * * * L V * * N * * A G * - - R * I D A N Y * I N L * - E R * Q
19 K * G * A V * * * * L V * * N * * A G * - - R * I D A N Y * I N F * - * R * Q
20 K * G * A V * * * * L V * * N * * A G * - - R * I D A N Y * I N F * - * R * Q
21 S * A * S V * D * * * L * * * * * A G * - - * * * N D N Y * I N L * - A R * Q
22 * * A * A V * D * * * L * * * * * A G * - - * * * N N N Y * I * L * - A R * Q
23 * * A * A V * D * * * L * * * * * A G * - - * * * N N N Y * I * L * - A R * Q
24 - * A * A V * D * * * L * * * * * G G * - - * * * D N N N Y I * L * - A R * Q
25 R * Q R * V * R * * A K * * E * * A R S - - * * Q N S N W * I N * * - A R * R
26 Q * * V * V * A * * F A * * K * * A Q E - - D * * P A * S * I N * G - E R * K
27 E * K * * * * * * * * T * V T * * A G * - - * * * D D L Y * I G M I - E - * H K
28 T * G R I V * D * * * T * * L * * G Q S - - D * I E G * * * I * * A - R F * Q
29 T * A R I V * D * * * T * * L * * G Q S - - D * I E R * * * I * * K - R H * R
30 T * A R V V * D * * * T * * L * * G Q S - - D * I E G * * * I * * A - R F * Q
31 T * A R I V * D * * * T * * * * * A A A - - * * * I N N F M I S S S - T H * Q
32 * * K K K V * E * * * V * * * * * A A S K H S * * - K N L - W R Y P S S K * T G T A E K C
33 Q - K K E L * E * * * V * * * * * A N N R N T * * - S N Y - F G H S S S K * T G T - E Q C
34 P * G T Q * * E * * K A * * * * * R D N I P S * D - N K Y - W L F P P * D * R Q E P E P C
35 P * G T Q * * E * * K A * * * * * R D N I P S * D - N K Y - W L F P P * D * R E E P E P C
36 W * Z Z Z * * Z * B * V * * Z * L R R B L S T * K - B Z Y - M F Y P D S R * R G P S E T C
37                                                       G
38 P)* L K E V * E * * L A * * * * * R D N L N T * D S K K Y - W M F P A * N * L E S E E P C
```

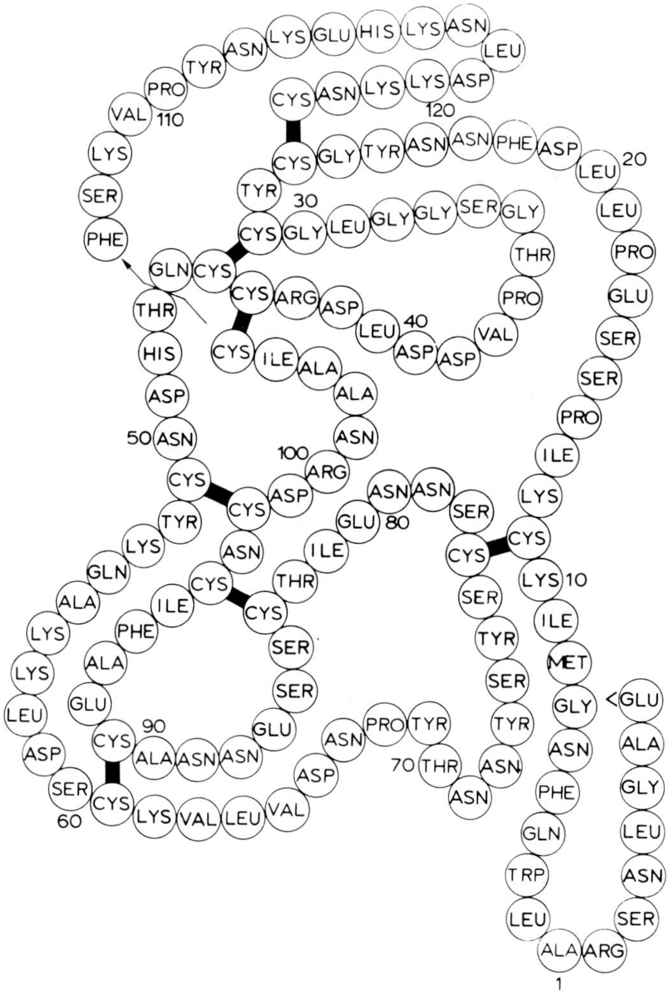

Fig. 1. Amino acid sequence of bovine prophospholipase A_2 and the connection of the disulphide bridges.

either they are catalytic residues and residues involved in binding of the cofactor Ca^{2+}: His-48; Asp-49; Asp-99; or they have an important structural function (e.g. the half-cysteines, five glycine residues).

Since it is known that upon binding of substrate (either monomers or aggregated substrate) hydrophobic interactions are involved, it is of interest to analyse the residues which surround the active site of bovine pancreatic PLA. Inspection of the X-ray model shows the astonishing fact that several hydrophobic side-chains surrounding the active site are not buried but point toward the surrounding water. This creates a large surface area with hydrophobic properties suitable for interactions with lipids. These surface residues are: Leu-2, Trp-3, Leu-19, Leu-20, Leu-31, Lys-56, Leu-58, (Val-Leu-Val-65), Tyr-69 and Thr-70. Table 1 shows that in all phospholipases these side-chains are highly variable (as would be expected for exposed residues), but mainly hydrophobic residues are present. Among the side-chains carrying a charge, only a single negatively-charged side-chain is found, although several arginine and lysine residues are present. This might suggest that interactions with lipid–water interfaces not only require a large hydrophobic surface area but also that a positive charge on the protein may add favourably to this interaction (see also 92a).

Two regions rich in lysine may be important for binding. In bovine pancreatic PLA the lysine residues 53, 56, 57 and 62 form a cluster which may be important for binding [93]. The C-terminal part of the sequence (residues 116–121) may also be important. Especially in venom PLA's the latter part contains a cluster of hydrophobic side chains (see Table 1). Since more than 10 residues contribute to the hydrophobicity of the protein surface, one might expect that substitution (or chemical modification) of only one of these side chains will not drastically alter the interaction with lipid–water interfaces per se.

Two proteins are reported to be devoid of phospholipase activity: notechis II-1 and taipoxin γ-chain. The former, which binds Ca^{2+} and does react with active site irreversible inhibitors, has a normal elapid phospholipase structure except for the substitution of Ser for the otherwise invariant Gly-30 [94]. Since this part of the main chain participates in Ca^{2+} binding, one might suppose that although the enzyme binds Ca^{2+} ions the Ca is not bound at the proper position. The taipoxin γ-chain has several salient structural features different from other PLA's: (1) at the N-terminus it contains 8 additional residues, as do the zymogens of the pancreatic PLA's; (2) if the cysteines present at positions 15 and 19 form a disulphide bridge, a short extra loop is present near the entrance of the active site; (3) it is the only sequence with Pro-31, in a part of the chain important for Ca binding; (4) there is no deletion between residues 55 and 68; and (5) a polysaccharide is attached to Asn-70, located at the entrance of the active site.

As early as 1972, it was suggested that the α-amino group of PLA forms an internal salt bridge, thereby stabilising the active site geometry [96]. This hypothesis has been supported by high (8.3–8.9) p*K* values of this group [97,98]. Also the finding that replacement of Ala-1 by other amino acids can have drastic effects (see Section 5, "Chemically modified enzymes") stresses the importance of this binding. Finally the refined X-ray structure of bovine PLA shows that Ala-1 is indeed buried in the interior of the enzyme. The α-amino group is linked via a water molecule to the side chain of Asp-99; moreover the α-ammonium group is hydrogen-bonded to the side chain of Gln-4 and to the main chain carbonyl carbon of Asn-71 (see also Section 8, "The 3-dimensional structure").

The precursors of the pancreatic enzymes, which are devoid of activity on micellar substrates but efficiently hydrolyse monomeric substrates, differ from the active enzymes only by the presence of a polar activation peptide at the N-terminus.

Activation peptides containing three, five or seven residues have been reported [5b,7,18,95] all containing an invariant arginine residue at the C-terminal end.

Despite a remarkable sequence homology of the enzymes isolated from pancreatic tissue and from the venoms of all classes of venomous snakes, their behaviour in solution is quite different. Whereas the enzymes from *C. adamanteus* and *C. atrox* only occur as dimers even at concentrations as low as 50 μg/ml [42,44], the enzyme from porcine pancreas exists as a monomer at concentrations as high as 5 mg/ml [99]. Several other phospholipases show a concentration-dependent association, generally in the concentration range 0.05–0.5 mg/ml. This equilibrium seems to be shifted to the monomeric form at low pH. Calcium ions display a more complex behaviour, showing either no influence on the monomer–dimer equilibrium or shifting it toward the monomeric or to the dimeric form [61,64,81,100]. Mal'tsev et al. [101] showed that Ca^{2+} ions alter the association-dissociation rate constants of the monomer–dimer equilibrium of *Naja naja oxiana* PLA but the equilibrium constant is hardly affected.

Since all extracellular PLAs are calcium-dependent it is not surprising that those PLAs that were tested are able to bind calcium ions. In general the observed dissociation constants fall in the range of 0.1–1 mM at pH 7–8. For a limited number of enzymes detailed studies pertaining to spectral and conformational changes as well as to amino acid side chains involved in the binding have been published (see Section 6, "Ligand binding").

4. *Kinetic data*

The kinetic behaviour of a large number of water-soluble enzymes acting on molecularly dispersed substrates (including esterases), has been analysed in detail. Usually these enzymes display classical Michaelis–Menten kinetics and important information has been obtained on the mechanism of action of these proteins. PLA (EC 3.1.1.4) belongs to a special group of esterases, the lipolytic enzymes, the specific activity of which strongly depends on the state of aggregation of the substrate. The rate of hydrolysis of phospholipids increases by several orders of magnitude on passing from the monomolecularly dispersed to micellar solutions. The analysis of the kinetic properties of this enzyme acting on monomolecularly dispersed substrates has provided a theory about the mechanism of catalysis (see Section 9, "Catalytic mechanism"). Attempts to reveal kinetic pathways for these enzymes acting on their biologically relevant aggregated substrates have so far not met with success, notwithstanding extensive efforts. To date, there is not even general agreement on the model of lipolysis from which the kinetic equations have to be derived.

As has been discussed in recent review papers [1b,102–105], the main difficulty in understanding lipolysis is our lack of information concerning the mechanisms leading to the observed enhanced rates as induced by certain organised lipid–water interfaces. Although it is evident that the physiochemical properties of the aggregated phospholipid systems play a predominant role in lipolysis, the effects of important factors such as steric environment and hydration of polar headgroups, chain-packing density and surface defects, surface charge and pH are still poorly understood. This results in the use of rather vague terms such as "quality of interface", "supersubstrate", etc.

Three speculative hypotheses have been forwarded to explain the burst in enzyme activity upon substrate aggregation.

(i) "Enzyme theory", which assumes a conformational change in the adsorbed enzyme, controlled by the micro-environment of the lipid–water interface and resulting in an optimisation of the active site.

(ii) "Substrate theory", which assumes a much higher susceptibility of substrate molecules toward the enzyme in the lipid–water interface.

(iii) "Product theory", which assumes that the rate-limiting step of product release, being very slow in water, is markedly increased in the hydrophobic lipid–water interface. The function of PLA's in vivo is a controlled degradation of aggregated long-chain phospholipids and our final aim should be the elucidation of the mechanism of action under these conditions. Based on the above-mentioned difficulties, we will try, however, to evaluate kinetic data obtained with other systems as well, and in the following order:

(a) Monomeric substrates
(b) Micellar substrates
 (i) micelles of short-chain lecithins
 (ii) mixed micelles of phospholipids with detergents
(c) Monomolecular surface films of medium-chain phospholipids
(d) Phospholipids present in bilayer structures

(a) Monomeric substrates

As early as 1961, Roholt and Schlamowitz [10] investigated the kinetics of crude PLA from *Crotalus durissus terrificus* on molecularly dispersed dihexanoyllecithin. The enzyme was found to act optimally at pH 8 and Ba^{2+} ions were shown to inhibit the hydrolysis by competition with the essential cofactor Ca^{2+} for binding to the protein. The highly water-soluble reaction products, hexanoic acid and 1-hexanoyl-lysolecithin, did not appear to influence the reaction rate. On the other hand a number of monoalkyl long-chain surfactants such as egg lysolecithin, sodium dodecylsulphate or Tween, strongly influenced the hydrolysis rate and it is now evident that these effects have to be attributed to the incorporation of the substrate in the detergent micelle (see Section 4b).

The first very detailed kinetic analysis of a highly purified PLA from *Crotalus adamanteus*, using as substrate monomeric 1,2-dibutyryl-lecithin, was reported in 1972 by Wells [106]. The pH-activity profile of this enzyme (optimum pH 8–8.5) is in agreement with the results of Roholt and Schlamowitz [10] and under no circumstances was it possible to find any cation which could replace Ca^{2+} in the enzymatic reaction. The pH-dependence of the reaction suggests that a group with pK 7.6 is involved in the catalytic step, as well as in Ca^{2+} binding [107]. Besides the important consequences of these studies for our understanding of the mechanism of catalysis by PLA, the author clearly demonstrated that his results are consistent with an ordered addition of ligands to the venom enzyme. Ca^{2+} adds first, followed by monomeric substrate. In addition the kinetic results point to an ordered release of products where fatty acid is released first from the enzyme, followed by the lysolecithin. It must be mentioned that the *Crotalus adamanteus* PLA has a strong tendency to form dimeric enzyme complexes in aqueous solution.

Using a series of homologous short-chain diacyllecithins varying in chain length between C_2 and C_5, Zhelkovskii et al. [108] also showed that a homogeneous preparation of PLA from the cobra *Naja naja oxiana* is able to hydrolyse these short-chain lecithins at concentrations far below their CMC. Although the individual kinetic constants k_{cat} and K_m could not be derived because the Michaelis constants are considerably higher than the CMC values, it is evident that the efficiency of the catalytic transformation of the substrate strongly depends on chain length of the hydrocarbon moiety of the substrate. From the results obtained it follows that the PLA molecule must possess an apolar region and most probably both acyl chains participate in the hydrophobic interaction between substrate and enzyme *.

Viljoen and Botes [109] investigated the kinetic properties of pure PLA from *Bitis gabonica* on monomeric dihexanoyllecithin as a function of pH. The authors confirmed the results of Wells [106] that these enzymes follow a kinetic mechanism of the ordered bi-ter type [106a] and found a pH-dependence of k_{cat} controlled by a group active in catalysis of p$K = 6.8$, being most probably a histidine residue. It is not clear why the authors used 0.5 mM lipid as highest substrate concentration taking into account the CMC of dihexanoyllecithin, which is about 10 mM. Although the value of k_{cat}/K_m can be determined in this way, the absolute values of k_{cat} and K_m could have been estimated with more accuracy by using higher substrate concentrations. The Michaelis constant, K_b, is pH-independent in the range 5.5–9.0, which would be in agreement with a predominantly hydrophobic interaction between enzyme and substrate. The comparison made by the authors between their present results (obtained with molecularly dispersed dihexanoyllecithin) and those reported previously by them (obtained with dihexadecanoyllecithin) should be re-evaluated (see Section 4d). Although the highly purified pancreatic (pro)phospholipases A_2 are also known to be able to hydrolyse molecularly dispersed short-chain lecithins [110,111], technical difficulties connected with the use of the titrimetric assay (see also [106]) have so far prevented more extensive kinetic analyses. Using short-chain lecithins containing thioester bonds, Volwerk et al. [12] reported kinetic data of porcine pancreatic PLA in the monomeric substrate region. In contrast to the venom enzymes, the initial velocity patterns of the pancreatic PLA

* Such an architecture would also explain the very bad substrate properties of glycolecithins [12,173] and 2-acyllysolecithins [161], which, because of their single chain could bind to the active site in an orientation unfavourable for catalysis.

are consistent with random addition of substrate and Ca^{2+} to the protein.

The V_{max}–pH profiles show that the activity of the pancreatic enzyme is controlled by a group of apparent pK 5.5, tentatively assigned to His-48.

(b) Micellar substrates

(i) Micelles of short-chain lecithins

The above-mentioned difficulties in obtaining detailed kinetic data on PLA with monomeric substrates, combined with the fact that lipolytic enzymes in vivo are acting on aggregated phospholipids, led various investigators to examine the kinetics of PLA acting on micellar short-chain lecithins. De Haas et al. [110] studied the action of porcine pancreatic PLA on a series of short-chain diacyllecithins varying in acyl chain length from C_6 to C_9. Large increases in reaction rates were observed upon passing the CMC and in the micellar region seemingly normal Michaelis curves were obtained describing the progressive adsorption of the enzyme at the surface of the micelles. Notwithstanding their slight differences in chemical structure, the various lecithins are degraded with very different rates, indicating the importance of the "quality" of the lipid–water interface for hydrolysis.

Initial rate measurements were interpreted to be consistent with a random addition of Ca^{2+} and substrate to the enzyme, which is in agreement with the results obtained for this enzyme in the monomeric substrate region [12]. These results would support the existence of separate and independent binding sites for substrate and metal activator on the enzyme, although Pieterson et al. [112] in direct binding studies reported a synergistic effect for Ca^{2+} and substrate binding between pH 5 and 8. The porcine pancreatic enzyme works optimally at a pH of about 6 but such values obtained with aggregated substrates have to be considered as apparent and are essentially uninterpretable [113,114].

A dramatic activation of the enzyme was found at high salt concentrations. No clear-cut explanation was provided but the concomitant decrease of the apparent K_m supports the idea that also micellar binding to this enzyme involves mainly hydrophobic forces. Detailed kinetic analyses of PLA from *Crotalus adamanteus* acting on dibutyryl-, dihexanoyl- and dioctanoyllecithin both below and above the CMC were reported by Wells [113]. Also for the venom enzyme a dramatic increase in catalytic efficiency was observed when the substrate concentration exceeded the CMC. In contrast to the pancreatic enzyme, this venom PLA requires an ordered addition of Ca^{2+} and substrate both in micellar and monomeric form. No activation of the venom enzyme was observed in the presence of high salt concentrations. Although the V_{max} of the phospholipase acting on *monomeric* dibutyryl lecithin is some 3000 times lower than the V_{max} measured on dioctanoyllecithin *micelles*, dibutyryl PC concentrations near the K_m of this substrate ($\sim$40 mM) were found to inhibit competitively the enzymic action of micellar dioctanoyl PC. This result was interpreted as a support for a mechanism of phospholipase A_2 in which the enzyme after each single encounter with the micellar interface and a catalytic cycle, returns to the aqueous phase. This argument, however, is valid only if diC_4-PC is not present in the diC$_8$-PC micelle. If part of the diC$_4$-PC is incorporated into mixed micelles together with diC$_8$-PC, the quality of the lipid–water interface will change and inhibition is to be expected. The observation that no hydrolysis of diC$_4$-PC occurs cannot be adduced as evidence that diC$_4$-PC does not partition between solvent and diC$_8$-PC micelles. Even if present in the micelle, the diC$_4$-PC monomer will hardly be able to compete for the monomer binding site on the enzyme with the monomeric diC$_8$-PC molecule. (Compare the monomer-E dissociation constants: K_m diC$_4$-PC ~ 40 mM; K_m diC$_6$-PC ~ 4 mM; K_m diC$_8$-PC ~ 0.4 mM.)

Indeed such a "single encounter mechanism" in which the enzyme "hops" up and down between bulk and micelle surface would not be fundamentally different from its interaction with monomeric substrate. The large rate enhancements attendant upon substrate aggregation were tentatively explained by assuming (a) marked increase in the rate of product release, (b) a much lower entropy of activation, or (c) conformational constraints placed on the glycerophosphocholine moiety of the substrate in the aggregated state. In an attempt to improve our understanding of the large rate enhancement observed with PLA when the substrate concentration exceeds the CMC, Pieterson et al. [111] compared the kinetic data of the "active" pancreatic enzyme with that of its natural zymogen using short-chain substrates below and above the CMC. Both proteins catalyse the hydrolysis of short-chain monomeric 3-*sn*-phosphatidylcholines with a similar, albeit low efficiency. Direct binding studies involving Ca^{2+} and monomeric substrate analoques and irreversible inactivation characteristics also point to a very similar architecture of the active centre in PLA and its zymogen [115]. The aggregated (micellar) form of the lecithins is hydrolysed effectively only by PLA and not by the zymogen. Apparently only the active form of the pancreatic enzyme recognizes certain organised lipid–water interfaces and hydrolyses such substrates in a very efficient way. These results together with a previous monolayer study [116; see also Section 4c] led to the hypothesis that "active" PLA, in contrast to its zymogen, contains a hydrophobic surface region, the Interface Recognition Site (IRS), through which the enzyme binds * to the lipid–water interface. Direct binding studies involving both active PLA and its zymogen with micellar substrates and analogues confirmed that only the "active" enzyme interacts with interfaces [111]. The fact that irreversible modification of the active centre in PLA does not impede the binding of the protein to interfaces [115] suggests a functional and topographic separation of IRS and active centre. Nuclear magnetic relaxation studies by Hershberg et al. [117,118] are in agreement with such topologically distinct sites. A similar conclusion was reached by Roberts et al. [119] for the *Naja naja* PLA. As shown in Fig. 2, two successive

* A comparable "hydrophobic head" or "interfacial affinity region" in lipolytic enzymes has been independently postulated by Brockerhoff [120]. Because the mode of interaction of the enzyme with the interface is still under discussion, "binding" is used in a rather loose sense and stands for different forms of interaction such as "adsorption", "penetration", "anchoring", etc.

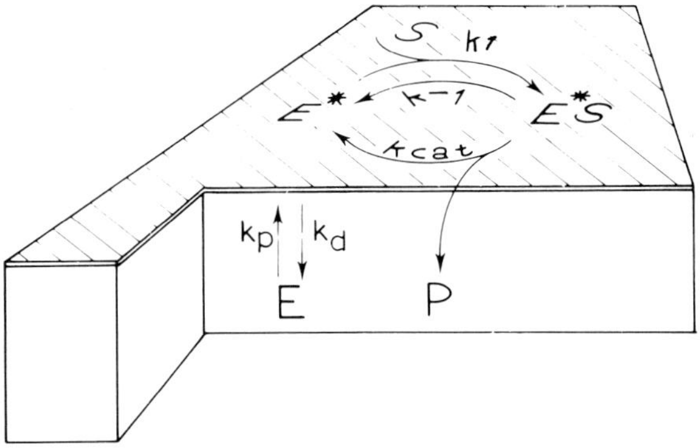

Fig. 2. Proposed model for the action of phospholipase A_2 (E) at an interface [116].

equilibria are supposed to exist; first a rate limiting, reversible penetration * of the enzyme into the interface ($E \rightleftharpoons E^\star$), followed by the formation of a "two-dimensional Michaelis complex" ($E^\star + S \rightleftharpoons E^\star S$). The dramatic rate enhancement observed for phospholipases A_2 from various sources when the substrate concentration exceeds the CMC and lipid–water interfaces are formed, has been atttributed to a conformational change in the bound protein ($E^\star$) resulting in an optimal alignment of the active site amino acid residues. This model could also explain why irreversible active-site inhibition of PLA by *p*-bromophenacylbromide is stimulated in the presence of certain micellar interfaces [115]. Although the apolar reagent is incorporated in various forms of lipid aggregates, such as micelles and lamellar structures, only those interfaces which allow binding of PLA to the interface gave rise to increased inhibition.

In a very interesting study, Allgyer and Wells [121] reanalysed the kinetics of *Crotalus adamanteus* PLA acting on monomeric and micellar diC_6-, diC_7- and diC_8-PC. The abnormal parabolic velocity dependence on substrate concentration near the CMC was tentatively explained by a thermodynamic model for micelle formation in which two species of micelles exist. In this formulation the first micelle is formed at lecithin concentrations near the CMC and the second micelle arises from the first at higher concentrations of lecithin. A satisfactory fit to the kinetic data was achieved assuming that the second micelle is the form of substrate responsible for the large rate enhancement observed above the CMC. In agreement with an early hypothesis of Brockerhoff [122] and with recent ^{13}C-NMR results of

* Penetration is used because of the multiple indications that at least for the pancreatic enzyme, hydrophobic interactions play a major role in the binding process [116]. Most probably an insertion of an apolar amino acid side chain in the hydrophobic lipid core is preceded by a looser adsorption process.

Schmidt et al. [123], the authors suggest that dehydration of the carbonyl groups in micelle II might be the main reason for the enhanced activity of PLA. The enzyme's extreme sensitivity to small changes in lipid hydration was noted earlier by Wells and colleagues [124–126]. Very recently, Johnson et al. [126a] reported a thermodynamic analysis of dihexanoyllecithin aggregation. For this lecithin the heat of dilution data for low lipid concentration could only fit by assuming the existence of premicellar aggregates, mainly dimers. The calorimetric measurements on the dihexanoyllecithin did not show the transition in micellar form proposed previously by Hershberg et al. [117]. No correlation has yet been reported between this self-association behaviour of the short-chain lecithin and phospholipase A kinetics.

From the foregoing it is clear that PLA's from different sources display dramatic rate enhancements when their substrates pass from the monomeric into the micellar form. Both for the *Crotalus* PLA and the pancreatic enzyme it has been demonstrated that substrate molecules at concentrations below their CMC are hydrolysed much more rapidly after incorporation into mixed micelles even with non-substrates or with competitive inhibitors! No agreement, however, exists on the origin of this interfacial activation. Wells et al. [106,113,127] prefer the "substrate" hypothesis: it is the lipid–water interface which confers a preferred conformation * on the substrate molecule which would allow for a higher fraction of productive single encounters with the enzyme. On the other hand, the investigators working with the pancreatic enzymes favour the "enzyme" theory in which PLA reversibly "binds" to the lipid–water interface, followed by a conformational change in the protein with increased catalytic activity. Although it could be argued that PLA's from various sources might follow different pathways, the high structural resemblance of these enzymes makes such an idea unattractive. In the reviewers' opinion the "enzyme" theory does not exclude the "substrate" hypothesis; both could be acting together to give the large rate enhancement observed. However, the assumption that the enzyme necessarily leaves the interface after each catalytic cycle is based on disputable arguments and it is not clear why such a mechanism would lead to accelerated catalysis.

(ii) Mixed micelles of phospholipids with detergents

Detergent solutions with a low CMC solubilise phospholipids by incorporation into mixed micelles. Such systems are attractive for kinetic investigations of lipolytic enzymes because, at least at first glance, they combine all the advantages of isotropy of micellar solutions with the possibility of investigating long-chain natural phospholipids by classical pH-stat assay techniques. In a series of papers Dennis and coworkers [130–134] extensively analysed the kinetic behaviour of PLA from *Naja naja naja* acting on lecithins (varying in chain length from C_6 to C_{16}) solubilised in

---

* Support for a change in monomer PL conformation/orientation occurring as the molecules become packed in an interface was obtained in ^{1}H- and ^{13}C-NMR studies of Roberts and colleagues [128,129], and by Plückthun and Dennis [129a] using ^{31}P-NMR.

the non-ionic detergent Triton X-100. Although this detergent is somewhat polydisperse, its neutral character constitutes a distinct advantage over charged amphiphiles such as bile salts, CTAB, SDS etc. in kinetic studies of PLA's which are dependent on metal cofactors. Biologically relevant phospholipids, such as the long-chain lecithins DMPC and DPPC form bilayer structures in water (liposomes, vesicles), interfaces which are hardly attacked by most PLAs (compare Section 4d). Addition of increasing amounts of Triton gradually transforms these lamellar structures into mixed micelles and at a molar ratio of Triton to lecithin of about 2:1, isotropic solutions are obtained which are optimally susceptible to the action of the cobra enzyme *. Higher mole fractions of the detergent gave rise to increasing "inhibition" of the PLA, a kinetic effect which has been ascribed to "surface dilution" of the substrate. To explain the observed surface dilution kinetics, Deems et al. [133,137] used a model of lipolysis comparable to the one shown in Fig. 2. By changing the lecithin concentration in the interface of the mixed micelle with Triton, they calculated approximate values of K_S^A ($=k_p/k_d$ in Fig. 2), the dissociation constant for the enzyme-mixed micelle complex and of K_M^B ($=K_M^{\star}$ in Fig. 2), the two-dimensional Michaelis constant for the catalytic step. Credit should be given to the authors for the originality of the idea to separate quantitatively the affinity constant of the enzyme for the interface and the binding to the substrate in the interface. Unfortunately the numerical values reported have to be considered as rather rough estimates, taking into account the simplifying assumptions which were required to apply the kinetic equations. As has been extensively discussed before [103], changes in the molar ratio of Triton to phospholipid probably induce differences in the quality of the lipid–water interface and thereby influence K_S^A. Such changes have been detected in fact by the authors [132]. On the other hand, reliable estimates of K_M^B are even more difficult to obtain. Under "saturating" conditions when all enzyme molecules were bound to the mixed micellar surface, the authors showed that the velocity remained linearly proportional with the amount of lecithin in the interface of the mixed micelle up to a mole fraction of 0.33 [130,131,133,137]. This implies that the two-dimensional lecithin concentration is far below K_M^B and even rough estimates of its absolute value become impossible.

In a similar attempt to separate $K_M^{\star}$ from k_p/k_d (Fig. 2) and to obtain a numerical value for the two-dimensional Michaelis constant, Slotboom et al. [209] used two enantiomeric 2-*sn*-lecithins containing fatty acids of different chain length in positions 1 and 3. By incorporating mixtures of both β-lecithins into Triton micelles, keeping *total* phospholipid concentrations and *total* amount of Triton constant, the enzyme activity could be followed as a function of the mole fraction of each of the β-lecithins. Because of the identical physicochemical properties of

* The authors demonstrated [135,136] that this formation of mixed micelles takes place only above the thermotropic phase transition temperature of the phospholipid. Formation of mixed micelles at temperatures below the transition temperature requires much higher ratios of Triton to phospholipid.

enantiomers the quality of the interface remains constant. Although this technique clearly showed that the $K_m^\star$ values for enantiomers are not identical, a quantitative relationship can be obtained only under interfacial saturation conditions (all E in form $E^\star$). Pancreatic PLA has a very low affinity for pure Triton micelles, as was also found for the cobra enzyme [119], and therefore the distribution of enzyme over bulk and interface ($E \rightleftharpoons E^\star$) will strongly depend on the total amount of β-lecithin incorporated into the mixed micelles. This implies for this detergent that interfacial saturation is difficult to reach. Using *n*-alkylphosphocholine as a carrier micelle for which the enzyme has a high affinity, k_{cat} and $K_m^\star$ values could be obtained for both stereoisomers. One must mention, however, that also in this case a simplifying assumption had to be made because the molecules of the carrier matrix are competitive inhibitors of the enzyme. In addition also in this study one might wonder whether the quality of the lipid–water interface remained rigorously constant upon incorporation of increasing amounts of β-lecithins.

Roberts et al. [119] proposed a new model for the interaction between *Naja naja* PLA and mixed micelles of Triton + phospholipid: two phospholipid molecules should be required, one to sequester the enzyme to the interface, the other for subsequent catalysis. Based on cross-linking experiments of the enzyme in the presence of excess substrate it was concluded that the substrate is essential for enzyme aggregation and that probably the resulting dimer unit is the active form of the enzyme. This "dual-phospholipid" model, however, was heavily based on the presumed "half-site reactivity" of this enzyme [138], which is now known to be incorrect [139]. Of course, the withdrawal of the "half-site reactivity" does not necessarily invalidate the proposal that the cobra enzyme aggregates to its enzymatically active dimer form in the presence of substrate. On the other hand the results of the cross-linking experiments, where under optimal conditions trimer formation is relatively more important than dimerisation, are not fully convincing. Perhaps the strongest evidence for the "dual-phospholipid" model has to be found in the "specificity reversal" of this enzyme (vide infra). An interesting observation in this study is that the cobra enzyme, just as the pancreatic PLA, has no affinity for pure Triton micelles. Only mixed micelles containing phospholipids (including sphingomyelin), bind to the enzyme in the presence of Ca^{2+} or Ba^{2+} ions. Also lysolecithin or free fatty acid incorporated in the Triton micelle enable the enzyme to bind to the mixed micelles and with these products no bivalent metal ions were required for binding. Although these findings might be interpreted as support for a mechanism in which PLA initially interacts with a *single* lipid molecule in the interface other explanations are also possible.

An interesting case of "specificity reversal" of the *Naja naja* PLA was described by Dennis and coworkers [15,140,141], which might bear direct relevance to the mechanism of action of this enzyme. Comparing the action of the enzyme on mixed micelles of Triton and long-chain lecithin with that on mixed micelles of Triton and long-chain PE, the cobra PLA hydrolyses the lecithin-containing micelles at a much higher rate. However, in Triton micelles containing both PE and PC in equimolar amounts, the enzyme was shown to possess a clear preference for PE as substrate.

The activating effect on PE hydrolysis appeared not to be limited to long-chain PC but several other phosphocholine-containing lipids showed a similar behaviour, such as lyso-PC, sphingomyelin and even dibutyryllecithin. These results can be tentatively explained by the possible existence of two binding sites on the enzyme molecule:

(i) an activator site which requires a lipid molecule containing the phosphocholine moiety and at least one fatty acyl chain and

(ii) a head-group non-specific catalytic site.

While it might be argued that activation of PLA toward PE by long-chain phosphocholine lipids could be caused by subtle changes in the lipid–water interface of the mixed micelle, the activating effect of the highly water-soluble dibutyryllecithin is supposed to constitute the strongest evidence for the proposed direct interaction of the PC molecule with the enzyme. Taking into account the relatively weak activating effect of dibutyryl PC (4 times), as compared to the two-fold activation by an aspecific, non-phosphocholine-containing lipid such as oleic acid, it is, however, of the utmost importance to be certain that dibutyryllecithin is not partially incorporated into the mixed micelle. The experimental techniques used by the authors, namely equilibrium gel filtration in the *absence* of PE and ^{31}P-NMR, would probably not detect a low incorporation of dibutyryl PC in the mixed micelle *. The activating effects of phosphocholine-containing lipids observed here on the rate of hydrolysis of more negatively charged phospholipids by venom PLA, are in agreement with previous reports on similar activation by *n*-alkylphosphocholine of *Crotalus adamanteus* PLA hydrolysis of negatively charged phospholipids such as cardiolipin, phosphatidylglycerol and phosphatidic acid [142,143]. The small size of a PLA molecule, however, dissuades one from postulating the presence of two binding sites for the relatively large phospholipid molecules. The previous suggestion of Dennis et al. that substrate might induce enzyme aggregation and that probably the resulting dimer is the active form of the enzyme, would solve the "sterical" problem but in that case the dimer structure should be asymmetric.

(c) Monomolecular surface films of medium-chain phospholipids

The principles, advantages and drawbacks of this attractive technique to investigate the kinetics of lipolytic enzymes have been discussed in considerable detail in two recent reviews [103,105]. We will, therefore, limit ourselves here to a discussion of a number of very recent papers.

The model of lipolysis proposed by Verger et al. [116] was recently checked by Pattus et al. [144–146] using two differently radioactively labelled preparations of

---

* Very recently Plückthun and Dennis [129a] reinvestigated the incorporaton of water-soluble short-chain phospholipids at concentrations below their CMC into detergent micelles. It was found that, in contrast to dihexanoylphosphatidylcholine, at most 5% of the dibutyryllecithin is incorporated into the micellar Triton X-100 phase.

porcine pancreatic PLA and a series of medium-chain lecithins containing C_8, C_9, C_{10} and C_{12} acyl chains. The lag time observed during pre-steady state kinetics reflects the rate-limiting step of the penetration of the enzyme into the monolayer. Film transfer experiments showed this penetration to be reversible but the desorption of the enzyme from the film is slow compared to the adsorption, which is in agreement with the results of Barque and Dervichian [147–149] (see, however, Momsen and Brockman [149a]). The kinetics of the penetration process is governed by the packing density of the substrate molecules and it seems that the polar headgroup of the phospholipid molecule and its hydration state play an important role. The steady state surface concentration of the enzyme decreases with increasing film pressure. However, this surface concentration increases with fatty acyl chain length of the substrate which is in agreement with the idea that hydrophobic interaction dominates the penetration process.

The influence of bulk pH on the pre-steady state kinetics of the porcine enzyme was investigated and it was found that at alkaline pH the penetration capacity markedly decreases (increase of induction time). In the presence of Ca^{2+}, the equilibrium surface concentration of the enzyme was found, however, to be pH-independent until the pH region where deprotonation of the α-NH_3^+ group of Ala-1 occurs. Deprotonation of this function results in a rapid desorption of the enzyme from the interface. At slightly acidic pH values ($\leqslant 6.0$), enzyme–substrate binding occurs in the absence of Ca^{2+}, but at higher pH only the E–Ca^{2+} complex is able to interact with the PC film. The rapid decomposition of the E–Ca^{2+}–PC complex at basic pH upon addition of EDTA again is a strong indication for the reversibility of the binding process.

Willman and Stewart-Hendrickson [150] investigated the influence of positive charge on the kinetics of hydrolysis of diC_{10}-PC monolayers by PLA from porcine pancreas and *Crotalus adamanteus*. Different insoluble long-chain amines were incorporated in the substrate PC film and hydrolysis rates were followed in a "zero-order" trough as function of pH and amine mole fraction. Because the amines possess very different apparent pK_a values in the mixed surface films it was possible to follow hydrolysis rates as a function of the surface charge of the monolayer. The authors conclude that the inhibition of both PLA's is caused exclusively by the positive surface charge of the film and not by changes in film packing. Unfortunately no use was made of radiolabelled enzymes so it is not clear whether the surface penetration step or the two-dimensional Michaelis parameters $K_m^\star$ and k_{cat} are modified by the positive charge of the film. Most probably more meaningful kinetics would have been obtained by the mixed-film technique (see below) which avoids a continuous change of the quality of the mixed film. Application of the "zero-order" trough [151] enabled Verger and colleagues to study the hydrolysis of mixed monomolecular films of triacylglycerol and lecithin by pancreatic lipase [152] and by pancreatic phospholipase A_2 [153]. Such studies are of particular relevance since lipolysis in vivo involves the participation of several classes of lipids (see also Burns and Roberts [153a]).

Mixed monolayer films of diC_{12}-PC and bovine brain sphingomyelin were used by Barenholz et al. [154]. They investigated two radiolabelled PLA's from porcine pancreas and from the venom of *Vipera berus* and studied the kinetics at different surface pressures and molar ratios of the phospholipids. Taking into account the complex thermotropic behaviour of natural sphingomyelins which are composed of various acyl chains (broad phase transition between 22° and 45°C), it can be expected that mixtures of this phospholipid with diC_{12}-PC will show non-ideal mixing in surface films (cf. Untracht and Shipley [155]). *V. berus* PLA, an enzyme characterised by a high penetrating power [156,157], is relatively insensitive to cracks * introduced in the surface film by increasing mole fractions of sphingomyelin. Its surface pressure–activity profile does not shift and the lower hydrolysis rates observed with increasing sphingomyelin content could be explained simply by substrate dilution. However, these experiments again demonstrate the high sensitivity of the weakly penetrating pancreatic PLA for surface defects. At low film pressures (10 dynes/cm) where the enzyme experiences no penetration problems, addition of sphingomyelin decreases enzymatic activity probably by substrate dilution. At high surface pressures, however, where the enzyme is unable to penetrate pure PC films, the insertion of sphingomyelin molecules in the film gives rise to phase separation and the resulting cracks are immediately recognised by the pancreatic enzyme, which enters the film and high hydrolysis rates are found. This results in a dramatic shift in the activity–surface pressure profile. It would be very interesting to repeat these experiments with a better defined synthetic sphingomyelin.

(d) Phospholipids present in bilayer structures

One of the earliest kinetic analyses of a pure PLA (*Bitis gabonica*) acting on DPPC was reported by Viljoen et al. [158]. Although the authors were under the impression that they studied *monomer* catalysis, the substrate concentrations applied in their assays were so far above the CMC reported by Tanford [159] for DPPC ($\sim 10^{-10}$ M), that we must assume that they worked with lipid *aggregates*, presumably bilayers. Using a somewhat obsolete enzyme assay technique in which proton release is followed by pH drop, they were able to measure initial hydrolysis rates at substrate concentrations ranging from 5 to 80 μM. The very low maximal velocity of the enzyme under these conditions (calculated from the figures to be about 0.5 $\mu mol \cdot min^{-1} \cdot mg^{-1}$ protein) is not in agreement with a 200 times higher V_{max} value given in Table 1 of the same paper.

Initial rate measurements in which substrate and Ca^{2+} concentrations were varied, confirm the mechanism proposed by Wells [106,113] for the *Crotalus adamanteus* PLA in which Ca^{2+} adds first to the enzyme, before the substrate molecule. Product inhibition experiments suggest that also in the *Bitis gabonica*

* Following a proposal of M.K. Jain, such ill-defined surface defects will occasionally be called "cracks".

enzyme the products are released in an obligatory order, fatty acid first and lysolecithin second. In summary, the results of Viljoen et al. might be interpreted by stating that the mechanisms of action of both venom PLA's are very similar, independent of the aggregation state of the substrate. On the other hand, the ill-defined physicochemical state of the substrate under the conditions used, together with the uncertainty about the maximal velocity, make such conclusions premature. Similar remarks have to be made on the kinetic experiments with PLA from *Naja mossambica mossambica* reported by Martin-Moutot and Rochat [63]. Long-chain diacylphospholipids such as PC which form aggregated bilayer structures in water, have for a long time been known to be very poor substrates for pancreatic PLA [160,161], and accurate kinetic analyses seemed to be impossible. However, following the initial reports of Op den Kamp et al. [162,163] that several fully saturated long-chain lecithins become very susceptible to hydrolysis by porcine pancreatic PLA at the thermotropic phase transition, renewed interest has arisen. At the transition temperature, domains of frozen molecules are separated from surface areas where the lipids are in the liquid-crystalline state, and most probably, surface defects exist at the borders, allowing penetration of the enzyme. Both above and below the phase transition the more regular and tighter packing of the phospholipid molecules prevents the anchoring of the enzyme to the interface and no hydrolysis is observed. One must mention that this sharp differentiation is found only with PLA's characterised by a weak penetrating power such as the pancreatic enzymes, β-bungarotoxin [164] or platelet phospholipase [165] in combination with multilayered liposomes of fully saturated lecithins. With increasing unsaturation of the lecithin acyl chains, resulting in looser packing of the phospholipid molecules in the interface, the more powerfully penetrating PLA's in particular are able to enter the bilayer to a certain extent and, at temperatures above the thermotropic phase transition, hydrolysis occurs. Similar results were reported recently by Goormaghtigh et al. [165a].

Wilschut et al. [166,167] extended the above studies and showed that sonicates of PC dispersions, especially those containing small unilamellar vesicles, are more susceptible to PLA hydrolysis than the multilamellar liposomes. They also observed that if sonication is carried out below the phase transition temperature, the resulting vesicles are hydrolysed over a much wider temperature range. Most probably the high curvature of the vesicles results in surface defects which facilitate penetration of the enzyme. These systems, however, are still of hardly any use in kinetic studies because of difficulties in determining initial rates and the variable effects of reaction products on the enzymatic velocity.

In order to overcome these difficulties, Jain and Cordes [168,169] proposed the incorporation of medium chain *n*-alkanols (C_6, C_8) in the aqueous dispersions of long-chain lecithins. By a number of different techniques including trapping experiments, they showed that the bilayers remained closed. They concluded that at optimal concentrations of activating alcohols, egg-PC liposomes and vesicles behave as excellent substrates for various PLAs and that normal Michaelis kinetics can be obtained. Most probably the alcohol chains inserted in the bilayer cause an increased spacing of the substrate molecules, allowing a facilitated penetration of the PLA molecule. However, effects of the alcohol molecule on the catalytic factors $K_m^{\star}$ and k_{cat} could not be excluded. In a subsequent study, Upreti and Jain [170] improved their assay system by using osmotic shock of the multilamellar vesicles before addition of the enzyme. A major disadvantage of the original substrate, phospholipid liposomes + alkanol, was the rather high apparent K_m of the lipolytic enzymes used. Because only the outer layer of the multilamellar vesicles is exposed to the enzyme, large amounts of substrates were required to obtain interfacial saturation. Moreover initial rate measurements were complicated because the rate of hydrolysis was increasing with time as successive bilayers were "opened" and more substrate became exposed. Due to a sudden decrease in ionic strength of the assay solution, the liposomes transiently "open" and such osmotically shocked bilayers offer an almost complete access of the enzyme to the substrate molecules. Because resealing of the liposomes is a rather slow process ($t_{\frac{1}{2}} \sim 10$ min), initial rate measurements were possible and the apparent K_m values were much lower. One must state that even with these osmotically shocked liposomes, the pancreatic PLA, in contrast to all venom enzymes tested, shows a lag phase at the beginning of hydrolysis and only after a certain induction time, τ, is a steady-state rate obtained [171].

This lag phase is strongly reminiscent of the behaviour of the pancreatic enzyme towards densely packed medium-chain PC monolayers [116]. Jain and Apitz-Castro showed that the lag period preceding the steady-state phase was not caused by increasing amounts of hydrolysis products. Moreover, the induction time appeared to be independent of concentrations of enzyme, substrate, alkanol and Ca^{2+}. These facts led the authors to a hypothetical kinetic mechanism for this enzyme, very similar to the model of Verger et al. [116] (cf. Fig. 2), in which the latency period is due to a slow, rate-limiting penetration of the enzyme into the lipid–water interface [144]. It is difficult to understand, however, how in this model τ could be independent of the concentration of the bilayer-perturbing alcohol. Moreover, the observation that calcium is not required for the slow penetration step is not in agreement with the monolayer results.

Recently, Upreti et al. [172] in a very detailed study, investigated the bilayer-perturbing capacity of an impressive series of different alkanols and the effect of the alcohol-modified bilayer on the kinetics of PLA. Whereas insertion of all alkanols into egg-PC liposomes resulted in an increase in free space in the substrate bilayer (surface defects), as evidenced by a higher accessibility to the enzyme and increasing velocities, estimation of the individual kinetic constants (cf. Fig. 2) remained impossible. The fact that increasing chain length of straight-chain *n*-alkanols results in a higher apparent K_m, whereas insertion of branched alcohols seems to have no influence on this parameter, suggests that the former alcohols might compete with substrate molecules for the hydrophobic binding site in the active centre. Jain and coworkers confirmed the original observation made by Bonsen et al. [173] that in mixtures of *sn*-3- and *sn*-1-lecithins having the same chain length, the D-isomer behaves as a pure competitive inhibitor characterised by the same binding constant to the enzyme. This makes the stereoisomeric *sn*-1-phospholipid the most ideal

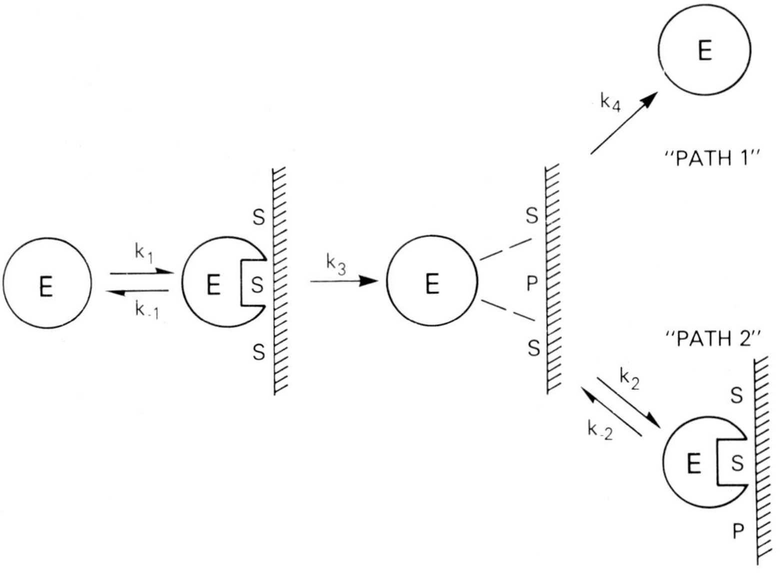

Fig. 3. Kinetic model for hydrolysis of phosphatidylcholine aggregates by *C. atrox* phospholipase A_2 [174].

phospholipid for determination of dissociation constants by direct binding experiments. Using *sn*-1-DPPC bilayers and radioactive PLA preparations from bee venom and porcine pancreas the authors clearly showed that addition of increasing amounts of alkanol to the PC-bilayer increases the amount of PLA bound to the lipid–water interface. Higher enzyme concentrations in the bilayer usually result in higher hydrolysis rates. The observed decrease in enzymatic activity at very high alcohol concentration, where even more enzyme was shown to be bound to the bilayer, is similar to the findings of Dennis [130,131] working with Triton-PC mixed micelles. Most probably this effect is caused by competitive inhibition and substrate dilution and/or is due to unfavourable effects of the microenvironment on k_{cat}.

It goes without saying that, at least for the venom PLA's, a more relevant approach to study the kinetics of the enzymes would be the use of an aqueous system containing only long-chain substrate, enzyme and Ca^{2+} ions.

Several groups investigated such systems using PLA's of different origin [114,174–176]. Tinker et al. [174], working with dispersions* of DPPC and of DMPC, analysed the kinetics of hydrolysis by *Crotalus atrox* PLA at different temperatures both below and above the phase transition temperature. They observed

* Unfortunately the authors prepared their vesicles by sonication below the phase transition temperature and no annealing was attempted. This procedure is known [177] to give unstable, very heterogeneous particles. The relatively low apparent K_m values reported by the authors (100–200 μM) suggest that most of the bilayers contained structural defects (cracks).

that the hydrolysis of gel-phase lecithins showed hyperbolic dependence of initial steady-state rates on bulk lipid concentration, which is in agreement with the results of Viljoen et al. [158] and of Martin-Moutot and Rochat [63]. However, hydrolysis of liquid-crystalline preparations showed a short initial burst of proton release, then a long lag period of very slow reaction, followed by a dramatic increase in the reaction rate. The accelerated proton release during the last stage is probably caused by the presence of considerable amounts of hydrolysis products in the interface. The lag period could indeed be abolished by pre-addition of the reaction products to the substrate bilayer before the reaction was started, an observation which was also reported by Roholt and Schlamowitz [10].

Based on these results the authors proposed a kinetic model of lipolysis which is quite different from that of Fig. 2, proposed by Verger et al. [116], Brockerhoff [120], Deems et al. [133] and Jain and Apitz-Castro [171]. As shown in Fig. 3, the key feature of this new model implies that the enzyme can only bind to the lipid–water interface by forming a 1:1 complex of enzyme and a single *substrate* molecule. This complex formation is supposed to involve a conformational change in the enzyme resulting in exposure of hydrophobic sites which subsequently penetrate the lipid surface. After the performance of *one* catalytic cycle, the enzyme molecule can either desorb from the surface and return to the aqueous phase ("hopping" *) or diffuse along the surface to an adjacent substrate molecule ("scooting" *). The authors proposed that the "hopping" model describes the rapid hydrolysis of the gel-phase phospholipids, whereas the slower hydrolysis of the liquid-crystalline phase would proceed by the "scooting" pathway. In a second paper, Tinker and Wei [175] worked out a mathematical treatment of the observed kinetics in the liquid-crystalline state and concluded "that the proposed model is consistent with current ideas on the mechanism of catalysis by this enzyme".

Very recently, Tinker et al. [178] analysed the hydrolysis of the gel-phase and studied the effects of reaction products on hydrolysis rates. Gel filtration experiments demonstrated that the enzyme binds to egg-PC bilayers even in the absence of Ca^{2+} and that incorporation of hydrolysis products in the bilayer *weakened* the enzyme binding. These observations together with the observed increase in hydrolysis rate at later stages of the reaction, where substantial amounts of lyso-PC and free fatty acids are present, were ascribed to a product-facilitated desorption of the enzyme from the surface. In this latter study both annealed and unannealed sonicated DPPC vesicles were used, but no attempt was made to separate the larger multilamellar structures from small unilamellar vesicles.

Kensil and Dennis [114] examined the action of *Naja naja naja* PLA on single-walled, sonicated vesicles of DPPC, DMPC and egg-PC as a function of temperature. They confirmed the observation of Tinker et al. [174] that the venom PLA hydrolyses the gel-phase phospholipids at a higher rate than the same substrate in

* "Hopping" and "scooting" are expressions used by Upreti and Jain [176] to differentiate between these pathways.

the liquid-crystalline state. In addition they also found an apparent stimulation of activity as the reaction proceeded above the phase transition temperature. This observation was tentatively attributed to an increase in phase transition temperature caused by increasing amounts of reaction products, whereby the enzyme could actually be hydrolysing *gel* state phospholipid, the preferred physical form. As a possible explanation for the enhanced hydrolysis of gel state phospholipids, the authors consider decreased hydration of head groups and better accessibility of the 2-ester function to the enzyme by a tilt of the acyl chains. In this study, well-characterised, annealed small unilamellar vesicles were used and consequently the apparent K_m values are about 30 times higher than reported by Tinker et al.

Finally, Upreti and Jain [176] reported on the kinetics of bee venom PLA acting on unmodified PC-bilayers. Packing alterations in the substrate aggregate were made by sonication, temperature change and osmotic shock. Again biphasic progress curves were found: after an initial rapid proton release in which less than 7% total available substrate is hydrolysed, the reaction slows down and only after production of a certain amount of lyso-PC and fatty acid, fast hydrolysis recommences. As a very attractive hypothesis to explain the observed kinetics, the authors propose that any treatment of the bilayer which introduces defect structures (cracks) and therefore free space, will enhance PLA activity. In terms of the model in Fig. 2 they do not preclude effects of the cracks on the catalytic parameters K_m^* and k_{cat}, but a highly important function of the surface defects is thought to be the shift of the equilibrium $E \rightleftharpoons E^\star$ to the right side. The specific influence on phosphatidylcholine bilayer packing exerted by the simultaneous presence of the hydrolysis products, lysolecithin and free fatty acid, has been demonstrated by Jain et al. [179] and Jain and De Haas [180]. While the PLA is unable to penetrate into the closely packed bilayers of pure lecithin, the presence of *both* lysolecithin + fatty acid results in surface defects (phase separation) and the enzyme displays a high affinity and catalytic power towards such "cracked" interfaces [181].

The hypothesis that cracks or irregularities in the lipid bilayer enhance PLA activity is furthermore illustrated by studies on a natural membrane using pancreatic PLA [182–184]. The *Acholeplasma laidlawii* membrane contains glycolipids (70%) and PG (30%), as the only substrate for PLA's. The physicochemical condition of the membrane can be manipulated by growth of the organisms in different fatty acids: e.g. palmitate addition yields membranes in which 80% of the esterified fatty acids consists of palmitate and the lipids undergo a phase transition between 15° and 40°C. At temperatures above the lipid phase transition PG is accessible for hydrolysis; below the lipid phase transition no PG is hydrolysed. In the latter condition proteins are aggregated, eliminating to a large extent the presence of irregularities in the gel state bilayer [182]. That membrane proteins may be responsible for irregularities in the membrane is illustrated by experiments on membranes which are enriched with branched-chain fatty acids. In this case protein aggregation does not occur upon a decrease in temperature and PG remains accessible also below the onset of the transition [184]. Another type of crack can be induced by binding the membranes at temperatures in between the onset and termination of the lipid phase transition. Now phase separation occurs between domains of gel-like lipids surrounded by liquid crystalline lipid molecules. Pancreatic PLA only has access to those PG molecules which are present in the fluid, protein-containing, areas of the lipid bilayer [183].

In a very recent study, Menashe et al. [185] reported on the action of porcine pancreatic PLA on annealed DPPC unilamellar vesicles. At or above the phase transition temperature long lag times were observed. Preincubation of the enzyme with substrate for a short period of time below the transition temperature followed by enzymatic assay at high temperature abolished the lag time. These results were explained by a slow substrate–enzyme organizational step above the phase transition, whereas this process is much more rapid with gel state phospholipids. The intrinsic activity of the enzyme is maximal when the substrate is in the liquid-crystalline state.

In a very recent paper Kupferberg et al. [185a] reported on the kinetics of *C. atrox* phospholipase A_2 hydrolysis of egg phosphatidylcholine in unilamellar vesicles. The time course of the reaction was analysed both in the absence and presence of bovine serum albumin, a protein which effectively traps the products of the enzymatic reaction. The authors conclude that during the enzymatic reaction only one of the products, lysolecithin, partially (40%) leaves the vesicle surface and inhibits the phospholipase competitively. In the presence of a large excess of serum albumin the product inhibition is relieved.

What is the additional information obtained from kinetic studies of PLA acting on intact PC-bilayers? One remarkable result seems to be the observation of Tinker et al. [174] and Kensil and Dennis [114] that gel-phase PC-bilayers are hydrolysed at a higher rate than the corresponding liquid-crystalline phase. These reports are in agreement with an early observation of Smith et al. [186]. It is clear, however, that independent of the physical structure of the PC-bilayers used (multilamellar liposomes, single-walled vesicles, annealed and unannealed), these systems are all characterised by similar, very complex progress curves. The reviewers feel that initial rate measurements with an acceptable accuracy are hardly possible and that therefore mathematical analyses of these systems using rate equations such as those developed by Gatt and Bartzai [187,188] are premature. On the other hand, the experimental results obtained by the various investigators appear to be in good agreement and therefore one should try, be it for the moment only in a rather qualitative and intuitive way, to explain the reported observations and try to fit them into a common and generalised model of lipolysis. At this moment two hypothetical models are under discussion:

(i) Model of Verger et al., cf. Fig. 2.

(ii) Model of Tinker et al., cf. Fig. 3.

It seems that in general investigators working with snake venom PLA's are more inclined to model (ii), whereas most people investigating the pancreatic enzyme prefer model (i).

Yet these two models are fundamentally different: while in the Verger model the enzyme is supposed to interact hydrophobically with the interface (penetration, anchoring) before Michaelis–Menten type E.S. formation and hydrolysis occurs, the prevailing pathway in the Tinker model ("hopping") implies initial formation by collision of an E.S. complex at the interface and a return of the enzyme into the aqueous bulk phase after each catalytic cycle. The generally observed accelerated hydrolysis of substrates in aggregated form is tentatively explained in the Verger model by a conformational change in the penetrated * enzyme with a concomitant optimisation of the active site. On the contrary, in the Tinker model, the high interface activity is attributed to a "hopping" of the enzyme from the interface to bulk solution and vice-versa and a prolonged stay of the enzyme at the surface of the aggregate ("scooting") is supposed to yield low hydrolysis rates. While the effective hydrolysis of gel-phase phospholipids and the observed rate increases upon product formation in the Tinker model are explained by product-facilitated *desorption* of enzyme from the interface, in the "Verger" model these phenomena are ascribed to a product-facilitated *adsorption* of enzyme to an interface containing more surface defects!

A frequently reported objection to the Verger model is that with several venom enzymes no indications could be found for initial adsorption to or penetration in the lipid–water interface using optical techniques such as ultraviolet difference spectroscopy or fluorescence spectroscopy. Most probably, however, these negative results are caused by the particular lipid–water aggregates used. In titration experiments with single-chain substrate, or product analogues such as lysolecithin, glycollecithins and *n*-alkylphosphocholines, for a number of venom PLAs ultraviolet and fluorescence signals were obtained [156,189,190], and saturation was usually observed. A second argument against this model could be the observation that the enzyme hydrolyses gel-phase phospholipids more rapidly than the liquid-crystalline phase. A priori, in the Verger model one would expect that adsorption of the enzyme and surface diffusion *in* the interface would be favoured by the more loosely packed liquid-crystalline phase and would result in increased hydrolysis rates. One should point out, however, that besides the difficulties mentioned in determining initial velocities with bilayer systems, comparison of the steady-state hydrolysis rates is hampered because of the unknown amounts of enzyme present at the interface. In addition, all investigators agree upon the fact that in phase-separated mixtures of lecithins, the most liquid component is hydrolysed more extensively. As regards the Tinker model the following points seem to be relevant.

(i) PLAs, independent of their origin, are known to possess an unusual affinity for all kinds of interfaces and adsorption occurs not only to lipid–water aggregates but also to glass, teflon, and many other surfaces, including the air–water interfaces. Therefore an ordered mechanism in which a Michaelis type E.S. complex would be required before hydrophobic interaction of the enzyme with the interface can occur, seems to be superfluous.

---

* Although the "penetration" process by various techniques has been shown to be reversible, the enzyme is thought to remain bound to the interface during a number of catalytic cycles.

(ii) A product(lyso-PC and/or fatty acid)-stimulated desorption of PLA from the lipid aggregate assumed to explain the observed higher hydrolysis rates, seems to be in contrast with the results of many direct binding studies. Several PLAs adsorb very well to micelles of single-chain detergents such as lyso-PC, fatty acid, *n*-alkylphosphocholines etc. Moreover the pancreatic PLA's which have no affinity for pure lecithin aggregates in bilayer form (liposomes or vesicles) strongly adsorb to these structures if low percentages of hydrolysis products are incorporated [181].

(iii) The "hopping" mechanism implies that desorption of PLA from the surface is a faster process than the formation of a new E.S. complex. This argument is based on a supposed slow surface diffusion of the enzyme in the lipid bilayer, a medium of higher viscosity than water, but does not take into account the well-known high mobility of free substrate molecules in the plane of the bilayer.

(e) Reversible inhibition of phospholipase A_2

Studies of inhibition kinetics have contributed to a large extent to our present knowledge of the mechanism of many enzymes. Unfortunately this approach has yielded only limited information on the mechanism of action of lipolytic enzymes. With the exception of the earlier work of Wells [106] in which product inhibition was successfully studied with *Crotalus adamanteus* PLA acting on monomeric substrate, similar studies on several other phospholipases A_2 were seriously impeded by unfavourable CMC/K_m ratios. An important problem is that inhibition studies of PLA acting on aggregated substrates, are plagued by even greater difficulties. Any incorporation of a possible inhibitor in an organised lipid–water interface will chance the quality of the interface and influence not only the Michaelis parameters $K_m^\star$ and k_{cat} (cf. Fig. 2) but also the amount of enzyme present *in* the interface (k_p/k_d in Fig. 2). In this way, several potential inhibitors of PLA act in fact as potent activators [10,173,181,191,192]. This subject has been discussed previously by Verger and De Haas [103] and up till now it has not been possible to separate the effects of inhibition in the classical chemical sense from purely physical effects.

(f) Monomeric or dimeric enzymes or higher aggregates?

The question whether PLAs are catalytically active as monomeric or dimeric proteins becomes particularly important after the reports of Wells [193] and Roberts et al. [138] that *Crotalus adamanteus* and *Naja naja naja* PLA's demonstrate "half-site" reactivity. Very recently, Smith and Wells [194] used "active enzyme ultracentrifugation" to demonstrate that it is the dimeric form of the enzyme which catalyses the hydrolysis of monomeric substrate. Although the suggestion of half-site reactivity for the *Naja naja naja* PLA has been withdrawn [139], this enzyme demonstrates a concentration-dependent aggregation in aqueous solution [137]: at concentrations below 50 μg/ml the enzyme exists predominantly in the monomeric form. However, additional evidence indicates that aggregated lipids shift this equilibrium to the dimeric state and that in fact the (asymmetric) dimer of this PLA is the catalytically active form of the enzyme.

A similar substrate-induced shift of monomeric into dimeric protein has been proposed for PLA from *Naja naja oxiana* [101,195]. Again the enzyme dimer is assumed to be organised asymmetrically but it is not clear why the enzyme should dimerise into asymmetric units in order to be able to hydrolyse monomeric diC_4-PC molecules.

Using equilibrium gel filtration, Van Eijk et al. [205a] showed a sigmoidal increase in apparent M_r of PLA from *Naja melanoleuca* when the protein was eluted through columns equilibrated with *monomeric* solutions of increasing tridecanylphosphocholine concentration. A maximal M_r-value of about 70000 was obtained at a lipid concentration of 0.25 mM (CMC of tridecanylphosphocholine = 0.28 mM) indicating the formation of aggregated protein in the presence of this single-chain substrate analogue. Similar observations have been made on PLA from *Naja naja naja* (E.A. Dennis, personal communication). With regard to the porcine pancreatic PLA, in aqueous solutions without lipids the enzyme exists as monomeric protein up to concentrations of several mg/ml. This is all the more remarkable taking into account the high number of hydrophobic amino acid side chains at the surface of the protein (cf. Section 3, "Structural aspects") and its well-known strong affinity for hydrophobic surfaces. Apparently stabilisation of the monomeric form of this PLA is caused by charge-charge repulsion of the molecules. Addition of monoacyl *phosphocholine*-containing substrate analogue in concentrations up to the CMC does not induce aggregation of this enzyme [12], suggesting that it is catalytically active as monomer. Using "active enzyme ultracentrifugation", Hille et al. [195a] indeed demonstrated that the catalytically active protein sediments as a monomer in substrate solutions below the CMC. Very recently, however, it was found in our laboratory that a strongly negatively-charged substrate, such as

$$\begin{array}{l} H_2C—S—CO—C_9H_{19} \\ | \\ H_2C—O—SO_3Na \end{array}$$

binds with high affinity to porcine pancreatic PLA in the presence of EDTA. At lipid concentrations far below the CMC this substrate induces enzyme aggregation and, at a lipid concentration of 100 μM, the resulting complex contains at least two enzyme molecules and several lipid monomers. Addition of Ca^{2+} in a concentration overcoming that of EDTA results in a highly effective hydrolysis. As one might expect, traces of sodium dodecyl sulphate behave as a very potent competitive inhibitor. If we assume that the charge-charge repulsions in aqueous PLA solutions, stabilising the monomeric protein structure, are caused mainly by the positively charged lysine and arginine cluster close to the hydrophobic IRS, it is understandable that both sodium dodecyl sulphate and the above-mentioned substrate have a high affinity for the enzyme. Such binding, relieving the charge repulsion and making the enzyme even more apolar, must result in a higher tendency of the protein to aggregate. The most remarkable fact, however, is the very high enzyme activity in the aggregated complex!

It appears, therefore, that the tendency of PLA to aggregate is a general property of this enzyme independent of its source. It is the hydrophobic/hydrophilic balance of any particular PLA which determines whether the enzyme has a strong or weak tendency to dimerise. At one side are the strongly aggregating enzymes from the *Crotalidae*, such as *C. atrox* and *C. adamanteus*, for which dissociation constants of $5 \cdot 10^{-10}$ M have been reported [185a]. Even at catalytic concentrations such enzymes always exist as dimers. An intermediate class are the *Naja* phospholipases and PLA from *Agkistrodon halys blomhoffii* which at catalytic concentrations occur as monomeric species. In the presence of lecithin solutions below the CMC they dimerise or aggregate into larger complexes but the functional role of enzyme aggregation in the hydrolysis is not yet clear. The other extreme class are the pancreatic PLA's. They possess a very low tendency to aggregate and even in monomeric lecithin solutions, they seem catalytically active as monomeric proteins. So far only the strongly adhering "alkyl sulphates" were found to induce enzyme aggregation which is correlated with high catalytic activity. It has to be stated, however, that direct binding studies of porcine pancreatic PLA with *micellar* phosphocholine-containing substrate analogues showed the presence of particles containing 2 or 3 enzyme molecules per 80–100 lipid monomers [99,196,197].

5. Chemically modified enzymes

(a) Specific amino acids

In the past decade a wide variety of more or less specific reagents have been used to modify almost all functional groups present in PLAs. As cited previously [198] one has to bear in mind that there exist no specific protein reagents, but only specific protein reactions. From this statement it may already be clear that it is necessary to first purify the modified protein to homogeneity before studying the effects produced by the modification. Obviously, the major goal of these studies is to pin-point active site residues in order to gain more insight into the mechanism of action of PLA. For some of these modifications it has been concluded that the residue modified is an active site residue, based almost exclusively on the observed loss of enzymatic activity toward substrate present as a lipid–water interface. Although this form of the substrate enables the enzyme to display its full enzymatic activity, PLA also has a distinct, though considerably lower activity toward the same substrate present as monomers. The enzymatic activity of PLA's on aggregated substrates can be completely lost by modification of a particular residue, while its active site remains intact. As a matter of fact such modifications lead to zymogen-like proteins. The loss of enzymatic activity toward aggregated substrates can be ascribed to the inability of the modified PLA to bind to lipid–water interfaces, or alternatively to bind non-specifically, preventing the formation of products. In these cases the residue modified is quite often termed "essential" without further proving its function. In order to avoid equivocal explanations it is therefore preferable to reserve the term "active site residues" to those residues directly involved in binding of the monomeric substrate and the essential Ca^{2+} ion, and to the residues performing the actual splitting of the ester bond. Modification of such residues will lead to loss of enzymatic activity of PLA toward substrate present as organised lipid–water interfaces *and* toward monomeric substrate. Residues which upon modification give rise to loss of PLA activity toward aggregated substrate, but which do not significantly affect enzymatic activity toward monomeric substrates are most likely involved in the binding to aggregated substrates.

(i) Sulphydryl groups and serine

Based on the absence of any free sulphydryl groups from all known PLA's it is generally agreed that no sulphydryl group is essential for activity or binding of PLA. It is now well established that various organo-phosphorus compounds do not cause inhibition of PLA's from different sources and that no Ser is present in the active site of this enzyme. In good agreement, no Ser residue close to the active site could be detected in the recently reported X-ray structure of bovine PLA [93].

(ii) Histidine

Studies by Volwerk et al. [115] revealed that the inactivation of porcine phospholipase A_2 and its zymogen by *p*-bromophenacyl bromide (BPB) follows similar pseudo first-order kinetics. When the residual enzymatic activity was less than 5%, amino acid analyses showed the loss of about one residue of His per mole of phospholipase A_2 or its zymogen in good agreement with the incorporation of 1.1–1.2 moles of [^{14}C]BPB per mole of protein. The [^{14}C]BPB incorporated was shown to be mainly localised on His-48, while 10% of the radioactivity was associated with His-115. Similar experiments with horse pancreatic phospholipase A_2 [199] lacking His-115 showed His-48 to be the only residue which reacted with BPB, demonstrating that His-48 is the primary site of modification and that alkylation of this residue produces a phospholipase A_2 inactive toward both micellar and monomeric substrate.

In agreement with the metal ion binding properties of the enzyme and its zymogen [110,112,200], both proteins are protected against BPB inactivation very efficiently by Ca^{2+} and Ba^{2+} while Mg^{2+} has no effect. In addition short-chain D-lecithins, the products of the phospholipase A_2 hydrolysis (lysolecithin and fatty acid), as well as the non-degradable substrate analogues (*n*-alkylphosphocholines), when present below their respective CMC's, all protect the enzyme and the zymogen efficiently against inactivation by BPB. The most effective protection was obtained when both Ca^{2+} and a monomeric D-lecithin were present. On account of the stoichiometric relationship between the loss of enzymatic activity and the incorporation of one mole of BPB/mole of protein and the effective protection by Me^{2+} and substrate analogues against the inactivation, His-48 was assigned to be an active site residue in phospholipase A_2.

From the effect of pH on the BPB inactivation of porcine phospholipase A_2, the apparent pK of His-48 was found to be 6.2 [115,173], while His-48 in the bovine phospholipase A_2 was shown to have a pK_{app} of 6.8 [201]. A group with an apparent pK of 6.3, corresponding most probably to a His residue, has been reported to control the rate of inactivation of human PLA by 1-bromo-octan-2-one [5b]. It should be emphasized that the protection against BPB inactivation with all lipids was observed *only below* their CMC's, thus as a result of the formation of the protein–monomer complex. Anomalous behaviour was observed when the rate of inactivation of PLA was studied with D-diC_6 or D-diC_7 lecithins in a concentration range above the respective CMC's.

The identical rates of inactivation of PLA and the zymogen, and their similar protection by divalent metal ions and monomeric substrate analogues suggest that the active site pre-exists at least partially in the zymogen. This idea is supported by the observation that the zymogen is capable of hydrolysing monomeric substrates [12,111], whereas it is inert towards micellar substrates. These results provide the strongest basis for the hypothesis that PLA contains an additional site for the interaction with lipid–water interfaces (IRS) which is absent in the zymogen.

From the inactivation of both porcine and equine PLA's with N-bromoacetyl-benzylamine it was established that exclusively the N-1 position of His-48 is alkylated, pointing to a specific orientation of the imidazole ring. This was confirmed by methylation of His-48 using methyl *p*-nitrobenzenesulphonate [199].

Although all data obtained from the BPB modification support the importance of His-48, which is conserved in the primary structure of all vertebrate PLA's, they do not specify its catalytic role. More conclusive evidence on this point was obtained recently by Verheij et al. [199] who used methyl *p*-nitrobenzenesulphonate to introduce a methyl group specifically on the N-1 position in His-48 of pancreatic PLA's. The methylated pancreatic PLA's have lost all their enzymatic activity toward both micellar and monomeric substrates, but still bind monomeric substrate analogues and Ca^{2+} with affinities comparable to the native enzymes. Binding of these ligands to the BPB or 1-bromo-octan-2-one-inhibited PLA's is, however, greatly impaired, most probably due to steric hindrance of these more bulky moieties [199]. Binding to lipid–water interfaces of pancreatic PLA inhibited with BPB, 1-bromo-octan-2-one or methyl *p*-nitrobenzenesulphonate is almost identical to that of the unmodified enzyme, thus indicating that the IRS and active site are topographically distinct [111]. Also, BPB-inactivated *Naja naja naja* PLA retained its affinity for mixed micelles [138].

Introduction of a [^{13}C]methyl group on His-48 enabled the determination of the pK value of the modified His residue by ^{13}C-NMR measurements. From the results obtained it was concluded that the proton on N-3 in the imidazole ring is involved in a strong interaction with a buried carboxylate group, thereby hindering rotation of the imidazole ring, and that the N-1 is involved in catalysis. Based on this result and other observations on the methylated phospholipase A_2 together with X-ray data, a catalytic mechanism for PLA was proposed (vide infra).

Since the publication on porcine PLA, several reports have appeared describing the selective modification of one His residue per protein molecule by BPB in various phospholipases A_2 and presynaptic snake venom neurotoxins [19a,36,41a,52,54,63,

64,68,71a,73,139,156,157,198,202–208 and 208a–c].

The His-residue modified with BPB has been positively assigned to be His-48 in a large number of PLAs and neurotoxic PLA's [54,64,68,198,202–206]. Both for β-bungarotoxin [203,204] and PLA from *Naja naja naja* [138], the His residue modified was shown to have a pK of 6.9. Ca^{2+} has been demonstrated to protect the inactivation by BPB for a number of these PLAs and neurotoxins [36,54,64,68,138,203–205]. Only for crotoxin B could no such protecting effect be demonstrated even at 25 mM Ca^{2+} [198]. Modified notexin [68] as well as modified β-bungarotoxin [36] have almost completely lost their Ca^{2+}-binding properties, like the modified pancreatic PLAs. In contrast, it has been reported that the BPB-inactivated PLAs from *Naja naja naja* [138], from *Naja nigricollis* [64] and from *Hemachatus haemachatus* [54] still bind Ca^{2+} with affinities comparable to corresponding native enzymes.

(iii) Tryptophan

The oxidation of two Trp residues per dimer in *Crotalus adamanteus* PLA [193] and of two Trp residues per subunit of PLA from venom of *Trimeresurus flavoviridis* (Habu snake) [73] by *N*-bromosuccinimide (NBS) renders the enzyme inactive toward micellar substrate [193]. It would be of interest to show whether this modified PLA possesses enzymatic activity toward monomeric substrates.

Reaction of 2-hydroxy-5-nitrobenzylbromide (HNB) with *Crotalus adamanteus* PLA also modifies two Trp residues per dimer [211]. In contrast to the NBS-oxidised PLA, the HNB-modified PLA retains full catalytic activity and also exhibits spectral perturbations in the presence of divalent cations.

Viljoen et al. [212] carried out Trp modification with NBS of PLA from *Bitis gabonica*. They were able to show that oxidation of Trp-31 was responsible for the observed loss of enzymatic activity toward substrate present as organised lipid–water interfaces. In addition these investigators found that Ca^{2+} or $diC_{16}PC$ (30 μM) do not, or only very weakly, protect against the oxidation. In contrast micelles of lyso PC, particularly in the presence of Ca^{2+}, do protect against oxidation of Trp-31. Although Viljoen et al. [212] claim that Trp-31 is an active-site residue, their second explanation that Trp-31 is involved in the binding to lipid–water interfaces seems more likely. This explanation is consistent with the fact that Trp-31 is variable in most PLA's. Moreover, Ca^{2+} ions alone do not protect against inactivation, whereas Ca^{2+} ions plus micelles do protect. Unfortunately, the enzymatic activity of the oxidised PLA toward monomeric substrate has not been tested. Apparently NBS is not incorporated in micelles of lyso PC, otherwise a more rapid modification would be expected. PLA from *Bitis gabonica* was also reacted with *o*-nitrophenylsulphenylchloride (NPC) [212], modifying predominantly Trp-70 with retention of full enzymatic activity.

Modification of the single Trp-3 residue in porcine pancreatic PLA with NPC did not affect the enzymatic activity when assayed on micellar L-diC_8PC [213]. In the egg-yolk assay the Trp-3-modified PLA possesses only half of the activity compared to the native enzyme.

Yoshida et al. [55] modified the single Trp at position 70 by NBS oxidation in one of the four iso PLAs isolated from the sea snake *Laticauda semifasciata* and found that the activity decreased considerably, becoming comparable to those of the other three isoenzymes lacking this Trp residue. Moreover, the authors reported the interesting observation that the Trp-modification changed the kinetic properties of this isoenzyme. NBS-oxidation of the Trp-containing enzyme produced a PLA which, like the native Trp-free isoenzymes, displayed biphasic kinetics.

NBS was reported by Howard and Truog [239] to oxidise Trp in β-bungarotoxin with loss of PLA activity and neurotoxicity.

Both NBS and 2-hydroxy-5-nitrobenzylbromide modified all of the tryptophan present in *Naja naja naja* PLA with the loss of almost all activity toward substrate present in lipid–water interfaces [138,215]. It is not certain whether all three Trp residues now known to be present in this PLA [139] were modified.

(iv) Methionine

PLA from *Crotalus adamanteus* venom was found to react slowly with 2-bromoacetamido-4-nitrophenol, which modified the single Met-10 residue [211]. When about 0.75 moles of *p*-nitrophenol groups were incorporated per subunit, all enzymatic activity was still present. No detectable spectral perturbations of the *p*-nitrophenol group were observed in the presence of divalent cations, demonstrating that these ions do not bind in the environment of Met.

Carboxymethylation of horse, bovine and pigiso- PLA's, all possessing only one Met residue at position 8, resulted in a rather slow loss of enzymatic activity [216,217]. When, however, 8 M urea is present, inactivation of porcine iso-PLA is fast [216]. The modified enzyme has lost its activity toward both micellar and monomeric substrates. Direct binding studies of this carboxymethylated iso-PLA showed that it no longer binds to lipid–water interfaces, but that it can still bind a monomeric substrate analogue and Ca^{2+}, albeit with a lower affinity than the native enzyme. Based on these observations, it was proposed that Met-8 was part of the IRS. The X-ray structure of bovine PLA [90, 218] indicates that Met-8 is buried in the interior of the protein. Apparently introduction of the zwitterionic group under rather vigorous conditions, considerably distorts part of the tertiary structure of the enzyme. Contrary to observation on native PLA, removal of urea does not result in proper refolding to the active conformation, resulting in the loss of enzymatic activity upon modification. Therefore the previous conclusion that Met-8 is part of the IRS is no longer tenable.

Porcine PLA, having an additional Met residue at position 20, is rapidly carboxymethylated in the absence of urea, under conditions where Met-8 of the iso-PLA is hardly reactive [217].

Although no inactivation was observed upon prolonged reaction of porcine PLA with methyliodide, the reagent slowly alkylated Met-20 as was demonstrated by incorporation of [^{14}C]methyliodide. Similarly, as observed for carboxymethylation, it was found that methylation of iso-PLA was considerably slower than that of normal porcine PLA. The observed differences in rates of alkylation of Met-8 and Met-20 in porcine PLA enabled Meyer [217] to prepare selectively both S-carboxymethyl Met-20- and S-methyl Met-20-porcine PLA's. Both modified proteins possess activities toward monomeric substrates similar to that of the native enzyme. Also, the affinities of both alkylated PLA's for monomeric and micellar substrate analogues, as well as for Ca^{2+}, were not affected. Furthermore, the specific activity of S-methyl Met-20-PLA with the egg-yolk assay was also found to be similar to that of native PLA, whereas that of the S-carboxymethyl Met-20 PLA was only about 50%. Monolayer experiments on these two modified PLA's revealed that the penetrating power was noticeably decreased, in particular for that of the carboxymethyl analogue. Most likely the more drastic effects on the properties of the enzyme upon carboxymethylation of Met-20, as compared to those upon methylation, are due to the additional introduction of a positive and a negative charge (carboxymethylation), or a positive charge only (methylation). The finding that the introduction of a positive charge on Met-20 has less influence on the properties of the pancreatic PLA is compatible with the occurrence of a positively charged Arg residue at this position in some snake venom PLA's (see Section 3, "Structural aspects"). These results, together with the 3-dimensional X-ray structure of the bovine PLA [218], suggest that Met-20 is part of the IRS.

(v) Lysine

Viljoen et al. [205] concluded that Lys is a residue essential for enzymatic activity of *Bitis gabonica* PLA, based on the observation that reaction of pyridoxal-5′-phosphate followed by reduction with sodium borohydride inactivated the enzyme toward substrate present at a lipid–water interface. The enzyme is protected against inactivation by micellar lysolecithin but not by Ca^{2+}. It is therefore very likely that the residue(s) modified are involved in some way in the binding to aggregated substrate. The loss of enzymatic activity was not due to modification of *one* particular Lys residue per enzyme molecule but to *four* different Lys residues, each modified by about 25%.

PLA (fraction DE-III) from *Naja melanoleuca* contains only 4 Lys residues. This prompted Van Eijk et al. [205a] to study modification of this protein with 4-chloro-3,5-dinitrobenzoic acid. Only Lys-6 readily reacted and Ca^{2+} ions enhanced the inactivation rate. The modified protein had only 1–2% residual activity when measured on micellar substrates and the activity toward monomeric dihexanoyl thiolecithin was also considerably lower. The affinity of the modified enzyme for Ca^{2+} ions increased 10-fold whereas the affinity for micellar substrates was not influenced. Yet Lys-6 cannot be considered as an active-site residue: reduction of the nitro groups to amino groups restored more than 50% of the original activity of the native enzyme measured with di-octanoyl lecithin. It was concluded that changes in the side chain of Lys-6 influence the conformation of the protein. This conformational change is reflected by altered k_{cat} values. A similar effect was found when Asn-6 in bovine AMPA was substituted by Arg [205b].

Due to a reduced reactivity of the α-NH_2 group in PLA from *Naja naja oxiana*, Apsalon et al. [205c] were able to modify selectively only the ϵ-NH_2 groups of all six

Lys residues. Blocking of all these ϵ-NH_2 groups with acetic anhydride, *o*-methylisourea, or the N-carboxyanhydride of *o*-nitrophenylsulphenylglycine does not lead to an appreciable decrease in enzymatic activity. When, however, all the ϵ-NH_2 groups were reacted with succinic anhydride, producing negatively charged groups, almost all catalytic activity was lost. Furthermore, Apsalon et al. [205c] found that blocking of the α-NH_2 group in addition to that of the ϵ-NH_2 groups with acetic anhydride or 2,4,6-trinitrobenzene sulphonic acid abolished the enzymatic activity of this PLA toward micellar substrate. These findings thus demonstrate that also in *N. naja oxiana* PLA, a free α-NH_2 group is essential for activity toward micellar substrate, as demonstrated previously for some other snake venom and pancreatic PLA's [96,189,213,245]. The reaction of *N. naja oxiana* PLA with pyridoxal phosphate led to an almost complete inactivation due to the incorporation of one pyridoxamino phosphate group per protein molecule as found after reduction of the Schiff base. The authors [205c] suggest that a Lys residue, not yet identified, has been covalently modified and is close to the anion-binding site of the enzyme. Although the secondary structure of the modified *N. naja oxiana* PLAs is retained, as judged from their CD spectra, the antigenic properties and the presynaptic activity of some of the modified PLA's were affected.

Pyridoxylation followed by reduction with ^{3}H-labelled sodium borohydride was used to label β-bungarotoxin radioactively [219]. The dissociation constant for binding to several tissue subfragments of nervous tissue was found to increase ten-fold upon pyridoxylation. No data were reported for loss of PLA activity.

(vi) Carboxylate groups

Recently, PLA from *Naja naja oxiana* has been modified with *N*-diazoacetyl-*N'*-(2,4-dinitrophenyl)-ethylenediamine (DBE) in the presence of Ca^{2+} [195,220]. When one carboxylate group per dimer was modified, the authors found complete inactivation of PLA using monomeric L-diC_4-PC as substrate. Their evidence, however, seems to be based heavily on the "half-site reactivity", previously observed by Dennis and coworkers [cf. 138] which is no longer valid [139]. Proflavin, a competitive inhibitor of this enzyme, and Ca^{2+} ions did not have any effect or even increased the incorporation. After reduction of the modified protein with sodium borohydride, indications were obtained for selective modification of an Asp residue which has not yet been identified.

In order to obtain information about the involvement of particular carboxylate groups in the active site and in Ca^{2+} binding of bovine pancreatic PLA, Fleer et al. [221] used the water-soluble 1-ethyl-3-(*N,N*-dimethyl)amino propyl carbodimide (EDC) and semicarbazide as the nucleophile. Depending on the conditions, they were able to block all carboxylates except one (Asp-99) or two (Asp-39 and Asp-99). Both modified proteins lost their enzymatic activity toward micellar and monomeric substrates and also lost their Ca^{2+}-binding properties. Repeating these experiments in the presence of Ca^{2+} ions, the carboxylate of Asp-49, in addition to those of Asp-39 and Asp-99, was not modified. This protein still possesses enzymatic activity. Its Ca^{2+}-binding properties were lost upon further modification in the absence of

Ca^{2+}, under conditions where only Asp-49 reacted. Therefore it was concluded that Asp-49 is the Ca^{2+}-binding ligand, which is in good agreement with the results from the X-ray structure of bovine pancreatic PLA [218]. From the pH dependence of the Ca^{2+}-binding to bovine PLA, a group with an apparent pK of 5.25 was found which was tentatively assigned to Asp-49.

Dinur et al. [221a] claimed to have modified one carboxylate group of porcine pancreatic PLA, with complete loss of activity toward egg-yolk lecithin, by reaction with *N*-ethyl-5-phenyl-isoxazolium-3′-sulphonate (Woodward's Reagent K), followed by reaction with radioactively labelled ethylglycinate. Unfortunately the investigators did not establish which carboxylate group was modified. The inactivation is protected by substrate analogues, viz. *n*-hexadecylphosphocholine and *N*-palmitoylaminoethylphosphocholine, present as micelles. However, in the presence of 30 mM $CaCl_2$, the rate of inactivation was enhanced two-fold. Taking into account the results obtained by Fleer et al. [221], it seems unlikely that the COOH of Asp-49, the proposed Ca^{2+}-binding ligand [90a], has been selectively modified by the Woodward's reagent. It is, therefore, a pity that Dinur et al. [221a] have not determined the Ca^{2+}-binding properties of their modified PLA. On the other hand, it does not seem very likely that the COOH of Asp-99, which is deeply buried in the interior of the bovine PLA, has been modified. The conflicting results obtained on COOH modification therefore require a thorough re-investigation.

(vii) Arginine

Recently, Vensel and Kantrowitz [222] reported the modification of an essential Arg residue in porcine pancreatic PLA by reaction with phenylglyoxal. It is known, however, that phenylglyoxal can transaminate α-amino groups even more rapidly than it modifies Arg residues [223]. Because the presence of a free α-amino group is essential for enzymatic activity and binding of porcine pancreatic PLA to lipid–water interfaces, Vensel and Kantrowitz [222] tried to prove by amino acid analysis and qualitative end-group analysis that the inactivation was not due to transamination. In the reviewers' opinion the methods used to show that transamination had not occurred are not sensitive enough. The effects of pH and micellar substrate analogues hold equally well for transamination of the α-amino group. Moreover, 2,3-butanedione and 1,2-cyclohexanedione, being more specific for Arg than phenylglyoxal, cause a much slower inactivation despite the large excess of each of these reagents used. From extensive model studies in our laboratory, it was determined that phenylglyoxal gives rise to excessive transamination of porcine pancreatic PLA with simultaneous modification of Arg residues, the number depending on reagent concentration. Using phenylglyoxal concentrations lower than those of Vensel and Kantrowitz complete inactivation of porcine PLA was observed. Then the protein was subjected to CNBr cleavage. After separation of the liberated N-terminal octapeptide from the remainder of the protein, it was found by amino acid analysis that, in addition to the disappearance of 80% of Arg-6, Ala-1 was almost completely absent.

Fleer et al. [224] preferred the use of [^{14}C]labelled 1,2-cyclohexanedione in the presence of borate to modify Arg residues in porcine PLA. Despite the formation of some transaminated PLA they were able to isolate a PLA modified exclusively at Arg-6. Extensive characterization revealed that the modification had almost no effect on the V_{max} values when assayed both on micellar and monomeric substrates and on the Ca^{2+}-binding properties as compared to unmodified PLA. The affinity of the modified PLA to micellar substrate analogues, as well as its penetrating capacity into monomolecular lecithin films were improved as compared to the unmodified PLA.

Upon reaction of *N. naja oxiana* PLA with 1,2-cyclohexanedione or acetylacetone, Apsalon et al. [224a] found little inhibition of activity unless borate was present. It has been shown that Arg-16 in this PLA was modified.

(viii) α-Amino group

Transamination of proteins by glyoxylic acid in the presence of Cu^{2+} is assumed to be specific for the α-amino group [225]. A rather rapid inactivation was observed for both porcine and equine PLA's, whereas bovine PLA was much more stable (Slotboom et al., to be published). Micellar substrate analogues almost completely protect porcine PLA against the modification. It was found that at a stage where PLA was approximately 80% inactivated about 15% of the potential activity of the zymogen was lost, indicative of some kind of side reaction. When the transamination reaction was performed in the presence of 6 M guanidine hydrochloride or 8 M urea complete inactivation of bovine, porcine and equine PLA's was observed within 30–60 min. After similar treatment of porcine ProPLA, all potential activity was recovered, indicating no additional inactivation. The transaminated porcine PLA had lost its enzymatic activity toward micellar substrate due to its considerably decreased affinity for lipid–water interfaces, but still retained its enzymatic activity toward monomeric substrate. In these respects the transaminated PLA thus very much resembles the zymogen. As a matter of fact the results of Photo CIDNP NMR spectroscopy [226] as well as the tentative 2.4 Å X-ray structure of transaminated bovine PLA (Dijkstra et al., to be published) support this conclusion. Subsequent treatment of a transaminated protein with *o*-phenylene diamine is reported [225] to remove selectively the N-terminal amino acid residue. This sequence of reactions was applied to the enzymatically inactive Ala^{-1}-AMPA *, which indeed produced in about 30% overall yield, enzymatically active AMPA having the same specific activity as authentic AMPA (Slotboom et al., to be published).

The use of glyoxylic acid to modify selectively the α-amino group is of particular interest for the snake venom PLA's, to study whether the effects on enzymatic activity and lipid-binding properties are similar to those observed for the pancreatic PLAs. Phospholipases A_2 from *Crotalus atrox, Vipera berus* and *Naja melanoleuca* were rapidly inactivated by glyoxylic acid in the presence of 4 M tetramethylurea [189]. After purification, the modified proteins have no enzymatic activity when

* AMPA in which an Ala residue has covalently been attached to the N-terminal Ala-1.

tested with micellar substrate but partially retained their activity toward substrate in monomeric form. Direct binding studies revealed that the affinity of the transaminated snake venom PLAs for lipid–water interfaces was decreased 5- to 10-fold, but in contrast to transaminated porcine PLA, a strong interaction was still observed. However, even though the modified venom PLAs do bind to lipid–water interfaces, no enhanced activity induced by the interface was observed. This was explained [189] by the assumption that PLA bound to lipid–water interfaces can occur in two conformations characterised by low and high turnover numbers, respectively, when acting on these aggregated substrates.

(ix) Tyrosine

Meyer et al. [227,228] nitrated Tyr residues in horse, porcine and bovine (pro)PLA's with tetranitromethane (TNM) giving rise to a rapid, partial loss of enzymatic activity, which is even more rapid in the presence of lysolecithin micelles and Ca^{2+}. This latter effect was attributed to the incorporation of the reagent into the lysolecithin micelles, thus enhancing the rate of nitration of those Tyr residues involved in the micellar binding site of PLA. The presence of lysolecithin also protects against polymerisation which was a side reaction in its absence. After purification of the mono- and di-NO_2 monomeric proteins it was found that in all three pancreatic PLAs Tyr-69 was always nitrated. In addition, Tyr-124 in porcine and Tyr-19 in horse PLA were also nitrated. All these mononitrated PLA's still possess 15–50% of the enzymatic activities of the respective unmodified enzymes when assayed on micellar substrates, indicating that the modified Tyr residues are not active-site residues. The NO_2-Tyr residues could be reduced to NH_2-Tyr residues by sodium dithionite. The various NH_2-Tyr PLA's are still enzymatically active and due to the low pK values of these NH_2 groups they could easily be transformed into the corresponding dansyl-NH_2-Tyr PLAs also possessing enzymatic activity.

From direct binding studies using ultraviolet difference spectroscopy, it was found that NO_2-Tyr-69-porcine as well as the dansyl-NH_2-Tyr-69-porcine, equine PLAs and in particular NO_2-Tyr-19- and dansyl-NH_2-Tyr-19-equine PLA, possess a higher affinity for lipid–water interfaces than the native enzymes. Upon interaction of the latter dansyl-NH_2-Tyr PLAs with micellar substrate analogues a considerable increase in fluorescence and a concomitant blue shift of the emission maximum of the dansyl group were observed. No such effects occurred for the corresponding dansyl-NH_2-Tyr-pro PLAs nor for dansyl-NH_2-Tyr-124-porcine PLA. It has, therefore, been concluded that Tyr-19 and Tyr-69 are part of the IRS in pancreatic PLA.

Monomer phospholipid binding at pH 6 as monitored by ultraviolet difference spectroscopy induces a strong hydrophobic perturbation of NO_2-Tyr-69 and -19. When measured at pH 8, monomer-binding decreased considerably, most probably due to charge repulsion between the phosphate moiety of the phospholipid analogue and the negatively charged NO_2-Tyr-69 residue which has a lower pK than Tyr.

Ca^{2+}-binding affects the NO_2-Tyr-69 residue as was shown by ultraviolet difference spectroscopy and the lowering of the pK of NO_2-Tyr-69, whereas no such effects were found for NO_2-Tyr-19 and -124.

The introduction of the NO_2 group and in particular of the dansyl-NH_2 group on Tyr-69 and Tyr-19 greatly enhances the penetrating power of these modified enzymes for monomolecular L-diC_{10}-PC films. When the pH is increased from 6 to 9, the penetrating power of the NO_2-Tyr-69-porcine and -equine PLAs, however, decreased considerably due to the introduction of a negative charge.

The availability of varous pure NO_2-Tyr PLAs was of great help for the identification of resonances in the ^{1}H-NMR spectrum of PLA originating from Tyr residues. Using the Photo CIDNP method it was possible to assign resonances corresponding to $H_{3,5}$ protons of Tyr-69 and Tyr-124 in porcine PLA [229].

Iodination of Tyr residues is a very attractive way to introduce a radioactive label, Reaction of bovine pancreatic (pro)-PLAs with an equimolar amount of iodine resulted for the bovine proteins in the exclusive monoiodination of Tyr-69, while in the porcine proteins in addition to extensive monoiodination of Tyr-69, Tyr-124 was also monoiodinated to a small extent [230]. As compared to the native enzyme, the iodinated enzyme has a higher specific activity in the egg-yolk assay, while similar V_{max} values were found using micellar diC_8-PC. The introduction of one atom of iodine on Tyr-69 in pancreatic PLA slightly increases the penetration capacity of the enzyme in monolayers of L-diC_{10}-PC, which is compatible with a better K_m found for monoiodinated PLA activity on micelles of diC_8-PC [144].

Crotalus adamanteus PLA upon reaction with iodine retained 88% of its activity when one mole of diiodotyrosine per protein molecule was present [193].

Bon et al. [231] also used iodination to label the subunits of crotoxin radioactively. Upon incorporation of one atom of iodine per mol of protein, the iodinated component B showed no significant decrease in PLA activity and retained full neurotoxic potential when tested after complexing with native component A.

Upon reaction of purified bee venom PLA with imidazolide derivatives of long-chain fatty acids, a single acyl residue is covalently coupled, presumably to a Tyr residue [191,232–234]. Kinetic analysis of the acylated enzyme shows an increase of the enzymatic activity which is almost entirely determined by enhancement of the V_{max} term (53-fold), with a small modification of the K_m value. Addition of free fatty acids has the same effect though to a lesser extent. Similar phenomena were observed for PLA's from *Vipera ammodytes* and *Naja naja* venoms. Of the possible explanations for this phenomenon given by the authors, the most attractive mechanism is that activation facilitates functional penetration of the lipid interface by the enzyme.

(b) Miscellaneous

(i) Modification of PLA with ethoxyformic acid anhydride (EOFA)

EOFA is a very reactive non-specific reagent which reacts in proteins with several amino acid side chains such as phenolates, imidazoles, carboxylates, sulphydryls, α- and ϵ-amines and guanidino groups [235–238]. Wells [193] used this reagent to identify whether a Lys or His residue might be important in the active site of *Crotalus adamanteus* PLA. Because no radioactive EOFA was used, the modification of His was determined by spectral changes at 230 nm. These measurements are not a reliable measure of the involvement of His when Tyr residues are simultaneously ethoxyformylated. The observation that EOFA modification is first-order with respect to dimeric enzyme and EOFA, led Wells to conclude that this modification is an example of "half-site reactivity". This hypothesis was supported by the findings that only one Lys residue/dimer is modified, that there were still detectable cation-induced optical effects and that there was recovery of the theoretically expected specific activities upon dissociation–reassociation of 50 and 100% inactivated PLA at pH 5.0. Based mainly on these observations, it was concluded that within the active site of *Crotalus adamanteus* PLA, a Lys residue was identified. Besides the observation that until now no Lys residue in any sequenced PLA has been reported on a position which in the tertiary structure of the bovine pancreatic PLA forms part of the active site (see "the 3D structure"), there are, in the reviewers' opinion, several reasons for re-evaluating this modification. It is now known that the *Crotalus adamanteus* PLA has a free α-NH_2 group which could also have reacted with EOFA. Moreover Tyr residue(s) are very likely to be simultaneously ethoxyformylated. Taking into account the large variety of possible sites for incorporation, a more direct determination of the residue(s) modified as well as of the number of residue(s) modified by radioactive EOFA should be considered.

Upon reaction of EOFA with *Naja naja naja* PLA, the group of Dennis [138] claimed that two amino groups, one Tyr and half a His per enzyme molecule were modified with retention of 15% of enzymatic activity. Based on this observation and the results obtained after consecutive EOFA/BPB and BPB/EOFA modifications respectively, it was concluded that EOFA also shows "half-site reactivity". Most probably the same arguments which led to the withdrawal of the "half-site reactivity" of BPB [139] also hold for EOFA modification.

EOFA and acetic anhydride have been reported to modify only NH_2 groups and no His or Tyr residues in crotoxin [198,231]. With a 50-fold excess, two NH_2 groups reacted in crotoxin with retention of all PLA activity and neurotoxicity, while higher concentrations of EOFA progressively modified more NH_2 groups with increasing losses of PLA activity and neurotoxicity. In this respect the separate crotoxin B-chain (basic PLA) behaves almost exactly as the complex.

Similarly, all PLA activity and neurotoxicity are lost upon reaction of EOFA with β-bungarotoxin, although no data were reported as to which amino acid residues were modified [239,240]. Ca^{2+} and diC_6-PC (above the CMC) were found to protect almost all PLA activity against inactivation by EOFA, whereas the neurotoxic properties were still lost. The authors suggest that there are possibly two sites on the protein: one responsible for PLA activity which can be protected; and another one for neurotoxicity which cannot be protected against EOFA modification.

Reaction of Notechis II-5 with EOFA showed the modification of one Tyr, one Lys and two His residues [208]. One of the His residues reacts slowly, the other fast. Although contradictory results were obtained as to whether PLA activity is lost or not, depending on the use of egg-yolk or *purified* egg-yolk PC, the authors claimed to have modified His-14 and His-21, which would mean that His-48 was not modified.

Most probably His-21 is involved in the binding of the enzyme to lipid–water interfaces. More extensive treatment with EOFA led to inactivation which could not be reversed with hydroxylamine. It was suggested that a Lys had been modified, although no supporting evidence was presented.

(ii) Cross-linking of PLA

In order to demonstrate cross-linking of *Naja naja naja* PLA under conditions in which the enzyme exists in an aggregated state, Lewis et al. [241] used various photoactivatable heterobifunctional aryl azides. The unpurified, cross-linked PLAs had all retained 20–80% of the enzymatic activity. Because this level of activity is significantly higher than can be explained by the presence of monomeric PLA in the mixture, the cross-linked proteins must retain some PLA activity.

To test the hypothesis that crotoxin A serves as a "chaperone" to enhance the specificity of crotoxin B, Hendon and Tu [242] cross-linked both polypeptide chains using the bifunctional cross-linking agent dimethyl-suberimidate. An average of three cross-links were introduced as found from the number of Lys residues blocked. Most likely two of these cross-links occur between the subunits A and B, while the third is presumably present as an intrapeptide cross-link on subunit B. No loss of PLA activity of the cross-linked crotoxin was observed, indicating that cross-linking does not interfere with the PLA active site present in the B-chain. In contrast, neurotoxicity of the cross-linked crotoxin is lost. Since the PLA activity of the cross-linked complex remains unaffected and since this activity is believed to be directly involved in presynaptic neurotoxicity, it appears that the loss of neurotoxicity occurs from some form of interference between the cross-linked complex and the target site, thus adding credence to the "chaperone" concept for crotoxin A.

(iii) Photoaffinity labelling

So far, only Huang and Law [243,243a] have used photoaffinity labelling to study the interaction of PLA (*Crotalus atrox*) with phospholipids. They synthesized a racemic 1,2-dihexanyl ether analogue of PE, containing in the polar head group an ethyl diazomalonyl group, which was found to be an effective substrate analogue. After photolysis of a mixture of the PLA and the photolabile PE analogue (present in a concentration of only 4 times its CMC), they observed covalent linkage of the enzyme with the PE by the photochemically generated carbene. From the amount of incorporated substrate analogue the ratio bound ligand to 14000 M_r polypeptide was 1.04. The radioactivity associated with the PE analogue, incorporated into the PLA, was found to be localised in two fragments viz. a large peptide comprising residues 43–97, and the N-terminal segment, residues 1–15. Undoubtedly important information for a better understanding of the architecture of the enzyme–substrate interaction can be expected upon further exploration of this attractive approach.

(iv) Semisynthesis of pancreatic phospholipase A_2

The α-helical N-terminal region of pancreatic PLA's has been shown to be directly involved in the binding of these enzymes to lipid–water interfaces [244]. Furthermore, the absence of micellar activity of the zymogen as well as of various α-amino blocked porcine AMPAs (vide infra) led Abita et al. [96] to conclude that the α-amino group stabilised the active geometry of the catalytic site. Semisynthesis was used to substitute various amino acid residues at the N-terminal region [213,245]. Such a semisynthetic approach requires that the ϵ-amino groups of Lys residues must be selectively protected, enabling removal and reintroduction of amino acid residues or peptides to take place exclusively at the free α-amino group. For the pancreatic PLAs this was done by amidination of the zymogens with methylacetimidate followed by tryptic activation. The resulting ϵ-amidinated PLAs (AMPAs) have about 70% of the enzymatic activity of native PLAs when assayed on micelles of L-diC$_8$-PC, and behave in all respects almost exactly as the unmodified PLA's. It is, therefore, not necessary to remove the protecting amidino groups afterwards. Using this procedure, Pattus et al. [144] prepared ^{3}H-labelled AMPA for monolayer studies (see Section 4, "Kinetic data"). Upon successive removal of N-terminal amino acid residues of porcine AMPA by the Edman procedure, des-Ala-1-, des-Ala-1.Leu-2-, and des-Ala-1.Leu-2.Trp-3-AMPA's were obtained which are devoid of enzymatic activity on micellar substrate. Although des-Ala-1-AMPA still possesses some activity toward monomeric substrate, removal of more than one amino acid residue further decreases this activity. Various amino acids were covalently coupled to des-Ala-1-AMPA, resulting in AMPA analogues always catalytically active on monomeric substrate. Whereas substitution of L-Ala-1 by Gly, β-Ala, L-Asn, L-Asp or L-NorLeu produced AMPA analogues catalytically active on micellar substrates, this was found not to be the case for AMPA analogues having N-terminally D-Ala, α-amino isobutyric acid, *N*-methyl-L-Ala, L-Leu or L-Phe. These latter analogues do not bind to lipid–water interfaces despite the availability of a free α-amino group [245; Slotboom et al., to be published]. Most likely this is due to the presence of a rather bulky, branched or D-amino acid residue, which for steric reasons prevents the proposed interactions shown later in Fig. 9 with concomitant distortion of the IRS [246]. Similarly various ^{13}C-enriched amino acids have been introduced at the N-terminal position of pancreatic AMPA's, enabling the determination of the pK values of the α-amino groups. A pK of 8.4 was found for the α-amino group of porcine AMPA, in good agreement with similar values (8.3 and 8.45, respectively) determined by proton titration [97] and by titration of protons released during tryptic activation of the zymogen [247]. Even higher pK values were found for the α-amino group of equine and bovine (L-[3-^{13}C]Ala-1)-AMPA viz. 8.8 and 8.9 respectively [98,229]. In contrast, (D-[3-^{13}C]Ala-1) porcine AMPA was found to have a normal pK value of 7.8 for its α-amino group [247]. These results together with the observation that introduction of an octan-2-one moiety on His-48 or addition of specific Ca^{2+} ions increase the pK of the α-amino group of (L-[3-^{13}C]Ala-1)-AMPA from 8.4 to 9.0 and not that of (D-[3-^{13}C]-Ala-1)-AMPA once more stresses the special environment of L-Ala-1 in pancreatic PLA's.

Using the same technique, but now coupling with the tripeptide Ala.Leu.Phe to des-Ala-1.Leu-2.Trp-3-AMPA, (Phe-3)-AMPA was obtained. This analogue was found to have about 40% of the enzymatic activity of AMPA, indicating that Trp-3 is not essential [213]. (Phe-3)-AMPA enabled the unambiguous assignment that in addition to Trp perturbation, one or more Tyr residues are also perturbed upon interaction with micellar substrate analogues [244].

Substitutions further on in the N-terminal region have been performed by covalent coupling of pre-assembled peptides to N-terminally shortened AMPA fragments prepared by selective proteolytic cleavage or CNBr splitting of tri-, hexa- and octa-peptides. It should be stated that these splittings caused the loss of all enzymatic activity, which could not be restored by non-covalent combining of peptide and protein fragments as observed for RNases. Similar findings were also reported for PLA from *Naja naja oxiana* [205c,206]. Recently, however, Kihare et al. [206a] reported that PLA from the venom of *Trimeresurus flavoridis* retained about 6% of its activity after CNBr cleavage of the N-terminal octapeptide, and that the N-terminally-shortened PLA occurs as a dimer like the native enzyme. Furthermore, the authors found evidence for the formation of a non-covalent complex of the octapeptide and the remainder of the protein, with a concomitant increase in catalytic activity up to 17% of the value of the native PLA. It has to be mentioned, however, that the amino acid sequence proposed for the N-terminal octapeptide [206a] deviates considerably from those of all other PLA sequenced, including that of *Trimeresurus okinavensis* [73b, Table 1). In particular, the presence of N-terminal pyroglutamic acid in this PLA seems curious, taking into account that pancreatic [96,213,245,246], as well as snake venom PLA [189] in which no α-NH_2 function is present, do not show enzymatic activity toward micellar substrates. Using N-terminally shortened porcine AMPA, Jansen [98] prepared [Gly-3] and [Glu-4] porcine AMPA's and showed that substitution of Trp-3, by Gly abolishes almost all micellar activity, most probably because of distortion of the α-helical structure. Although Gln-4 is absolutely conserved in all PLA's sequenced, [Glu-4]-AMPA possesses about 40% of the activity of AMPA. Interestingly, the penetrating power of [Gly-4]AMPA into monolayers of L-diC_{10}-PC was decreased, whereas that of [Glu-4]AMPA was increased as compared to that of unmodified AMPA. Recently Van Scharrenburg et al. [248] substituted Asn-6 in the bovine AMPA by Arg, which occurs at this position in the porcine enzyme. This substitution was found to increase both the low affinity for lipid–water interfaces and the low penetrating capacity of the bovine AMPA for monolayers to values comparable with those for porcine AMPA. Substitution of the absolutely conserved Phe-5, located in the hydrophobic wall around the active site cleft (see Fig. 13), by a Tyr residue in bovine AMPA causes the loss of almost all catalytic activity. The affinity of [Tyr-5] AMPA for micellar lipid–water interfaces is identical to that of native AMPA and the observed loss of activity is therefore very likely due to a distortion of the active site [205b]. When the absolutely conserved Gln-4 is substituted by norleucine in bovine AMPA, about 25% of the original activity toward monomeric substrate is retained. Toward micellar substrate, however, all catalytic activity is lost, because [Nle-4]-AMPA does not bind to micellar lipid–water interfaces [205b]. The substitution of Nle for Gln-4 most probably perturbs the extended system of H-bridges between Ala-1 and Gln-4 and between Ala-1 and the active site Asp-99 (Fig. 9), [90a] thereby preventing the formation of a functional IRS. In this respect it is of interest to note that a pure PLA from porcine intestinal mucosa has recently been shown to possess an Asn residue at position 4 instead of a Gln (R. Verger, personal communication). This latter PLA only displays catalytic activity towards phosphatidylglycerol when present as a monomolecular layer. Also β-bungarotoxin has been shown to have an Asn at position 4 [38]. It can thus be concluded that these substitutions may yield valuable information on the role of the N-terminal amino acid residues on enzymatic activity and lipid-binding properties of pancreatic PLA's, but more work has to be done to explain the observed findings correctly.

6. *Ligand binding*

(a) Binding of Ca^{2+}

(i) Pancreatic phospholipases A_2

Equilibrium gel filtration studies demonstrated that both porcine PLA and its zymogen possess only one high-affinity Ca^{2+}-binding site per protein molecule [112,200,247]. Binding of Ca^{2+} to porcine PLA and pro-PLA induces ultraviolet difference spectra which are characterised by a large peak at 242 nm and two small peaks at 282 and 288 nm. It was tentatively concluded that the observed difference spectrum originates from a shift of a Tyr residue to a more polar environment and a charge effect on a His residue. Qualitatively identical difference spectra were obtained for both proteins with Ba^{2+} and Sr^{2+}. Both from ^{1}H-NMR and fluorescence titration studies using native and His-48-modified pancreatic PLA's, it was demonstrated that Ca^{2+}-binding decreases the pK value of His-48 from about 7 to 5.7 [199,263]. Ca^{2+} does not influence the fluorescence spectra of PLA and pro-PLA. However, addition of Ca^{2+} enhances the ANS fluorescence induced by PLA and its zymogen, enabling the determination of the metal ion dissociation constants [112]. A similar conclusion was reached by Brittain et al. [264] who used Tb^{3+} as a luminescent probe of Ca^{2+} sites in proteins. Ca^{2+} dissociation constants were also derived from inactivation of PLA by BPB [112,115].

The dissociation constants for the porcine PLA-Ca^{2+} and the pro-PLA-Ca^{2+} complexes are similar. Values obtained by the various techniques showed good agreement. The dissociation constants of the Ba^{2+} and Sr^{2+} complexes do not differ substantially from those obtained for Ca^{2+}. Values were found ranging from 100 mM at pH 4, 2.5 mM at pH 6 to 0.2 mM at pH 10, and the pH dependency suggests that the metal ion binding site contains one or more carboxylates. Recently similar values were reported for human PLA [5b].

For the bovine PLA the pH dependency of $K_{Ca^{2+}}$ was shown to be controlled by a single carboxylate group with an apparent pK of 5.2, which by chemical modification studies was tentatively assigned to Asp-49 [221]. Obviously no Ca^{2+}-binding could be detected for the Asp-49-modified bovine PLA, whereas Ca^{2+}-binding to BPB-modified pancreatic PLA is greatly impaired, probably due to steric hindrance

[199]. A similar pK value was very recently reported by Andersson et al. [277] for porcine pro-PLA using ^{43}Ca-NMR. With this technique, the authors found a dissociation rate constant of 2.5×10^3/s. Together with the reported $K_{Ca^{2+}}$ value (0.4 mM at pH 7.5) it was concluded that the Ca^{2+}-binding site of porcine pro-PLA is more rigid or generally less accessible to an incoming Ca^{2+} ion, as observed for rabbit skeletal muscle troponin C.

To date, Gd^{3+} is the only metal ion found which can substitute for Ca^{2+} with retention of some enzymatic activity. Dissociation constants for PLA and pro-PLA were evaluated from water proton relaxation (PRR) titrations. The K_{Me} for Ca^{2+}, Eu^{3+} and Tb^{3+} were determined by competition of these cations with Gd^{3+}. The $K_{Ca^{2+}}$ values determined in this way agreed very well with those obtained directly, whereas K_{Me} for Eu^{3+} and Tb^{3+} for PLA were 0.07 and 0.08 mM respectively, at pH 5.8 [118].

Finally it has to be mentioned that the affinity of the enzyme for Ca^{2+} is considerably enhanced at neutral pH by micellar substrate analogues [112,118,247]. This synergistic effect explains the discrepancies observed between Ca^{2+}-dissociation constants determined directly and those obtained from kinetic analysis.

(ii) Venom phospholipases A_2

Binding of Ca^{2+} to notexin [202], notechis II-1 [68] and taipoxin [207] induced almost identical ultraviolet difference spectra to those observed for porcine PLA. Somewhat lower $K_{Ca^{2+}}$ values were reported for these proteins as compared to the value obtained for porcine PLA. In addition it was concluded that one Ca^{2+} was bound per protein molecule, except for taipoxin which binds two Ca^{2+} ions. In this latter protein one Ca^{2+} is bound to the α-subunit and one to the γ-subunit, while β-subunit has no affinity for Ca^{2+}. Although it appears very likely that indeed one Ca^{2+} is bound per polypeptide chain, this conclusion is based on the assumption that the maximal absorbance is due to the binding of one Ca^{2+} per protein molecule. It was concluded that BPB modified notexin is still able to bind one Ca^{2+} per protein molecule, although its $K_{Ca^{2+}}$ value (25 mM at pH 7.4) was 178-fold higher than that found for native notexin.

Abe et al. [36] demonstrated by equilibrium dialysis that β-bungarotoxin binds one mole of Ca^{2+} per mole of protein and a $K_{Ca^{2+}}$ of 0.15 mM was found at pH 8. Similarly, as found for porcine PLA, this Ca^{2+} binding induces a conformational change as detected by fluorescence measurement in the presence of the dye ANS. Comparable K_{Me} values for Ca^{2+}, Ba^{2+} and Sr^{2+} were obtained as determined by equilibrium dialysis, whereas Mg^{2+} and Mn^{2+} do not bind. Fluorescence experiments with BPB-modified β-bungarotoxin showed that Ca^{2+} up to a concentration of 5 mM induced only a very small effect on the fluorescence of the dye–toxin complex. These fluorescence studies indicate that BPB-modified β-bungarotoxin has lost its Ca^{2+}-binding properties.

Using equilibrium dialysis, Wells [211] showed for the *Crotalus adamanteus* PLA the presence of two cation-binding sites per dimer with a dissociation constant of about 5×10^{-5} M at pH 8 for the alkaline earth cations. Ultraviolet difference spectroscopy revealed that Ca^{2+}, Ba^{2+} and Sr^{2+} bind to this PLA.

Although *Crotalus atrox* PLA like all other PLA's requires Ca^{2+} for activity, no ultraviolet difference spectrum was produced up to 20 mM Ca^{2+} at pH 7.4 [265]. The observed effects of Ca^{2+} on the CD spectrum, the enhancement of fluorescence of ANS-PLA complex by Ca^{2+} and the heat effect in microcalorimetry suggest that the enzyme binds Ca^{2+}. So far only a kinetically determined $K_{Ca^{2+}}$ value (1.1×10^{-3} M at pH 7.5) has been reported. Taking into account the very similar amino acid sequences of the *Crotalus adamanteus* and *Crotalus atrox* PLA in which all aromatic residues are conserved (see Section 3, "Structural aspects"), it is remarkable that the metal ion-induced difference spectra are so different.

Binding of Ca^{2+} to *Bitis gabonica* PLA produces an ultraviolet difference spectrum rather similar to that observed for *Crotalus adamanteus* PLA [266]. The difference spectrum of the *Bitis gabonica* PLA was ascribed to both solvent- and charge-induced perturbations of predominantly Trp. Moreover, Ca^{2+} binding to *Bitis gabonica* PLA also shows a red-shifted peak with a maximum at 240–245 nm, which was not reported for *Crotalus adamanteus* PLA, and which was used to determine the dissociation constant. Similarly Viljoen et al. [266] also observed pH-dependent spectral perturbations both in the absence and presence of Ca^{2+}. More recently, Viljoen and Botes [109] found from the pH dependency of spectral changes in the presence of Ca^{2+}, three transition zones from which pK values of 5.66, 6.75 and 9.15 (at 25°C) were calculated. Based on the heats of ionisation of groups associated with these various pK values, the group with pK 5.66 was assigned to a carboxylate involved in Ca^{2+}-binding. The other two groups with pK values of 6.75 and 9.15 were assigned to a His and a Tyr residue, respectively. From the observation that Ca^{2+} induces a difference spectrum in BPB-modified PLA, Viljoen and Botes [109] conclude that Ca^{2+} is still able to bind, but no dissociation constant is reported. From kinetic data the group involved in Ca^{2+}-binding was found to have a pK value of 6.4.

At basic pH, Ca^{2+}-binding to *Naja naja naja* PLA induces a blue-shifted ultraviolet difference spectrum with minima at 292 and 283 nm, due to charge-induced perturbation of Trp. In contrast, at acid pH, Ca^{2+} induces a red-shifted ultraviolet difference spectrum with maxima at 290.5 and 282 nm due to solvent-induced perturbation of Trp and possibly Tyr [215]. Binding constants for Ca^{2+} in the pH range 3.5–8.5 were thus determined and were found to be in good agreement with those obtained from quenching effects of Ca^{2+} on the fluorescence intensity. The binding of Ca^{2+} to the enzyme is pH-dependent with a pK of 5.9 and a $K_{Ca^{2+}}$ of 0.15 mM for the unprotonated form of the enzyme. The difference spectrum induced by Ca^{2+} at acidic pH is similar to the titration difference spectrum observed in the absence of Ca^{2+}. The latter spectrum shows a pH-dependency controlled by a group with a pK of about 7. It has been concluded that Ca^{2+} binding to *Naja naja naja* PLA triggers a conformational change lowering the pK of a critical residue, probably the active site His. Ca^{2+}-binding also affects the monomer–dimer equilibrium. The ultraviolet difference spectrum induced by Ca^{2+} with BPB-modified enzyme was consistent with Trp perturbation and perturbation of the newly added chromophore.

The binding constant for Ca^{2+} was not changed.

The Ca^{2+}-induced difference spectra of PLA's from *Naja nigricollis* [64] and from *Hemachatus haemachatus* [54], are negative, with minima at 290 and 283 nm, and are interpreted to be primarily charge-induced perturbations of Trp. In addition, a positive peak at 260 nm was also observed which upon titration enabled the authors to determine the dissociation constants. Similar binding constants were obtained for both BPB-modified enzymes, although the Ca^{2+}-induced difference spectra drastically changed. Both PLA's from *Naja nigricollis* and *Hemachatus haemachatus* also markedly enhance the emission intensity of ANS, but in contrast to pancreatic PLA and β-bungarotoxin, Ca^{2+} decreases the fluorescence of the complex.

From the tryptophan fluorescence of *N. naja siamensis, N. naja kaouthia* and *N. naja atra* PLAs Teshima et al. [266a] found $K_{Ca^{2+}}$ values similar to those reported by Roberts et al. [215] for *N. naja naja* PLA. Teshima and coworkers reported perturbation of the pK value of an ionisable group from 7.55 to 7.25 and that protonation of another group with a pK value of 5.4 competed with the Ca^{2+}-binding to the three *Naja* PLA's. On the basis of the X-ray data for bovine PLA [90a,199], the former group was assigned to His-48 and the latter to Asp-49. Recently, Ikeda and Samejima [266b] reported the Ca^{2+}-binding constants for PLA-II from *A. halys blomhoffii* to be larger than those for porcine pancreatic PLA but smaller than those for the cobra enzymes, under similar conditions.

(b) Binding of monomeric zwitterionic substrate analogues

A pre-requisite for these studies is the availability of suitable phospholipids fulfilling at least the conditions (i) that they are not hydrolysed by the enzyme, (ii) that they must behave as competitive inhibitors, and (iii) that they must possess a large enough monomer concentration range together with a good affinity. Similarly, as discussed already (see Section 4, "Kinetic data") for monomer kinetics, direct binding studies are also hampered by the phenomenon that quite often the dissociation constants exceed the CMC values. Because short-chain 1-*sn*-phosphatidylcholines like D-diC_6- or D-diC_7-PCs have been shown to be competitive inhibitors, these lecithins have been used as suitable substrate analogues to study monomer binding. Although it could not strictly be proven that lysolecithins are indeed competitive inhibitors, results similar to those with D-lecithins were obtained. However, the use of either D-lecithins or 1-acyl lyso-PCs has the drawback that particularly in the presence of Ca^{2+} ions, a slow non-specific hydrolysis might occur due to the rather high enzyme concentrations used as compared to kinetic studies. It is, however, possible to substitute Ca^{2+} by Ba^{2+} or Sr^{2+} ions which are competitive for Ca^{2+}. Alternatively, one can use non-hydrolysable substrate analogues. *n*-Alkyl phosphocholines having alkyl moieties of 10, 12 or 14 carbon atoms and CMC values of about 10, 1 and 0.1 mM, respectively, proved to be most useful. Just as for lysolecithins, no evidence is yet available that these substrate analogues are competitive inhibitors. Nevertheless their interaction behaviour with PLA is in all respects similar to that of monomeric short-chain D-lecithins or 1-acyl lysolecithins.

Binding of monomers of short-chain D-lecithins or 1-acyl lyso-PC's to porcine PLA or pro-PLA induces similar red-shifted ultraviolet difference spectra, having peaks at 282 and 288 nm caused by perturbation of Tyr residue(s) [111,200]. In agreement with this observation hardly any perturbation of the unique Trp residue at position 3 was observed in fluorescence spectroscopy with these or other monomeric substrate analogues [244]. Equilibrium gel filtration was also used to study monomer binding of D-lecithins to porcine PLA and pro-PLA. Both techniques enabled the determination of the dissociation constants for binding of monomeric D-diC_7-PC to porcine PLA. The assumption that a 1:1 complex occurs was confirmed recently by Volwerk et al. [12] using equilibrium dialysis *. It was found from ultraviolet difference spectroscopy and from BPB inactivation that the dissociation constant of monomeric 1-acyl lyso-PCs decreases from 43 to 0.06 mM when the acyl moiety increases from 7 to 14 carbon atoms. It was concluded therefore that monomer binding is mainly due to hydrophobic interactions [111,115]. The affinity of monomers of D-diC_7-PC or *n*-dodecanylphosphocholine for porcine PLA remains constant between pH 4 and 7 and is not much affected by Ca^{2+}. In particular in the absence of Ca^{2+}, the affinity decreases above pH 7 [200]. Methyl-His-48-porcine and -equine PLAs bind monomers of *n*-decanylphosphocholine with the same affinities as their respective native enzymes [199]. In contrast, no detectable binding was observed for monomers of D-diC_7-PC to BPB-inhibited porcine PLA using equilibrium gel filtration [111]. This lack of binding is probably due to steric hindrance.

Using *Naja melanoleuca* (fraction DE-III) PLA, Van Eijk et al. [205a] studied the binding of this enzyme to monomers of *n*-alkyl phosphocholines by ultraviolet and fluorescence spectroscopy. The fluorescence data showed sigmoid binding curves. At the CMC of the phospholipids no abrupt change in the signal was observed. In fact the signal slowly increased from about 80% of the maximal value at the CMC to a maximum value which was reached at a lipid concentration of about twice the CMC. Gel filtration studies showed that the M_r of phospholipase increased to values of about 70000 at concentrations of C_{13}-PC equal to its CMC. Again a sigmoidal dependence was observed.

Recently the group of Ikeda [266b,c] determined the affinity of various *n*-alkylphosphocholines (alkyl = octanyl, decanyl, dodecanyl and tetradecanyl, respectively), for the cobra venom PLAs of *N. n. siamensis, N.n. kaouthia* and *N. n. atra* as well as for PLA-II of *Agkistrodon halys blomhoffii* using aromatic circular dichroism or ultraviolet difference spectroscopy. For all three cobra venom PLA's the dissociation constants (2.4 mM for *n*-decanylphosphocholine) were found to be almost constant in the pH range 2.5–8.5 and were hardly affected by Ca^{2+}. Comparison of the dissociation constants of the various complexes of the three cobra venom PLAs

* This 1:1 stoichiometry is found only when the pancreatic PLA binds phosphocholine-containing substrate analogues such as *n*-alkylphosphocholines. Substitution of the zwitterionic head group by strongly negatively-charged groups, e.g. *n*-dodecanylsulphate, results in cooperative binding of several detergent molecules.

with the homologous *n*-alkylphosphocholines showed that monomer binding increases with increasing chain length as has been found previously for the porcine pancreatic PLA [111,115]. This latter effect was also observed for PLA from *A. halys blomhoffii* [266b]. In contrast to the cobra venom PLAs, Ikeda and Samejima [266b] found for PLA-II from *A. halys blomhoffii* that Ca^{2+} or an increase of the pH lowers the affinity for monomer binding. It has to be mentioned, however, that the Japanese investigators assume that only one phospholipid molecule binds per enzyme molecule and that no aggregation occurs. Although this assumption has been verified for monomer binding of the zwitterionic phospholipid analogues to porcine pancreatic PLA [12], it has been shown recently that monomeric snake venom PLAs do aggregate in the presence of these phospholipid analogues [205a; E.A. Dennis, personal communication]. Therefore, the results of these direct binding experiments [266b,c] should be interpreted cautiously.

(c) Binding to aggregated lipids

As discussed extensively already (vide supra), a number of theories have been developed in the last decade to explain the high catalytic activity of PLA toward substrate present in organised lipid–water interfaces, as compared to its low activity on the same substrate present in monomeric form. Irrespective of whatever model we adopt, it is obvious that investigations providing detailed information on the protein–lipid interaction are of the utmost importance. Unfortunately, direct binding studies consume rather large quantities of enzyme, and probably this is the main reason that until now most attention has been paid to the pancreatic PLA's. Although most of these studies so far are limited to micellar substrate analogues there is a growing interest to extend these investigations to bilayer structures.

(i) Pancreatic PLA

Binding of micelles of D-diC_7-PC, lyso-PC or *n*-alkylphosphocholines to porcine PLA further increases the peaks in the ultraviolet difference spectrum produced already by monomer phospholipid binding, while a concomitant shift of the maximal difference absorption from 288 to 292 nm is observed, indicative of both Tyr and Trp perturbation [111,244]. Binding of micelles to PLA can also be monitored by fluorescence spectroscopy where a large increase in fluorescence intensity and a blue shift of about 10 nm of the emission maximum is observed [244]. No such effects are observed for pro-PLA [244]. Elution of a mixture of PLA and pro-PLA in the presence of lysolecithin micelles on Sephadex G-75 showed that only PLA elutes at the void volume bound to the lipid micelles, whereas pro-PLA elutes at its normal position according to its M_r [200]. These observations are in agreement with the presence of a binding site for aggregated lipids on the enzyme in addition to the monomer binding site. A similar conclusion was reached by Hershberg et al. [118] from PRR studies.

Equilibrium gel filtration studies using either micelles of C_{14} lyso-PC or mixed micelles of D-diC_{10}-PC and C_{14} lyso-PC were performed by Pieterson et al. [111] to obtain quantitative data on the binding. It was concluded that one molecule of porcine PLA was bound to about 35 lipid monomers in the mixed micelle and to about 15 in the lysolecithin micelle. The affinity of porcine PLA was found to be higher for the mixed micelles ("K_d"$=2.1\times10^{-5}$ M) at pH 6 than for the C_{14} lyso-PC micelles ("K_d"$=1.6\times10^{-4}$ M) (J.C. Vidal, unpublished results). The bovine PLA, although it has the same PLA-phospholipid ratio in the complex as the porcine PLA, possesses a lower affinity ("K_d"$=1.0\times10^{-4}$ M) for the mixed micelles. BPB-inactivated porcine PLA was found to have a similar capacity to that of the native PLA to interact with these lipid–water interfaces, and it was concluded that the recognition site for interfaces is not only functionally but also topographically distinct from the monomer-binding and catalytic site.

More recently, Soares de Araujo et al. [99], Hille et al. [196] and Donné-Op den Kelder et al. [197] used equilibrium gel filtration and light scattering to study the complex formation of porcine PLA with micelles of various *n*-alkylphosphocholines and lysolecithins. From the results obtained it turned out that the binding is not a simple additive process but rather an insertion of two enzyme molecules into the micelle, followed by a reorganisation of the detergent monomers.

Soares de Araujo et al. [99] found from micro–calorimetry that the binding of PLA to micelles of *n*-hexadecanylphosphocholine is a rapid, exothermic process. Using non-linear regression analysis of binding data it is possible from these measurements to determine the enthalpy changes (ΔH), the number of lipid molecules complexed with one PLA molecule (N) and the dissociation constant (K_d). The

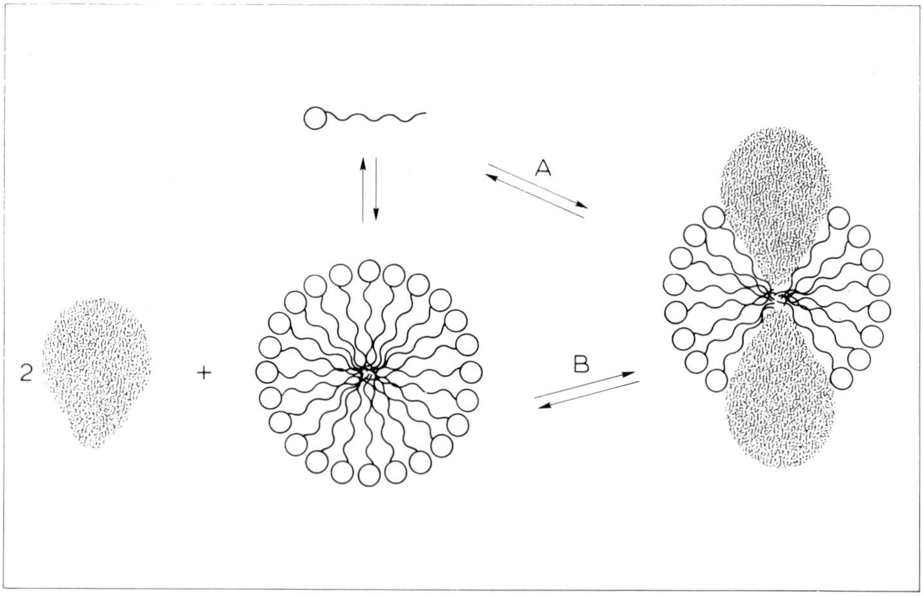

Fig. 4. Schematic view of the pathways for the formation of a complex between phospholipase A_2 and micelles of *n*-hexadecanylphosphocholine [99].

low ΔH values, the positive ΔS changes and the negative value of the heat capacity ΔC_p, support the idea that mainly hydrophobic interactions determine the stability of the PLA–lipid complex. A highly schematic drawing of the complex formation in agreement with the stoichiometry found by the various techniques is given in Fig. 4. At least two possible pathways (A and B) can be considered [267] via which the final complex is constructed. The co-micellisation mechanism (pathway A) has been proposed for some water-soluble proteins containing several high-affinity lipid-binding sites [268–270]. For the pancreatic PLA, Soares de Araujo et al. [99] strongly favoured insertion of the protein into the micelle (pathway B). The authors emphasized that the dimeric structure of pancreatic PLA in the complex shown in Fig. 4 should not be interpreted to mean that an enzyme dimer is functionally active in catalysis.

Although these physico-chemical techniques provide valuable information, the measurements are rather time-consuming and require large quantities of protein. It is, therefore, more advantageous to use fluorescence or ultraviolet difference spectroscopy. These techniques were used by Van Dam-Mieras et al. [244] to study the

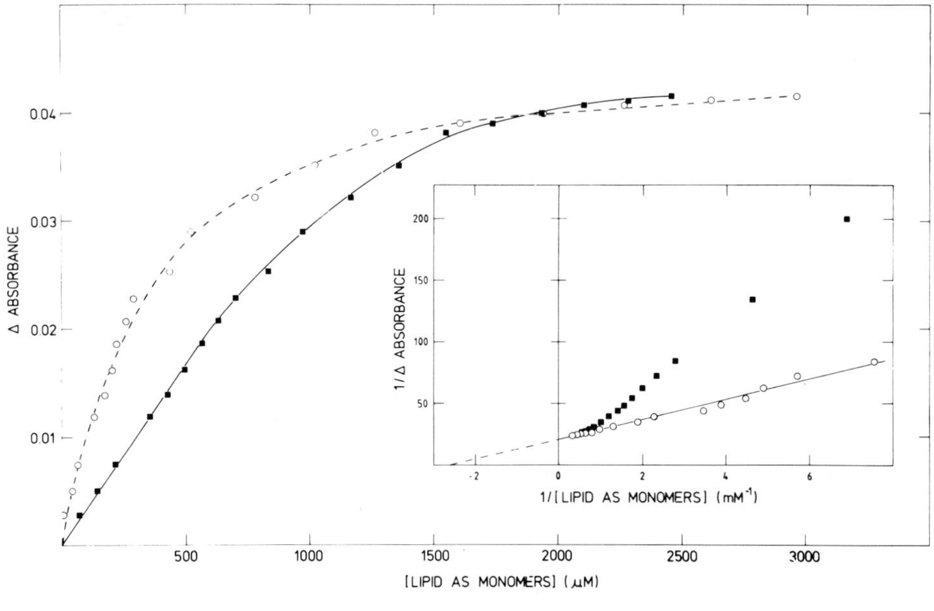

Fig. 5. A direct plot of the ultraviolet absorption difference spectroscopy signal at 292 nm relative to the *n*-octadecanylphosphocholine concentration expressed as monomers. The difference signal at 292 nm relative to total lipid concentration (■) is shown, the solid curve through these points represents the result of the computer fit. In addition, the observed signal is plotted as a function of free lipid (○). The broken curve gives the calculated difference signal relative to free lipid monomers. Inset: a double reciprocal plot of the observed difference signal at 292 nm as a function of total lipid (■) and free lipid (○), respectively. The concentration of PLA is 27.4 μM. All measurements were made at 25°C and pH 4.0 [196].

binding of porcine PLA to *n*-hexadecanylphosphocholine micelles. In this study, dissociation constants were calculated from total lipid concentrations. However, recently this method has been shown to be incorrect, since it leads to excessively high apparent K_d values (Fig. 5) [196]. As shown in Fig. 5, plotting of the ultraviolet absorption difference signals relative to free lipid concentration (expressed as monomers) requires non-linear regression analysis to obtain quantitative data. When the signal is plotted versus free lipid concentration the direct plot fits a hyperbola. Consequently the corresponding double-reciprocal plot is a straight line, whereas it is curved when lipid *total* is plotted. Donné-Op den Kelder et al. [197] showed that only when complex formation is measured by titrating enzyme to lipid, can K_d and the number of lipid molecules complexed with one PLA molecule (N) be obtained graphically without the use of a computer. However, this latter procedure requires large amounts of enzyme. Using both techniques, the authors determined the K_d values as well as the stoichiometry of the porcine PLA complexes formed with a series of saturated and unsaturated *n*-alkylphosphocholines and lysolecithins. In good agreement with the results obtained from microcalorimetry, they found that all the PLA–lipid complexes formed with the saturated phospholipid analogues consisted of 2 PLA molecules and about half the number of monomers present in the original pure micelle. The PLA–lipid complexes formed with the unsaturated phospholipid analogues were found to contain 3 PLA molecules and about 70% of the monomers present originally in the pure micelles. The dissociation constants were found to be dependent on the chain-length of the phospholipid analogue and range from 23 μM for *n*-tetradecanylphosphocholine micelles to 6.6 μM for *n*-octadecanylphosphocholine micelles at pH 6, whereas the affinity for lyso PC's was 2–6-fold lower. These observations further support the conclusion of Soares de Araujo et al. [99] that the stability of the PLA–lipid complex is predominantly due to hydrophobic interactions. Determination of the M_r of the protein part in the enzyme–*n*-octadecanylphosphocholine complex, using the sedimentation equilibrium centrifugation method described by Reynolds and Tanford [271], gave a value of 30000 in good agreement with the model proposed [196].

Studying the pH-dependency of the stability of the PLA *n*-octadecanylphosphocholine complex, Donné-Op den Kelder et al. [197] found that a protonated group with a p*K* of 6.25 controls this binding, and it has been suggested that the active-site residue His-48 and/or Asp-49 are the most likely candidates involved in the lipid-binding process. In particular at basic pH, Ca^{2+} is required for binding of PLA to micellar compounds, by stabilising the conformation of the enzyme that has optimum micelle-binding properties. Similar studies but now using methyl-His-48- and octan-2-one-His-48-modified PLA's showed that the micelle-binding of these proteins is now controlled by a group with p*K* 4.6, while addition of Ca^{2+} at high pH values again restores the micelle-binding properties of these modified PLA's. Therefore, most probably the group having p*K* 4.6 should be assigned to Asp-49. Apparently, upon alkylation of the N-1 atom of His-48, the rather higher p*K* value of Asp-49 drops from 6.25 to 4.6, the latter value being normal for a carboxylate group in a protein.

(ii) Snake venom PLA

Prigent-Dachary et al. [190] used fluorescence spectroscopy to study binding of various snake venom PLAs to vesicles of long-chain phospholipids. They found that strong inhibitors of blood clotting (PLAs from *Naja nigricollis, Naja mossambica mossambica* and *Vipera berus orientale*) interact with PC, PC + PS and PS vesicles, although a higher affinity was found for the PS-containing vesicles than for the pure PC vesicles. Poor inhibitors of blood coagulation (PLAs from *Bitis gabonica, Crotalus adamanteus, Crotalus atrox* and *Naja melanoleuca DE II*) do not or only weakly bind to these vesicles. Using the "non-hydrolysable" diC_{16}-ether-PC it was demonstrated that Ca^{2+} promotes the complex formation which can occur whenever the lipids are in the crystal or fluid phase. Inactivation of the anti-coagulant PLA from *Naja nigricollis* with BPB decreased the affinity of the enzyme for the phospolipids two-fold.

Very recently Jain et al. [181] compared the binding of porcine and *Naja melanoleuca* PLAs to long-chain phospholipid dispersions (vesicles) using various techniques. Qualitatively, gel filtration, differential scanning calorimetry and freeze-fracture electron microscopy showed binding of *Naja melanoleuca* PLA to vesicles of pure diC_{14}-ether-PC. Similar experiments with porcine PLA did not reveal any binding to the diC_{14}-ether-PC vesicles alone. However, only when vesicles of the ternary system PC + lyso-PC + FA were used did the porcine PLA show affinity for the bilayer phospholipids. More quantitative data about the binding of these two PLAs to bilayer structures were obtained from fluorescence and ultraviolet difference spectroscopy. Binding of *Naja melanoleuca* PLA to pure diC_{14}-ether-PC vesicles causes an increase in fluorescence intensity and in parallel a blue shift of the emission maximum, which for the porcine PLA again occur exclusively in the ternary bilayer system. Using the curve-fitting procedure for lipid binding as described by Soares de Araujo et al. [99] and Hille et al. [196] it was found that the K_d values for *Naja melanoleuca* PLA were lower than for the porcine PLA for the same ternary system and that the number of phospholipid molecules contributing to the binding is lower for the *Naja melanoleuca* PLA than for the porcine PLA. The results thus suggest that the binding of pig PLA is regulated by the organisation of the bilayer and the factors favouring phase separation in bilayers also favour the binding of the pancreatic PLA to bilayers.

Recently, Verheij et al. [156,189] using ultraviolet difference spectroscopy determined the dissociation constants and the stoichiometry of the PLA-*n*-hexadecanylphosphocholine complexes for a number of snake venom PLAs in the presence of Ca^{2+} (*Vipera berus, Naja melanoleuca* and *Crotalus atrox*). The dissociation constants were found to be in the range, 1.6–8 μM, comparable to that of porcine PLA, but the ratio lipid to protein (N) is considerably lower for the snake venom PLA's than for the porcine PLA. BPB-inactivated *Vipera berus* PLA also binds to micelles, though with a two-fold lower affinity as compared to the native enzyme.

In the absence of Ca^{2+}, Wells [211] did not observe an ultraviolet difference spectrum of *Crotalus adamanteus* PLA with micelles of D-diC_6-PC. Similar observations have been reported by Tinker for *Crotalus atrox* PLA (personal communication).

In direct binding studies of *Bitis gabonica* PLA with diC_{16}-PC, lyso-PC or fatty acid, Viljoen et al. [266] found ultraviolet difference spectra originating from perturbation of Trp residues, both in the presence and absence of Ca^{2+}. It was assumed that Ca^{2+} is necessary for producing an active conformation of the enzyme, allowing the productive binding of substrate, and that in the absence of Ca^{2+} unproductive binding gives rise to the observed difference spectrum.

Roberts et al. [119] and Adamich et al. [137] used equilibrium gel filtration to study binding of native and BPB-modified *Naja naja naja* PLA's to mixed micelles of Triton X-100 and long-chain PC's (and other phospholipids). They found binding only when divalent metal ions were present. In contrast, no metal ions were required for binding of *Naja naja naja* PLA to mixed micelles of Triton X-100 and fatty acid or lyso-PC. The reported K_d values [137] have no physical meaning since it was assumed that the complex formed is additive (vide supra).

7. Immunology

Ouchterlony's double immunodiffusion showed that only cow and sheep pancreatic PLA gave precipitin lines of complete identity to both antisera. Horse PLA only partially cross-reacts with pig PLA using anti-horse PLA serum, whereas pig PLA shows a partial cross-reaction with horse, cow and sheep PLA towards anti-pig serum [217,272]. With the exception of a partial immunological identity between human and porcine enzymes [5a], no line of precipitation could be visualised between human PLA and the antisera to the other mammalian PLA's, nor between these various mammalian homologous enzymes with the antiserum to human prophospholipase A_2 [5b].

Similar results were obtained from the micro-complement fixation assay. With this technique in particular, horse and cow PLA show considerable immunological differences, whereas the pig enzyme takes an intermediate position between these phospholipases. Ouchterlony's immunodiffusion did not discriminate between the enzyme and its zymogen since a complete cross-reaction toward anti-PLA serum was observed. However, the complement fixation assay detects a considerable difference. Using this assay iso-porcine PLA could be clearly distinguished from porcine PLA although there are only four substitutions in their sequences [251]. Moreover, with the micro-complement fixation assay it turned out that the N-terminal sequence Ala^1-Arg^6 is most probably part of an antigenic determinant of phospholipase A_2. Radioimmunoassay, using monovalent phospholipase A_2-specific F_{ab} fragments revealed a maximum number of three antigenic sites of PLA that can simultaneously be occupied by antibody. The F_{ab} fragments were separated into three fractions, using three immunoadsorbent columns in series. These F_{ab} fragments showed different inhibitory properties toward binding of PLA to micellar substrate. One of these F_{ab} fragments turned out to protect PLA effectively against BPB modification [272].

8. *The 3-dimensional structure*

Not all PLA's crystallize readily to yield crystals suitable for X-ray analysis. The enzyme from porcine pancreas never yielded suitable single crystals despite numerous attempts, while its precursor produced crystals of poor quality which allowed calculation of an electron-density map only at a resolution of 3 Å [89]. The revised sequence of porcine PLA [88] could, however, not be incorporated into this electron density map. This observation and the absence of regular α-helices and β-pleated sheets suggest that the crystals contained denatured protein. In the meantime it was found that both the active enzyme and the precursor of bovine pancreatic PLA crystallised as high quality single crystals. Using these crystals and three heavy-atom derivatives, the three-dimensional structure was determined to a resolution of 2.4 Å [90]. Subsequently, diffraction data to 1.7 Å resolution were collected and the phospholipase model was crystallographically refined at this resolution to a final R-factor of 17.1% [90a,218].

PLA's from *Crotalus adamanteus* and *Crotalus atrox* also yield crystals suitable for X-ray analysis. In both cases one dimer per asymmetric unit was present [273]. Interpretation of the electron density map at a resolution of 2.5 Å shows that the main chain folding of *Crotalus atrox* phospholipase A_2 is very similar to that of bovine phospholipase [91a] (see Fig. 10). Furthermore it was found that the C-terminal appendage is linked indeed via a disulphide bridge to Cys-50 (see also Section 3, "Structural aspects"). In the dimer both active sites are shielded from the surrounding water. It is difficult to visualize how Ca^{2+} ions and substrate molecules can enter the heavily shielded active site (Fig. 10).

Notexin, a neurotoxic PLA, forms crystals diffracting to a resolution of 1.8 Å. There are 6 molecules in the unit cell [274]. No further data obtained with this phospholipase have been published so far.

Phospholipase A_2 from *Naja naja naja* venom has been crystallized under a variety of conditions [274a]. Several different crystal forms were obtained depending upon pH and the presence of calcium ions. The best characterized crystals contain two PLA molecules in the asymmetric unit.

In the absence of 3-dimensional structures of other PLA's we assume that the 3-D structure of the bovine pancreatic and the rattlesnake PLAs can also be compared to other (venom) PLA's. For this reason we will give a somewhat detailed description of the structure of bovine PLA. The molecule is kidney-shaped with the dimensions 22 Å × 30 Å × 42 Å; it has a high content of secondary structure with about 50% α-helix and 10% β-structure (Fig. 6). The structure is stabilised by a large number of hydrogen bridges. In addition the loops are held together by seven disulphide bridges. For example, the two long antiparallel α-helices corresponding to residues 40–58 and 90–108 are connected by two disulphide bridges (Cys-44 to Cys-105 and Cys-51 to Cys-98). In these helices the active centre residues His-48, Asp-49, Tyr-52 and Asp-99 are bound tightly together.

Fig. 7 shows a 3-dimensional view of the active centre of bovine PLA including the backbones of residues 28–33, 48–52 and 98–99 and some of the side chains.

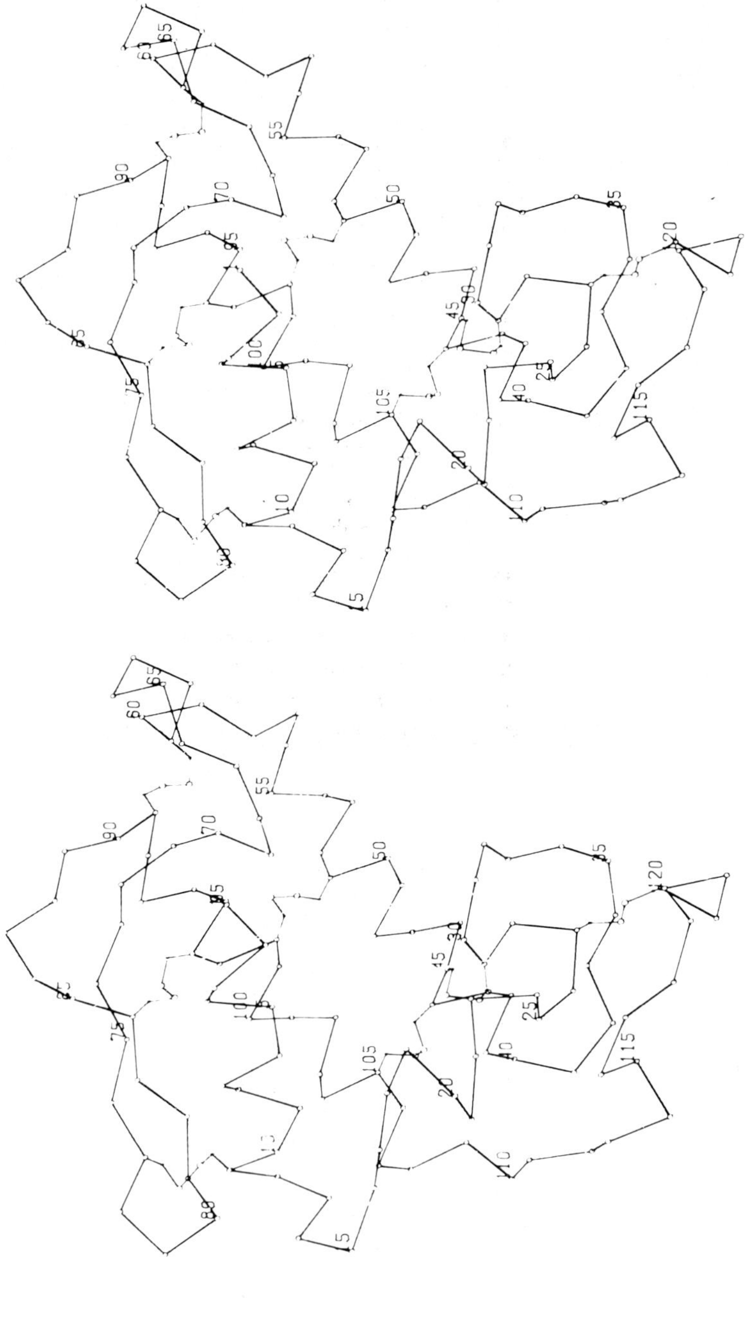

Fig. 6. Stereo diagram showing the conformation and disulphide bridges of the bovine pancreatic phospholipase molecule [90].

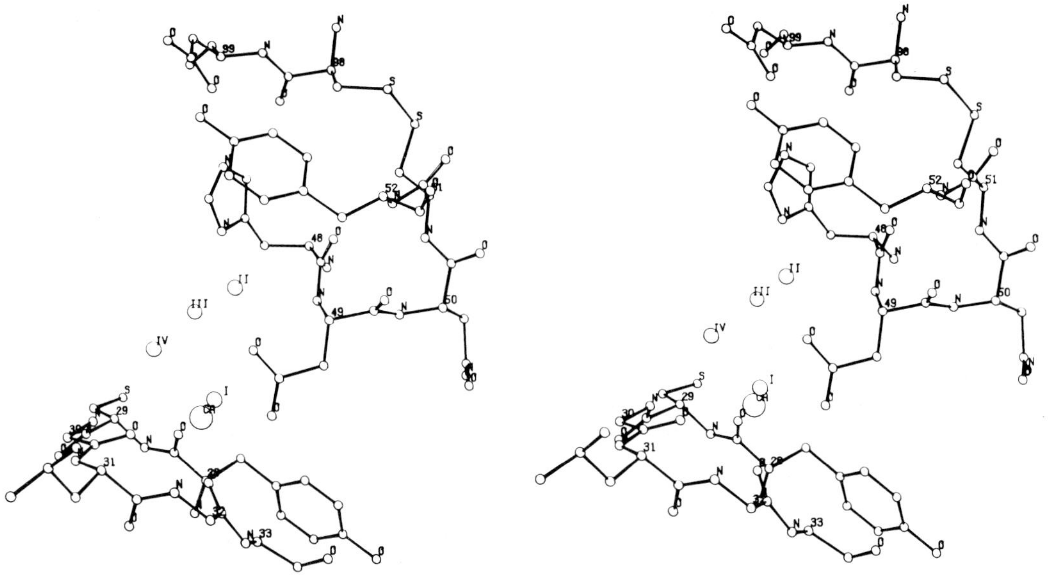

Fig. 7. Stereo picture of the active site of phospholipase A_2 including the calcium ion and several water molecules [218].

Note that the amino acids in this part of the sequence are invariant in all phospholipases except residues 31 and 50 (see Table 1). The main chain of residues 28–33 is part of the calcium-binding loop which runs from residues 25 to 42 and contains the five glycines conserved in all phospholipases. When the folding pattern of bovine PLA is summarized in a Ramachandran plot, these 5 glycine residues are found in regions disallowed for other amino acids. Substitution of these glycines for

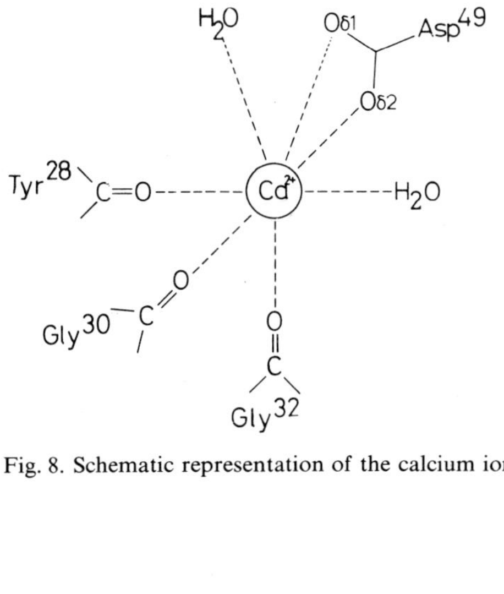

Fig. 8. Schematic representation of the calcium ion and its ligands [218].

other amino acids, while maintaining the chain folding pattern, would be energetically highly unfavourable [218].

The calcium ion is located in the active site surrounded by 7 oxygen ligands (Fig. 8), viz. 3 carbonyl oxygens, the δ^1 and δ^2 oxygens of Asp-49 and two water molecules [218,221]. Six of these ligands are found at the corners of an octahedron. The Ca^{2+} ion can be replaced by a Ba^{2+} ion although Ba^{2+} does not orient itself into exactly the same position, probably due to its larger size (B.W. Dijkstra, personal communication).

The imidazole ring of His-48 is in close proximity to the side chains of Asp-99, Tyr-52 and a water molecule (Fig. 9). The N-3 atom of His-48 is at hydrogen-bonding distance (2.8 Å) of one of the carboxylate oxygens of Asp-99. Close to the N-1 of His-48 (about 3 Å) a water molecule is found (water molecule I in Fig. 7). This water molecule could very well perform the nucleophilic function in the ester hydrolysis by analogy with the active centre serine in the serine esterases. The carbonyl oxygens of Asp-99 are also hydrogen-bonded to the hydroxyl groups of Tyr-52 (2.55 Å) and Tyr-73 (2.50 Å). Both tyrosine residues are invariant in all phospholipases. *Via* a water molecule these residues are also hydrogen-bonded to the α-amino group, the side-chain of Gln-4, and the carbonyl oxygens of Pro-68 and Asn-71. Gln-4 is invariant in all phospholipases and the interactions with the α-amino group and the main chain carbonyl oxygens do not necessarily depend on the side-chains present. One might, therefore, predict that in all phospholipases such an extended proton relay system does exist. This system probably has a structural rather than a catalytical function, since proteins devoid of the α-amino group (e.g. precursor)

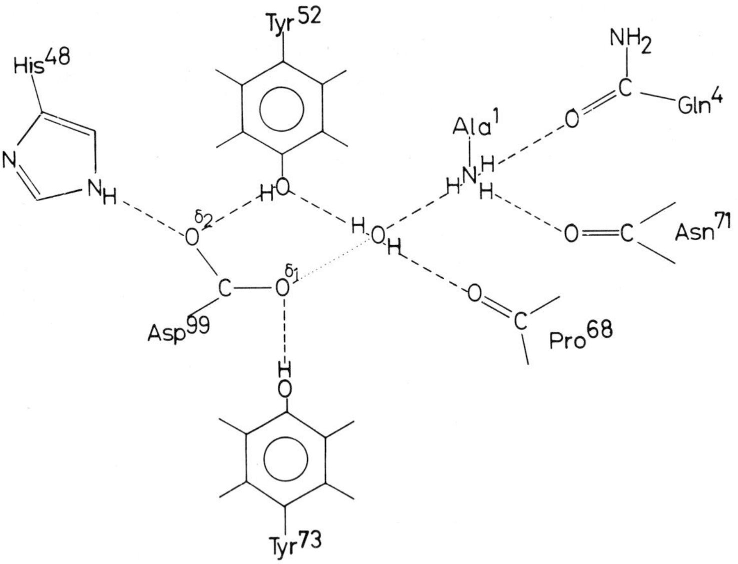

Fig. 9. Proton relay system of phospholipase [218].

effectively hydrolyse monomeric substrates. The system is buried in the interior of the protein and the Asp^{99}-His^{48} couple is shielded from the surrounding solvent by a number of invariant hydrophobic residues: Phe-5, Ile-9, Ala-102, Ala-103, Phe-106 and the disulphide bridge between Cys-29 and Cys-45. In addition Phe-22 (Tyr in most venom enzymes) is part of this hydrophobic active site wall.

Whereas the hydrophobic residues forming the active site wall are mostly invariant, the situation at the surface surrounding the active site is quite different. As amply discussed already (see Section 3, "Structural aspects") the entrance of the active site (IRS) is composed of highly variable, mainly hydrophobic amino acid side-chains. The fact that the surface does not impose strict spatial requirements upon the size of the side-chains has apparently given rise to a great variety of in general hydrophobic residues.

If we finally try to predict how the primary structure of about 30 venom PLAs (Table 1) would fit the three-dimensional structure of the bovine pancreatic PLA, we come to the following conclusions. In all PLAs the residues around the Asp^{99}-His^{48} couple and the potential Ca^{2+} ligands are invariant (or highly conserved). There is no obvious reason why all PLAs could not form an extended proton relay system as depicted in Fig. 9. The residues around the entrance to the active site (IRS) are variable but with few exceptions they are hydrophobic. The large deletion between residues 57 and 68 found in the venom PLA's shortens two external loops around the disulphide bridge between Cys-61 and Cys-91 without affecting the gross shape of the molecule. Therefore we tentatively predict that the PLA's from the different sources not only show a high degree of sequence homology but also have very similar three-dimensional properties. This conclusion is supported by the X-ray analysis of *Crotalus atrox* phospholipase (Fig. 10). The shape of the molecule is similar to that of the pancreatic enzyme. The dimeric form of the enzyme seems to be stabilised by ionic interaction between Lys-64 (Lys-69 in Table 1) of one protomer and Asp-49 from the other protomer (Fig. 11). The Ca^{2+}-binding loop (residues 27–34) also seems to contribute to the stability of the dimer *via* interactions with the N-terminal region of the other protomer.

Another X-ray determination deals with the structure of the precursor of bovine pancreatic PLA. Good crystals of this protein have been obtained and the results show that the structure is nearly identical to that of the active PLA, except for the N-terminal region and around Tyr-69. In the precursor, these residues show a high mobility, whereas they are fixed in the active PLA. Because the N-terminal residues and Tyr-69 are part of the IRS this observation is of the utmost interest (J. Drenth, personal communication).

9. *Catalytic mechanism*

In this section we will try to compare data emerging from chemical modifications, direct binding studies, and X-ray crystallography and see how these data fit a proposed catalytic model for bovine pancreatic PLA. Kinetic analyses of the

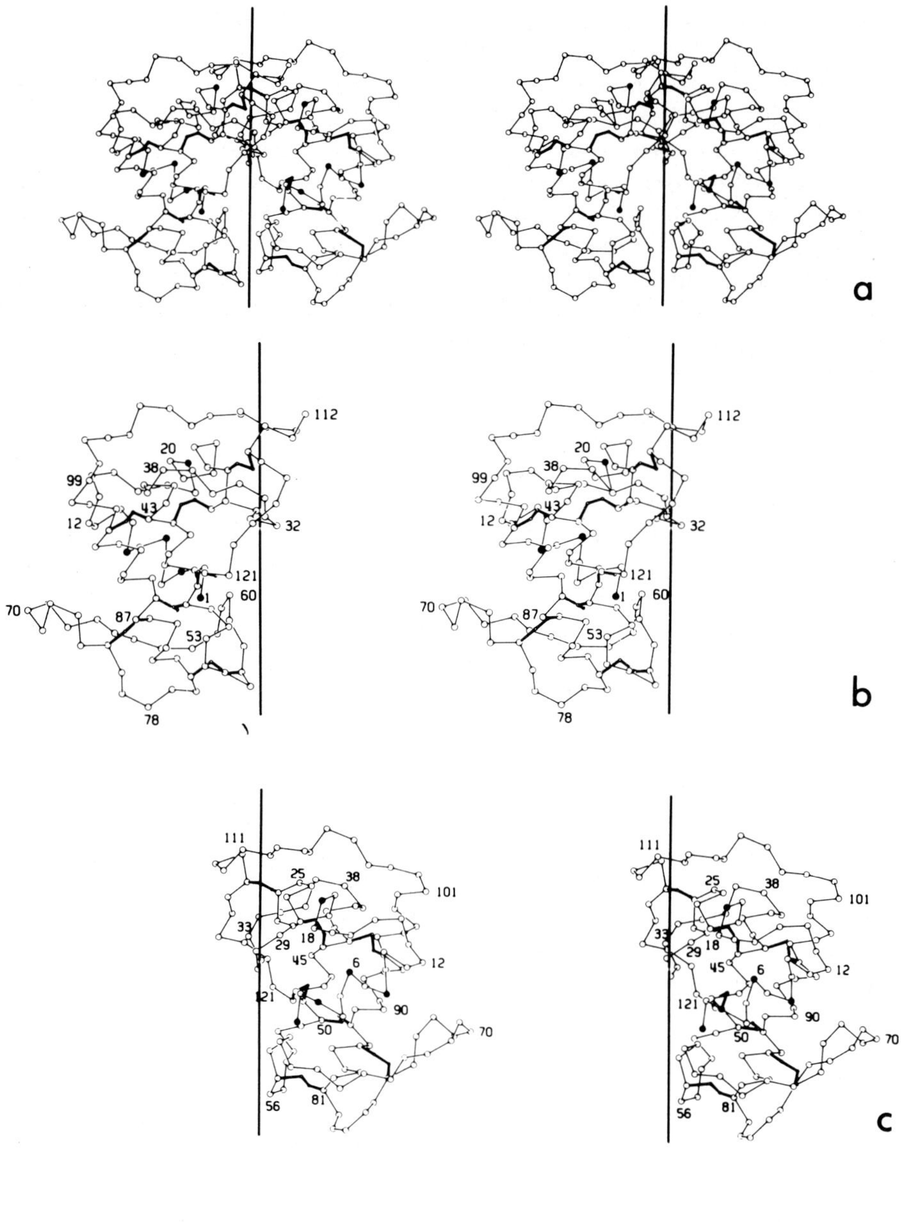

Fig. 10. Stereo views of the unrefined C_a positions of: (a) the entire phospholipase A_2 molecule from the venom of *C. atrox*; (b) the left (L) protomer alone (enlarged scale); and (c) the right protomer alone (scale as in L). The numbering system has not been adjusted to fit a homologous scheme, but proceeds without additions or deletions from the NH_2 terminus to the COOH terminus. The sequence positions corresponding to the residues that constitute the putative interfacial recognition surface are indicated by *darkened atoms*.

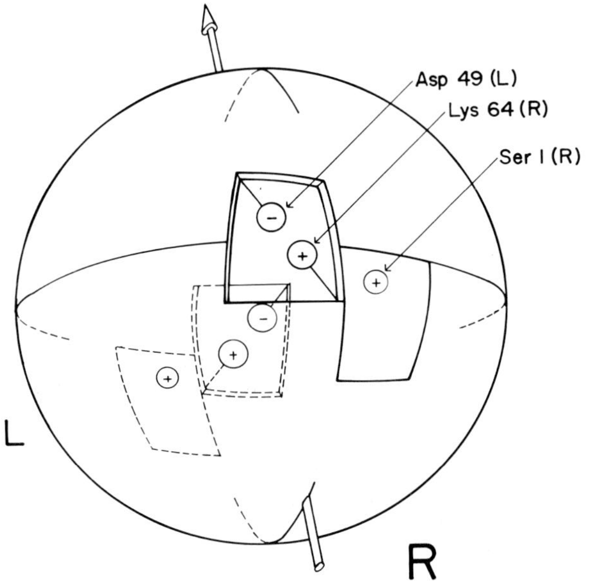

Fig. 11. A schematic representation of the surface of the phospholipase A_2 dimer from *C. atrox*. The large *large arrow* is the local dyad that skewers the oblate ellipsoid. The *path* to the *front right* is the region on the right (R) protomer corresponding to the interfacial recognition surface of the mamalian enzyme and is designated by a + indicating the approximate position of the NH_2 terminus. The adjacent *window*, slightly *above* and immediately to the *left* of the interfacial recognition surface, appears tp be the likely portal of access to the cavity that houses the catalytic and cofactor-binding sites of both protomers. A salt bridge between Lys-64 of the R protomer and the Asp-49 from the L protomer lies across this portal. The symmetry-related regions are shown to the *left rear* in *broken lines* [91a].

hydrolysis of aggregated substrate require a binding step of the enzyme to the lipid–water interface prior to the Michaelis–Menten complex formation. It has been shown (see Section 4, "Kinetic data") that such an additional binding step complicates the interpretation of kinetic data in terms of well-defined rate and binding constants. Only by using monomeric short-chain phospholipids can interpretable kinetic data be obtained [10,12,106]. As already pointed out in the previous sections, we know that: (1) hydrolysis requires an ester bond 5 or 6 atoms separated from a negative charge and the ester bond must be present in a specific stereochemical orientation; (2) Ca^{2+} ions are required for the reaction; Ba^{2+} and Sr^{2+} ions are competitive inhibitors. They bind in a 1 : 1 ratio to the enzyme in a pocket formed by 3 backbone carbonyl groups and the side-chain of Asp-49; (3) monomeric substrates or substrate analogues bind in a 1 : 1 ratio; in this binding process hydrophobic interactions predominate; (4) His-48 is involved in catalysis with its N-1 group oriented toward the solvent. The pK of this group is about 6.5, a value that drops to about 5.5 in the presence of Ca^{2+} ions; (5) although the enzyme hydrolyzes esters it is not a classical serine esterase. It does not react with organophosphates and no results have been obtained in favour of the existence of an acyl enzyme. Therefore, Wells [211] proposed that a water molecule must be the nucleophile attacking the ester bond.

The catalytic mechanism described here heavily depends on the X-ray structure of bovine pancreatic PLA. We assume that this structure does not differ significantly from the structure of any PLA (from pancreas or venom). Such an assumption is not unrealistic since we have seen that venom and pancreatic PLA's show a high degree of homology. In the X-ray structure, His-48 is located in a cleft near the absolutely conserved side-chains of Asp-49, Tyr-52 and Asp-99 (Table 1). The wall of the cleft is constituted of residues with highly conserved, hydrophobic side-chains. Based on the chemical evidence (vide supra) and the spatial arrangement of the side-chains, a mechanism has been proposed [199] which is described in Fig. 12.

The presence of the Asp[99]-His[48] couple suggests a comparison with the serine esterases. The serine residue found in the serine esterases is lacking in PLA but instead a water molecule about 3 Å away from the N-1 nitrogen of His-48 is supposed to perform the nucleophilic function in the ester hydrolysis by analogy with the active centre serine in the esterases. When this water molecule attacks the substrate carbonyl carbon atom, the imidazole ring of His-48 picks up a proton from the water molecule, thereby facilitating the reaction. This proton is subsequently

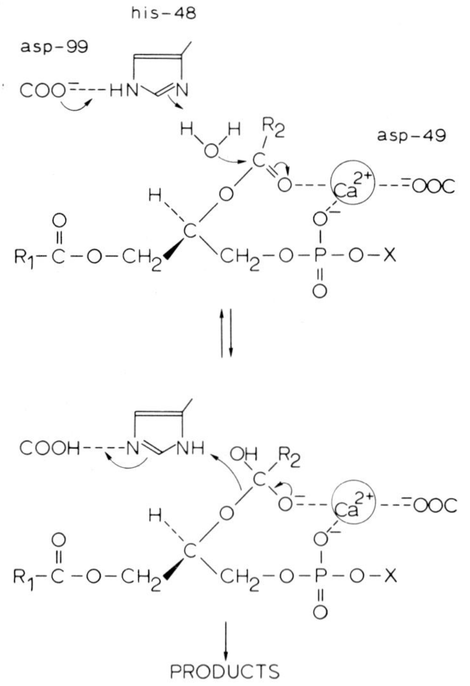

Fig. 12. Proposed catalytic mechanism [199].

donated by the imidazole ring to the alkoxy oxygen, just as in the serine enzymes where the proton from serine is transferred by His to the leaving group [275,276].

The function of the Ca^{2+} ion may be to bind the negative phosphate group. If this were the only role of the Ca^{2+} ion it is not clear why in the presence of the slightly larger Ba^{2+} ions (1.34 Å vs. 0.99 Å) a ternary complex is formed but not hydrolysed. A possible explanation is that because Ca^{2+} is a stronger Lewis acid than Ba^{2+} it can more easily polarise the ester carbonyl function and stabilise the tetrahedral intermediate in concert with the backbone NH group of residue 30.

No X-ray crystallographic data of an enzyme-substrate (analogue) complex are yet available. However, it is possible to fit a substrate molecule in the active centre with the susceptible ester bond in the required position relative to the attacking water molecule, the phosphate group close to the Ca^{2+} ion and the remaining part of the polar head group (e.g. choline) pointing towards the solvent. The two acyl chains, while running parallel to each other, can be fitted into a shallow cleft on the enzyme surface in between the apolar side-chains of Leu^{2}, Leu^{19}, Leu^{20} and Leu^{31} (Fig. 13).

How does this mechanism fit data from PLA's other than the bovine pancreatic PLA? The side-chains of the calcium ligand Asp-49, the Asp^{99}-His^{48} couple and Tyr-52 are invariant in all PLAs and most probably fulfil a similar role. The role of Tyr-52 is not very clear although it is at hydrogen bridge distance from Asp-99 and

Fig. 13. The space-filling model of bovine pancreatic phospholipase.

may help to stabilise the charge of the Asp^{99}-His^{48} couple. Albeit somewhat variable, the residues forming the wall of the active site cavity are very hydrophobic in all phospholipases (Section 8). Consequently, we must assume that in all phospholipases, the Asp^{99}-His^{48} couple is accommodated in a hydrophobic micro-environment. Despite this similarity, the reported pK values of the group controlling catalysis — and according to Fig. 10 this must be histidine — vary between 5.5 and 7.6 [12,106,109] and may suggest that subtle changes near the Asp^{99}-His^{48} couple change its pK drastically. For all pancreatic enzymes, the active site histidine shows a "normal" pK value of about 6.5 and this value is lowered to about 5.5 in the presence of Ca^{2+} ions [199,201,263]. Also in *Naja naja naja* PLA the pK of the active centre histidine is lowered upon addition of Ca^{2+} [100]. A further increase in k_{cat} values above pH 7 observed in pancreatic as well as venom PLAs might be ascribed to a conformational change induced by deprotonation of a residue with a pK value around 8. The nature of this group has not yet been elucidated although it has been suggested to be a lysine [193] or the α-amino group [12].

The binding of monomeric substrate analogues to pancreatic, *Naja n. oxiana* and *C. adamanteus* PLA's has been shown to be a mainly hydrophobic process resulting in a 3-fold better binding for each additional methylene group [12,106,108,113,115, 266b and c]. Also, modification of His-48 with alkylating reagents is only successful when the reagents possess an apolar part [100,199]. Indeed, if the side-chains of residues 2, 19, 20 and 31 contribute predominantly to the binding of monomers, we may expect from Table 1 that this hydrophobic interaction plays an important role in all phospholipases. These residues are also an integral part of the larger hydrophobic surface (IRS) (see Section 3) that is supposed to interact with lipid–water interfaces. Therefore, one expects a somewhat different orientation of the substrate molecule bound to the active site when the enzyme becomes embedded in a lipid–water interface. Whether this conformational change alone is responsible for the fact that aggregated substrates are hydrolysed with high velocity compared to monomeric substrates is not yet clear. Other factors like the conformation and the hydration of the substrate [120] and the entropy loss upon binding [113] may also play an important role. Finally, it is also conceivable that in the hydrolysis of monomers the release of products is slow, whereas *in* the interface the product is replaced rapidly by a new substrate molecule by lateral diffusion. This diffusion is rapid enough to allow turnover numbers at least one order of magnitude higher than the observed maximal turnover numbers (about 7000/s).

10. Prospects

Despite the availability of many primary structures and a high-resolution X-ray structure, our understanding of the mechanism of action of PLA is limited. Undoubtedly this is due to the fact that PLA acts on substrates that are "insoluble" in water. In the presence of phospholipid aggregates no meaningful interpretation of the effects of inhibitors can be made. Also the hydrolysis of monomeric substrates only yields limited information due to the fact that the affinities of PLA for these substrates are too low to allow for detailed kinetic studies using inhibitors. Also, the observed aggregation of phospholipases with phospholipids at concentrations well below the CMC is a complication for kinetic studies.

The high degree of homology of venom and pancreatic PLAs suggests a common mode of action for all phospholipases. However, even relatively simple questions like: "is the active enzyme acting as monomer or dimer?" cannot be easily answered. The great variation in amino acid side-chains located at the surface of the protein will certainly induce large differences in properties of the phospholipases. Results obtained with phospholipases from one source should be treated with great care and no generalised conclusions should be drawn from these. Therefore it seems important that comparative studies are carried out.

Much information has been obtained from chemical modification studies and it can be expected that the vast amount of information obtained from sequence analysis will promote more modification studies. Also for this reason the elucidation of the three-dimensional structure of more venom PLAs as well as (a) PLA-inhibitor complex(es) is highly desirable.

At present, our knowledge of the apoenzyme exceeds that of PLA–lipid complexes. Further studies on the interactions of PLA with aggregated phospholipids are required to obtain detailed information about the lipid–protein complexes. Only by combining our knowledge on phospholipid orientation, hydratation, conformation and motion in the interface (see e.g. two recent reviews by Hauser et al. [278] and Büldt and Wohlgemüt [279]), and the conformational changes of the protein in the complex, may one expect to understand how the fine structure of the lipid–water interface determines the activity of lipolytic enzymes.

Acknowledgements

Dr. M.R. Egmond is gratefully acknowledged for critically proof-reading the manuscript, and for his help in the preparation of Table 1. The authors would like to express their appreciation to colleagues for making available manuscripts prior to publication: E.A. Dennis, B.W. Dijkstra, J. Drenth, D. Eaker, R.L. Heinrikson, P. Lind, S. Nishida, J.A.F. Op den Kamp, B.W. Shen, P.B. Sigler, R. Verger, C.C. Viljoen, M.A. Wells, T. Wieloch, C.C. Yang and H. Yoshida.

We thank Drs. B.W. Dijkstra, J. Drenth, R. Verger and D.O. Tinker for generously supplying various figures.

Thanks are due to Miss E.J.G. de Haas and Miss R.G. Obbink for typing the manuscript.

References

1 Shen, B.W. and Law, J.H. (1979) in A.M. Scanu, R.W. Wissler and G.S. Getz (Eds), The Biochemistry of Atherosclerosis, Dekker Inc., New York, pp. 275–291.
1a Van den Bosch, H. (1980) Biochim. Biophys. Acta 604, 191–246.
1b Dennis, E.A., Darke, P.L., Deems, R.A., Kensil, C.R. and Plückthun, A. (1981) Mol. Cell Biochem. 36, 37–45.
2 Van den Bosch, H. and Aarsman, A.J. (1979) Agents and Actions 9, 382–389.
3 Magee, W.L., Gallai-Hatchard, J., Saunders, H. and Thompson, R.H.S. (1962) Biochem. J. 83, 17–25.
4 Uthe, J.F. and Magee, W.L. (1971) Can. J. Biochem. 49, 776–784.
5 Figarella, C., Clemente, F. and Guy, O. (1971) Biochim. Biophys. Acta 227, 213–217.
5a Grataroli, R., De Caro, A., Guy, O., Amic, J. and Figarella, C. (1981) Biochimie 63, 677–684.
5b Grataroli, R., Dijkman, R., Dutilh, C.E., Van der Ouderaa, F., De Haas, G.H. and Figarella, C. (1982) Eur. J. Biochem. 122, 111–117.
6 Nieuwenhuizen, W., Kunze, H. and De Haas, G.H. (1974) Methods Enzymol. 32B, 147–154.
7 Evenberg, A., Meyer, H., Verheij, H.M. and De Haas, G.H. (1977) Biochim. Biophys. Acta 491, 265–274.
8 Salach, J.I., Turini, P., Hauber, J., Seng, R., Tisdale, H. and Singer, T.P. (1968) Biochem. Biophys. Res. Commun. 33, 936–941.
9 Salach, J.I., Turini, P., Seng, R., Hauber, J. and Singer, P. (1971) J. Biol. Chem. 246, 331–339.
10 Roholt, O.A. and Schlamowitz, M. (1961) Arch. Biochim. Biophys. 94, 364–379.
11 Aarsman, A.J., Van Deenen, L.L.M. and Van den Bosch, H. (1976) Bioorg. Chem. 5, 241–247.
12 Volwerk, J.J., Dedieu, A.G.R., Verheij, H.M., Dijkman, R. and De Haas, G.H. (1979) Recl. Trav. Chim. Pays-Bas 98, 214–220.
12a Wolf, C., Sagaert, L. and Bereziat, G. (1981) Biochem. Biophys. Res. Commun. 99, 275–283.
13 Henderson, T.O., Kruski, A.W., Davis, L.G., Glonek, T. and Scanu, A.M. (1975) Biochemistry 14, 1915–1920.
14 Brasure, E.B., Henderson, T.O., Glonek, T., Pattnaik, N.M. and Scanu, A.M. (1978) Biochemistry 17, 3934–3938.
15 Roberts, M.F., Adamich, M., Robson, R.J. and Dennis, E.A. (1979) Biochemistry 18, 3301–3308.
16 Tsao, F.H.C., Cohen, H., Snyder, W.R., Kézdy, F.J. and Law, J.H. (1973) J. Supramol. Struct. 1, 490–497.
17 Van Wezel, F.M. and De Haas, G.H. (1975) Biochim. Biophys. Acta 410, 299–309.
18 Dutilh, C.E., Van Doren, P.J., Verheul, F.E.A.M. and De Haas, G.H. (1975) Eur. J. Biochem. 53, 91–97.
19 Wittich, K.A. and Schmidt, H. (1969) Enzym. Biol. Clin. 10, 477–486.
19a Castle, A.M. and Castle, J.D. (1981) Biochim. Biophys. Acta 666, 259–274.
20 Wells, M.A. (1975) Biochim. Biophys. Acta 380, 501–505.
21 Rock, C.O. and Snyder, F. (1975) J. Biol. Chem. 250, 6564–6566.
22 Louw, A.I. and Carlsson, F.H.H. (1979) Toxicon 17, 193–197.
23 Apsalon, U.R., Shamborant, O.G. and Miroshnikov, A.I. (1977) Bioorgh. Khim. 3, 1553–1559.
24 Gubenšek, F. and Žunič, D. (1978) Toxicon 16, 419.
25 Gritsuk, V.I., Meshcheryakova, E.A., Okhanov, V.V., Efremov, E.S. and Miroshnikov, A.I. (1979) Bioorgh. Khim. 5, 1222–1232.
26 Kawauchi, S., Iwanaga, S., Samejima, Y. and Suzuki, T. (1971) Biochim. Biophys. Acta 236, 142–160.
27 Kawauchi, S., Samejima, S., Iwanaga, Y. and Suzuki, T. (1971) J. Biochem. 69, 433–437.
28 Hanahan, D.J., Joseph, M. and Morales, R. (1980) Biochim. Biophys. Acta 619, 640–649.
29 Augustyn, J.M. and Elliott, W.B. (1970) Biochim. Biophys. Acta 206, 98–108.
30 Shipolini, R.A., Callewaert, G.L., Cottrell, R.C., Doonan, S., Vernon, C.A. and Banks, B.E.C. (1971) Eur. J. Biochem. 20, 459–468.

30a Miroshnikov, A.I., Gritsuk, V.I., Meshcheryakova, E.A., Okhanov, V.V., Tuichibaev, M.U. and Tashmukhamedov, B.A. (1981) Bioorgh. Khim. 7, 494–501.
31 Howard, N.L. (1975) Toxicon 13, 21–30.
32 Botes, D.P. and Viljoen, C.C. (1974) Toxicon 12, 611–619.
33 Ferlan, I. and Gubenšek, F. (1978) Period. Biol. 80 (Suppl. 1), 31–36.
34 Alagón, A.C., Molinar, R.R., Possani, L.D., Fletcher, Jr., P.L., Cronan, Jr., J.E. and Juliá, J.Z. (1980) Biochem. J. 185, 695–704.
35 Vidal, J.C. and Stoppani, A.O.M. (1971) Arch. Biochem. Biophys. 147, 66–76.
36 Abe, T., Alemà, S. and Miledi, R. (1977) Eur. J. Biochem. 80, 1–12.
37 Moody, T.W. and Raftery, M.A. (1978) Arch. Biochem. Biophys. 189, 115–121.
38 Kondo, K., Toda, H., Narita, K. and Lee, Ch.Y. (1982) J. Biochem. 91, 1531–1548.
38a Kondo, K., Toda, H. and Narita, K. (1981) J. Biochem. 89, 37–47.
39 Hanley, M.R., Eterovic, V.A., Hawkes, S.P., Hebert, A.J. and Bennett, E.L. (1977) Biochemistry 16, 5840–5848.
40 Tobias, G.S., Donlon, M.A., Catravas, G.N. and Shain, W. (1978) Biochim. Biophys. Acta 537, 348–357.
41 Wernicke, J.F., Oberjat, T. and Howard, B.D. (1974) J. Neurochem. 22, 781–788.
41a Kondo, K., Toda, H. and Narita, K. (1981) J. Biochem. 89, 29–36.
42 Wells, M.A. and Hanahan, D.J. (1969) Biochemistry 8, 414–424.
43 Wu, T.W. and Tinker, D.O. (1969) Biochemistry 8, 1558–1568.
44 Hachimori, Y., Wells, M.A. and Hanahan, D.J. (1971) Biochemistry 10, 4084–4089.
45 Slotta, K.H. and Fraenkel-Conrat, H.L. (1938) Chem. Ber. 71, 1076–1081.
46 Habermann, E. and Breithaupt, H. (1978) Toxicon 16, 19–30.
47 Breithaupt, H., Rübsamen, K. and Habermann, E. (1974) Eur. J. Biochem. 49, 333–345.
48 Breithaupt, H., Omori-Satoh, T. and Lang, J. (1975) Biochim. Biophys. Acta 403, 355–369.
49 Cate, R.L. and Bieber, A.L. (1978) Arch. Biochem. Biophys. 189, 397–408.
50 Gopalakrishnakone, P., Hawgood, B.J., Theakston, R.D.G. and Reid, A.H. (1979) Toxicon 17 (Suppl. 1), 57.
51 Nair, B.C., Nair, C. and Elliott, W.B. (1979) Toxicon 17, 557–569.
52 Fohlman, J. and Eaker, D. (1977) Toxicon 15, 385–393.
53 Joubert. F.J. (1975) Eur. J. Biochem. 52, 539–554.
54 Yang, C.C. and King, K. (1980) Toxicon 18, 529–547.
55 Yoshida, H., Kudo, T., Shinkai, W. and Tamiya, N. (1979) J. Biochem. 85, 379–388.
56 Possani, L.D., Alagón, A.C., Fletcher, Jr., P.L., Varela, M.J. and Juliá, J.Z. (1979) Biochem. J. 179, 603–606.
57 Chang, W.C., Hsu, H.P. and Lo, T.B. (1976) Toxicon 14, 409–410.
58 Deems, R.A. and Dennis, E.A. (1981) Methods Enzymol. 71C, 703–710.
59 Andreasen, T.J., Doerge, D.R. and McNamee, M.G. (1979) Arch. Biochem. Biophys. 194, 468–480.
60 Karlsson, E. and Pongsawasdi, P. (1980) Toxicon 18, 409–419.
61 Joubert, F.J. and Van der Walt, S.J. (1975) Biochim. Biophys. Acta 379, 317–328.
62 Joubert, F.J. (1977) Biochim. Biophys. Acta 493, 216–227.
63 Martin-Moutot, N. and Rochat, H. (1979) Toxicon 17, 127–136.
64 Yang, C.C. and King, K. (1980) Biochim. Biophys. Acta 614, 373–388.
65 Evans, H.J., Franson, R.C., Qureshi, G.D. and Moo-Penn, W.F. (1980) J. Biol. Chem. 255, 3793–3797.
66 Halpert, J. and Eaker, D. (1975) J. Biol. Chem. 250, 6990–6997.
67 Halpert, J. and Eaker, D. (1976) J. Biol. Chem. 251, 7343–7347.
68 Halpert, J. and Eaker, D. (1976) FEBS Lett. 71, 91–95.
69 Fohlman, J., Eaker, D., Karlsson, E. and Thesleff, S. (1976) Eur. J. Biochem. 68, 457–469.
70 Fohlman, J. (1979) Toxicon 17, 170–172.
71 Leonardi, T.M., Howden, M.E.H. and Spence, I. (1979) Toxicon 17, 549–555.
71a Vaughan, G.T., Sculley, T.B. and Tirrell, R. (1981) Toxicon 19, 95–101.

72 Mebs, D. and Samejima, Y. (1980) Experientia 36, 868–869.
73 Ishimara, K., Kihara, H. and Ohno, M. (1980) J. Biochem. 88, 443–451.
73a Ouyang, C., Teng, C.M., Chen, Y.C. and Lin, S.C. (1978) Biochim. Biophys. Acta 541, 394–407.
73b Joubert, F.J. and Haylett, T. (1981) Z. Physiol. Chem. 362, 997–1006.
74 Aleksiev, B. and Shipolini, R. (1971) Z. Physiol. Chem. 352, 1183–1188.
75 Aleksiev, B. and Tchorbanov, B. (1976) Toxicon 14, 477–485.
76 Tchorbanov, B., Aleksiev, B., Bukolova-Orlova, T., Burstein, E. and Atanasov, B. (1977) FEBS Lett. 76, 266–268.
77 Šket, D., Gubenšek, F., Pakin, R. and Lebez, D. (1973) Toxicon 11, 193–196.
78 Boffa, G.A., Boffa, M.C., Zakin, M.M. and Burstein, M. (1971) Protides Biol. Fluids, Proc. Colloq. 19, 85–90.
79 Delori, P.J. (1973) Biochimie 55, 1031–1045.
80 Boffa, G.A., Boffa, M.C. and Winchenne, J.J. (1976) Biochim. Biophys. Acta 429, 828–838.
81 Shiloah, J., Klibansky, C., De Vries, A. and Berger, A. (1973) J. Lipid Res. 14, 267–278.
82 Simon, T., Bdolah, A. and Kochva, E. (1980) Toxicon 18, 249–259.
83 Simon, T. and Bdolah, A. (1980) Toxicon 18, 369–373.
84 Shipolini, R.A., Callewaert, G.L., Cottrell, R.C. and Vernon, C.A. (1974) Eur. J. Biochem. 48, 465–476.
85 Shipolini, R.A., Doonan, S. and Vernon, C.A. (1974) Eur. J. Biochem. 48, 477–483.
86 De Haas, G.H., Slotboom, A.J., Bonsen, P.P.M., Van Deenen, L.L.M., Maroux, S., Puigserver, A. and Desnuelle, P. (1970) Biochim. Biophys. Acta 221, 31–53.
87 De Haas, G.H., Slotboom, A.J., Bonsen, P.P.M., Nieuwenhuizen, W., Van Deenen, L.L.M., Maroux, S., Dlouha, V. and Desnuelle, P. (1970) Biochim. Biophys. Acta 221, 54–61.
88 Puijk, W.C., Verheij, H.M. and De Haas, G.H. (1977) Biochim. Biophys. Acta 492, 254–259.
89 Drenth, J., Enzing, C.M., Kalk, K.H. and Vessies, J.C.A. (1976) Nature 264, 373–377.
90 Dijkstra, B.W., Drenth, J., Kalk, K.H. and Vandermaelen, P.J. (1978) J. Mol. Biol. 124, 53–60.
90a Dijkstra, B.W., Kalk, K.H., Hol, W.G.J. and Drenth, J. (1981) J. Mol. Biol. 147, 97–123.
91 Heinrikson, R.L., Krueger, E.T. and Keim, P.S. (1977) J. Biol. Chem. 252, 4913–4921.
91a Keith, C., Feldman, D.S., Deganello, S., Glick, J., Ward, K.B., Jones, E.O. and Sigler, P.B. (1981) J. Biol. Chem. 256, 8602–8607.
92 Van Scharrenburg, G.J.M., De Haas, G.H. and Slotboom, A.J. (1980) Z. Physiol. Chem. 361, 571–576.
92a Sukhanov, V.A., Sidorov, O.Yu., Okhanov, V.V., Basharuli, V.A., Shvets, V.I. and Miroshnikov, A.I. (1981) Mol. Biol. (Moscow) 15, 139–144.
93 Dijkstra, B.W., Drenth, J. and Kalk, K.H. (1981) Nature 289, 604–606.
94 Lind, P. and Eaker, D. (1980) Eur. J. Biochem. 111, 403–409.
95 Nieuwenhuizen, W., Steenbergh, P. and De Haas, G.H. (1973) Eur. J. Biochem. 40, 1–7.
96 Abita, J.P., Lazdunski, M., Bonsen, P.P.M., Pieterson, W.A. and De Haas, G.H. (1972) Eur. J. Biochem. 30, 37–47.
97 Janssen, L.H.M., de Bruin, S.H. and De Haas, G.H. (1972) Eur. J. Biochem. 28, 156–160.
98 Jansen, E.H.J.M. (1979) Ph.D. Thesis, State University of Utrecht, The Netherlands.
99 De Araujo, P.S., Rosseneu, M.Y., Kremer, J.M.H., Van Zoelen, E.J.J. and De Haas, G.H. (1979) Biochemistry 18, 580–586.
100 Deems, R.A. and Dennis, E.A. (1975) J. Biol. Chem. 250, 9008–9012.
101 Mal'tsev, V.G., Zimina, T.M., Kurenbin, O.I., Belen'kii, B.G., Aleksandrov, S.L., Pavlova, N.P., Dyakov, V.L. and Antonov, V.K. (1979) Bioorgh. Khim. 5, 1710–1719.
102 Brockerhoff, H. and Jensen, R.G. (1974) in Lipolytic Enzymes, Academic Press, New York.
103 Verger, R. and De Haas, G.H. (1976) Annu. Rev. Biophys. Bioeng. 5, 77–117.
104 Sémériva, M. and Desnuelle, P. (1978) Adv. Enzymol. 46, 319–370.
105 Verger, R. (1980) Methods Enzymol. 64B, 340–392.
106 Wells, M.A. (1972) Biochemistry 11, 1030–1041.
106a Cleland, W.W. (1963) Biochim. Biophys. Acta 67, 104–137.

107 Wells, M.A. (1974) Biochemistry 13, 2265–2268.
108 Zhelkovskii, A.M., Dyakov, V.L. and Antonov, V.K. (1978) Bioorgh. Khim. 4, 1665–1672.
109 Viljoen, C.C. and Botes, D.P. (1979) Toxicon 17, 77–87.
110 De Haas, G.H., Bonsen, P.P.M., Pieterson, W.A. and Van Deenen, L.L.M. (1971) Biochim. Biophys. Acta 239, 252–266.
111 Pieterson, W.A., Vidal, J.C., Volwerk, J.J. and De Haas, G.H. (1974) Biochemistry 13, 1455–1460.
112 Pieterson, W.A., Volwerk, J.J. and De Haas, G.H. (1974) Biochemistry 13, 1439–1445.
113 Wells, M.A. (1974) Biochemistry 13, 2248–2257.
114 Kensil, C.R. and Dennis, E.A. (1979) J. Biol. Chem. 254, 5843–5848.
115 Volwerk, J.J., Pieterson, W.A. and De Haas, G.H. (1974) Biochemistry 13, 1446–1454.
116 Verger, R., Mieras, M.C.E. and De Haas, G.H. (1973) J. Biol. Chem. 248, 4023–4034.
117 Hershberg, R.D., Reed, G.H., Slotboom, A.J. and De Haas, G.H. (1976) Biochim. Biophys. Acta 424, 73–81.
118 Hershberg, R.D., Reed, G.H., Slotboom, A.J. and De Haas, G.H. (1976) Biochemistry 15, 2268–2274.
119 Roberts, M.F., Deems, R.A. and Dennis, E.A. (1977) Proc. Natl. Acad. Sci. USA 74, 1950–1954.
120 Brockerhoff, H. (1973) Chem. Phys. Lipids 10, 215–222.
121 Allgyer, T.T. and Wells, M.A. (1979) Biochemistry 18, 4354–4361.
122 Brockerhoff, H. (1968) Biochim. Biophys. Acta 159, 296–303.
123 Schmidt, C.F., Barenholz, Y., Huang, C. and Thompson, T.E. (1977) Biochemistry 16, 3948–3954.
124 Misiorowski, R.L. and Wells, M.A. (1974) Biochemistry 13, 4921–4927.
125 Poon, P.H. and Wells, M.A. (1974) Biochemistry 13, 4928–4936.
126 Wells, M.A. (1974) Biochemistry 13, 4937–4942.
126a Johnson, R.E., Wells, M.A. and Rupley, J.A. (1981) Biochemistry 20, 4239–4242.
127 Wells, M.A. (1978) in C. Galli, G. Galli and G. Porcelatti (Eds.), Advances in Prostaglandin and Thromboxane Research, Vol. 3, Raven Press, New York, pp. 39–45.
128 Roberts, M.F., Bothner-By, A.A. and Dennis, E.A. (1978) Biochemistry 17, 935–941.
129 Burns, R.A. and Roberts, M.F. (1980) Biochemistry 19, 3100–3106.
129a Plückthun, A. and Dennis, E.A. (1981) J. Phys. Chem. 85, 678–683.
130 Dennis, E.A. (1973) J. Lipid Res. 14, 152–159.
131 Dennis, E.A. (1973) Arch. Biochem. Biophys. 158, 485–493.
132 Dennis, E.A. (1974) J. Supramol. Struct. 2, 682–699.
133 Deems, R.A., Eaton, B.R. and Dennis, E.A. (1975) J. Biol. Chem. 150, 9013–9020.
134 Roberts, M.F., Otnaess, A.B., Kensil, C.A. and Dennis, E.A. (1978) J. Biol. Chem. 253, 1252–1257.
135 Robson, R.J. and Dennis, E.A. (1979) Biochim. Biophys. Acta 573, 489–500.
136 Dennis, E.A. (1974) Arch. Biochem. Biophys. 165, 764–773.
137 Adamich, M., Roberts, M.F. and Dennis, E.A. (1979) Biochemistry 18, 3308–3314.
138 Roberts, M.F., Deems, R.A., Mincey, T.C. and Dennis, E.A. (1977) J. Biol. Chem. 252, 2405–2411.
139 Darke, P.L., Jarvis, A.A., Deems, R.A. and Dennis, E.A. (1980) Biochim. Biophys. Acta 626, 154–161.
140 Adamich, M. and Dennis, E.A. (1978) Biochem. Biophys. Res. Commun. 80, 424–428.
141 Adamich, M., Roberts, M.F. and Dennis, E.A. (1979) Biochemistry 18, 3308–3314.
142 Van Deenen, L.L.M. and De Haas, G.H. (1963) Biochim. Biophys. Acta 70, 538–553.
143 De Haas, G.H., Bonsen, P.P.M. and Van Deenen, L.L.M. (1966) Biochim. Biophys. Acta 116, 114–124.
144 Pattus, F., Slotboom. A.J. and De Haas, G.H. (1979) Biochemistry 18, 2691–2697.
145 Pattus, F., Slotboom, A.J. and De Haas, G.H. (1979) Biochemistry 18, 2698–2702.
146 Pattus, F., Slotboom, A.J. and De Haas, G.H. (1979) Biochemistry 18, 2703–2707.
147 Dervichian, D.G. and Barque, J.P. (1979) J. Lipid Res. 20, 437–446.
148 Barque, J.P. and Dervichian, D.G. (1979) J. Lipid Res. 20, 447–455.
149 Barque, J.P. and Dervichian, D.G. (1979) J. Lipid Res. 20, 599–606.
149a Momsen, W.E. and Brockman, H.L. (1981) J. Biol. Chem. 256, 6913–6916.

150 Willman, C. and Stewart-Hendrickson, H. (1978) Arch. Biochem. Biophys. 191, 298–305.
151 Verger, R. and De Haas, G.H. (1973) Chem. Phys. Lipids 10, 127–136.
152 Pieroni, G. and Verger, R. (1979) J. Biol. Chem. 254, 10090–10094.
153 Pieroni, G. and Verger, R. (1983) Eur. J. Biochem. to be published.
153a Burns, Jr., R.A. and Roberts, M.F. (1981) J. Biol. Chem. 256, 2716–2722.
154 Barenholz, Y., Pieroni, G. and Verger, R. (1982) Biochim. Biophys. Acta, in press.
155 Untracht, S.H. and Shipley, G.G. (1977) J. Biol. Chem. 252, 4449–4457.
156 Verheij, H.M., Boffa, M.C., Rothen, Ch., Bryckaert, M.C., Verger, R. and De Haas, G.H. (1980) Eur. J. Biochem. 112, 25–32.
157 Boffa, M.C., Rothen, Ch., Verheij, H.M., Verger, R. and De Haas, G.H. (1980) in D. Eaker and T. Wadström (Eds.), Natl. Toxins, Proc. 6th Int. Symp. on Animal, Plant and Microbial Toxins, pp. 131–138.
158 Viljoen, C.C., Schabort, J.C. and Botes, D.P. (1974) Biochim. Biophys. Acta 360, 156–165.
159 Tanford, C. (1976) in The Hydrophobic Effect: Formation of Micelles and Biological Membranes, J. Wiley, New York.
160 Van Deenen, L.L.M., De Haas, G.H. and Heemskerk, C.H.Th. (1963) Biochim. Biophys. Acta 67, 295–304.
161 De Haas, G.H., Postema, N.M., Nieuwenhuizen, W. and Van Deenen, L.L.M. (1968) Biochim. Biophys. Acta 159, 103–117.
162 Op den Kamp, J.A.F., De Gier, J. and Van Deenen, L.L.M. (1974) Biochim. Biophys. Acta 345, 253–256.
163 Op den Kamp, J.A.F., Kauerz, M.Th. and Van Deenen, L.L.M. (1975) Biochim. Biophys. Acta 406, 169–177.
164 Strong, P.N. and Kelly, R.B. (1977) Biochim. Acta 469, 231–235.
165 Kainagi, R. and Koizumi, K. (1979) Biochim. Biophys. Acta 556, 423–433.
165a Goormaghtigh, E., Van Campenhoud, M. and Ruysschaert, J.M. (1981) Biochem. Biophys. Res. Commun. 101, 1410–1418.
166 Wilschut, J.C., Regts, J., Westenberg, H. and Scherphof, G. (1976) Biochim. Biophys. Acta 433, 20–31.
167 Wilschut, J.C., Regts, J., Westenberg, H. and Scherphof, G. (1978) Biochim. Biophys. Acta 508, 185–196.
168 Jain, M.K. and Cordes, E.H. (1973) J. Membr. Biol. 14, 101–118.
169 Jain, M.K. and Cordes, E.H. (1973) J. Membr. Biol. 14, 119–134.
170 Upreti, G.C. and Jain, M.K. (1978) Arch. Biochem. Biophys. 188, 364–375.
171 Jain, M.K. and Apitz-Castro, R.C. (1978) J. Biol. Chem. 253, 7005–7010.
172 Upreti, G.C., Rainier, S. and Jain, M.K. (1980) J. Membr. Biol. 55, 97–112.
173 Bonsen, P.P.M., De Haas, G.H., Pieterson, W.A. and Van Deenen, L.L.M. (1972) Biochim. Biophys. Acta 270, 364–382.
174 Tinker, D.O., Purdon, A.D., Wei, J. and Mason, E. (1978) Can. J. Biochem. 56, 552–558.
175 Tinker, D.O. and Wei, J. (1979) Can. J. Biochem. 57, 97–106.
176 Upreti, G.C. and Jain, M.K. (1980) J. Membr. Biol. 55, 113–121.
177 Szoka, Jr., F. and Papahadjopoulos, D. (1980) Annu. Rev. Biophys. Bioeng. 9, 467–508.
178 Tinker, D.O., Law, R. and Lucassen, M. (1980) Can. J. Biochem. 58, 898–912.
179 Jain, M.K., Van Echteld, C.J.A., Ramirez, F., De Gier, J., De Haas, G.H. and Van Deenen, L.L.M. (1980) Nature 284, 486–487.
180 Jain, M.K. and De Haas, G.H. (1981) Biochim. Biophys. Acta 642, 203–211.
181 Jain, M.K., Egmond, M.R., Verheij, H.M., Apitz-Castro, R., Dijkman, R. and De Haas, G.H. (1982) Biochim. Biophys. Acta, 341–348.
182 Bevers, E.M., Singal, S.A. and Op den Kamp, J.A.F. (1977) Biochemistry 16, 1290–1295.
183 Bevers, E.M., Op den Kamp, J.A.F. and Van Deenen, L.L.M. (1978) Eur. J. Biochem. 84, 35–42.
184 Bouvier, P., Op den Kamp, J.A.F. and Van Deenen, L.L.M. (1981) Arch. Biochem. Biophys. 208, 242–247.

185 Menashe, M., Lichtenberg, D., Guttierrez-Merino, C. and Biltonen, R.L. (1981) J. Biol. Chem. 256, 4541–4543.
185a Kupferberg, J.P., Yokoyama, S. and Kézdy, F.J. (1981) J. Biol. Chem. 256, 6274–6281.
186 Smith, A.D., Gul, S. and Thompson, R.H.S. (1972) Biochim. Biophys. Acta 289, 147–157.
187 Gatt, S. and Bartfai, T. (1977) Biochim. Biophys. Acta 488, 1–12.
188 Gatt, S. and Bartfai, T. (1977) Biochim. Biophys. Acta 488, 13–24.
189 Verheij, H.M., Egmond, M.R. and De Haas, G.H. (1981) Biochemistry 20, 94–99.
190 Prigent-Dachary, J., Boffa, M.C., Boisseau, M.R. and Dufourq, J. (1980) J. Biol. Chem. 255, 7734–7739.
191 Drainas, D. and Lawrence, A.J. (1978) Eur J. Biochem. 91, 131–138.
192 Rosenthal, A.F. and Ching-Hsien Han, S. (1970) Biochim. Biophys. Acta 218, 213–220.
193 Wells, M.A. (1973) Biochemistry 12, 1086–1093.
194 Smith, C.M. and Wells, M.A. (1981) Biochim. Biophys. Acta 663, 687–694.
195 Zhelkovskii, A.M., Dyakov, V.D., Ginodman, L.M. and Antonov, V.K. (1978) Bioorgh. Khim. 4, 1665–1672.
195a Hille, J.D.R. et al., to be published.
196 Hille, J.D.R., Donné-Op den Kelder, G.M., Sauve, P., De Haas, G.H. and Egmond, M.R. (1981) Biochemistry 20, 4068–4073.
197 Donné-Op den Kelder, G.M., Hille, J.D.R., Dijkman, R., De Haas, G.H. and Egmond, M.R. (1981) Biochemistry 20, 4074–4078.
198 Jeng, T.W. and Fraenkel-Conrat, H. (1978) FEBS Lett. 87, 291–296.
199 Verheij, H.M., Volwerk, J.J., Jansen, E.H.J.M., Puijk, W.C., Dijkstra, B.W., Drenth, J. and De Haas, G.H. (1980) Biochemistry 19, 743–750.
200 Pieterson, W.A. (1973) Ph.D. Thesis, State University of Utrecht.
201 Dutilh, C.E. (1976) Ph.D. Thesis, State University of Utrecht.
202 Halpert, J., Eaker, D. and Karlsson, E. (1976) FEBS Lett. 61, 72–76.
203 Kondo, K., Toda, H. and Narita, K. (1978) J. Biochem. 84, 1291–1300.
204 Kondo, K., Toda, H. and Narita, K. (1978) J. Biochem. 84, 1301–1308.
205 Viljoen, C.C., Visser, L. and Botes, D.P. (1977) Biochim. Biophys. Acta 483, 107–120.
205a Van Eijck, J., Verheij, H.M. and De Haas, G.H. (1982) to be published.
205b Van Scharrenburg, G.J.M., Puijk, W.C., Egmond, M.R., Van der Schaft, P.H., De Haas, G.H. and Slotboom, A.J. (1982) Biochemistry 21, 1345–1352.
205c Apsalon, U.R., Aianyan, A.E., Meshcheryakova, E.A., Surina, E.A., Miroshnikov, A.I., Gotgil'f, I.M. and Magazanik, L.G. (1980) Bioorgh. Khim. 6, 1068–1078.
206 Magazanik, L.G., Gotgil'f, I.M., Slavnova, T.I., Miroshnikov, A.I. and Apsalon, U.R. (1979) Toxicon 17, 477–488.
206a Kihara, H., Ishimaru, K. and Ohno, M. (1981) J. Biochem. 90, 363–370.
207 Fohlman, J., Eaker, D., Dowdall, M.J., Lüllmann-Rauch, R., Sjödin, T. and Leander, S. (1979) Eur. J. Biochem. 94, 531–540.
208 Eaker, D. (1978) in Ch. H. Li (Ed.), Versatility of Proteins, Academic Press, New York, pp. 413–431.
208a Menashé, M., Rochat, H. and Zlotkin, E. (1981) Insect Biochem. 11, 137–142.
208b Condrea, E., Fletcher, J.E., Rapuano, B.E., Yang, C.-C. and Rosenberg, P. (1981) Toxicon 19, 61–71.
208c Ouyang, C., Jy, W., Zan, Y.-P. and Teng, C.-M. (1981) Toxicon 19, 113–120.
209 Slotboom, A.J., Verger, R., Verheij, H.M., Baartmans, P.H.M., Van Deenen, L.L.M. and De Haas, G.H. (1976) Chem. Phys. Lipids 17, 128–147.
210 Mebs, D. and Samejima, Y. (1980) Toxicon 18, 443–454.
211 Wells, M.A. (1973) Biochemistry 12, 1080–1085.
212 Viljoen, C.C., Visser, L. and Botes, D.P. (1976) Biochim. Biophys. Acta 438, 424–436.
213 Slotboom, A.J. and De Haas, G.H. (1975) Biochemistry 14, 5394–5399.
214 Joubert, F.J. and Taljaard, N. (1980) Eur. J. Biochem. 112, 493–499.

215 Roberts, M.F., Deems, R.A. and Dennis, E.A. (1977) J. Biol. Chem. 252, 6011–6017.
216 Van Wezel, F.M., Slotboom, A.J. and De Haas, G.H. (1976) Biochim. Biophys. Acta 452, 101–111.
217 Meyer, H. (1979) Ph.D. Thesis, State University of Utrecht.
218 Dijkstra, B.W. (1980) Ph.D. Thesis, State University of Groningen.
219 MacDermot, J., Westgaard, R.H. and Thompson, E.J. (1978) Biochem. J. 175, 281–288.
220 Zhelkovsky, A.M., Apsalon, U.R., Dyakov, V.L., Ginodman, L.M., Miroshnikov, A.I. and Antonov, V.K. (1977) Bioorg. Khim. 3, 1430–1432.
221 Fleer, E.A.M., Verheij, H.M. and De Haas, G.H. (1981) Eur. J. Biochem. 113, 283–288.
221a Dinur, D., Kantrowitz, E.R. and Hajdu, J. (1981) Biochem. Biophys. Res. Commun. 100, 785–792.
222 Vensel, L.A. and Kantrowitz, E.R. (1980) J. Biol. Chem. 255, 7306–7310.
223 Takahashi, K. (1968) J. Biol. Chem. 243, 6171–6179.
224 Fleer, E.A.M., Puijk, W.C., Slotboom, A.J. and De Haas, G.H. (1981) Eur. J. Biochem. 116, 277–284.
224a Apsalon, U.R. and Miroshnikov, A.I. (1980) Bioorgh. Khim. 6, 773–779.
225 Dixon, H.B.F. and Fields, R. (1972) Methods Enzymol. 25B, 409–419.
226 Egmond, M.R., Slotboom, A.J., De Haas, G.H., Dijkstra, K. and Kaptein, R. (1980) Biochim. Biophys. Acta 623, 461–466.
227 Meyer, H., Verhoef, H., Hendriks, F.F.A., Slotboom, A.J. and De Haas, G.H. (1979) Biochemistry 18, 3582–3588.
228 Meyer, H., Puijk, W.C., Dijkman, R., Foda-Van der Hoorn, M.M.E.L., Pattus, F., Slotboom, A.J. and De Haas, G.H. (1979) Biochemistry 18, 3589–3597.
229 Jansen, E.H.J.M., Meyer, H., De Haas, G.H. and Kaptein, R. (1978) J. Biol. Chem. 253, 6346–6347.
230 Slotboom, A.J., Verheij, H.M., Puijk, W.C., Dedieu, A.G.R. and De Haas, G.H. (1978) FEBS Lett. 92, 361–364.
231 Bon, C., Changeux, J.P., Jeng, T.W. and Fraenkel-Conrat, H. (1979) Eur. J. Biochem. 99, 471–481.
232 Drainas, D., Moores, G.R. and Lawrence, A.J. (1978) FEBS Lett. 86, 49–52.
233 Lawrence, A.J. and Moores, G.R. (1975) FEBS Lett. 49, 287–291.
234 Lawrence, A.J. (1975) FEBS Lett. 58, 186–189.
235 Larroquère, J. (1964) Bull. Soc. Chim. Fr. 1543–1551.
236 Melchior, Jr. W.B. and Fahrney, D. (1970) Biochemistry 9, 251–258.
237 Mühlràd, A., Hegyi, G. and Toth, G. (1967) Acta Biochim. Biophys. Acad. Sci. Hung. 2, 19–29.
238 Burstein, Y., Walsh, K.A. and Neurath, H. (1974) Biochemistry 13, 205–210.
239 Howard, B.D. and Truog, R. (1977) Biochemistry 16, 122–125.
240 Ng, R.H. and Howard, B.D. (1978) Biochemistry 17, 4978–4986.
241 Lewis, R.V., Roberts, M.F., Dennis, E.A. and Allison, W.S. (1977) Biochemistry 16, 5650–5654.
242 Hendon, R.A. and Tu, A.T. (1979) Biochim. Biophys. Acta 578, 243–252.
243 Huang, K.-S. and Law, J.H. (1978) in S. Gatt, L. Freysz and P. Mandel (Eds.), Adv. Exp. Med. Biol. Enzymes of Lipid Metabolism, Vol. 101, Plenum Press, New York, pp. 177–183.
243a Huang, K.-S. and Law, J.H. (1981) Biochemistry 20, 181–187.
244 Van Dam-Mieras, M.C.E., Slotboom, A.J., Pieterson, W.A. and De Haas, G.H. (1975) Biochemistry 14, 5387–5394.
245 Slotboom, A.J., Jansen, E.H.J.M., Pattus, F. and De Haas, G.H. (1978) in R.E. Offord and C. Dibello (Eds.), Semisynthetic Peptides and Proteins, Academic Press, London, pp. 315–349.
246 Slotboom, A.J., Van Dam-Mieras, M.C.E. and De Haas, G. (1977) J. Biol. Chem. 252, 2948–2951.
247 Slotboom, A.J., Jansen, E.H.J.M., Vlijm, H., Pattus, F., Soares de Araujo, P. and De Haas, G.H. (1978) Biochemistry 17, 4593–4600.
248 Van Scharrenburg, G.J.M., Puijk, W.C., Egmond, M.R., De Haas, G.H. and Slotboom, A.J. (1981) Biochemistry 20, 1584–1591.
249 Evenberg, A., Meyer, H., Gaastra, W., Verheij, H.M. and De Haas, G.H. (1977) J. Biol. Chem. 252, 1189–1196.
250 Fleer, E.A.M., Verheij, H.M. and De Haas, G.H. (1978) Eur. J. Biochem. 82, 261–269.
251 Puijk, W.C., Verheij, H.M., Wietzes, P. and De Haas, G.H. (1979) Biochim. Biophys. Acta 580, 411–415.

252 Lind, P. and Eaker, D. (1981) Toxicon 19, 11–24.
253 Joubert, F.J. (1975) Biochim. Biophys. Acta 379, 345–359.
254 Joubert, F.J. (1975) Biochim. Biophys. Acta 379, 329–344.
255 Ovchinnikov, Yu.A., Miroshnikov, A.I., Nazimov, I.V., Apsalon, U.R. and Soldatova, L.N. (1979) Bioorgh. Khim. 5, 805–813.
256 Tsai, T.H., Wu, S.H. and Lo, T.B. (1981) Toxicon 19, 141–152.
257 Lind, P. and Eaker, D. (1982) Eur. J. Biochem. 124, 441–447.
258 Botes, D.P. and Viljoen, C.C. (1974) J. Biol. Chem. 249, 3827–3837.
259 Randolph, A., Sakmar, T.P. and Heinrikson, R.L. (1980) in Frontiers in Protein Chemistry, Elsevier, pp. 297–322.
260 Fraenkel-Conrat, H., Jeng, T.W. and Hsiang, M. (1980) in D. Eaker and T. Wadström (Eds.), Natural Toxins, Proceedings of the 6th International Symposium on Animal, Plant and Microbial Toxins.
261 Samejima, Y., Iwanaga, S. and Suzuki, T. (1974) FEBS Lett. 47, 348–350.
262 IUPAC-IUB Commission on Biochemical Nomenclature, Eur. J. Biochem. 5 (1968) 151–153.
263 Aguiar, A., De Haas, G.H., Jansen, E.H.J.M., Slotboom, A.J. and Williams, R.J.P. (1979) Eur. J. Biochem. 100, 511–518.
264 Brittain, H.G., Richardson, F.S. and Martin, R.B. (1976) J. Am. Chem. Soc. 98, 8255–8260.
265 Purdon, A.D., Tinker, D.O. and Spero, L. (1977) Can. J. Biochem. 55, 205–214.
266 Viljoen, C.C., Botes, D.P. and Schabort, J.C. (1975) Toxicon 13, 343–351.
266a Teshima, K., Ikeda, K., Hamaguchi, K. and Hayashi, K. (1981) J. Biochem. 89, 13–20.
266b Ikeda, K. and Samejima, Y. (1981) J. Biochem. 89, 1175–1184.
266c Teshima, K., Ikeda, K., Hamaguchi, K. and Hayashi, K. (1981) J. Biochem. 89, 1163–1174.
267 Robinson, N.C. and Tanford, C. (1975) Biochemistry 14, 369–378.
268 Makino, S., Reynolds, J.A. and Tanford, C. (1973) J. Biol. Chem. 248, 4926–4932.
269 Haberland, M.E. and Reynolds, J.A. (1975) J. Biol. Chem. 250, 6636–6639.
270 Rosseneu, M.Y., Soetewij, F., Middelhoff, G., Peeters, H. and Brown, W.V. (1976) Biochim. Biophys. Acta 441, 68–80.
271 Reynolds, J.A. and Tanford, C. (1976) Proc. Natl. Acad. Sci. USA 73, 4467–4470.
272 Meyer, H., Meddens, M.J.M., Dijkman, R., Slotboom, A.J. and De Haas, G.H. (1978) J. Biol. Chem. 253, 8564–8569.
273 Pasek, M., Keith, C., Feldman, D. and Sigler, P.B. (1975) J. Mol. Biol. 97, 395–397.
274 Kannan, K.K., Lövgren, S., Cid-Dresdner, H., Petef, M. and Eaker, D. (1977) Toxicon 15, 435–439.
274a Wang, A.H.J. and Yang, C.C. (1981) J. Biol. Chem. 256, 9279–9282.
275 Kraut, J. (1977) Annu. Rev. Biochem. 46, 331–358.
276 Komiyama, M. and Bender, M.L. (1979) Proc. Natl. Acad. Sci. USA 76, 557–560.
277 Andersson, T., Drakenberg, T., Forsén, S., Wieloch, T. and Lindström, M. (1981) FEBS Lett. 123, 115–117.
278 Hauser, H., Pascher, I., Pearson, R.H. and Sundell, S. (1981) Biochim. Biophys. Acta 650, 21–51.
279 Büldt, G. and Wohlgemuth, R. (1981) J. Membr. Biol. 58, 81–100.

Abbreviations

PLA: phospholipase A_2 (EC 3.1.1.4)
pro-PLA: prophospholipase A_2
AMPA: ε-amidinated phospholipase A_2
des-Ala-1-AMPA: ε-amidinated phospholipase A_2 from which the N-terminal Ala-1 has been removed.
PL: phospholipid
FA: fatty acid

PC (phosphatidylcholine, L-lecithin, di-C_n-PC, 1,2-diacyllecithin, *sn*-3-lecithin): 1,2-diacyl-*sn*-glycero-3-phosphocholine
D-Lecithin (D-diC_n-PC, *sn*-1-lecithin): 2,3-diacyl-*sn*-glycero-1-phosphocholine
β-Lecithin (*sn*-2-lecithin): 1,3-diacyl-*sn*-glycero-2-phosphocholine
Lysolecithin (lyso-PC, 1-acyllysolecithin): 1-acyl-*sn*-glycero-3-phosphocholine
DMPC: 1,2-dimyristoyl-*sn*-glycero-3-phosphocholine
DPPC: 1,2-dipalmitoyl-*sn*-glycero-3-phosphocholine
DiC_n ether PC: 1,2-dialkyl-*rac*-glycero-3-phosphocholine
PE: 1,2-diacyl-*sn*-glycero-3-phosphoethanolamine
PS: 1,2-diacyl-*sn*-glycero-3-phospho-L-serine
PG: 1,2-diacyl-*sn*-glycero-3-phospho-1′-glycerol
ANS: 1-anilinonaphthalene-8-sulphonic acid
Boc: *t*-butyloxycarbonyl
BPB: *p*-bromophenacyl bromide
CNBr: cyanogen bromide
Dansyl: 5-(dimethylamino)naphthalene-1-sulphonyl
DBE: *N*-diazoacetyl-*N′*-(2,4-dinitrophenyl)ethylenediamine
EDC: 1-ethyl-3-(*N*,*N*-dimethyl)aminopropyl carbodiimide
EDTA: ethylene diamine tetracetic acid
EOFA: ethoxyformic acid anhydride
HNB: 2-hydroxy-5-nitrobenzylbromide
NBS: *N*-bromosuccinimide
NPC, *o*-nitrophenylsulphenylchloride
NPS: *o*-nitrophenylsuccinimide
TNM: tetranitromethane
CTAB: cetyl trimethylammonium bromide
SDS: sodium dodecylsulphate
Triton X-100: *p*-(1,1,3,3-tetramethylbutyl)phenoxypolyoxyethylene glycol
Tween: polyoxyethylenesorbitol fatty acid ester
CMC: critical micelle concentration
IRS: interface recognition site
CD circular dichroism
NMR: nuclear magnetic resonance
Photo-CIDNP: photochemically-induced dynamic nuclear polarisation
PRR: proton relaxation rate
IEP: iso-electric point

CHAPTER 11

Genetic control of phospholipid bilayer assembly

CHRISTIAN R.H. RAETZ

Department of Biochemistry, College of Agricultural and Life Sciences, University of Wisconsin-Madison, Madison, WI 53706, U.S.A.

1. Introduction

As documented throughout this volume, all biological membranes contain a great diversity of lipid substances. Prokaryotic membranes, as illustrated by *Escherichia coli* consist mainly of phospholipids, such as phosphatidylethanolamine, phosphatidylglycerol and cardiolipin [1,2]. Eukaryotic systems are characterized by the additional presence of sterols, sphingolipids, plasmalogens and an abundance of choline and inositol-linked glycerophospholipids [3–6]. Considering only the phospholipids, the membranes of *E. coli* contain about ten major molecular species (i.e. chemically distinct combinations of polar headgroups and fatty acids), while eukaryotic systems possess approx. ten times as many [1–6]. If minor phospholipids and metabolic intermediates are also counted, then prokaryotic membranes contain about 100 phospholipid structures, whereas eukaryotic membranes have about 1000 [1–6].

All growing cells possess enzyme systems capable of generating a certain degree of lipid heterogeneity [1–6]. In bacteria, this occurs on the inner surface of the cytoplasmic membrane [1,2], while in eukaryotic systems it takes place largely on the cytoplasmic face of the endoplasmic reticulum [3,3a,6]. As a rule, the enzymes of phospholipid biogenesis are minor integral membrane proteins, although their proper functioning is critical to membrane assembly [1,3]. Relatively few of the phospholipid enzymes have been purified to homogeneity and studied chemically, but considerable progress toward this goal has been made over the past decade [1,3].

The metabolic control and biological significance of lipid heterogeneity are not well understood. Several fundamental questions remain unresolved in this area (and in all organisms). These include the following: (1) What mechanisms regulate (or set) the total membrane phospholipid content of cells? (2) What regulates the ratios of polar headgroups and fatty acid species? (3) What determines or sets the cellular level of each of the phospholipid enzymes? (4) What coordinates phospholipid synthesis with membrane protein and macromolecular syntheses? (5) By what mechanisms do phospholipids move from one side of a membrane to the other or between two membranes, especially from a membrane which can generate its own phospholipids to another that cannot? (6) What are the functions of the individual phospholipid species?

Hawthorne/Ansell (eds.) Phospholipids

The isolation of biochemically defined mutants altered in phospholipid biogenesis represents a relatively unexploited [1–3] and powerful empirical approach to this area. In addition to providing new information that could lead to the solution of the problems outlined above, the genetic dissection of phospholipid synthesis is essential for the following: (1) demonstration that biosynthetic and catabolic pathways deduced from studies in vitro are physiologically significant; (2) elucidation of the function of ancillary enzymes, illustrated by studies of diacylglycerol kinase of *E. coli,* which could not be explained by enzymatic studies alone [1,7,8]; (3) identification of the genes controlling and coding for the enzymes of lipid bilayer assembly, permitting the use of gene cloning techniques for enzyme overproduction, gene isolation and DNA sequencing [1,9–12]; (4) modification in vivo of membrane lipid composition, facilitating studies of membrane biogenesis and function [1,2,13–15].

In this chapter we will examine recent progress with the isolation of *biochemically defined* mutants altered in phospholipid biogenesis, especially with the use of enzyme-specific autoradiography of colony preparations immobilized on filter paper [16–18]. Most of the existing mutants have been isolated from the bacterium *E. coli* [1], although recent extensions of the colony autoradiography techniques to mammalian cells grown in tissue culture [17,18] will also be considered. In addition, some choline and inositol auxotrophs of lower eukaryotes have been available for many years, and these have recently received renewed attention as vehicles for membrane lipid modification [19–22].

2. *Approaches to the isolation of* Escherichia coli *mutants defective in phospholipid metabolism*

Amongst the prokaryotes, *E. coli* is the organism of choice for most genetic studies, since approx. 1000 genes out of about 5000 have now been identified [23]. *E. coli* genes are especially easy to isolate using the recently developed techniques of molecular cloning, especially if a mutant defective in the gene of interest is already available [24,25]. Since the nucleotide sequence of isolated DNA fragments can be determined very rapidly [26,27], it is likely that the sequence of the entire *E. coli* chromosome will be known by the end of the decade.

E. coli is the best characterized prokaryote with respect to its membrane lipid biogenesis [1,2]. Hence, most current genetic studies have utilized this system. However, the general approaches to the isolation of lipid mutants, outlined below, are not limited to *E. coli.*

(a) Isolation of auxotrophs and supplementation of phospholipids by fusion

A classical approach to obtaining defined mutants in a metabolic pathway involves the isolation of auxotrophs requiring certain end products or intermediates for growth [28]. In the case of *E. coli* phospholipid metabolism, it is possible to feed cells early precursors, such as fatty acids (reviewed in [13–15]) or glycerol-3-phosphate

[1,2]. Mutants requiring these substances have been extensively characterized. In contrast, supplementation of intact phospholipids has not yielded the desired mutants in the late steps of phospholipid metabolism [1,2]. Two reasons for this are: (1) fusion of exogenous phospholipid vesicles with membranes of growing Gram-negative bacteria is relatively inefficient and occurs only to a limited extent in "deep rough" mutants that are partially defective in their lipopolysaccharide [29–31]; and (2) there are a variety of endogenous phospholipases and lysophospholipases in bacteria which are potentially capable of degrading phospholipid molecules [1]. Consequently, no one has isolated auxotrophs of *E. coli* (or other bacteria) requiring intact lipids, although methods for this may eventually be developed.

(b) Analogs or inhibitors of metabolism

The genetic dissection of DNA, RNA and protein synthesis has been aided considerably by the use of specific inhibitors, many of which are antibiotics. For instance, the drug rifampicin inhibits DNA-dependent RNA polymerase of *E. coli,* and mutants resistant to this drug provided the first clues to the location of genes coding for RNA polymerase [23,28]. Similarly, protein synthesis inhibitors such as chloramphenicol and streptomycin act on defined ribosomal proteins of *E. coli* [23,28]. Comparable specific inhibitors do not exist for probing the synthesis of membrane phospholipids. Nevertheless, such compounds may eventually be discovered, particularly since existing genctic evidence demonstrates that certain phospholipids are essential for cell growth [1,2]. The recent discovery of a compound (globomycin) that blocks the processing of the outer membrane lipoprotein of *E. coli* illustrates the potential for this approach [32].

Certain analogs of lipid precursors do exist, which could provide an avenue for the isolation of mutants. A methylene analog of glycerol-3-phosphate (3,4-dihydroxy-butyl-1-phosphonate) developed by Tropp and collaborators [33–36] is an effective false substrate for phosphatidylglycerophosphate synthase in *E. coli.* Since this is not its sole site of action, mutants with defined metabolic lesions have not been obtained.

(c) Radiation suicide

Exposure of microbial cells to certain tritiated metabolites (amino acids, sugars, nucleosides) results in the uptake and incorporation of these radioactive substances into macromolecules [28]. Storage of cells treated in this manner leads to a loss of viability because of radiation damage. Mutants unable to incorporate the labeled precursors may survive preferentially. This approach has been especially successful for enriching some kinds of mutants defective in protein synthesis [28].

The possibility of isolating mutants in phospholipid synthesis by radiation suicide was first investigated by Cronan et al. [37]. [2-^{3}H]Glycerol-3-phosphate was utilized as the suicide reagent in the hope of finding *E. coli* mutants blocked in the acylation of glycerol-3-phosphate [37]. Despite the identification of many temperature-sensitive organisms amongst the surviving cells [37,38], mutants with definitely characterized lipid lesions were not obtained [38]. As reviewed elsewhere [1], mutants initially thought to be defective in the glycerol-3-phosphate acyltransferase [37,39] (*plsA*) were subsequently found to be defective in all macromolecular synthesis due to a lesion in adenylate kinase [40,41].

Despite the risk of obtaining mutants blocked in energy-generating systems, radiation suicide protocols deserve renewed consideration for the enrichment of phospholipid mutants. For instance, Cronan, Silbert, and collaborators [42] have found many strains altered in fatty acid synthesis amongst the survivors of an acetate suicide procedure. A serine suicide enrichment has been reported for the isolation of one *E. coli* mutant blocked in phosphatidylserine synthase [43,44], although 300 colonies were examined individually and no second isolates were obtained.

(d) "Brute force"

Because of the above restrictions, "brute force" screening for enzymatically defined lipid mutants has been utilized [1]. The feasibility of the so-called "brute force" approach was convincingly demonstrated by DeLucia and Cairns [45], who isolated mutants of *E. coli* lacking DNA polymerase I. In this procedure, cells are exposed to a potent chemical mutagen, and cloned on agar from single cells without any prior enrichment procedures. Subsequently, each colony serves as an inoculum from which a culture is grown and a cell-free extract is prepared. If a sufficient number of such extracts are assayed (usually several thousand), a few mutants lacking the enzyme of interest are obtained. Weiss and Milcarek [46] have devised a partially automated procedure to generate such lysates, and have isolated various nuclease mutants in this manner. The Weiss and Milcarek method could be adapted without modification to phospholipases.

To further facilitate the "brute force" approach, Hirota and coworkers [47] have established a bank of several thousand *E. coli* strains—each derived from a separate mutagenesis—which carry random temperature-sensitive lesions. These organisms can be screened one at a time for the desired biochemical alterations. This collection has already provided many mutants in penicillin-binding proteins [48], membrane enzymes [49], ribosomal proteins [50], and other cellular components [51]. The success of the "brute force" strategy demonstrates that chemical mutagenesis induces biochemically identifiable lesions with a very high frequency, and that selection techniques are not inevitably necessary for the isolation of mutants [48–51]. In the case of *E. coli* lipid metabolism, phosphatidylserine decarboxylase [52] and cardiolipin synthase mutants [53] have been isolated by the "brute force" strategy.

(e) Enzymatic colony sorting on filter paper

The success of "brute force" as a means of isolating defined mutants is undisputable, but the actual assay and mapping of mutants by "brute force" are extremely tedious, especially if more than one or two enzymes are to be examined. For this reason our laboratory has developed rapid screening assays [7,16,54–56,56a] for detecting lipid

TABLE 1

Labeling schemes for the detection of phospholipid enzymes in *Escherichia coli* colony preparations immobilized on filter paper

Enzyme	Gene designation [a]	Labeled precursor	Other required substrates or additions	Ref.
Glycerol-3-P acyltransferase	*plsB*	[U-^{14}C]glycerol-3-P	Palmitoyl coenzyme A; Mg^{++}	110a [b]
CDP-diacylglycerol synthase	*cds*	[α-^{32}P]dCTP	Phosphatidic acid; Mg^{++}	56
Diacylglycerol kinase	*dgk*	[γ-^{32}P]ATP	1, 2 Diolein; Mg^{++}	7, 8
Phosphatidylserine synthase	*pss*	[3-^{14}C]-L-serine	CDP-diacylglycerol	54
Phosphatidylglycero-phosphate synthase	*pgsA*	[U-^{14}C]glycerol-3-P	CDP-diacylglycerol; Mg^{++}	55
CDP-diacylglycerol hydrolase	*cdh*	[α-^{32}P]CTP	Phosphatidic acid; Mg^{++}; EDTA	110a [b]

[a] See also Figs. 5 and 6.
[b] See text.

enzymes directly in immobilized colony preparations, and we have found that this approach is applicable to many reactions involved in lipid metabolism (Table 1). Colony autoradiography can be used with bacteria [16,57,58], yeasts [59], or animal cells [17,18,60,60a], and has yielded most of the lipid mutants presently available.

The details of the rapid colony screening assays have been published [7,16,54–56]. Briefly, a disc of filter paper is pressed down on an agar plate on which several hundred colonies of mutagen-treated cells are present. Following this, the paper is lifted off, and in the process most of the material from each colony is transferred to the paper. Enough cells remain on the plate to keep growing and reform the original pattern. The colonies attached to the filter paper can be rendered permeable by treatment with lysozyme and EDTA, coupled with freezing and thawing. Recently, we have found that drying of the filter paper after the freezing-thawing cycle [110a] dramatically improves cell lysis without requiring exposure of the paper to elevated temperatures (50–70°C), as in the published methods [7,16,54–56]. Colonies treated in this manner remain immobilized on the paper and can carry out reactions of phospholipid synthesis in vitro, for example, the conversion of [α-^{32}P]dCTP to [α-^{32}P]dCDP-diacylglycerol dependent on phosphatidic acid (Fig. 1). The lipid generated around each colony lysate is precipitated with trichloroacetic acid after about 30 min. Unreacted radioactive precursor is washed away on a Buchner funnel. The lipid generated in situ is detected qualitatively by autoradiography, and following this, the colonies on the paper are stained with a protein dye, such as Coomassie blue, to locate all colonies including mutants (Fig. 1). Superimposition of the

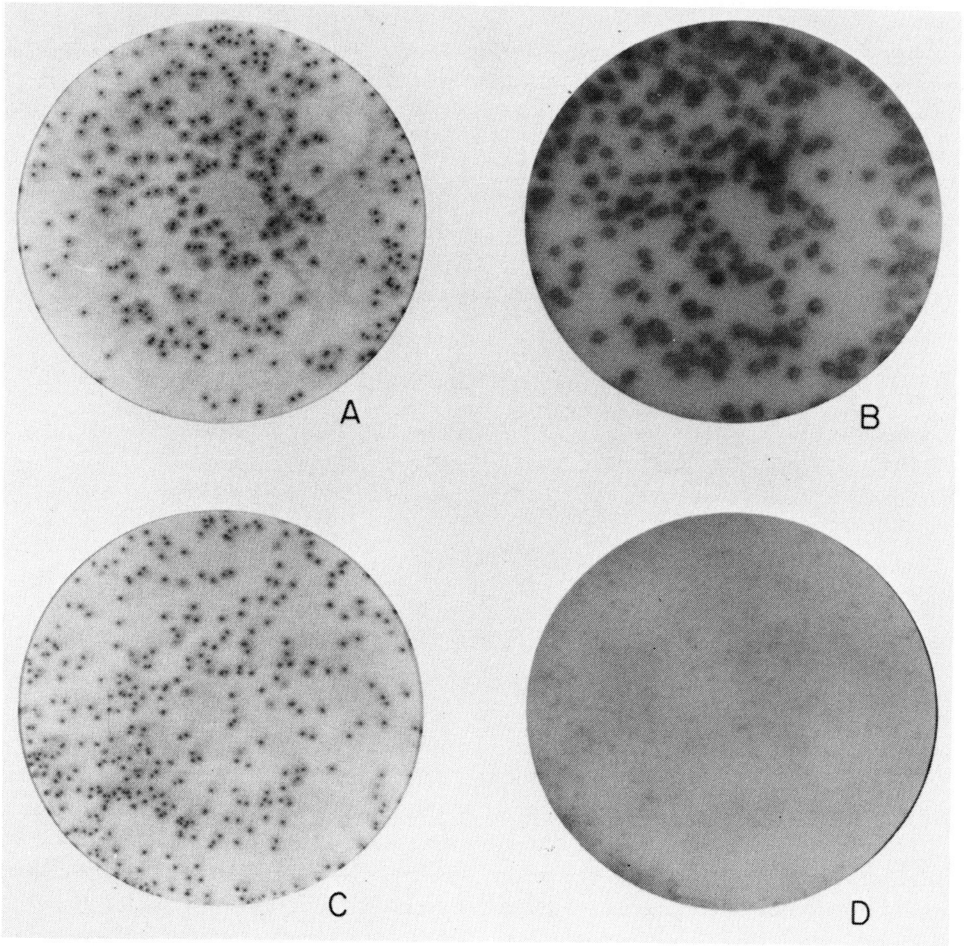

Fig. 1. Autoradiographic detection of CDP-diacylglycerol synthase in *E. coli* colony preparations immobilized on filter paper. Colonies were labelled with [α-^{32}P]dCTP as reported previously [56]. Panels A and B are the filter paper and autoradiograms, respectively, of colony preparations incubated in the presence of phosphatidic acid, while C and D show a different paper and autoradiogram derived from an incubation in the absence of phosphatidic acid. The intense halos of Panel B represent [α-^{32}P]dCDP-diacylglycerol around each colony.

autoradiogram on the stained paper allows identification of mutants as blue colonies lacking a black overlay (Fig. 2). In our experience with the various assays listed in Table 1, one mutant is found in every 1000 to 10000 colonies screened [7,16,54–56]. No assumptions are made about the potential properties or phenotypes of such mutants, allowing for the isolation of absolutely defective as well as conditionally defective strains. A maximum of 40000 colonies can be assayed per day if necessary.

It is important to note that colony autoradiography can be employed as a final, definitive screening technique, even if the desired mutants are first enriched from a larger population by the selective methods outlined above.

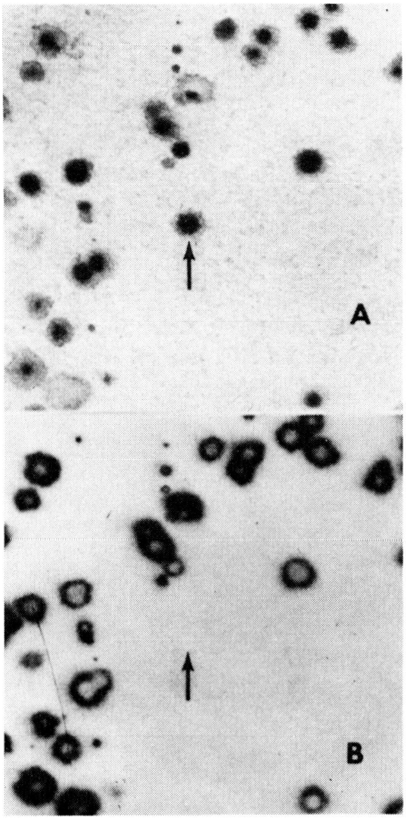

Fig. 2. Identification of an *E. coli* mutant in phosphatidylglycerophosphate synthase [16,55] by comparison of a stained filter paper with the corresponding autoradiogram. Panel A shows an area of a stained filter paper (about 3 cm × 3 cm in the original) in which a mutant was found. This was done by comparing it to the corresponding autoradiogram shown in Panel B. The arrow indicates the position of a colony that did not synthesize any radioactive phosphatidylglycerophosphate under these conditions (see also Table 1 and [16,55]). This strain was subsequently found to lack phosphatidylglycerophosphate synthase when cell-free extracts were assayed by conventional methods [16,55]. In practice, mutants can be identified more rapidly by superimposing the autoradiogram (Panel B) on the blue paper colony copy (Panel A). Reprinted from reference [16] with permission of the publisher.

3. Genetic approaches to phospholipid metabolism in yeasts and fungi

Among the earliest chemically defined mutants reported in the genetic literature were the auxotrophs of *Neurospora crassa* requiring choline or inositol for growth

[61–63]. Strains of this kind have recently been obtained from *Saccharomyces cerevisiae* as well [20,22]. Fatty acid-dependent strains of yeast have also been studied extensively and have helped to clarify the structure of the fatty acid synthase complex in this system [5]. Within the past 5 years the choline and inositol auxotrophs have been utilized for the purpose of membrane lipid modification [19,21].

As with *E. coli,* no one has developed methods for supplementing yeasts or fungi with intact phospholipid molecules. Antibiotics that block the late stages of lipid synthesis are not available, and radiation suicide protocols—for instance, using tritiated choline or inositol—have not been used to obtain lipid mutants. "Brute force" screening of yeast colonies for biochemical variants has been very successful in the case of polyamine metabolism [64] and could certainly be adapted to study lipid synthesis. The feasibility of colony autoradiography has been documented both with *Neurospora crassa* and with *S. cerevisiae* [59], but the necessary selective assays in situ (as in Table 1) have not been developed. Studies of membrane lipid genetics would be especially fruitful with *S. cerevisiae,* since methods for DNA-mediated genetic transformation [65] and molecular cloning in yeast are already available [66].

4. Genetic approaches to phospholipid metabolism in higher mammalian cells

As indicated above, the molecular complexity of lipids in higher eukaryotic membranes is an order of magnitude greater than in prokaryotic systems [1–6]. The structure and organization of membranes within eukaryotic cells is radically different, suggesting that there must be unique mechanisms of metabolic control and compartmentalization. Higher eukaryotic cells also perform a variety of physiologically unique functions, some of which may involve lipid metabolism, including stimulus-triggered responses [67] and biogenesis of enveloped viruses [68]. Mutants in phospholipid, sphingolipid and sterol synthesis might help explain how lipids participate in these processes.

(a) Transfer of animal cell colonies to filter paper and its application to somatic cell genetics

Permanent lines from a variety of animal and human sources can be propagated from single cells [69]. When diluted appropriately and allowed to attach to a plastic surface bathed in a liquid growth medium, such cells divide every 12–24 h and generate macroscopic colonies after 8–16 days. The isolation of biochemically defined mutants derived from such mammalian cells is well documented and has been reviewed elsewhere [69,70].

In contrast to *E. coli* or yeast, methods for the analysis of animal cell colonies are very limited. Unlike microorganisms, tumor cells do not grow well on solid agar surfaces, excluding the use of classical replica plating for mutant analysis [17,18].

The obvious need for a simple replica-plating procedure applicable to animal cells led us to explore unconventional conditions for cultivating macroscopic animal cell colonies and for transferring them from one surface to another. In 1978, we made the provocative observation that single CHO cells can proliferate extremely well when sandwiched between a plastic surface and a piece of smooth Whatman paper (No. 50) weighed down with glass beads to assure even contact [17]. This maneuver allowed replacement of the growth medium when necessary without disturbing the colony pattern. Some cells from each developing colony invade the overlaying paper fibers, while other cells remain attached to the plate. The spread of loose cells into regions between colonies, which tends to obscure the pattern, is eliminated by the overlay, presumably because convection currents are reduced (Fig. 3). The cloning of animal cells between paper and plastic facilitates the growth of much larger colonies

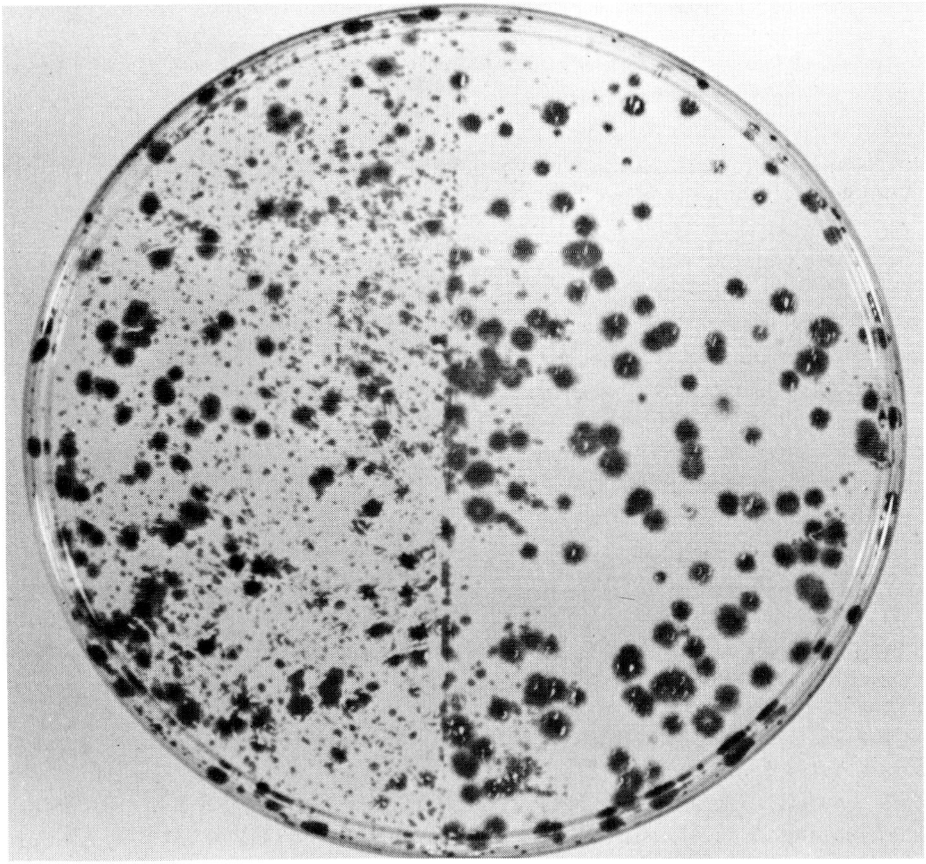

Fig. 3. Reduction of secondary animal cell colonies by filter paper overlay. 1-day-old cells were overlayed on the right side, but not on the left, with filter paper and beads. After 9 days the plate was stained with Coomassie Blue.

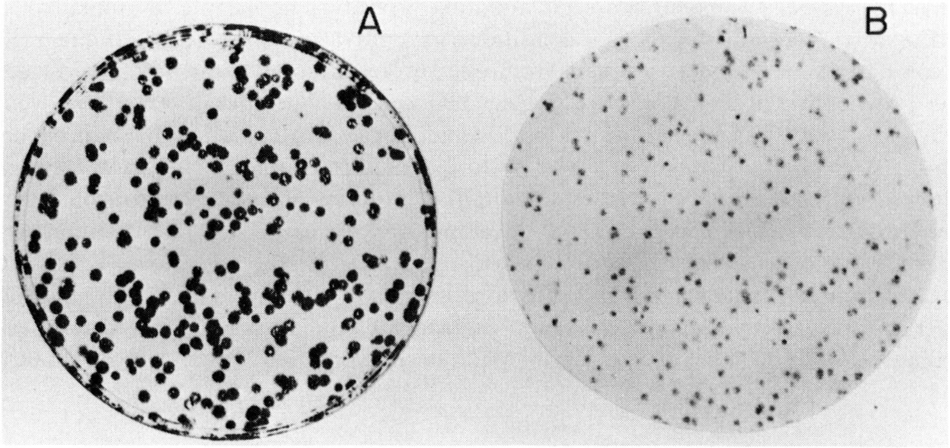

Fig. 4. Transfer of a 9-day-old animal cell colony pattern from master plate (left) to filter paper (right), both stained with Coomassie Blue.

(0.3–0.5 cm in diameter) than previously possible. When the paper is removed from the plate, a high-resolution copy of the colony pattern is available both on the plate and on the paper (Fig. 4).

The colonies on the paper can be rendered permeable (by freezing and thawing) and used for autoradiographic enzyme screenings in situ [17,18], as described above for *E. coli.* Alternatively, the immobilized cells can be left intact and labeled directly with specific precursors, such as [^{14}C]choline, thymine or leucine [17,18,60]. The advantage of working with intact cells is that an entire pathway can be examined in one step [17,18], but mutants defective in transport or energy generation will also be recovered. As many as 10^4–10^5 animal cells can be screened for specific biochemical alterations with this method [17,18].

Cells attached to paper can further be utilized to propagate one or more true replica plates by placing the paper into a fresh plastic dish [17,18,71]. As in the original transfer of the colonies to the paper, the glass beads create even contact with the replica plate, which takes an additional 3–6-days to form, depending on temperature. In all colony screening schemes, viable cells are retrieved at a later date (after the mutants have been located) from the master plate, which is stored at 28°C [17,18].

At present, this procedure for cloning animal cells on filter paper has been used to isolate inositol auxotrophs [71], temperature-sensitive mutants defective in CDP-choline synthase (cholinephosphate cytidylytransferase, EC 2.7.7.15) [60], and ethanolamine phosphotransferase (EC 2.7.8.1) mutants [71a], using the Chinese hamster ovary (CHO) cell line as the parent. Robbins has recently described CHO mutants defective in a lysosomal α-mannosidase [72], Glaser et al. have isolated UV-sensitive CHO mutants [73], and Hirschberg et al. [74] have obtained CHO mutants with altered glycoprotein synthesis using the filter-paper approach. The applicability of the overlay and copying technique to other cell lines has also been documented for mouse L cells [17,18], and certain hormone-sensitive pituitary tumor lines [18]. Recently, we have found that polyester cloth is preferable to filter paper in certain settings [74a].

To date, filter paper or polyester screening appears to be the most effective method available for isolating somatic cell lipid mutants of defined biochemistry. Other possible approaches include radiation suicide (for instance, with tritiated choline of high specific radioactivity) and the use of metabolic analogs. With regard to the latter, mammalian cells effectively incorporate analogs of choline into their phospholipids [75]. In general, this results in the inhibition of growth, but no one has attempted to isolate animal cell mutants resistant to such compounds. The isolation of animal cell lines dependent on intact lipid molecules may also be feasible, and will be considered in detail below.

5. General properties of E. coli *phospholipid mutants*

Fig. 5 presents the enzymatic reactions for membrane phospholipid synthesis in *E. coli* [1,2]. Genetic symbols indicate sites at which mutants are available. The lysophosphatidic acid acyltransferase is the only enzyme which has not been subjected to any genetic analysis. The enzymes involved in the assembly of the membrane-derived oligosaccharides (MDO) have not been identified [76,77], but it is possible to inhibit MDO synthesis nonspecifically by using mutants defective in gluconeogenesis or UDP-glucose synthesis [78,79].

Fig. 6 indicates the locations of the phospholipid genes on the chromosome of *E. coli* (which is circular and has an M_r-value of about $2 \cdot 10^9$). Many of the genes identified so far appear to be structural (i.e., coding for the polypeptide chains of enzymes) rather than regulatory.

The extent to which mutations in the phospholipid genes allow modifications of cellular lipid composition is shown in Table 2. Certain phospholipid perturbations are compatible with cell growth, while others are not [1]. The least flexible parameter is the total phospholipid content, which cannot be lowered by more than 40% without inhibiting cell growth [81,82]. The ratio of zwitterionic to anionic phospholipids is also an important factor, although a two to three-fold deviation from wild type in either direction still permits growth under most conditions (Table 2, see *pss*-21, *psd*-4 and *pgsA*-444). Least critical is the level of cardiolipin, which varies considerably in wild-type cells and can be reduced further by the *cls* mutation without obvious adverse effects [53]. In addition, many of the existing mutations cause partial (rather than complete) metabolic blocks (*psd*-4 at 30°C or *cds*-8 at pH 6, Table 2), leading to massive accumulations of intermediates which are ordinarily present at extremely low levels. Depending on the metabolite, the *E. coli* membrane is capable of accepting virtually any "abnormal" lipid in the range of 10–20% of the total phospholipid content. Some of these extraneous lipids have useful phenotypic manifestations. For instance, *cds*-8 at pH 7 renders the cells partially resistant to

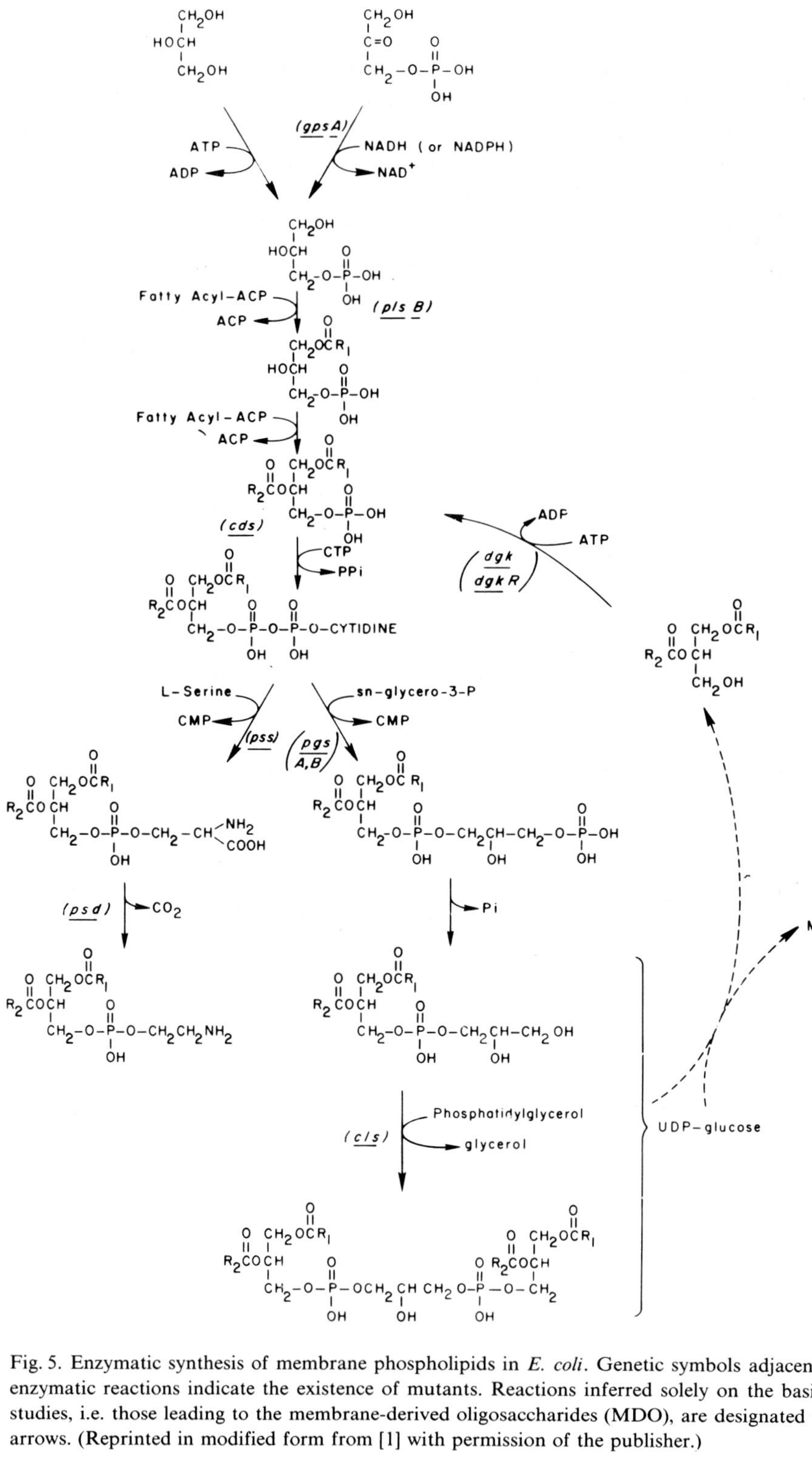

Fig. 5. Enzymatic synthesis of membrane phospholipids in *E. coli*. Genetic symbols adjacent to specific enzymatic reactions indicate the existence of mutants. Reactions inferred solely on the basis of genetic studies, i.e. those leading to the membrane-derived oligosaccharides (MDO), are designated with dashed arrows. (Reprinted in modified form from [1] with permission of the publisher.)

erythromycin [56a], while *dgk*-6 makes the bacteria hypersensitive [7,8] to low osmolarities (below 20 mM).

Two additional classes of mutations (not listed in Table 2), altering lipid metabolism, have no effect on growth under ordinary laboratory conditions. These are: (1) mutants (*pldA*) lacking the outer membrane phospholipase A [88,89]; and (2) mutants blocked indirectly in the formation of the membrane-derived oligosaccharides (such as glucosephosphate isomerase mutants (*pgi*) grown on casamino acids [78,79]).

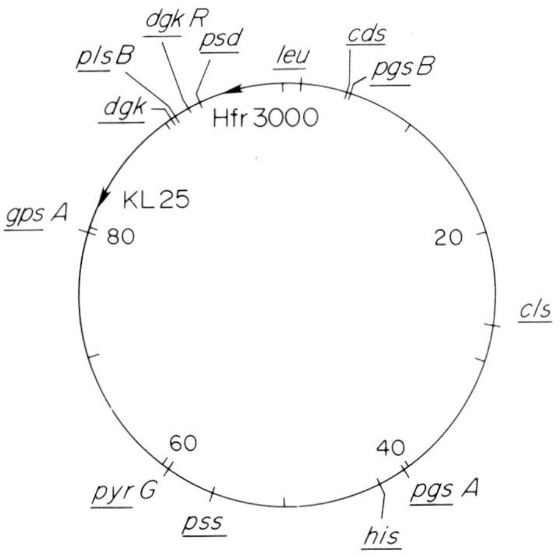

Fig. 6. Locations of mutations responsible for defects in the enzymatic synthesis of membrane phospholipids on the chromosome of *E. coli.* See Fig. 5 (and text) to correlate genetic symbols with enzymatic reactions. Not shown in Fig. 5 is *pyrG* (CTP synthase, EC 6.3.4.2). Mutants defective in diacylglycerol kinase (*dgk*) appear to define the structural gene, while regulatory mutants with elevated levels of the kinase are designated (*dgkR*). The leucine (*leu*) and histidine (*his*) genes, as well as the origins of Hfr3000 and KL25 [23], are provided as reference points.

The results of Table 2 must be considered in any model of phospholipid function deduced from physical, chemical, or reconstitution studies.

6. E. coli *mutants in phosphatidic acid synthesis*

(a) Glycerol-3-phosphate acyltransferase (EC 2.3.1.15) K_m *mutants* (plsB)

Existing mutants defective in glycerol-3-phosphate acyltransferase (*plsB*) have been isolated by penicillin enrichment for glycerol-3-phosphate auxotrophs [80]. These mutants have a 10-fold higher than normal K_m for glycerol-3-phosphate [80], and therefore require supplementation with exogenous glycerol-3-phosphate (or glycerol) to increase the internal glycerol-3-phosphate pool. While the existing *plsB* mutants are not temperature-sensitive for growth, such variants could presumably be isolated

TABLE 2

Extreme modifications of *E. coli* K-12 lipid composition caused by mutations in polar headgroup synthesis

Defective gene	Phenotype	Conditions of Cell Growth	Resulting lipid composition [a]						Comments and Ref.
			PE	PG	CL	PA	PS	DG	
			% of phospholipid phosphorus [b]					% [c]	
None (*E. coli* K-12)	None	30–42°C, log phase	70–80	15–25	1–10	0.1–0.5	<0.5	0.2–0.6	Not pH or osmotically sensitive
plsB-26	glycerol auxotroph	37°C, omit glycerol for 30 min	75	23	2	– [d]	–	–	Lipid: protein ratio is reduced by 40% [80–82]
dgk-6	osmotic sensitivity	37°C, log phase	80	19	1	–	–	9	Higher DG level at low osmolarity [7,8]
cds-8	pH-sensitive	37°C, pH 6, log phase	80		[12] [e]	8	–	–	Partial erythromycin resistance [56,56a] [f]
cds-8	pH-sensitive	37°C, pH 8 for 2 h	62		[10] [e]	28	–	–	Partial erythromycin resistance [56,56a] [f]
pyrG-51	Cytidine auxotroph	37°C, deplete cytidine for 2 h	58		[11] [e]	31	–	–	Lipid : protein ratio is increased 3 fold [56a,83] [f]

pss-21	TS-42°C [g]	30°C, log phase	45	41	14	0.4	–	–	Hypersensitivity to antibiotics [54,84,85]
pss-21	TS-42°C	42°C, 4.5 h	24	26	46	–	–	–	Filamentation in some cases [54,84]
psd-4	TS 42°C	30°C, log phase	67	10	3	–	20	–	[52,86]
psd-4	TS 42°C	42°C, 4 h	31	7	14	–	48	–	Filamentation [86]
pgsA-444	None	42°C, log phase	83	13	1	1	–	–	[55]
pgsA-444, *pgsB*-1	TS-42°C	42°C, 3 h	97	1	1	1	–	–	Accumulation of two lipid A precursors [55,87]
pgsB-1	None	42°C, log phase	72	24	3	0.7	–	–	[87]
cls^{-}	None	37°C, log phase	79	21	<0.4	–	–	–	Normal growth of f1 bacteriophage [53]

[a] Abbreviations: PE, phosphatidylethanolamine; PG, phosphatidylglycerol; CL, cardiolipin; PA, phosphatidic acid; PS, phosphatidylserine; DG, diacylglycerol. Gene designations are explained in Fig. 5 and text.

[b] Values for wild-type cells indicate ranges observed under different growth conditions and with the different parental strains used in the references of this table. All values given are estimates of the true chemical composition. Unless otherwise indicated the ratio of lipid to protein is unaltered by the polar headgroup modifications. We estimate the phospholipid content at 6–8% of the dry weight of *E. coli.*

[c] Percent diacylglycerol is calculated relative to the total, chloroform-soluble esterified fatty acid.

[d] Blanks indicate that accurate values have not been determined, but are expected to be in the normal range.

[e] Values in brackets represent the sums of PG and CL.

[f] See text.

[g] TS, temperature-sensitive for growth.

as well. When glycerol-3-phosphate is withheld from the growth medium, the rate of endogenous phospholipid synthesis in a *plsB* mutant drops abruptly by 90–95% [81,82]. Cell growth gradually ceases over the course of about 30 min, but levels of ATP and other nucleotides remain high for several hours [81,82]. As growth stasis sets in, the ratio of phospholipid to protein drops about 40% below wild type [81,82]. Inner and outer membranes isolated from glycerol-3-phosphate starved *plsB* mutants are denser than wild-type membranes and are easily separable from them on sucrose gradients [81].

Studies of macromolecular and membrane protein synthesis in the *plsB* mutants have made it necessary to revise earlier suggestions regarding the coupling of membrane protein and membrane lipid synthesis [90,91]. It appears that the synthesis of membrane proteins continues during glycerol-3-phosphate and phospholipid limitation [81]. Continuous lipid synthesis is not required for membrane protein insertion [92,93] and proteolytic processing [94–96].

In contrast to macromolecular and membrane protein syntheses, the inhibition of phospholipid synthesis in *plsB* mutants causes a rapid cessation of fatty acid synthesis de novo [97]. This apparent coupling of phospholipid and fatty acid synthesis may be very significant, and is also observed in the Gram-positive mutants [98]. In both instances, the mechanism is unknown. Perhaps the accumulation of completed fatty acid chains on acyl carrier protein prevents further fatty acid synthesis when all possible acyl carrier protein molecules are occupied. The levels of acyl acyl carrier protein and other fatty acid precursors have not been measured in this setting.

It is not possible to bypass the glycerol-3-phosphate requirement of these acyltransferase mutants by feeding cells lysophosphatidic acid [31]. Although a considerable amount of this substance is bound by *E. coli,* only some of it is converted to phospholipid [31]. The rate at which this occurs appears insufficient to support cell growth.

Prototrophic revertants of glycerol-3-phosphate acyltransferase mutants able to grow without supplementation are also very interesting [99]. Some of them are revertants in the structural gene which have regained a normal enzyme with a low K_m [99]. However, other phenotypic revertants still have the high K_m characteristic of the defective acyltransferase, yet the cells have lost their requirement for glycerol-3-phosphate [99]. This occurs because of a secondary mutation in the biosynthetic glycerol-3-phosphate dehydrogenase (EC 1.1.1.8), which has lost its normal mechanism of feedback inhibition by glycerol-3-phosphate, and hence allows the synthesis of much more endogenous glycerol-3-phosphate [99,100].

In summary, glycerol-3-phosphate acyltransferase mutants of *E. coli* have proved especially useful for studies of membrane biogenesis and for examining the coordination of membrane lipid and membrane protein synthesis [1,2]. An advantage of using glycerol-3-phosphate acyltransferase mutants for this purpose is the fact that during glycerol-3-phosphate starvation, endogenous glycerol-3-phosphate is still generated, though at the wild-type levels, which are insufficient for lipid synthesis. Therefore, other reactions involving glycerol-3-phosphate can continue, and phospholipid synthesis is blocked selectively.

Further genetic studies aimed at isolating other types of acyltransferase mutants, for instance, temperature-sensitives or regulatory mutants, deserve consideration. A rapid autoradiographic screening assay for this enzyme has recently been developed (Table 1). It should be possible to identify the genes that regulate the level of acyltransferase, for instance, the promoter adjacent to the structural gene, by examining a larger number of revertants of the *plsB* mutants able to grow without glycerol-3-phosphate supplementation. This assumes that overproduction of the K_m-defective enzyme is functionally equivalent to a structural gene reversion.

Recently, the *plsB* gene has been isolated and has been found in very close proximity to the *dgk* gene by chemical as well as genetic methods [12]. The possibility of a common control region which includes a gene for phosphatidic acid synthesis de novo (*plsB*) together with a gene for salvage of phosphatidic acid synthesis (*dgk*) is discussed further below under "Molecular cloning".

(b) Mutants in the biosynthetic glycerol-3-phosphate dehydrogenase (EC 1.1.95.5; gpsA)

In glucose-grown cells, the glycerol-3-phosphate which serves as a precursor for phospholipid synthesis is usually generated by a soluble enzyme (Fig. 5), termed the biosynthetic glycerol-3-phosphate dehydrogenase. Mutants of this enzyme have been isolated in conjunction with the acyltransferase mutants discussed above [80]. In view of the selection scheme employed, existing mutant isolates defective in the dehydrogenase (*gpsA*) are glycerol-3-phosphate (or glycerol) auxotrophs. When glycerol-3-phosphate is removed from the medium, the endogenous pool of glycerol-3-phosphate drops below that present in wild-type cells. The effects on phospholipid synthesis closely resemble those observed with *plsB*, and the cells acquire membranes with a greater than normal density. The properties of *E. coli gpsA* mutants also resemble those of similar mutants isolated earlier from Gram-positive bacteria [98]. The *gpsA* locus [101] is far removed from the *plsB* gene (Fig. 6).

(c) Mutants in diacylglycerol kinase (dgk)

Pieringer and Kunnes [102] first established that membranes of *E. coli* contain diacylglycerol kinase which generates phosphatidic acid from ATP and 1,2-diacylglycerol. The role of this enzyme in membrane lipid biogenesis remained uncertain [103,104], until the recent isolation of diacylglycerol kinase mutants [7,8]. Since mutants defective in the acyltransferases (*plsB*) are inhibited by 90–95% in their capacity to generate phospholipids [80–82], a major role for the kinase in phospholipid synthesis de novo was excluded. Very little diacylglycerol (and no triacylglycerol) is present in wild-type *E. coli* [7,8], the former representing 0.2–0.6% of the esterified fatty acid (Table 2). A priori, this small amount of diacylglycerol could have arisen from the chemical breakdown of a minor, unstable lipid. Also, the diacylglycerol pool does not turn over rapidly like that of true de novo intermediates, for instance, phosphatidic acid and CDP-diacylglycerol [105,106].

The isolation of mutants lacking diacylglycerol kinase has been achieved with the use of colony autoradiography [7,8], as shown in Table 1. Mutants deficient in the kinase accumulate substantial amounts of 1,2-diacylglycerol (7–11%), about 15- to 30-fold higher than the wild type (Table 2). While kinase mutants are not temperature-sensitive for growth, they do not divide on media of low osmolarity [7]. Spontaneous revertants containing the kinase can be selected on this basis [7]. The finding that substantial amounts of diacylglycerol can accumulate in certain *E. coli* mutants argues that diacylglycerol is the true substrate for this enzyme in vivo and, further, that there must be a significant source of diacylglycerol in *E. coli* not previously appreciated.

The primary source of diacylglycerol in *E. coli* is likely to be phosphatidylglycerol, which has recently been shown to act as a donor of its glycerol-1-phosphate headgroup in the formation of the membrane-derived oligosaccharides [76–79]. The transfer of glycerol-1-phosphate to MDO precursors (or analog disaccharides) has recently been demonstrated in vitro (E.P. Kennedy, personal communication), and this should generate 1,2-diacylglycerol as a by-product. Since most of the energy expended in lipid synthesis is consumed during the formation of the hydrocarbon fatty acid chains, it is reasonable that there should be a mechanism for salvaging diacylglycerol moieties, whatever the source. A minor salvage pathway for phosphatidic acid amounting to 5–10% of the total made at any given time is not incompatible with the studies of the acyltransferase mutants discussed above.

The suggestion of the diacylglycerol cycle [8] (see Fig. 5) is further supported by the construction of double mutants defective both in diacylglycerol kinase and in gluconeogenesis, specifically in phosphoglucose isomerase [8]. When MDO synthesis is inhibited in these organisms by growing them on amino acid as the carbon source, there is little phosphatidylglycerol turnover and much less accumulation of diacylglycerol (only 2.4% of the total lipid) when the kinase is defective [8]. Addition of glucose to such double mutants activates MDO synthesis and consequently causes diacylglycerol to reaccumulate [8] to levels observed in single-step dgk^- mutants (i.e. 7%). The possibility of minor sources of diacylglycerol besides MDO synthesis cannot be eliminated. Reports of a phospholipase C in *E. coli* exist, but this work has not been substantiated [107,108].

7. E. coli *mutants in CDP-diacylglycerol synthesis*

(a) CDP-diacylglycerol synthase (phosphatidate cytidylyl transferase, EC 2.7.7.41; cds)

Recently, our laboratory has developed in situ screening assays for detecting the conversion of phosphatidic acid to CDP-diacylglycerol [56] by following the incorporation of [α-^{32}P]ribo- or deoxycytidine triphosphate to trichloroacetic acid-precipitable material (see Table 1) dependent on phosphatidic acid (Fig. 1). In an initial screening, six strains were identified (out of 20000 colonies) in which the specific activity of CDP-diacylglycerol synthase was below 5% of wild type [56]. Synthesis of both deoxy-CDP-diacylglycerol and ribo-CDP-diacylglycerol was similarly affected, suggesting that the same enzyme synthesizes both liponucleotides [56]. All mutations responsible for CDP-diacylglycerol synthase deficiency mapped near minute 4 on the *E. coli* chromosome (Fig. 6).

None of the mutants isolated in this particular screening are temperature-sensitive for growth [56]. However, some of the mutants accumulate 10- to 50-fold more phosphatidic acid than is present ordinarily, or as much as 5% of the total lipid. The phospholipid to protein ratio remains unchanged by this modification, although the increase in phosphatidic acid occurs preferentially at the expense of phosphatidylglycerol and cardiolipin [56]. Presumably, any residual CDP-diacylglycerol synthase activity present in these mutants is enough to allow for a normal rate of phospholipid synthesis in vivo, especially with the compensatory elevation of the phosphatidic acid pool. Despite the increased levels of phosphatidic acid, this material still behaves as a precursor to the major phospholipids [56]. In short-term $^{32}P_i$ pulses, the enlarged phosphatidic acid pool is labeled preferentially, and turnover studies reveal that this label subsequently becomes incorporated into the major lipids, as in the wild type [56].

In an attempt to obtain mutants blocked more tightly at the CDP-diacylglycerol synthase step, we have examined alternative methods for selecting additional mutants [56a]. In a survey of antibiotic sensitivity patterns, we have observed that mutants containing 5% phosphatidic acid in their membranes are partially resistant to the antibiotic, erythromycin. When plated on agar containing 100 μg/ml of erythromycin, the frequency with which the CDP-diacylglycerol synthase mutants are recovered is increased by 20- to 40-fold. With this strategy we have recently isolated 40 new CDP-diacylglycerol synthase mutants [56a]. In several cases genetic mapping has been attempted, giving the same results as in the initial studies (Fig. 6). A subpopulation of the erythromycin-resistant *cds*$^-$ mutants grow poorly at pH 8 [56a]. This is due to additional phosphatidic acid accumulation at high pH in these mutants, amounting to as much as 25–30% of the total membrane lipid at the point of growth inhibition (Table 2). The maximum level of phosphatidic acid compatible with log phase growth of *E. coli* appears to be about 10%.

The studies of the CDP-diacylglycerol synthase mutant illustrate an important principle, associated with the isolation of enzymatically defined mutants by "brute force" or colony autoradiography. Although a series of mutants obtained in this way may have no detectable enzyme activity as measured in vitro, only a subset of such mutants will possess corresponding metabolic alterations or growth phenotypes in vivo. Thus, there is no simple correlation between residual enzymic activity and the extent of lipid modification in vivo. This anomaly can be rationalized because the conditions used to assay an enzyme in vitro may not adequately reflect intracellular circumstances. Despite this limitation, "brute force" screening and colony autoradiography yield enzymatically defined lesions, which, at the very least, permit localization of the genes coding for the enzyme of interest, and may suggest further methods for the isolation of additional mutants.

(b) Cytidine auxotrophs (pyrG)

E. coli mutants defective in the conversion of UTP to CTP are cytidine auxotrophs [83]. We have recently examined the effects of cytidine starvation of such mutants on the formation of membrane phospholipids [56a]. Although RNA and protein syntheses cease abruptly, membrane phospholipids continue to be made at a reduced rate. The phospholipid synthesized upon cytidine starvation consists primarily (over 85%) of phosphatidic acid. As in the case of the CDP-diacylglycerol synthase mutants, phosphatidic acid accumulates to a final level of 25–30% of the total lipid, but the lipid to protein ratio is about twice normal (Table 2) because protein synthesis is preferentially inhibited.

The use of cytidine auxotrophs to perturb membrane lipid synthesis may prove useful, since the biosynthetic phospholipid enzymes presumably remain intact. It may be possible to achieve the conversion of phosphatidic acid to phosphatidylethanolamine and phosphatidylglycerol in membranes isolated from cytidine auxotrophs, avoiding the non-physiological detergents generally employed to stimulate these conversions. Perhaps regulatory effectors can be identified and isolated with such a system. The 100-fold accumulation of phosphatidic acid in cytidine auxotrophs and CDP-diacylglycerol synthase mutants argues that the glycerol-3-phosphate acyltransferase and the lysophosphatidic acid acyltransferase are not feedback-inhibited by this intermediate.

(c) CDP-diacylglycerol hydrolase (cdh)

The membrane-bound CDP-diacylglycerol synthase of *E. coli* acts with approximately equal efficiency on CTP and dCTP [109]. However, a separate membrane-bound enzyme (CDP-diacylglycerol hydrolase or CDP-diacylglycerol pyrophosphatase, EC 3.6.1.26) in *E. coli* hydrolyzes CDP-diacylglycerol, though not deoxy-CDP-diacylglycerol [110]. Both ribo- and deoxy-CDP-diacylglycerol can be detected in vivo [106]. The biological function of the CDP-diacylglycerol hydrolase is unknown.

In an attempt to analyze this problem, we have developed an enzymatic assay for CDP-diacylglycerol hydrolase in colonies immobilized on paper [110a]. For this purpose, the colonies are allowed to generate [α^{32}P]ribo-CDP-diacylglycerol in situ for 40 min. Following this, EDTA is introduced into the reaction mixture, which inhibits the synthase but not the hydrolase. After an additional 40 min, any ribo-CDP-diacylglycerol synthesized during the first 40 min is degraded by the CDP-diacylglycerol hydrolase (Table 1). With this assay mutants defective in the hydrolase have dark halos, while the surrounding wild-type colonies are pale. The *cdh* mutation maps at a distinct site near minute 88 on the *E. coli* chromosome (not shown in Fig. 6).

8. E. coli *mutants in phosphatidylethanolamine synthesis*

(a) Phosphatidylserine synthase (*EC 2.7.8.8*; pss)

A total of six phosphatidylserine synthase mutants have been reported, and they appear to define the structural gene for the enzyme [43,44,54,84]. Several of them have been characterized in considerable detail with regard to their membrane lipid composition [44,84]. As expected from the pathway of Fig. 5, inhibition of phosphatidylserine synthase by genetic means leads to increased utilization of CDP-diacylglycerol for phosphatidylglycerol and cardiolipin synthesis [44,84]. The lipid to protein ratio remains the same [84]. The elevation of cardiolipin levels in the *pss* mutants is especially striking and occurs both in the inner and in the outer membrane [84].

Strains harboring *pss*8 [54] or *pss*21 [84] contain a reduced level of phosphatidylethanolamine (45–55%) at 30°C, and this drops to about 30% after 4–6 h at 42°C (Table 2). Even at 42°C, however, some residual phosphatidylethanolamine continues to be made. Whether this material originates from residual enzymatic activity or arises by a separate mechanism has not been ascertained. It would be desirable to isolate additional phosphatidylserine synthase mutants to define more precisely the extent to which this enzyme is responsible for the synthesis of phosphatidylethanolamine. In any case, it is clear that CDP-diacylglycerol-dependent phosphatidylserine synthesis represents the major pathway in *E. coli.*

Further characterization of phosphatidylserine synthase mutants has revealed that the gross membrane protein composition is unaffected by modification of the polar headgroups [85]. Furthermore, the lipopolysaccharide [85] appears to be the same as in *pss*$^{+}$ parental strains, and the fatty acid composition [84] is not greatly altered, particularly at 30°C [84]. These findings are of interest in view of the extreme antibiotic hypersensitivity of *pss* mutants at all temperatures [85], especially towards hydrophilic antibiotics such as gentamycin and streptomycin. The antibiotic hypersensitivity of *pss* mutants suggests that inhibitors of the phosphatidylserine synthase would potentiate the action of numerous antibiotics already in clinical use for the treatment of Gram-negative infections. As yet, specific inhibitors have not been designed for this enzyme. Obvious serine analogs, such as α-methylserine or serine methyl ester are ineffective (Raetz, C.R.H., unpublished).

All existing phosphatidylserine synthase mutants are stabilized by the addition of salts or divalent cations to the growth medium [44,84]. However, in the best available mutant (*pss*21), the cells are not able to grow at 42°C even under optimized ionic conditions [84]. Suppression of the temperature-sensitive phenotype of these mutants by supplementation with lipids or lysolipids also has not been possible [111]. It may be that the excess of polyglycerophospholipids rather than the absence of phosphatidylethanolamine inhibits cell growth.

The further characterization of membrane functions in *E. coli pss* mutants would be of considerable interest, and methods for the selection of additional mutants might become obvious through such studies.

(b) Phosphatidylserine decarboxylase (EC 4.1.1.65; psd)

Hawrot and Kennedy [52,86,112] have examined the gene for phosphatidylserine decarboxylase designated *psd*. The initial mutants in this locus were obtained by "brute force" screening [52], permitting determination of the chromosomal location [112] of the *psd* gene (Fig. 6). Following this, mutagenesis of a localized region of the chromosome in the vicinity of *psd* led to the isolation of additional mutants [112] with a significantly altered lipid composition (Table 2). Many of these organisms are temperature-sensitive for growth [86,112], particularly when phosphatidylethanolamine drops below 50% and is replaced by phosphatidylserine (Table 2). As in the case of phosphatidylserine synthase mutants, the lipid that accumulates (i.e. phosphatidylserine) is found in the inner and in the outer membrane (Hawrot, E. and Kennedy, E.P., personal communication). The decarboxylase itself, like the other phospholipid enzymes, is associated primarily with the inner membrane [113,114]. If a *psd* mutant is shifted from 42°C back to 30°C, most of the excess phosphatidylserine is decarboxylated, suggesting free flow of lipids between inner and outer membranes. The association of the decarboxylase with the inner membrane exerts an additional stabilizing effect on the enzyme [50].

Although not studied in great detail, both the *pss* [54] and *psd* [86] mutants form long filaments at non-permissive temperatures under some conditions. Whether or not there is a direct relationship of lipid modification to cell division has not been determined. Filamentation is a relatively non-specific response, observed with many mutants in macromolecular synthesis [115].

9. E. coli mutants in polyglycerophospholipid synthesis

(a) Phosphatidylglycerophosphate synthase (EC 2.7.8.5; pgsA *and* pgsB)

Unlike the relatively straightforward genetic analysis of phosphatidylethanolamine synthesis, the characterization of mutants in phosphatidylglycerol synthesis has revealed some intriguing complexities [16,55,87]. Mutants unable to generate anionic phospholipids under nonpermissive conditions (42°C) have recently been isolated in our laboratory [55,87]. With these strains it is possible to reduce the phosphatidylglycerol content to about 1% (Table 2). Unlike the other lipid modifications discussed above, a two-step mutagenesis is required to achieve this [53,87].

In an initial screening of 250000 colonies, 25 mutants deficient in phosphatidylglycerophosphate synthase (glycerophosphate phosphatidyltransferase, EC 2.7.8.5) were isolated by colony autoradiography [55]. In many cases the residual enzymatic activity determined in extracts was $<5\%$ of the wild type. The location of the gene responsible for this enzymatic defect [55], designated *pgsA,* is shown in Fig. 6. Since the residual activity in many of the isolates is inactivated at 70°C, whereas the wild type is not [55], it is likely that the *pgsA* locus is the structural gene for the enzyme, which consists of a single polypeptide [116].

Biochemical and phenotypic analysis of all 25 *pgsA* mutants has revealed that none of them are temperature-sensitive for growth or show reductions in the level of phosphatidylglycerol corresponding to the enzymatic lesions measured by assay in vitro [16,55]. This anomaly can be explained two ways. On the one hand, the phosphatidylglycerophosphate synthase may not represent the sole route for the synthesis of phosphatidylglycerol in vivo. However, there is no evidence for isoenzymes or alternate mechanisms. Another possibility is that the phosphatidylglycerophosphate synthase is present in great excess, or that the residual activity is somehow stabilized in vivo, as shown above with the *cds* mutants. We favor the latter explanation, since it has not been possible to isolate insertion or deletion mutants of *pgsA,* in which there is no possibility of residual enzymatic activity (Raetz, C.R.H., unpublished).

As an approach to this problem, we have isolated second-step mutants, starting with one of the 25 available, partially defective *pgsA* strains as the parent [55]. In this manner, a temperature-sensitive mutant has been generated, which stops synthesizing phosphatidylglycerol at 42°C and in which the level can be reduced from about 15% at 30°C to 1% after 3 h at 42°C (Table 2). Unexpectedly, the second step mutation introduced into this strain, which is termed *pgsB,* is not a second alteration in the *pgsA* structural gene, but rather maps at a distinct site [87], near minute 4 on the chromosome (see Fig. 6).

The *pgsB* mutation confers several interesting features on strains harboring lesions in *pgsA* [55,87]. These are: (1) temperature-sensitive growth and defective phosphatidylglycerol synthesis, at 42°C; (2) temperature-sensitive net synthesis of the phosphatidylglycerophosphate synthase enzyme, upon shifting of the cells to 42°C [55]; and (3) accumulation at 42°C of two novel glycolipids, which are partially acylated lipopolysaccharide precursors lacking KDO or other sugars [55,117]. All three abnormalities are corrected by introduction of either the *pgsA*$^+$ or the *pgsB*$^+$ gene into the double mutant [55,87], despite the considerable distance between these genes on the *E. coli* linkage map (Fig. 6).

The necessity that both genes be defective for the expression of the above traits implies that products specified by these genes may interact in normal cells. While the biochemical basis of the *pgsB* lesion remains unknown, its existence implies a previously unrecognized link between phosphatidylglycerol and lipopolysaccharide synthesis. This connection could be indirect, such as a common requirement for a processing step needed for the insertion of two biosynthetic enzymes into the cytoplasmic membrane. The double mutant (of which there is just one isolate) represents the only method for eliminating phosphatidylglycerol from *E. coli* membranes [55,87]. As in most other instances, techniques for the selective enrichment of *pgs* mutants would be very desirable.

Using [3-^{32}P]glycerol-3-phosphate and CDP-diacylglycerol, we have recently developed a coupled, two-step assay for phosphatidylglycerophosphate (Icho, T. and Raetz, C.R.H.), analogous in principle to that described above for CDP-diacylglycerol hydrolase. There appears to be more than one phosphatidylglycerophosphate in the membranes of *E. coli.*

(b) Cardiolipin synthase (cls)

A careful study by Pluschke et al. [53] reported the isolation of one mutant lacking cardiolipin in vivo. This strain was found using the "brute force" approach and the random mutant collection of Hirota and coworkers [53].

The absence of cardiolipin in vivo is correlated with a deficiency of the cardiolipin synthase, assayed by the method of Hirschberg and Kennedy [118]. Depending on growth conditions, the mutant contains 10–50 times less of this lipid than the parental strains [53]. There is a slight compensatory rise in the amount of phosphatidylglycerol. Whether cardiolipin is altogether nonessential, or whether the small residual level maintains membrane functions dependent on cardiolipin is uncertain. To resolve this it will be necessary to isolate deletion mutants missing the *cls* gene entirely. This may not be too difficult, since the location of the *cls* gene has been accurately determined (Fig. 6).

In addition to characterizing the genetics and biochemistry of their mutation, Pluschke et al. [53] examined the growth of the bacteriophage fl and found it to be unaffected. In wild-type cells this virus causes considerable accumulation of cardiolipin [119], while in the *cls*$^-$ mutants it causes a build-up of phosphatidylglycerol [53]. This observation implies that the elevated level of cardiolipin associated with fl infection of wild-type cells results from increased phosphatidylglycerol synthesis rather than decreased cardiolipin turnover.

10. E. coli mutants in membrane lipid turnover and catabolic enzymes

(a) Mutants unable to generate membrane-derived oligosaccharides

In *E. coli* and related Gram-negative bacteria, the polyglycerophosphatides turn over much more rapidly than phosphatidylethanolamine [1]. The turnover of phosphatidylglycerol, in part, is due to its conversion to cardiolipin [1]. Transfer of the glycerol-1-phosphate headgroup of phosphatidylglycerol to precursors of the membrane-derived oligosaccharide (MDO) is the second major cause of turnover (see Fig. 5) [76–79]. Phosphatidylglycerol also serves as a precursor to the glycerol moiety of the Braun lipoprotein [120,121], but this represents a minor route.

The MDO are a family of periplasmic substances with M_r-values of approx. 2000 [76,77]. The carbohydrate portion consists entirely of glucose [76,77,122]. Charge heterogeneity is caused by differential modification with glycerol-1-phosphate and succinate [76,77]. The MDO represent about 1% of the dry weight of wild-type *E. coli,* and their function is unknown [1], although their synthesis is inhibited by high osmolarity (Kennedy, E.P., personal communication). The enzymes responsible for the assembly of MDO, especially the transfer of glycerol-1-phosphate moieties from phosphatidylglycerol, have not been characterized.

That the glycerol-1-phosphate transfer to MDO should account for a major portion of phosphatidylglycerol turnover can be deduced from the relative amount of MDO and phosphatidylglycerol [1,76], and from studies with mutants unable to generate the membrane-derived oligosaccharides [78,79]. Mutants blocked in the formation of UDP-glucose or strains defective in gluconeogenesis (discussed above) can be used for this purpose [78,79]. Although cell growth is not inhibited by these lesions, the cells are unable to synthesize MDO, and the turnover of phosphatidylglycerol is reduced considerably [78,79]. The origin of the slow residual rate of phosphatidylglycerol turnover (also observed with phosphatidylethanolamine) in this setting is uncertain (Raetz, C.R.H., unpublished). In any event, MDO synthesis and the rapid phosphatidylglycerol turnover it produces are not essential for cell division or growth [1,86–79].

(b) Mutants in catabolic enzymes (pldA)

At least nine enzymes catalyze some form of phospholipid degradation in *E. coli* [1]. This includes two phospholipases and two lysophospholipases [1]. Mutants defective in the predominant, detergent-resistant phospholipase (*pldA*) associated with the outer membrane have no obvious phenotype [88,89], although the fatty acid release usually associated with T4 or λ infection does not occur [123,124]. As in the case of cls^- mutants, deletions or insertion mutations of *pldA* have not been studied. Also, multiple mutants lacking all four of the major lipases have not been constructed.

The elucidation of the function of catabolic enzymes and their possible relationship to the slower phases of phospholipid turnover (noted above) deserve further study. Mutant screening schemes based on colony autoradiography or histochemistry in situ could probably be developed for the lipases. Genetic studies of phospholipid catabolism in eukaryotic systems are non-existent but might be fruitful, since all membrane fractions, including lysosomes, have some hydrolytic capacity [125,126] (See also Chapter 9). Despite the initial, discouraging results obtained with the *pldA* mutants of *E. coli* [1], the possibility that certain lipases or combinations of lipases act in concert to control phospholipid composition cannot be eliminated.

11. Molecular cloning of E. coli *genes coding for the lipid enzymes*

During the past 5 years, the rapid development of molecular cloning technology has permitted the construction of bacteria containing multiple copies of specific genes or gene clusters [24,25,127–128]. Especially convenient as a bridge between genetic and biochemical studies is a collection of 2000 *E. coli* strains prepared by Clarke and Carbon [25], each of which carries a hybrid colE1 plasmid into which a unique fragment of *E. coli* DNA has been inserted. Such hybrid plasmids are maintained at 5–20 copies per cell [25]. The average M_r-value of the *E. coli* inserts in this particular collection is about $8 \cdot 10^6$ (approx. 0.25% of the *E. coli* chromosome). Each strain in the collection contains a different *E. coli* fragment, representing virtually every gene [25].

TABLE 3

Overproduction of phospholipid enzymes by gene cloning techniques

Cloned gene	Original plasmid vectors	M_r of *E. coli* DNA insert ($\cdot 10^6$)	Multiplication factor for enzyme overproduction	Trivial designation for hybrid plasmid	Ref.
gpsA	pMB9	3	60	pDC2	11
plsB	colE1	approx. 8	4–8	pLC9-28	12
	pACYC184	3	8–12	pVL1	
dgk	colE1	approx. 8	3–14	pLC9-28	12
	pACYC184	0.96	17	pVL1P4	
pss	colE1	11	17	pLC34-44	9.129
	pBR322/λNOP	2.2	80–140	pPS3155λ	
pgsA	pSC101	5	8	pPG2	130, 131
	pBR322	1.0	18–20	pPG2-10	
pgsB	colE1	approx. 8	1	pLC26-43	87
psd	colE1	approx. 8	30–50	pLC8-47	10
cdh	colE1	approx. 8	4–6	pLC16-4	25

An extremely useful application of the phospholipid mutants is the fact that they facilitate the identification of those few hybrid plasmids in the Clarke and Carbon colony bank which carry the *E. coli* DNA coding for the phospholipid enzymes. In practice, this is done by transferring the individual hybrid plasmids (using replica-plating techniques) from the Clarke and Carbon collection to a specific lipid mutant as the recipient [9,10,12,87]. Restoration of an enzymatic defect and correction of the associated phenotype (if any) in the recipient indicates that the inserted DNA of a particular plasmid (each of which is identified by a number) represents the desired phospholipid gene.

Distinct hybrid plasmids bearing *plsB, dgk, pss, psd, pgsA, pgsB cdh* and *gpsA* have been identified (Table 3). Most of these have been found in the Clarke and Carbon collection, although some have been constructed separately (Table 3). When such strains are assayed for the enzyme carried on the hybrid plasmid, specific overproduction is generally observed, corresponding to the maintenance of the hybrid plasmids in multiple copies per chromosome [9–12,87,129–131]. Depending on the plasmid vector, conditions of cell growth, and properties of the DNA insert, specific overproductions as high as 150-fold [129] have been achieved (Table 3).

To demonstrate that the hybrid plasmids are causing enzyme overproduction (rather than activation), some lipid enzymes such as the biosynthetic glycerol-3-phosphate dehydrogenase (*gpsA*) and the phosphatidylserine synthase (*pss*) have been purified to homogeneity from normal and plasmid-bearing cells [9,11,129]. As shown for phosphatidylserine synthase in Table 4, the specific activity of crude extracts is much higher in the case of the plasmid-bearing strains, while the specific activities of the homogenous enzymes are virtually identical [9]. With phosphati-

TABLE 4

Purification of phosphatidylserine synthase to homogeneity from wild-type (A324) and *pss*$^+$ hybrid plasmid-bearing strain (RA324)

Step	A324 [a]		RA324 [b]	
	Specific activity, units/mg	Yield %	Specific activity, units/mg	Yield %
1. Broken cells	7.1 [c]	100	120 [c]	100
2. Cell supernatant	7.3	85	140	97
3. Streptomycin sulfate	27	88	400	83
4. Polymer partitioning	21	42	320	50
5. Ammonium sulfate	47	33	820	42
6. Phosphocellulose	–	20	–	19
7. DEAE-Sephadex	34000±15% [d]	16	39000±15% [d]	13
Protein yield	1.9 mg		8.5 mg	

[a] Started with 320 g wet weight of cells.
[b] Started with 150 g wet weight of cells.
[c] For the purification of the synthase, the enzymatic activity was determined at 30°C under the conditions described by Larson and Dowhan [136].
[d] Standard deviation of four determinations of the specific activity. These data are taken from reference [9] with permission of the publisher.

dylserine synthase overproduction now as high as 150-fold [129], it is possible to isolate hundreds of mg of this enzyme, while previously even a few mg were difficult to obtain. Thus, a variety of chemical and physical studies, including X-ray crystallography, can be contemplated for these enzymes in the near future.

Molecular cloning has also facilitated the purifications of the glycerol-3-phosphate acyltransferase (*plsB*) [132] and the phosphatidylserine decarboxylase (*psd*) [10]. The former has never actually been purified to homogeneity from plasmid-free cells. The latter can be overproduced as much as 45-fold by brief isoleucine starvation of cells in late exponential phase [10]. Under these conditions of extreme overproduction, about half of the decarboxylase is recovered in the soluble fraction rather than in the membrane [10]. Perhaps membrane binding sites have become limiting, or alternatively, an essential processing step required for membrane insertion cannot keep up with the large amounts of polypeptide made from the hybrid plasmids. With all other genes examined so far (*dgk, plsB, pss* and *pgsA*) the overproduced enzyme has the same subcellular localization as in plasmid-free cells [9,11,12,129–131].

Since it is possible to isolate chemical amounts of hybrid plasmid DNA, free of chromosomal DNA [24], the structure and function of the DNA coding for each of the lipid enzymes can be examined directly. The technology for base sequencing of DNA has become so simple [26,27] that it is preferable to infer the amino acid sequence of the lipid enzymes from their DNA sequence than to analyze them directly. For example, the nucleotide sequences of the closely linked diacylglycerol kinase (*dgk*) and glycerol-3-phosphate acyltransferase (*plsB*) genes (Fig. 6) have recently been completed (Lightner, V.A., Bell, R.M. and Modrich, P., personal communication). Transcription of these two genes occurs bidirectionally from a central starting region, which is about 170 base pairs in length. The nucleotide sequence of the *plsB* gene is in excellent agreement with the M_r-value [132] and the amino acid composition of the isolated protein. The *dgk* gene, which is the likely structural gene for the kinase, codes for a putative polypeptide that is exceptionally hydrophobic, consistent with the observed in butan-1-ol solubility of this enzyme from *E. coli* [138].

In addition to providing the primary structure of the mature protein, the nucleic acid sequencing will also reveal the presence (or absence) of leader peptides which may be required for membrane insertion [133,134] and will facilitate identification of regulatory sites, such as promoters, which are adjacent to the structural genes [133,134]. The isolated DNA can also be used to direct the in vitro synthesis of the phospholipid enzymes (Dowhan, W., personal communication), which could reveal protein factors required for transcription and translation.

E. coli strains with elevated enzyme levels provide new tools for studies of phospholipid metabolism. Surprisingly, examination of cloned *plsB, pss* and *pgsA* genes has failed to reveal any major perturbations of cellular lipid composition corresponding to the extent of enzyme overproduction [9,129–131,135]. The enzymes may already be present in excess normally, or they may be down-regulated in vivo by mechanisms which cannot be assessed in vitro.

12. Further genetic approaches to the control of E. coli *phospholipid gene expression*

Despite the progress in defining the gene coding for the enzymes of phospholipid synthesis (Fig. 6), the fundamental questions posed at the beginning of this chapter concerning the regulation of phospholipid metabolism remain unanswered. From the study of mutants (Table 2), it is clear that genetic defects in the biosynthetic enzymes can render any step of phospholipid synthesis rate-limiting, but this does not necessarily pinpoint the actual regulatory mechanisms that are operative in wild-type cells. Perhaps, mutants with altered lipid compositions, but not defective in the biosynthetic enzymes, might provide some insight into regulation. Conversely, phenotypic revertants of mutants defective in biosynthetic enzymes which retain the enzymatic lesion but are bypassed by some secondary mutation would be useful.

The mechanisms which set the cellular level of each phospholipid enzyme are also unknown. Though not extensively studied, the synthesis of the phospholipid enzymes in wild-type *E. coli* is not dramatically altered by changing the conditions of culture [1]. As indicated above (Table 3), introduction of multiple gene copies invariably leads to enzyme overproduction. Those enzymes which have been purified to homogeneity each represent between 0.01 and 0.1% of the total protein, approximately equivalent to 1000 polypeptide chains per enzyme per wild-type cell [116,132,136–138]. Since the phospholipid enzymes are generally integral membrane proteins, it is conceivable that there is post-translational modification or proteolysis in conjunction with membrane insertion. In view of the relatively fixed cellular demand for phospholipids, the level of gene expression could be determined solely by the efficiency of the promoter adjacent to each structural gene.

To determine whether or not regulatory signals other than promoters exist for the enzymes in this system, we have recently developed a general strategy for detecting *E. coli* mutants with elevated levels of phospholipid gene expression [138a]. To do this we have used the rapid in situ autoradiographic procedure described above [16], except that short-term assays have been employed to locate colonies with greater than normal enzymatic activity. Out of 20000 colonies derived from a stock of mutagen-treated cells, we have recently identified four strains in which the level of diacylglycerol kinase is 5–10 times higher than normal [138a]. Other phospholipid enzymes (Fig. 5) are unaffected. In some of these mutants the selective elevation has been shown to result from an alteration in a new gene, designated *dgkR*1 (see Fig. 6), which maps at a distinct site several minutes away from the structural gene, termed *dgk*. The considerable genetic and physical distance between these two genes excludes the possibility that the *dgkR* locus represents a promoter. Further, the product of the *dgkR*1 gene appears to act in a *trans* fashion, since introduction of hybrid colE1 plasmids carrying the *dgk* structural gene [12] into a mutant harboring *dgkR*1 results in a specific multiplicative overproduction of the kinase (Table 5). This means that the *dgkR*-1 mutation acts on each copy of the *dgk* structural gene present in these strains. The resulting specific activity of the kinase is about 73 times higher than normal, a factor of 12.9 being contributed by the presence of the multiple *dgk* structural genes and a factor of 5.6 by the presence of the regulatory mutation.

TABLE 5

The *dgkR*-1 mutation acts in trans on multiple copies of the *dgk* structural gene cloned on a hybrid colE1 plasmid

Cell-free extracts were prepared from fresh overnight cultures grown on LB broth, as described elsewhere [56]. Precision of duplicate determinations is approx. ±10%. The number in brackets (bottom row, right side) indicates the value expected for a perfect multiplicative interaction between the hybrid dgk^+ plasmids and the *dgkR*-1 mutation (i.e. 12.9×5.6).

Strain	Chromosomal genotype	Plasmid genotype	Diacylglycerol kinase specific activity nmol/min/mg protein	Observed ratio
R477	dgk^+ $dgkR^+$	None	3.4	1.0
R477/pLC9-28	dgk^+ $dgkR^+$	hybrid colE1-dgk^+	44	12.9
GK1H	dgk^+ *dgkR*-1	None	19	5.6
GK1H/pLC9-28	dgk^+ *dgkR*-1	hybrid colE1-dgk^+	250	73.4 (72.2)

Although the biochemistry of the *dgkR*-1 lesion has not been defined, it is very likely that there are novel proteins (or metabolites) which control the expression of at least some of the phospholipid enzymes. Such regulation might involve transcriptional, translational, or even post-translational mechanisms. To explore the generality of such regulatory loci we have recently isolated a new mutant with 4–6 fold elevation of phosphatidylserine synthetase (Sparrow, C.P. and Raetz, C.R.H.). The latter is designated *pssR,* is also *trans*-acting, and causes actual polypeptide overproduction as judged by purification.

An additional, unexploited approach would involve the isolation of mutants that are temperature-sensitive for the synthesis of specific enzymes. The *pgsB* mutation fits this description [55,87], though it was not specifically isolated for this purpose. To find mutants that are temperature-sensitive for the synthesis of phospholipid enzymes, one could make a replica plate of mutagen-treated colonies grown at 30°C and shift the replica plate to 42°C for a period of 6–8 h. Following this it is still possible to perform in situ enzymatic assays by the filter paper procedure. During the 6–8 h at 42°C any normal enzyme made at 30°C would be diluted out by continued cell growth. Loci involved in essential processing or modification steps could certainly be detected in this way.

13. Choline and inositol auxotrophs of fungi and yeasts

(a) Neurospora crassa

The classical genetic studies of Beadle and Tatum [61] in the early 1940s demonstrated that single gene mutations of *Neurospora crassa* could lead to defects in the synthesis of defined chemical substances. They isolated 380 mutants by examining 68000 ascospores derived from a mutagen-treated stock. Some of these variants required specific amino acids, B group vitamins or nucleic acid bases for growth. By 1944, mutants dependent on choline or inositol had already been identified in this collection, and provided sensitive microbiological assays for the quantitation of these substances in biological samples [62,63]. The inositol-less mutants died especially rapidly in the absence of supplementation [139], but simultaneous inhibition of protein synthesis prevented cell death. This sparing phenomenon has been used to enrich for mutants defective in many other metabolic processes starting with an inositol-less parent [140]. The molecular basis and primary cause of inositol-less death are unknown.

Biochemical analyses of the choline auxotrophs by Scarborough and Nyc has revealed the existence of two subclasses [141,142]. One type is defective in the phosphatidylethanolamine methyltransferase (EC 2.1.1.7), while the other is lacking the phosphatidylmonomethylethanolamine (phosphatidyldimethylethanolamine)-methyltransferase. Absence of this major de novo synthetic mechanism causes a choline dependency, since choline can still be utilized for lecithin synthesis in *Neurospora* via the CDP-choline pathway of Kennedy and Weiss [143]. The inositol auxotrophs have not been studied as extensively as the choline requiring strains but almost certainly are defective in the generation of *myo*-inositol from glucose-6-phosphate [19].

The use of choline and inositol auxotrophs to perturb the phospholipid composition of the membranes of *Neurospora crassa* has received little attention until recently. In a careful study, Hubbard and Brody [19] have analyzed the composition of zwitterionic and anionic phospholipids in wild-type and auxotrophic cells, subjected to choline or inositol limitation. In the choline auxotrophs, extensive replacement of phosphatidylcholine by phosphatidyldimethylethanolamine, phosphatidylmonomethylethanolamine or phosphatidylethanolamine is possible, depending on the supplement added to the medium [19]. Dimethylethanolamine can replace choline almost entirely as the major polar head group, without inhibiting growth, and other extensive modifications are also possible [19]. Despite the considerable variation in the relative proportions of different zwitterionic phospholipid species caused by these substitutions, the sum of all phospholipid species and the ratio of total zwitterionic to total anionic species remain virtually constant [19]. Similar lipid modification studies with the inositol auxotrophs demonstrate that the phosphatidylinositol depletion in this sytem is correlated with the accumulation of phosphatidylserine, possibly because CDP-diacylglycerol is their common precursor [19,144]. These studies suggest the existence of compensatory mechanisms to maintain certain critical parameters of lipid composition (i.e., total content and charge). Various membrane functions have not been examined in the *Neurospora* membranes subjected to lipid modification.

In contrast to *Neurospora crassa,* higher eukaryotic cells like mouse fibroblasts do not generate phosphatidylcholine by methylation of phosphatidylethanolamine [60,75,145]. Instead, they require choline for growth and utilize the CDP-choline pathway [60,75]. Work by Glaser and collaborators has shown that replacement of choline in the growth medium by either ethanolamine, monomethylethanolamine or dimethylethanolamine leads to extensive incorporation of these headgroups into the phospholipids [75,145]. As in *Neurospora,* dimethylethanolamine can replace choline almost entirely as the major headgroup, yet the cells can proliferate normally [75]. Although nonphysiological choline analogs are also incorporated, they do not support prolonged cell growth [75,145].

(b) Saccharomyces cerevisiae *and other yeasts: inositol auxotrophs*

The isolation of inositol-requiring mutants of *Saccharomyces cerevisiae* and their detailed study by mapping, complementation and biochemical analysis have been reported by Culbertson and Henry [20]. A series of 52 inositol auxotrophs of *S. cerevisiae* are available, and these consist of ten independently segregating loci [20]. A total of 36 representatives exist for the *ino*-1 region alone. Complementation studies suggest the possible existence of 13 further subclasses within *ino*-1, but the possibility of intracistronic versus intercistronic complementation has not been resolved [20].

Enzymatic analyses of mutants in all 10 inositol loci have revealed 100 to 10000-fold reductions in the conversion in vitro of glucose-6-phosphate to inositol phosphate and inositol [146]. Considering the relatively large number of genes identified in this study compared to the few enzymatic steps required for the conversion of glucose-6-phosphate to inositol, it may be that structural as well as regulatory genes exist for the inositol system [20,146]. Essentially all of the inositol-less mutants of yeast manifest the same phenomenon of inositol-less death noted above [20,147], and as in the case of the *Neurospora* inositol auxotrophs, the inclusion of inhibitors of protein synthesis such as cycloheximide during inositol starvation provides complete protection against loss of cell viability [148].

The inositol-containing lipids of *S. cerevisiae* constitute about one-third of the total membrane lipid [21]. Phosphatidylinositol is a major component of this material, but in addition, diphosphoinositides and triphosphoinositides have been detected [21,149,150]. Phosphoinositol-containing sphingolipids are also present in yeast [151–154] but are generally absent in higher eukaryotes (see also Chapter 7).

Using the *ino*-1-13 auxotroph of *S. cerevisiae,* Becker and Lester [21] have studied the changes of phospholipid composition resulting from inositol deprivation. After 20 h of starvation, there is about a 10-fold decrease in the content of phosphatidylinositol, accompanied by a massive accumulation of phosphatidic acid and CDP-diacylglycerol [21], which are presumably precursors of this material [144]. The inositol-containing sphingolipids continue to be made, suggesting that they may be derived from phosphatidylinositol [21]. As in the case of *Neurospora,* the ratio of anionic to zwitterionic lipids remains relatively constant [21]. Whether or not the changes in membrane lipid composition are directly responsible for inositol-less death remains uncertain, since other as yet unidentified inositol metabolites could also play a critical role in cell physiology. For this reason it would be desirable to isolate mutants specifically defective in the enzymatic conversion of CDP-diacylglycerol to phosphatidylinositol.

Various yeasts isolated from natural sources, such as *Saccharomyces carlsbergensis,* are inherently inositol auxotrophs (without mutagenesis). Though not as extensively studied, the metabolism of lipids is also altered in these yeasts upon inositol deprivation [155,156]. In *S. carlsbergensis* the accumulation of neutral lipids, including triacylglycerols, is especially prominent. The metabolic basis for this response is not fully understood [156].

(c) Choline auxotrophs of S. cerevisiae

Choline-requiring mutants of *S. cerevisiae* were isolated by Atkinson et al. [22,157]. These investigators treated *S. cerevisiae* with the mutagen ethylmethanesulfonate and identified colonies by replica plating dependent on 1 mM choline for growth [22]. Three independently isolated strains were obtained, and the mutation causing the choline requirement was recessive [22]. The three isolates failed to complement each other [22], suggesting identical lesions in all cases [22].

Unlike the choline auxotrophs of *Neurospora* discussed above, the choline auxotrophs of *S. cerevisiae* isolated by Atkinson et al. [22,157] can grow in the presence of either choline or ethanolamine. This eliminates the possibility of defects in the methyltransferase. The phospholipid compositions of these strains are interesting, since phosphatidylserine is largely absent, even when the cells are grown in the presence of choline [22,157]. This suggests that the primary defect in these mutants is a deficiency in the formation of phosphatidylserine, which in *S. cerevisiae* is probably generated from CDP-diacylglycerol and serine [144], as it is in *E. coli* [1]. Higher eukaryotes do not possess this CDP-diacylglycerol-dependent mechanism for phosphatidylserine formation [3,71].

As a consequence of phosphatidylserine depletion, the cells are unable to generate sufficient endogenous phosphatidylethanolamine by phosphatidylserine decarboxylation [157] and hence require either ethanolamine or choline for growth in order to generate adequate amounts of phosphatidylcholine. A defect in the generation in vitro of phosphatidylserine in these mutants has also been documented [157]. These results demonstrate that in *S. cerevisiae* the normal levels of phosphatidylserine (i.e. approximately 6% of the total lipid) are not essential for maintenance of functions involved in cell growth or division [26,157].

As yet, mutants defective in the methylation of phosphatidylethanolamine have not been identified in *S. cerevisiae.* However, Yamashita and Oshima have recently observed an interesting interaction between inositol and choline metabolism in *Saccharomyces* [158]. They found that inclusion of inositol in the growth medium of wild-type yeasts reduces the specific activity of the phosphatidylethanolamine methyltransferase by 4- to 10-fold [158]. Removal of inositol from the growth medium results in a restoration of high levels of this enzymatic activity [158]. In this setting, they have isolated a mutant of yeast which is unable to grow in the presence of inositol unless also supplemented with choline [158]. Presumably, the mutant is more sensitive to the repression of the phosphatidylethanolamine methyltransferase than wild type [158]. The exact biochemical nature of the mutation is unknown [158], but Yamashita and Oshima have also reported *S. cerevisiae* mutants defective in choline transport and choline kinase derived from the above [158] by a second round of mutagenesis [159]. As yet, these mutants have not been subjected to intensive biochemical investigations [159].

In summary, the genetic modifications of phospholipid metabolism in yeasts and fungi have been confined to studies of inositol and choline auxotrophs, since these precursors are rapidly taken up from the medium. The supplementation of such lower eukaryotes with intact phospholipids has not been attempted, and it is unlikely to work in view of the thick cell walls which surround these organisms. Mutants defective in the synthesis and processing of the intermediates of the phospholipid pathways have not been obtained, nor are mutants in regulation of phospholipid synthesis available in these systems. The scope of phospholipid genetics in lower eukaryotes could be extended considerably by the use of colony autoradiography [16], which is applicable to both *Neurospora* and *Saccharomyces* [16,59]. Further, the techniques for gene cloning are well advanced with *S. cerevisiae* [65,66] and would greatly facilitate the isolation of the phospholipid genes.

14. Genetic modification of membrane phospholipid synthesis in mammalian cells

(a) Characterization of inositol auxotrophs of CHO cells

Using *myo*-inositol auxotrophs of CHO cells, we have studied the effects of *myo*-inositol depletion on phospholipid metabolism in this system [17,71]. After three days of inositol starvation, the phosphatidylinositol level in the mutant decreases from about 7.1% to 0.8%, while there is a compensatory rise in the amount of phosphatidylglycerol from 0.7% in the presence of *myo*-inositol, to 10% in its absence. The amount of cardiolipin remains unaltered during this modification. Similar, but less dramatic, lipid alterations also occur upon inositol starvation of parental cells, which retain viability and continue to grow at a normal rate.

The phospholipid modifications resulting from inositol depletion of this *myo*-inositol CHO auxotroph differ strikingly from those observed with *myo*-inositol auxotrophs of lower eukaryotes, which accumulate either phosphatidic acid and CDP-diacylglycerol or phosphatidylserine, as discussed above [19,21]. Phosphatidylglycerol accumulation in these CHO cells occurs to the same extent in crude mitochondrial and microsomal membranes, despite the predominant localization of phosphatidylglycerophosphate synthase in mitochondria and phosphatidylinositol synthase in microsomes [71]. These results imply that phosphatidylglycerol and CDP-diacylglycerol are able to flow in vivo between subcellular organelles and that the latter is accessible to both biosynthetic enzymes. Evidence for phosphatidylglycerol and CDP-diacylglycerol translocation between organelles in vitro has also recently been described [3,160,161]. It will be of great interest to examine various membrane functions in strains depleted of phosphatidylinositol.

(b) Autoradiographic detection of CHO mutants defective in phosphatidylcholine synthesis

We have recently screened 20000 colonies of CHO cells derived from a mutagen-treated stock for variants unable to incorporate [methyl-^{14}C]choline into

Fig. 7. [Methyl-^{14}C]choline autoradiography of CHO cell colonies immobilized on filter paper. Mutagen-treated cells (Experimental Procedures) were dispersed with trypsin [60] and placed in 100-mm diameter tissue culture dishes to yield approx. 200 colonies/plate at 33°C. After 1 day the cells were overlayed with a disc of Whatman No. 50 filter paper and glass beads [17,60] and incubated at 33°C for another 16 days. After aspirating the medium and decanting the beads the filter paper was removed from the dish with sterile tweezers and placed cell-side-up on a sterile metal or glass pan tilted at a 60° angle [17,18]. The surface tension of the residual medium held the disc firmly against the pan. The paper was then rinsed *vigorously* with a 30-ml stream of medium lacking serum to remove loose cells. Next, it was placed cell-side-up on top of an even layer of glass beads in a dish filled with enough medium (containing 10% dialyzed serum) to keep the paper moist. After 16 h at 40°C the disc was placed on an absorbent paper towel to remove excess moisture and was transferred to another dish containing 1 ml of growth medium supplemented with 0.1 mM [methyl-^{14}C]choline (10 μCi/μmol). After 4 h of further incubation at 40°C, the paper was treated with 1 ml of 10% trichloroacetic acid, which resulted in the precipitation of

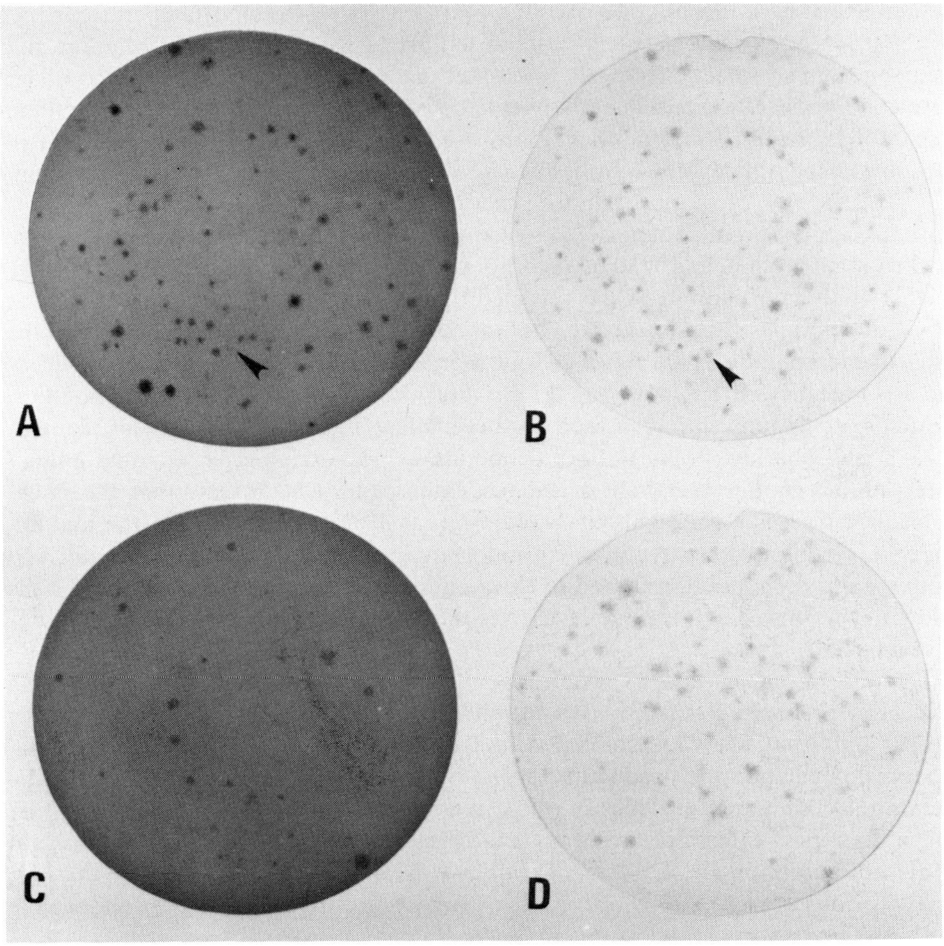

the choline-linked phospholipids. Unincorporated radioactive precursor was removed by placing each disc in a Buchner funnel and passing five 50-ml volumes of 2% trichloroacetic acid through the paper under vacuum [60]. The papers were then dried overnight at room temperature and exposed to Kodak XR-5 X-ray film for 4 days. After autoradiography the papers were stained overnight with 0.05% Coomassie Brilliant Blue G in 10% acetic acid to visualize all the colonies. To destain the papers they were stacked in a beaker containing 300 ml of methanol–water–acetic acid (45 : 45 : 10, v/v) and stirred at 37°C for about 1 h. Several changes of destaining solution resulted in the appearance of bright blue colonies on a virtually white background. Throughout these manipulations, the master plate was stored at 28°C under otherwise normal growth conditions in medium supplemented with 10% serum, 20 U/ml Mycostatin, and 2.5 μg/ml Fungizone. Mutants identified as blue-staining colonies lacking an autoradiographic halo (arrow indicates position of mutant 58) were retrieved with glass cloning cylinders [60, 162] from the master plates. All candidates were passed through the above cloning procedure one more time to achieve their complete purification. (A) and (C) represent the autoradiograms from the original mutant screening and the subsequent repurification, respectively, while (B) and (D) are the corresponding stained filter papers. (Reprinted from [60] with permission of the publisher.)

trichloroacetic acid-precipitable phospholipid (Fig. 7) at elevated temperatures, i.e. 40°C [60]. [Methyl-^{14}C]choline is metabolized primarily to phosphatidylcholine and sphingomyelin by intact CHO cells, and other macromolecules are not labeled under these conditions [60]. Consequently, viable cells immobilized on paper (rather than preparations made permeable) have been used in these experiments, permitting all steps of phosphatidylcholine synthesis to be screened simultaneously [17,18]. Mutant 58, identified by this approach, is specifically defective in phosphatidylcholine synthesis while several other isolates also obtained from the same screening are blocked in thymidine and leucine incorporation as well as choline metabolism [60]. Further analysis of mutant 58 has revealed that the strain grows almost normally at 33°C, the permissive temperature, but divides only once or twice at 40°C, the restrictive temperature [60]. After 20 h of incubation at 40°C, the phosphatidylcholine level declines from 41% to 21% in the mutant, while other phospholipids, including sphingomyelin, continue to be made [60,60a]. Parental cells contain 50–58% phosphatidylcholine at both temperatures. Ion-exchange chromatography of water-soluble choline metabolites isolated from mutant 58 reveals that the phosphocholine level is elevated about 3-fold, both at 33°C and at 40°C in the mutant, while CDP-choline decreases from 0.4 nmol/mg protein to less than 0.07 nmol/mg protein when the mutant is shifted to elevated temperatures [60,60a]. Wild-type cells maintain the same CDP-choline level (0.5–0.6 nmol/mg protein) at both temperatures [60,60a].

To confirm that mutant 58 is defective in the synthesis of CDP-choline, extracts have been prepared from mutant and wild-type cells. A 40-fold reduction in the specific activity of the CDP-choline synthase is observed in mutant 58, and mixing experiments exclude the production of inhibitors of CDP-choline synthesis by the mutant [60,60a]. Other enzymes of phosphatidylcholine synthesis are unaffected by this mutation. Temperature-resistant revertants derived from mutant 58 regain nearly normal levels of CDP-choline synthase [60a]. These studies provide the first genetic evidence that CDP-choline is the primary precursor of the phosphocholine head group of phosphatidylcholine in any mammalian system.

The availability of mutants of this kind provides new approaches to studies of the regulation of the membrane phosphatidylcholine content and creates the possibility of eventually isolating and mapping the genes involved in phosphatidylcholine metabolism by analogy to the gene cloning studies already in progress with *E. coli* (see above). The continued synthesis of sphingomyelin under conditions of CDP-choline limitation [60] suggests that CDP-choline is not the direct precursor of sphingomyelin, but that a reaction involving lecithin as the donor of the phosphorylcholine head group is more likely [163,164].

The observation that mutant 58 is temperature-sensitive for growth in the presence of 10% bovine fetal serum (which is a component of the growth medium) is especially intriguing. This level of serum contributes approx. 20–50 μM choline-linked phospholipids to the growth medium, particularly phosphatidylcholine and lysophosphatidylcholine bound to lipoproteins (Esko, J.D. and Raetz, C.R.H., unpublished). If receptor-mediated uptake of lipoproteins [165,165a] could deliver some of this material intact to the appropriate subcellular membranes, the temperature-sensitive phenotype of mutant 58 should be bypassed. Since phenotypic suppression does not occur, it appears that CHO cells do not possess adequate mechanisms for lipoprotein uptake, or alternatively that phospholipids incorporated during lipoprotein endocytosis are extensively degraded. Mammalian lysosomes are known to contain a phospholipase C [126], and all sub-cellular membranes have phospholipase A activity [125].

Because of the inadequacy of serum, it is of interest that phosphatidylcholine dispersions added to the growth medium effectively suppress the phenotype of mutant 58 at 40°C (Fig. 8). Even colony formation from single cells is restored (data not shown). This finding suggests that there are mechanisms for the functional utilization of certain preparations of exogenous phospholipids during membrane biogenesis, and it demonstrates that the temperature-sensitive phenotype of mutant 58 can be attributed to the lesion in CDP-choline synthase [60,60a] and not some secondary mutation.

In addition to bovine liver and egg lecithin dispersions, phenotypic bypass can be achieved by chemically synthesized dipalmitoyl lecithin and by lysolecithin (Esko, J.D., Nishijima, M. and Raetz, C.R.H., unpublished). Serum lipoproteins can be removed by KBr flotation [167] without affecting the ability of either lecithin or lysolecithin to correct the growth defect. The lysolecithin bypass demonstrates that the acyltransferases specific for lysophospholipids, originally described by Lands [168,169], can be sufficient to support cellular growth, when de novo synthesis is blocked by mutation. The chemically detected phosphatidylcholine content of mutant

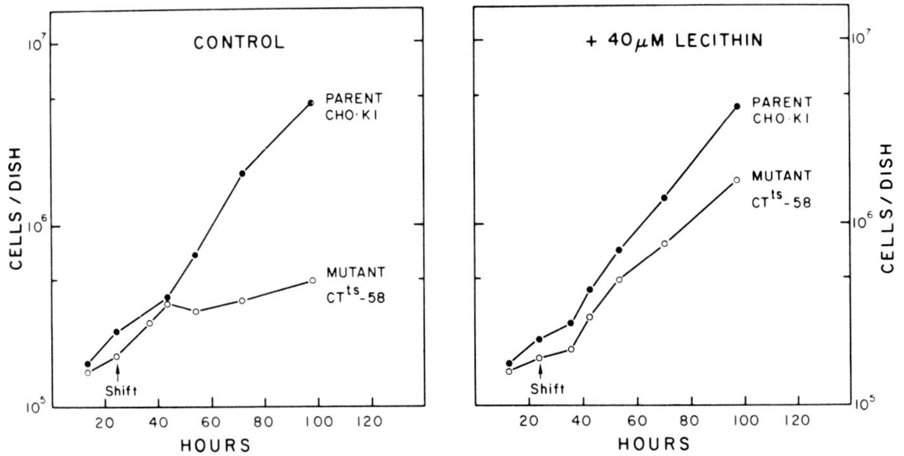

Fig. 8. Growth of parental CHO·Kl and mutant 58 cells at 40°C in the absence and presence of exogenous lecithin. Multiple 60 mm dishes were inoculated with about $2 \cdot 10^5$ cells and incubated at 33°C for 24 h. Thereafter cultures were shifted to 40°C in the absence (left panel) or in the presence (right panel) of 40 μM egg lecithin, added from a concentrated sonic dispersion. At indicated times cells from duplicate dishes were dispersed with trypsin [166] and counted on a Coulter Model B Counter. The author thanks J.D. Esko and M. Nishijima for providing the above data.

58 at 40°C is raised to about 80% of the wild-type level by the inclusion of lysolecithin in the medium (data not shown).

The possibility of bypassing phospholipid synthesis de novo by exogenous supplementation in CHO cells will be especially useful for the isolation of additional mutants in this process, since the frequency of observed mutations may be much higher if synthesis de novo is rendered non-essential. Conditionally lethal mutations are less frequent than lesions that cause an absolute defect independent of temperature. The possibility of introducing exogenous phospholipids into CHO cells in a functionally useful state differs from similar attempts to correct lipid lesions in mutants of *E. coli* [31]. As noted above, McIntyre and Bell [31] supplemented mutants defective in glycerol-3-phosphate acyltransferase with lysophosphatidic acid and observed extensive binding, but phenotypic bypass could not be demonstrated.

The mechanisms by which exogenous lipids enter cells to cause phenotypic bypass also deserve further study. Whether it is a simple fusion process [170] or is mediated by specific proteins (or surface receptors) remains to be determined. It is probable that the mechanisms for lysolecithin uptake will be different from those for lecithin incorporation. While mutant cells are capable of increasing their chemical phosphatidylcholine content, by utilizing some fraction of the supplement, it appears that wild-type cells do not do this. The inclusion of lecithin or lysolecithin in the growth medium of parental CHO cells does not alter their phosphatidylcholine content, suggesting that cells may regulate the net uptake of the exogenous lipid depending on their need for it (Esko, J.D., Nishijima, M. and Raetz, C.R.H., in preparation).

(c) Other assays in situ for detection of lipid enzymes in CHO colonies

Mutant 58 was isolated by permitting intact cells to incorporate [methyl-^{14}C]choline in vivo [60]. It is also possible to assay some of the mammalian phospholipid enzymes directly in situ by rendering the cells permeable and labeling with appropriate precursors in vitro [17,18]. For instance, the CDP-ethanolamine and CDP-choline phosphotransferase reactions can be assessed in colonies, and a variant with an altered ethanolamine phosphotransferase has recently been obtained [71a]. Excellent autoradiographic assays for the microsomal glycerol-3-phosphate acyltransferase, phosphatidylinositol synthase (CDP-diacylglycerol inositol phosphatidyltransferase, EC 2.7.8.11) [17] and phosphatidylglycerophosphate synthase have also been developed (Raetz, C.R.H., unpublished), but mutants are not yet available.

15. Summary

The identification by empirical methods of the genetic material coding for the phospholipid enzymes is beginning to provide a vast, new base of information on which to formulate hypotheses regarding membrane biogenesis and function. Features of phospholipid structure which are essential or non-essential for growth are becoming clearer, and this has implications for the role of phospholipid asymmetry and phospholipid physical properties in biological systems.

While most of the genetic studies to date have provided physiological verification of the metabolic schemes derived from earlier enzymological investigations, many new biochemical findings have been uncovered by this work. In *E. coli*, the study of the *dgk* locus [7,8] has explained the function of the kinase in a diglyceride recycling system, and the studies of the *dgkR* regulatory mutants (Table 5) have led to the inescapable conclusion that there are regulatory proteins (or metabolites) that help to determine the level of phospholipid gene expression. The study of phosphatidylglycerol genetics has revealed two interacting genes, a structural gene (*pgsA*) and a secondary gene (*pgsB*) which may provide a link between phosphatidylglycerol and lipopolysaccharide formation [55,87]. Two novel glycolipid species have been isolated in the course of this work, which have implications for the order of lipopolysaccharide assembly [117].

Gene cloning [9–12,129–131] has been especially productive in bridging biochemical and genetic studies and can be expected to provide a wealth of additional structural information in the near future. Cloning of the *psd* gene has revealed the existence of a soluble form of the decarboxylase, possibly an intermediate in maturation and membrane insertion [10]. Mapping and cloning studies of *plsB* and *dgk* have revealed a very close linkage of these two phospholipid genes, necessitating a search for common regulatory elements. The perturbation of the antibiotic resistance spectrum of *E. coli* cells harboring either the *pss* [85] or the *cds* [56,56a] mutations may prove useful for the design of new drugs and the treatment of gram-negative infections. Studies with inositol and choline auxotrophs of lower eukaryotes [19,21] have suggested the existence of precise mechanisms for the control of the phospholipid content and the polar headgroup charge.

Most exciting is the feasibility of extending phospholipid genetics to complicated, higher eukaryotic systems [17,18,60,71]. The characterization of CDP-choline synthase mutants [60] suggests that there are mechanisms for phospholipid uptake which can support cellular growth and which are enhanced by phospholipid depletion due to mutation. Work in higher eukaryotic systems is only beginning and should be complemented by similar studies of lower eukaryotic organisms such as *S. cerevisiae*. Fundamental mechanisms for the regulation of membrane biogenesis are certain to emerge.

Acknowledgements

I am indebted to Jeffrey Esko, Barry Ganong and William Dowhan for their critical reading of the preliminary version of this manuscript. I thank Sarah Green for her assistance and patience during the preparation of this article.

This work was supported in part by United States Public Health Service grants AM 21722, AM 19551 and 1KO4-AM00584.

References

1 Raetz, C.R.H. (1978) Microbiol. Rev. 42, 614–659.
2 Cronan Jr., J.E. (1978) Annu. Rev. Biochem. 47, 163–189.
3 Bell, R.M. and Coleman, R.E. (1980) Annu. Rev. Biochem. 49, 459–487.
3a. Bell, R.M., Ballas, L.M. and Coleman, R.A. (1981) J. Lipid Res. 22, 391–403.
4 Van den Bosch, H. (1974) Annu. Rev. Biochem. 43, 243–277.
5 Bloch, K. and Vance, D. (1977) Annu. Rev. Biochem. 46, 263–298.
6 Snyder, F. (Ed.) (1977) Lipid Metabolism in Mammals, Vols. 1, 2, Plenum, New York.
7 Raetz, C.R.H. and Newman, K.F. (1978) J. Biol. Chem. 253, 3882–3887.
8 Raetz, C.R.H. and Newman, K.F. (1979) J. Bacteriol. 137, 860–868.
9 Raetz, C.R.H., Larson, T.J. and Dowhan, W. (1977) Proc. Natl. Acad. Sci. USA 74, 1412–1416.
10 Tyhach, R.J., Hawrot, E., Satre, M. and Kennedy, E.P. (1979) J. Biol. Chem. 254, 627–633.
11 Clark, D., Lightner, V., Edgar, R., Modrich, P., Cronan Jr., J.E. and Bell, R.M. (1980) J. Biol. Chem. 255, 714–717.
12 Lightner, V.A., Larson, T.J., Tailleur, P., Kantor, G.D., Raetz, C.R.H., Bell, R.M. and Modrich, P. (1980) J. Biol. Chem. 255, 9413–9420.
13 Silbert, D.F., Cronan Jr., J.E., Beacham, I.R. and Harder, M.E. (1974) Fed. Proc. 33, 1725–1732.
14 Silbert, D.F. (1975) Annu. Rev. Biochem. 44, 315–339.
15 Cronan Jr., J.E. and Gelmann, E.P. (1975) Bacteriol. Rev. 39, 232–256.
16 Raetz, C.R.H. (1975) Proc. Natl. Acad. Sci. USA 72, 2274–2278.
17 Esko, J.D. and Raetz, C.R.H. (1978) Proc. Natl. Acad. Sci. USA 75, 1190–1193.
18 Esko, J.D. and Raetz, C.R.H. (1982) The Enzymes, in press.
19 Hubbard, S.C. and Brody, S. (1975) J. Biol. Chem. 250, 7173–7181.
20 Culbertson, M.R. and Henry, S.R. (1975) Genetics 80, 23–40.
21 Becker, G.W. and Lester, R.L. (1977) J. Biol. Chem. 252, 8684–8691.
22 Atkinson, K.D., Jensen, B., Kolat, A.I., Storm, E.M., Henry, S.A. and Fogel, S. (1980) J. Bacteriol. 141, 558–564.
23 Bachmann, B.J. and Low, K.B. (1980) Microbiol. Rev. 44, 1–56.
24 Sinsheimer, R.L. (1977) Annu. Rev. Biochem. 46, 415–438.
25 Clarke, L. and Carbon, J. (1976) Cell 9, 91–99.
26 Maxam, A.M. and Gilbert, W. (1977) Proc. Natl. Acad. Sci. USA 74, 560–564.
27 Wu, R. (1978) Annu. Rev. Biochem. 47, 607–634.
28 Miller, J.H. (1972) Experiments in Molecular Genetics, Cold Spring Harbor Laboratory, Cold Spring Harbor, New York.
29 Jones, N.P. and Osborn, M.J. (1977) J. Biol. Chem. 252, 7398–7404.
30 Jones, N.P. and Osborn, M.J. (1977) J. Biol. Chem. 252, 7405–7412.
31 McIntyre, T.M. and Bell, R.M. (1978) J. Bacteriol. 135, 215–226.
32 Inukai, M., Takeuchi, M., Shimizu, K. and Arai, M. (1979) J. Bacteriol. 140, 1098–1101.
33 Kaback, J., DeFillippe, L., Engel, R. and Tropp, B.E. (1972) J. Med. Chem. 15, 1074–1075.
34 Shopsis, C.S., Engel, R. and Tropp, B.E. (1972) J. Bacteriol. 112, 408–412.
35 Cheng, P.-J., Nunn, W.D., Tyhach, R.J., Goldstein, S.L., Engel, R. and Tropp, B.E. (1975) J. Biol. Chem. 250, 1633–1639.
36 Tyhach, R.J., Engel, R. and Tropp, B.E. (1976) J. Biol. Chem. 251, 6717–6723.
37 Cronan Jr., J.E., Ray, T.K. and Vagelos, P.R. (1970) Proc. Natl. Acad. Sci. USA 65, 737–744.
38 Godson, G.N. (1973) J. Bacteriol. 113, 813–824.
39 Cronan, Jr., J.E. and Godson, G.N. (1972) Mol. Gen. Genet. 116, 199–210.
40 Glaser, M., Nulty, W. and Vagelos, P.R. (1975) J. Bacteriol. 123, 128–136.
41 Esmon, B.E., Kensil, C.R., Cheng, C.C.-H. and Glaser, M. (1980) J. Bacteriol. 141, 405–408.
42 Harder, M.E., Beacham, I.R., Cronan Jr., J.E., Beacham, K., Honegger, J.L. and Silbert, D.F. (1972) Proc. Natl. Acad. Sci. USA 69, 3105–3109.
43 Ohta, A., Okonogi, K., Shibuya, I. and Maruo, B. (1974) J. Gen. Appl. Microbiol. 20, 21–32.

44 Ohta, A. and Shibuya, I. (1977) J. Bacteriol. 132, 434–443.
45 DeLucia, P. and Cairns, J. (1969) Nature 224, 1164–1166.
46 Weiss, B. and Milcarek, C. (1974) Methods Enzymol. 29, 180–193.
47 Matsuhashi, M., Takagaki, Y., Maruyama, I.N., Tamaki, S., Nishimura, Y., Suzuki, H., Ogino, U. and Hirota, Y. (1977) Proc. Natl. Acad. Sci. USA 74, 2976–2979.
48 Suzuki, H., Nishimura, Y. and Hirota, Y. (1978) Proc. Natl. Acad. Sci. USA 75, 664–668.
49 Matsuhashi, M., Maruyama, I.N., Takagaki, Y., Tamaki, S., Nishimura, Y. and Hirota, Y. (1978) Proc. Natl. Acad. Sci. USA 75, 2631–2635.
50 Isono, K. (1980) in Ribosomes (Chambliss, G., Craven, G.R., Davies, J., Davis, K., Kahan, L. and Nomura, M., Eds.), pp. 641–669, University Park Press, Baltimore.
51 Hirota, Y., Suzuki, H., Nishimura, Y. and Yasuda, S. (1977) Proc. Natl. Acad. Sci. USA 74, 1417–1420.
52 Hawrot, E. and Kennedy, E.P. (1975) Proc. Natl. Acad. Sci. USA 72, 1112–1116.
53 Pluschke, G., Hirota, Y. and Overath, P. (1978) J. Biol. Chem. 253, 5048–5055.
54 Raetz, C.R.H. (1976) J. Biol. Chem. 251, 3242–3249.
55 Nishijima, M. and Raetz, C.R.H. (1979) J. Biol. Chem. 254, 7837–7844.
56 Ganong, B.R., Leonard, J. and Raetz, C.R.H. (1980) J. Biol. Chem. 255, 1623–1629.
56a. Ganong, B.R. and Raetz, C.R.H. (1982) J. Biol. Chem. 257, 389–394.
57 Mukherjee, P.K. and Paulus, H. (1977) Proc. Natl. Acad. Sci. USA 74, 780–784.
58 Lehmann, V., Rupprecht, E. and Osborn, M.J. (1977) Eur. J. Biochem. 76, 41–49.
59 Cramer, C.L. and Davis, R.H. (1979) J. Bacteriol. 137, 1437–1438.
60 Esko, J.D. and Raetz, C.R.H. (1980) Proc. Natl. Acad. Sci. USA 77, 5192–5196.
60a. Esko, J.D., Wermuth, M.M. and Raetz, C.R.H. (1981) J. Biol. Chem. 256, 7388–7393.
61 Beadle, G.W. and Tatum, E.L. (1941) Proc. Natl. Acad. Sci. USA 27, 499–506.
62 Horowitz, N.H. and Beadle, G.W. (1943) J. Biol. Chem. 150, 325–333.
63 Beadle, G.W. (1944) J. Biol. Chem. 156, 683–689.
64 Cohn, M.S., Tabor, C.W. and Tabor, H. (1980) J. Bacteriol. 142, 791–799.
65 Hinnen, A., Hicks, J.B. and Fink, G.R. (1978) Proc. Natl. Acad. Sci. USA 75, 1929–1933.
66 Petes, T.D. (1980) Annu. Rev. Biochem. 49, 845–876.
67 Michell, R.H. (1975) Biochim. Biophys. Acta 415, 81–147
68 Leonard, J. (1978) Annu. Rev. Biophys. Bioeng. 7, 139–174.
69 Puck, T.T. (1972) The Mammalian Cell as a Microorganism—Genetic and Biochemical Studies in vitro, Holden-Day, California.
70 Baker, R.M. and Ling, V. (1978) in Methods in Membrane Biology (Korn, E.D., Ed.), Vol. 9, Plenum, New York, pp. 337–384.
71 Esko, J.D. and Raetz, C.R.H. (1980) J. Biol. Chem. 255, 4474–4480.
71a. Polokoff, M.A., Wing, D.C. and Raetz, C.R.H. (1981) J. Biol. Chem. 256, 7687–7690.
72 Robbins, A.R. (1979) Proc. Natl. Acad. Sci. USA 76, 1911–1915.
73 Busch, D.B., Cleaver, J.E. and Glaser, D.A. (1980) Som. Cell. Genet. 6, 407–418.
74 Hirschberg, C.B., Baker, R.M., Spencer, L.A., Watson, D., Averbuch, T. and Perez, M. (1980) Fed. Proc. 39, 2003.
74a. Raetz, C.R.H., Wermuth, M.M., McIntyre, T.M., Esko, J.D. and Wing, D.C. (1982) Proc. Natl. Acad. Sci. USA 79, 3223–3227.
75 Glaser, M., Ferguson, K.A. and Vagelos, P.R. (1974) Proc. Natl. Acad. Sci. USA 71, 4072–4076.
76 Van Golde, L.M.G., Schulman, H. and Kennedy, E.P. (1973) Proc. Natl. Acad. Sci. USA 70, 1368–1372.
77 Kennedy, E.P., Rumley, M.K., Schulman, H. and Van Golde, L.M.G. (1976) J. Biol. Chem. 251, 4208–4213.
78 Schulman, H. and Kennedy, E.P. (1977) J. Biol. Chem. 252, 6299–6303.
79 Schulman, H. and Kennedy, E.P. (1977) J. Biol. Chem. 252, 4250–4255.
80 Bell, R.M. (1974) J. Bacteriol. 117, 1065–1076.
81 McIntyre, T.M. and Bell, R.M. (1975) J. Biol. Chem. 250, 9053–9059.

82 McIntyre, T.M., Chamberlain, B.K., Webster, R.E. and Bell, R.M. (1977) J. Biol. Chem. 252, 4487–4493.
83 Friesen, J.D., An, G. and Fiil, N.P. (1978) Cell 15, 1187–1197.
84 Raetz, C.R.H., Kantor, G.D., Nishijima, M. and Newman, K.F. (1979) J. Bacteriol. 139, 544–551.
85 Raetz, C.R.H. and Foulds, J. (1977) J. Biol. Chem. 252, 5911–5915.
86 Hawrot, E. and Kennedy, E.P. (1978) J. Biol. Chem. 253, 8213–8220.
87 Nishijima, M., Bulawa, C.E. and Raetz, C.R.H. (1981) J. Bacteriol. 145, 113–121.
88 Ohki, M., Doi, O. and Nojima, S. (1972) J. Bacteriol. 110, 864–869.
89 Abe, M., Okamoto, N., Doi, O. and Nojima, S. (1974) J. Bacteriol. 119, 543–546.
90 Hsu, C.C. and Fox, C.F. (1970) J. Bacteriol. 103, 410–416.
91 Wilson, G. and Fox, C.F. (1971) J. Mol. Biol. 55, 49–60.
92 Nunn, W.D. and Cronan Jr., J.E. (1974) J. Biol. Chem. 249, 724–731.
93 Weisberg, L.J., Cronan Jr., J.E. and Nunn, W.D. (1975) J. Bacteriol. 123, 492–496.
94 Wickner, W., Mandel, G., Zwizinski, C., Bates, M. and Killick, T. (1978) Proc. Natl. Acad. Sci. USA 75, 1754–1758.
95 Wickner, W. (1979) Annu. Rev. Biochem. 48, 23–46.
96 Wickner, W. (1980) Science 210, 861–866.
97 Nunn, W.D., Kelley, D.L. and Stumfall, M.Y. (1977) J. Bacteriol. 132, 526–531.
98 Mindich, L. (1972) J. Bacteriol. 110, 96–102.
99 Bell, R.M. and Cronan Jr., J.E. (1975) J. Biol. Chem. 250, 7153–7158.
100 Edgar, J.R. and Bell, R.M. (1978) J. Biol. Chem. 253, 6354–6363.
101 Cronan Jr., J.E. and Bell, R.M. (1974) J. Bacteriol. 118, 598–605.
102 Pieringer, R.A. and Kunnes, R.S. (1965) J. Biol. Chem. 240, 2833–2838.
103 Schneider, E.G. and Kennedy, E.P. (1973) J. Biol. Chem. 248, 3739–3741.
104 Schneider, E.G. and Kennedy, E.P. (1976) Biochim. Biophys. Acta 441, 201–212.
105 Chang, Y.-Y. and Kennedy, E.P. (1967) J. Biol. Chem. 242, 516–519.
106 Raetz, C.R.H. and Kennedy, E.P. (1973) J. Biol. Chem. 248, 1098–1105.
107 Proulx, P.R. and Van Deenen, L.L.M. (1967) Biochim. Biophys. Acta 144, 171–174.
108 Okuyama, H. and Nojima, S. (1969) Biochim. Biophys. Acta 176, 120–124.
109 Langley, K.E. and Kennedy, E.P. (1978) J. Bacteriol. 136, 85–95.
110 Raetz, C.R.H., Dowhan, W. and Kennedy, E.P. (1976) J. Bacteriol. 125, 855–863.
110a. Bulawa, C.E., Ganong, B.R., Sparrow, C.P. and Raetz, C.R.H. (1981) J. Bacteriol., 148, 391–393.
111 Homma, H., Nishijima, M., Kobayashi, T., Okuyama, H. and Nojima, S. (1981) Biochim. Biophys. Acta, 663, 1–13.
112 Hawrot, E. and Kennedy, E.P. (1976) Mol. Gen. Genet. 148, 271–279.
113 White, D.A., Albright, F.A., Lennarz, W.J. and Schnaitman, C.A. (1971) Biochim. Biophys. Acta 249, 636–642.
114 Bell, R.M., Mavis, R.D., Osborn, M.J. and Vagelos, P.R. (1971) Biochim. Biophys. Acta 249, 628–635.
115 Hirota, Y., Ryter, A. and Jacob, F. (1968) Cold Spring Harbor Symp. Quant. Biol. 33, 677–693.
116 Hirabayashi, T., Larson, T.J. and Dowhan, W. (1976) Biochemistry 15, 5205–5211.
117 Nishijima, M. and Raetz, C.R.H. (1981) J. Biol. Chem. 256, 10690–10696.
118 Hirschberg, C.B. and Kennedy, E.P. (1972) Proc. Natl. Acad. Sci. USA 69, 648–651.
119 Woolford Jr., J.L., Cashman, J.S. and Webster, R.E. (1974) Virology 58, 544–560.
120 DiRienzo, J.M., Nakamura, K. and Inouye, M. (1978) Annu. Rev. Biochem. 47, 481–532.
121 Chattopadhyay, P.K. and Wu, H.C. (1977) Proc. Natl. Acad. Sci. USA 74, 5318–5322.
122 Schneider, J.E., Reinhold, V., Rumley, M.K. and Kennedy, E.P. (1979) J. Biol. Chem. 254, 10135–10138.
123 Sakakibaru, Y., Doi, O. and Nojima, S. (1972) Biochem. Biophys. Res. Commun. 46, 1434–1440.
124 Bradley, W.E.C. and Astrachan, L. (1971) J. Virol. 8, 437–445.
125 Brockerhoff, H. and Jensen, R.G. (1974) Lipolytic Enzymes, Academic Press, New York.
126 Matzuzawa, Y. and Hostetler, K.Y. (1980) J. Biol. Chem. 255, 646–652.
127 Cohen, S.N., Chang, A.C.Y., Boyer, H.W. and Helling, R.B. (1973) Proc. Natl. Acad. Sci. USA 70, 3240–3244.

128 Hershfield, V., Boyer, H.W., Yanofsky, C., Lovett, M.A. and Helinski, D.R. (1974) Proc. Natl. Acad. Sci. USA 71, 3455–3459.
129 Ohta, A., Waggoner, K., Louie, K. and Dowhan, W. (1981) J. Biol. Chem. 256, 2219–2225.
130 Ohta, A. and Dowhan, W. (1979) Abstracts of the XIth International Congress of Biochemistry, p. 376.
131 Ohta, A., Waggoner, K., Radominska-Pyrek, A. and Dowhan, W. (1981) J. Bacteriol. 147, 552–562.
132 Larson, T.J., Lightner, V.A., Green, P.R., Modrich, P. and Bell, R.M. (1980) J. Biol. Chem. 255, 9421–9426.
133 Nakamura, K. and Inouye, M. (1979) Cell 18, 1109–1117.
134 Movva, N.R., Nakamura, K. and Inouye, M. (1980) Proc. Natl. Acad. Sci. USA 77, 3845–3849.
135 Snider, M.D. (1979) J. Biol. Chem. 254, 7197–7202.
136 Larson, T.J. and Dowhan, W. (1976) Biochemistry 15, 5212–5218.
137 Dowhan, W., Wickner, W.T. and Kennedy, E.P. (1974) J. Biol. Chem. 249, 3079–3084.
138 Bohnenberger, E. and Sanderman, H., Jr. (1979) Eur. J. Biochem. 94, 401–407.
138a. Raetz, C.R.H., Kantor, G.D., Nishijima, M. and Jones, M.L. (1981) J. Biol. Chem. 256, 2109–2112.
139 Shatkin, A.J. and Tatum, E.L. (1961) Am. J. Bot. 48, 760–771.
140 Lester, H.C. and Gross, S.R. (1959) Science 129, 572.
141 Scarborough, G.A. and Nyc, J.F. (1967) J. Biol. Chem. 242, 238–242.
142 Scarborough, G.A. and Nyc, J.F. (1967) Biochim. Biophys. Acta 146, 111–119.
143 Kennedy, E.P. and Weiss, S.B. (1956) J. Biol. Chem. 222, 193–214.
144 Steiner, M.R. and Lester, R.L. (1972) Biochim. Biophys. Acta 260, 222–243.
145 Esko, J.D., Gilmore, J.T. and Glaser, M. (1977) Biochemistry 16, 1881–1890.
146 Culbertson, M.R., Donahue, T.F. and Henry, S.A. (1976) J. Bacteriol. 126, 243–250.
147 Henry, S.A., Donahue, T.F. and Culbertson, M.R. (1975) Mol. Gen. Genet. 143, 5–11.
148 Henry, S.A., Atkinson, K.D., Kolat, A.I. and Culbertson, M.R. (1977) J. Bacteriol. 130, 472–484.
149 Prottey, C., Seidman, M.M. and Ballou, C.E. (1971) Lipids 5, 463–468.
150 Lester, R.L. and Steiner, M.R. (1968) J. Biol. Chem. 243, 4889–4893.
151 Smith, S.W. and Lester, R.L. (1974) J. Biol. Chem. 249, 3395–3405.
152 Wagner, H., and Zofcsik, W. (1966) Biochem. Z. 346, 343–350.
153 Steiner, S., Smith, S.W., Waechter, C.J., and Lester, R.L. (1969) Proc. Natl. Acad. Sci. USA 64, 1042–1048.
154 Steiner, S. and Lester, R.L. (1972) J. Bacteriol. 109, 81–88.
155 Shafari, T. and Lewin, L.M. (1968) Biochim. Biophys. Acta 152, 787–790.
156 Hayashi, E., Hasegawa, R. and Tomita, T. (1976) J. Biol. Chem. 251, 5759–5769.
157 Atkinson, K., Fogel, S. and Henry, S.A. (1980) J. Biol. Chem. 255, 6653–6661.
158 Yamashita, S. and Oshima, A. (1980) Eur. J. Biochem. 104, 611–616.
159 Hosaka, K. and Yamashita, S. (1980) J. Bacteriol. 143, 176–181.
160 Stuhne-Sekalec, L. and Stanacev, N.Z. (1979) Can. J. Biochem. 57, 618–624.
161 Van Golde, L.M.G., Oldenborg, V., Post, M., Batenburg, J.T., Poorthuis, B.J.H.M. and Wirtz, K.W.A. (1980) J. Biol. Chem. 255, 6011–6013.
162 Jacobs, L. and DeMars, R. (1977) in Handbook of Mutagenicity Test Procedures (Kilbey, B.J., Legator, M., Nichols, W. and Ramel, C., Eds.), Elsevier, New York, pp. 193–220.
163 Diringer, H., Marrgraf, W.D., Koch, M.A. and Anderer, F.A. (1972) Biochem. Biophys. Res. Commun. 47, 1345–1352.
164 Ullman, M.D. and Radin, N.S. (1974) J. Biol. Chem. 249, 1506–1512.
165 Goldstein, J.L., Anderson, R.G.W. and Brown, M.S. (1979) Nature 279, 679–685.
165a. Pearse, B.M.F. and Bretscher, M.S. (1981) Annu. Rev. Biochem. 50, 85–101.
166 Litwin, J. (1973) in Tissue Culture—Methods and Applications (Kruse, P.F. and Patterson, M.K., Eds.), Academic Press, New York, pp. 188–192.
167 Innerarity, T.L., Pitas, R.E. and Mahley, R.W. (1979) J. Biol. Chem. 254, 4186–4190.
168 Lands, W.E.M. (1960) J. Biol. Chem. 235, 2233–2237.
169 Lands, W.E.M. and Crawford, C.G. (1976) in Enzymes of Biological Membranes (Martinosi, A., Ed.), Vol. 2, Plenum, New York, pp. 3–85.
170 Pagano, R.E. and Weinstein, J.N. (1978) Annu. Rev. Biophys. Bioeng. 7, 435–468.

Subject index